I0056249

The Editor

Prof. K.V. Peter is basically a horticulturist, a plant breeder and a University Professor. He is an acknowledged and decorated scientist and science manager. A post graduate from G.B. Pant University of Agriculture and Technology, he did post doctoral research at BARC Beltsville Maryland USA and worked at Laboratories at AVRDC Tainan and Guadeloupe (French West Indies). He was associated with development and release of improve and high yielding varieties/hybrids in tomato, brinjal, chilli, bittergourd melons, amaranths and cowpea. Sources of resistance to bacterial wilt in tomato, chilli and brinjal, aphids in cowpea and viral leaf curl in chilli were located by him and are now used in breeding programmes. Prof. Peter provided managerial support to the Indian Institute of Spices Research, Calicut to possess the World's largest germplasm of black pepper and cardamom. A prolific writer and academic editor he authored/edited 74 books published both in India and abroad. He received 7 scholarships and fellowships at various stages of education. He is member in the Board of Examiners in 12 Universities, Chairman/member in 130 committees of Government /research institutes, Fellow in 3 National Academies and 3 Scientific societies, member in 18 Scientific societies. His publishers include Taylor and Francis, Elsevier, National Book Trust, I.C.A.R., New India Publishing Agency, Astral International Pvt. Ltd., Studium Press (USA), Universities Press, Hyderabad and Kerala Language Institute, Trivandrum.

Prof. Peter is recipient of several awards like Rafi Ahmed Kidwai Award 1996-1998 for outstanding research in Horticulture-ICAR, New Delhi; Recognition Award 2000–National Academy of Agricultural Sciences, New Delhi; Dr. M.H. Marigowda National Award for the Best Horticulturist–2000; Silver Jubilee Medal for outstanding contributions in Vegetable Research; Dr. Harbhajan Singh Award 1993; Silver Jubilee Memonto; Biotech Product and Process Development Award 2003; Shiva Sakthi-HSI National Award for Life Time Achievement in Horticulture-2008 and Dr. K. Ramiah Memorial Award-National Academy of Agricultural Sciences, New Delhi for outstanding contribution to Crop Improvement-2009. He was Director when the Best Institution Award was conferred to IISR, Calicut by ICAR, New Delhi. He was Vice-Chancellor when Sardar Patil Award 2003 for the best ICAR institution was conferred to Kerala Agricultural University.

Family: Vimala is wife, Anvar and Ajay sons, Anu and Cynara Daughters-in-law and Antony Ajay Peter, grand son and Anna Vimala Anvar, grand daughter. Parents are Late Kuruppacharil Devassey Varkey and Late Rosa Varkey.

Horticulture for Nutrition Security

Horticulture for Nutrition Security

Editor
Professor K.V. Peter

2015
Daya Publishing House®
A Division of
Astral International Pvt. Ltd.
New Delhi – 110 002

© 2015 EDITOR

Publisher's note:

Every possible effort has been made to ensure that the information contained in this book is accurate at the time of going to press, and the publisher and author cannot accept responsibility for any errors or omissions, however caused. No responsibility for loss or damage occasioned to any person acting, or refraining from action, as a result of the material in this publication can be accepted by the editor, the publisher or the author. The Publisher is not associated with any product or vendor mentioned in the book. The contents of this work are intended to further general scientific research, understanding and discussion only. Readers should consult with a specialist where appropriate.

Every effort has been made to trace the owners of copyright material used in this book, if any. The author and the publisher will be grateful for any omission brought to their notice for acknowledgement in the future editions of the book.

All Rights reserved under International Copyright Conventions. No part of this publication may be reproduced, stored in a retrieval system, or transmitted in any form or by any means, electronic, mechanical, photocopying, recording or otherwise without the prior written consent of the publisher and the copyright owner.

Cataloging in Publication Data--DK
Courtesy: D.K. Agencies (P) Ltd. <docinfo@dkagencies.com>

Horticulture for nutrition security / editor, Professor K.V. Peter.
pages   cm
Includes bibliographical references and index.
ISBN: 978-93-5130-677-1 (International Edition)

1. Horticulture--India. 2. Food security--India. 3. Nutrition--India.
I. Peter, K. V., editor.

DDC 338.1750954   23

Published by   : **Daya Publishing House®**
A Division of
Astral International Pvt. Ltd.
– ISO 9001:2008 Certified Company –
4760-61/23, Ansari Road, Darya Ganj
New Delhi-110 002
Ph. 011-43549197, 23278134
E-mail: info@astralint.com
Website: www.astralint.com

Laser Typesetting   : **Classic Computer Services**, Delhi - 110 035

Printed at   : **Replika Press Pvt. Ltd.**

PRINTED IN INDIA

Devotion

The book HORTICULTURE FOR NUTRITION SECURITY is devoted to Prof. M.S. Swaminathan, Father of Green Revolution for his commitment to make India hunger free by 2030. He coined the much quoted "There is a horticulture remedy for every nutritional malady". Voted one among the three great Asians-Mahathma Gandhi, Rabindranath Tagore and M.S. Swaminathan- he becomes history and a household name.

Acknowledgement

I acknowledge Prof. V.L. Chopra, Former Member, Planning Commission, GOI for writing the FOREWORD. I bow before him for qualities of excellence, patience and whole heartedness. I acknowledge all the 35 scientists who contributed chapters to the book. Mahtab S. Bamji supported me to get permission for reproducing from the INSA Report and the Current Science. Prof. P.I. Peter, Chairman NoniBiotech, Chennai and Dr. Kirti Singh, Chairperson, World Noni Research Foundation, Chennai provided me support and facilities. My wife Vimala, sons Anvar and Ajay, daughters-in-law Anu and Cynara; granddaughter Anna and grand son Antony; their smiles and faces motivated me in loneliness.

I acknowledge FAO of UN, Rome for permission to reproduce preamble-I: The State of Food and Agriculture-2014 in Brief: Innovations in Family Farming.

I acknowledge Mr. Anil Mittal of Astral International Pvt. Ltd., New Delhi for publishing the book well in time in a near error free format.

Foreword

The National Food Security Bill-2013 was enacted by the Parliament of India 'to provide for food and nutritional security in human life cycle approach by ensuring access to adequate quantity of quality food at affordable prices to people to live a life with dignity and for matters connected therewith or incidental there off'. India is also signatory to the UN resolution of Millennium Development Goals (MDGs-8) which range from halving extreme poverty rates and eliminating child mortality to halting the spread of HIV/AIDS and providing universal primary education by 2015.Despite great strides in food production (52 million tones in 1951 to 236 million tons in 2014), 26 per cent of Indian population is still below poverty line (per capita income lesser than US $ 1.5/capita/day) and has limited access to nutritious food. India also houses 60 per cent of the worlds anemic. Every third child is born with low birth weight and is condemned to poor mental and physical development and immunity unless rehabilitated within the first year of life.

"There is a horticultural remedy for every nutritional malady' says Prof. M.S. Swaminathan, the Father of Green Revolution in India. Horticulture has emerged as a core sector with a crop coverage of 21 million ha and an annual production of 240.5 million tones. This includes vegetables, fruits, root and tuber crops,mushrooms, ornamentals, medicinal and aromatic plants, nuts and plantation crops(industrial crops).Though these crops occupy hardly 11 per cent of cropped area, they contribute over 30.4 per cent to the gross agricultural output in the country. Fruits and vegetables are also rich sources of vitamins, minerals, protein, anti-oxidants, lipids, fiber and several beneficial phyto-chemicals. The per capita consumption of fruits (65g/day) and vegetables (120g/day) in India is very low against the recommended minimum of 120g and 280g respectively. With the present level of population, the annual requirement of horticulture produce will be 360 million tones by 2020-21 against the present 240.5 million tones. Family Farming, Nutrition gardens at each homestead, urban horticulture, vertical gardens and protected cultivation are now promoted to attain self sufficiency in horticulture produce. The FAO of UN declared the year 2014 as year of Family Farming and 2015 as year of Soils. The National Academy of Agricultural Sciences, New Delhi has released two Policy Papers- No. 7 on' Diversification of Agriculture for Human Nutrition' and No. 30 on ' Organic Farming: Approaches and possibilities in the context of Indian Agriculture'. As former President of the Academy, I am happy that Prof. K V Peter, a Fellow of the Academy, has made use of these policy papers as Preambles to the book. The Indian National Science Academy, after wide consultations, issued a policy paper on 'Micro-nutrient Security for India-Priorities for Research and Action-2011'. Viable water and soil are pre-requisites to successful horticulture crop production for freedom from pesticide and other undesirable residues. The issue of hidden hunger leading to loss of working hours and causing debility disorders can be addressed by encouraging family farming and nutrition garden in each homestead.

The present book HORTICULTURE FOR NUTRITION SECURITY has 21 chapters covering both basic and applied aspects by 35 active working scientists. I congratulate the Editor, Prof. K V Peter, and the publisher, The Astral International Pvt. Ltd New Delhi, for undertaking this project. I co- edited the book 'HANDBOOK OF INDUSTRIAL CROPS' with Prof. K V Peter which was published by Taylor and Francis. The book was received very well. Likewise, it is my conviction that the present book will be an asset for students, teachers and policy planers.

V.L. Chopra

Food and Agriculture Organization of the United Nations-2014. The State of Food and Agriculture 2014 in Brief: Innovation in Family Farming

More than 500 million family farms manage the majority of the world's agricultural land and produce most of the world's food. We need family farms to ensure global food security, to care for and protect the natural environment and to end poverty, undernourishment and malnutrition. But these goals can be thoroughly achieved if public policies support family farms to become more productive and sustainable; in other words policies must support family farms to innovate within a system that recognizes their diversity and the complexity of the challenges faced.

The State of Food and Agriculture 2014: Innovation in family farming analyses family farms and the role of innovation in ensuring global food security, poverty reduction and environmental sustainability. It argues that family farms must be supported to innovate in ways that promote sustainable intensification of production and improvements in rural livelihoods. Innovation is a process through which farmers improve their production and farm management practices.

This may involve planting new crop varieties, combining traditional practices with new scientific knowledge, applying new integrated production and post-harvest practices or engaging with markets in new, more rewarding ways. But innovation requires more than action by farmers alone. The public sector- working with the private sector, civil society and farmers and their organizations - must create an innovation system that links these various actors, fosters the capacity of farmers and provides incentives for them to innovate.

Family farms are very diverse in terms of size, access to markets and household characteristics, so they have different needs from an innovation system. Their livelihoods are often complex, combining multiple natural-resource-based activities, such as raising crops and animals,fishing, and collecting forest products, as well as off-farm activities, including agricultural and non- agricultural enterprises and employment. Family farms depend on family members for management decisions and most of their workforce, so innovation involves gender and intergenerational considerations. Policies will be more effective if they are tailored to the specific circumstances of different types of farming households within their institutional and agro-ecological settings. Inclusive research systems, advisory services, producer organizations and cooperatives, as well as market institutions are essential.

The challenges of designing an innovation system for the twenty-first century are more complex than those faced at the time of the Green Revolution. The institutional framework is different due to a declining role of the public sector in agricultural innovation and the entry of new actors, such as private research companies and advisory services, as well as civil society organizations. At the same time, farmers are having to address globalization, increasingly complex value chains, pressures on natural resources, and climate change.

Family Farms: Size and Distribution

There are more than 570 million farms in the world.Although the notion of family farming is imprecise, most definitions refer to the type of management or ownership and the labour supply on the farm. More than 90 percent of farms are run by an individual or a family and rely primarily on family labour. According to these criteria, family farms are by far the most prevalent form of agriculture in the world. Estimates suggest that they occupy around 70 - 80 percent of farm land and produce more than 80 percent of the world's food in value terms.

. The vast majority of the world's farms are small or very small, and in many lower-income countries farm sizes are becoming even smaller. Worldwide, farms of less than 1 hectare account for 72 percent of all farms but control only 8 percent of all agricultural land. Slightly larger farms between 1 and 2 hectares account for 12 percent of all farms and control 4 percent of the land, while farms in the range of 2 to 5 hectares account for 10 percent of all farms and control 7 percent of the land. In contrast, only 1 percent of all farms in the world are larger than 50 hectares, but these few farms control 65 percent of the world's agricultural land. Many of these large, and sometimes very large, farms are family-owned and operated.

The highly skewed pattern of farm sizes at the global level largely reflects the dominance of very large farms in high-income and upper-middle-income countries and in countries where extensive livestock grazing is a dominant part of the agricultural system. Land is somewhat more evenly distributed in the low-and lower-middle-income countries where more than 95 percent of all farms are smaller than 5 hectares. These farms occupy almost three-quarters of all farm land in the low-income countries and almost two-thirds in the lower- middle income group. In contrast, farms larger than 50 hectares control only 2 percent and 11 percent, respectively, of the land in these income groups.

Exactly what can be considered a small farm- below 0.5 or 1 hectare, or some other size - will depend on agro-ecological and socio-economic conditions, and their economic viability will depend on market opportunities and policy choices. Below a certain level, a farm may be too small to constitute the main means of support for a family. In this case, agriculture may make an important contribution to a family's livelihood and food security, but other sources of income through off-farm employment, transfers or remittances are necessary to ensure the family lives a decent life. On the other hand, many small or medium- sized family farms in the low- and middle-income countries could make a greater contribution to global food security and rural poverty alleviation, depending on their productive potential, access to markets and capacity to innovate. Through a supportive agricultural innovation system these farms could help transform world agriculture.

Family Farms, Food Security and Poverty

In most countries, small and medium-sized farms tend to have higher agricultural crop yields per hectare than larger farms because they manage resources and use labour more intensively. This means that the share of small and medium-sized farms in national food production is likely to be even larger than the share of land they manage.

A large proportion of family farmers with small land holdings also depend on other natural resources, especially forests, pastureland and fisheries. The intensive resource use on these farms may threaten sustainability of production. These small and medium-sized farms are central to global natural resource management and environmental sustainability as well as to food security.

While smaller farms tend to achieve higher yields per hectare than larger farms, they produce less per worker. Labour productivity - or output per worker - is also much lower **in** low-income countries than in high-income countries. Increased labour productivity is a precondition for sustained income growth, so enabling farming families in low- and middle-income countries to raise their labour productivity is essential if we are to boost farm incomes and make inroads into reducing rural poverty.

Although smaller farms tend to have higher yields than larger farms within the same country, cross-country comparisons show that yields per hectare are much lower in poorer countries, where smaller farms are more prevalent, than in richer countries. This seeming paradox simply reflects the fact that yields in low-income countries are far lower, on average, than in richer countries and far lower than they could be if existing technologies and management practices were appropriately adapted and more widely adopted in low- income countries. Innovation aimed at increasing yields in developing countries could have significant impacts in terms of expanding agricultural production, increasing farm incomes and lowering food prices, thereby reducing poverty and enhancing food security by making food more affordable and accessible to both rural and urban populations.

The potential to improve labour productivity and yields can only be realized if family farmers are able to innovate. There are two main, but interrelated, pathways through which farmers' productivity may be increased: the development, adaptation and application of new technologies and farm management practices; and the wider application of existing technologies and practices. The first expands the potential for more productive use of existing resources by pushing out the production possibility frontier. The second allows farmers to achieve more of this potential.

Innovation Systems for Family Farming

Innovation happens when individuals and groups adopt new ideas, technologies or processes that, when successful, spread through communities and societies. The process is complex, involving many actors, and it cannot function in a vacuum. It is furthered by the presence of an effective innovation system. Among other things, an agricultural innovation system includes the general enabling economic and institutional environment required by all farmers. Other key components are research and advisory services and effective agricultural producers' organizations. Innovation often builds on and adjusts local knowledge and traditional systems in combination with new sources of knowledge from formal research systems.

One fundamental driver for all innovators- including family farmers - is access to markets that reward their enterprise. Farmers with access to markets, including local markets, for their produce - whether it be food staples or cash crops – have a strong incentive to innovate. Technologies help farmers to enter the market by allowing them to produce marketable surpluses. Innovation and markets depend on, and reinforce, each other. However, investments in physical and institutional market infrastructure are essential to allow farmers to access markets both for their produce and for inputs. Efficient producers' organizations and cooperatives can also play a key role in helping farmers link to input and output markets.

Because family farms are so diverse in terms of size, access to markets and other characteristics, general policy prescriptions are unlikely to meet the needs of all of them. Public support for innovation should take into consideration the specific structure of family farming in each country and setting, as well as the policy objectives for the sector.

Some family farmers manage large commercial enterprises and require little from the public sector beyond agricultural research to ensure long-term production potential and the enabling environment and infrastructure that all farmers need to be productive, although they may require regulation, support and incentives to become more sustainable. Other, very small, family farms engage in markets primarily as net food buyers. They produce food as an essential part of their survival strategy, but they often face unfavourable policy environments and have inadequate means to make farming a commercially viable enterprise.

Many such farmers supplement both income and nutrition from other parts of the landscape, through forests, pastures and fisheries and from off-farm employment. For these farmers, diversification and risk spreading through these and other livelihood strategies will be necessary. While agriculture and agricultural innovation can improve livelihoods, they are unlikely to be the primary means of lifting this group of farmers out of poverty. Helping such farmers escape poverty will require broad-based efforts, including overall rural development policies and effective social protection. In between these two extremes are the millions of small and medium-sized family farms that have the potential to become economically viable and environmentally sustainable enterprises. Many of these farms are not well integrated into effective innovation systems and lack the capacity or incentives to innovate.

Public efforts to promote innovation in agriculture for family farms must focus on providing inclusive research, advisory services, market institutions and infrastructure that the private sector is typically unable to provide. For example, applied agricultural research for crops, livestock species and management practices of importance to smallholders are public goods and should be a priority. A supportive environment for producer organizations and other community-based organizations can also help promote innovation among family farms.

Promoting Sustainable Productivity on Family Farms

Demand for food is growing while land and water resources are becoming ever more scarce and degraded. Climate change will make these challenges yet more difficult. Over the coming decades, tanners need to produce significantly larger amounts of food, mostly on land already in production. The large gaps between actual and potential yields for major crops show that there is significant scope for increased production through productivity growth on family farms. This can be achieved by developing new technologies and practices or through overcoming barriers and constraints to the adaptation and adoption of existing technologies and practices. Overcoming poverty in low- and middle-income countries also means boosting labour productivity through innovation on family farms as well as providing farming families with other opportunities for employment.

It is not enough to produce more. If societies are to flourish in the long term, they must produce sustainably. The past paradigm of input-intensive production cannot meet the challenge. Productivity growth must be achieved through sustainable intensification. That means, inter alia, conserving, protecting and enhancing natural resources and ecosystems, improving the livelihoods and well- being of people and social groups and bolstering their resilience - especially to climate change and volatile markets.

The world must rely on family farms to grow the food it needs and to do so sustainably. For this to happen, family farmers must have the knowledge and economic and policy incentives they need to provide key environmental services, including watershed protection, biodiversity conservation and carbon sequestration.

Overcoming Barriers to Sustainable Farming

Smaller family farms tend to rely on tried and trusted methods because one wrong decision can jeopardize an entire growing season; but they readily adopt new technologies and practices that they perceive to be beneficial in their specific circumstances. Nevertheless, several obstacles often stand in the way of farmers adopting innovative practices that combine productivity increases with preservation and improvement of natural resources. Key impediments include the absence of physical and marketing infrastructure, financial and risk management instruments, and secure property rights.

Farmers often face high initial costs and long pay- off periods when making improvements. This can prove to be a prohibitive disincentive, especially in the absence of secure land rights and of access to financing and credit. Farmers are also unlikely to undertake costly activities and practices that generate public goods (such as environmental conservation) without compensation or local collective action. Furthermore, improved farm practices and technologies often only work well in the agro-ecological and social contexts for which they were designed, and if solutions are not adapted to local conditions, this can be a serious impediment to adoption.

Local institutions, such as producers' organizations, cooperatives and other community-based organizations, have a key role to play in overcoming some of these barriers. The effective functioning of local institutions and their coordination with the public and private sectors and with farmers themselves, both men and women, can determine whether or not small family farms can introduce innovative, sustainable improvements suited to their needs and local conditions.

Agricultural Research and Development: Focusing on Family Farms

Investing in agricultural research and development (R&D) is important for boosting agricultural productivity, preserving the environment and eradicating poverty and hunger. A large body of evidence confirms that there are high returns to public investments in agricultural R&D. In many countries such investment is currently insufficient. Private-sector research is increasingly important, especially in high-income countries, but it cannot replace public research. Much agricultural research can be considered a public good, where the benefits of the knowledge generated cannot be appropriated by a private company and is therefore unlikely to attract the private sector. Returns to agricultural R&D often take a long time to materialize and, in addition, research is cumulative, with results building up over time. In this context, a continuous long-term public commitment to agricultural research is fundamental. Innovative forms of more short-term financing can help, but stable institutional funding is needed to maintain a core long-term research capacity.

All countries need a certain level of domestic research capacity because technologies and practices can rarely be imported without some adaptation to local agro-ecological conditions.

However, countries need to consider carefully what research strategy is best suited to their specific needs and capacities. Some countries, particularly those with too few funds to run strong national research programmes, may need to focus on adapting the results of international research to conditions at home. Others, with bigger research budgets, may also want to devote resources to more basic research. The establishment of international partnerships and a careful division of labour between international research with broader applications and national research geared to domestic needs is a priority. There is also scope for South-South cooperation between large countries with major public research programmes and countries with less national research capacity facing similar agro-ecological conditions.

Research that meets the needs of family farms in their specific agro-ecological and social conditions is essential. Combining farmer-led innovation and traditional knowledge with formal research can contribute to sustainable productivity. Involving family farmers in defining research agendas and engaging them in participatory research efforts can improve the relevance of research for them. This may include working closely with producers' organizations and creating incentives for researchers and research organizations to interact with family farms and their different members, including women and youth, and to undertake research tailored to their specific circumstances and needs.

Promoting Inclusive Rural Advisory Services

While investments in agricultural R&D are needed in order to expand the potential for sustainable production, sharing knowledge about technologies and innovative practices among family farmers is perhaps even more important for closing existing gaps in agricultural productivity and sustainability between developing and developed countries. Agricultural extension and advisory services are critical for this challenge, but far too many farmers, and especially women, do not have regular access to such services. Modern extension features many different kinds of advisory services as well as service providers from the public, private and non-profit sectors. While there is no standard model for delivery of extension services, governments, private businesses, universities, NGOs, and producer organizations can play the role of service providers for different purposes and for different approaches. Strengthening the various types of service providers is an important component of promoting innovation.

Governments still have a strong role to play in the provision of agricultural advisory services. Like research, agricultural advisory services generate benefits for society that are greater than the value captured by individual farmers and commercial advisory service providers. These benefits- increased productivity, improved sustainability, lower food prices, poverty education, etc.- constitute public goods and call for the involvement of the public sector in the provision of agricultural advisory services. In particular, the public sector has a clear role in providing services to small family farms, especially in remote areas, who are unlikely to be reached by commercial service providers and who may have a strong need for neutral advice and information on suitable farming practices. Other areas include the provision of advisory services relating to more sustainable agricultural practices, or for climate change adaptation or mitigation through reduced greenhouse gas emissions or increased carbon sequestration. The public sector is also responsible for ensuring that the advisory services provided by the private sector and civil society are technically sound and socially and economically appropriate.

For rural advisory services to be relevant and have the necessary impact, the needs of different types of family farms as well as different household members in farming families need to be addressed. Engaging women and youth effectively and ensuring that they have access to advisory services that take into account their needs and constraints are central to ensuring effectiveness. Participatory approaches, e.g. farmer field schools in which farmers learn from other farmers, peer- learning mechanisms and knowledge-sharing activities, provide effective means for achieving these aims. More information and evidence is needed on experiences with different extension models and their effectiveness. Efforts to gather and share such information should be promoted at the national and international levels.

Developing Capacity for Innovation in Family Farming

Innovation presupposes a capacity to innovate at the individual, collective, national and international levels. The skills and capacities of individuals involved in all aspects of the agricultural innovation system - farmers, extension service providers, researchers, etc. - must be upgraded through education and training at all levels. Special attention needs to be given to women and girls based on their needs and roles in agriculture and rural livelihood strategies. A further focus must also be on youth in general, who tend to have a greater inclination to innovate than elder farmers and represent the future of agriculture. If youth perceive agriculture as a potential profession with scope for innovation, this can have major positive implications for the prospects for the sector.

Collective innovation capacity depends on effective networks and partnerships among the individuals and groups within the system. Producers' organizations and cooperatives are of particular importance. Strong, effective and inclusive organizations can facilitate the access of family farms to markets for inputs and outputs, to technologies and to financial services such as credit.

They can serve as a vehicle for closer cooperation with national research institutes; provide extension and advisory services to their members; act as intermediaries between individual family farms and different information providers; and help small farmers gain a voice in policy-making to counter the often prevailing influence of larger, more powerful interests. Furthermore, family farmers who depend on other resources, such as forests, pastures and fisheries can benefit by linking with producer organizations within these sectors. Linking producer organizations across these sectors can further strengthen the case for clear tenure rights and better coordination between policies and service providers.

At national and international levels, the right environment and incentives for innovation are created by good governance and sound economic policies, secure property rights, market and other infrastructure, and a conducive regulatory framework. Governments must support the development of effective and representative producers' organizations and ensure that they participate in policy-making processes.

Micro-nutrient Security for India: Priorities for Research and Action

Report by Indian National Science Academy, New Delhi

Recently, the Indian National Science Academy (INSA) brought out a position paper on "Nutrition Security for India–issues and the way forward" based on deliberations in a symposium and subsequent discussions. This policy paper has been prepared as an off-shoot of the earlier effort, to focus specifically on priorities for research and action, for addressing the issue of micronutrients (MN) deficiencies – the hidden hunger. (INSA, 2009 www.insa.ac.in). It is beamed at researchers, funding agencies and policy makers and planners.

Malnutrition in India, particularly among women, children and adolescents is an emergency needing immediate attention if the country has to have inclusive growth and development. Health portfolio tends to concentrate on infectious diseases (vaccination, oral rehydration, treatment of infections), and noncommunicable diseases like cancer and cardiovascular diseases. Important as these are, nutrition cannot be subsumed under these. Nutrition has to be the basis of judging national development. Without good nutrition, neither communicable nor non-communicable diseases can be controlled. Malnutrition is the worst form of noncommunicable disease and is an important risk factor for chronic diseases at a later date. Maternal malnutrition has multigenerational adverse effects on human health and development.

Nutrition Security implies 'Physical, economic and social access to balanced diet, clean drinking water, safe environment, and health care'. Nutrition literacy and leadership at all levels is needed to understand and act. After 63 years of independence, India has among the highest incidence of under-nutrition in the world1. Almost 50 per cent of children under 5 are under weight (weight for age) and stunted (height for age). Over 30 per cent of adults are also undernourished. Besides deficiency of calories and protein, deficiency of micronutrients (MN) (vitamins and minerals) is rampant2,3,4. MN deficiency is referred to as the hidden hunger since often times it is not an obvious killer or crippler, but extracts heavy human and economic cost. Though anthropometric deficits are attributed to protein calorie malnutrition, MN deficiencies contribute significantly, because MNs are needed for utilisation of proteins and calories and to fight infections from a young age.

MN deficiency has a complex aetiology. Besides poor diet (due to poverty, ignorance, low agricultural productivity, and cultural factors); inadequate access to safe drinking water, clean disease-free environment, and health- care outreach also contribute. Infections result in loss of appetite, impaired absorption and utilisation of nutrients, particularly micronutrients. Environmental xenobiotics also impair MN utilisation. The present paper specifically focuses on feasible approaches and research needed to improve dietary access to MN.

Magnitude of the Problem of Micro-nutrient Deficiencies in India

Among the MN deficiencies, iron deficiency anaemia (IDA) is the most serious public health problem2,3,4. Estimates of IDA in women and children have varied from 50-70 per cent ; pregnant women being particularly susceptible. Iodine deficiency disease (IDD) is another worrisome public health problem[3]. Though, its magnitude has declined in recent years after the introduction of iodised salt, the problem still persists, and not confined to the Sub Himalayan regions as earlier thought. Fortunately, some of the severe vitamin-deficiency diseases such as beri beri (thiamine-vitamin B1 deficiency), pellagra (niacin deficiency), and scurvy (vitamin C deficiency) have disappeared. Blindness due to vitamin A deficiency and rickets due to vitamin D deficiency remain as clinical rather than public health problems. However, milder clinical manifestations and biochemical (sub-clinical) evidence of these deficiencies is rampant. Also osteoporosis in adults, particularly women after menopause due to calcium and vitamin D deficiency is common.[5]

Functional significance of sub-clinical MN deficiencies needs to be established.

Dietary Aetiology of MN Deficiency in India

Repeated diet surveys done by the National Nutrition Monitoring Bureau (NNMB) (National Institute of Nutrition, ICMR) in 9 states of India and some other surveys, indicate, the following[2].

Cereal-pulse based Indian diets are qualitatively deficient in micronutrients particularly iron, calcium, vitamin A, riboflavin and folic acid (hidden hunger), due to low intake of income-elastic protective foods such as pulses, vegetables particularly green leafy vegetables (GLV), fruits, and foods of animal origin. In recent years, there has been substantial erosion of area under cultivation of coarse grains and millets and share of these nutritious grains in total cereals produced and consumed[6].

More than 70 per cent of preschool children consume less than 50 per cent RDA of iron, vitamin A and riboflavin.

Within a family dietary deficits are more marked for preschool children due to inequitable distribution of food. This is because of lack of awareness of children's nutritional needs, and inability of child to articulate. While income cannot be blamed if the family has enough food for adults, time constraint on the mother who has to go out to work to supplement the family income, is a factor. In recent years concern has been expressed about the inadequate intake of other micronutrients such as zinc, vitamin D, calcium and vitamin B12.

More research is needed to establish the extent of dietary deficiency and requirement of these nutrients.

Consequences of Micro-nutrient Deficiencies

Apart from human suffering due to morbidity and mortality, malnutrition in general and MN deficiencies in particular have a high economic cost. 'Productivity losses due to poor nutrition are estimated to be more than 10 per cent of lifetime earnings for individuals, and 2-3 per cent of GDP to the nation. Cost of treating malnutrition is 27 times more than the investment required for its prevention'[7]. According to a panel of Nobel laureates, of the top 10 priorities selected for advancing global welfare using methodologies based on the theory of welfare economics, in Copenhagen Consensus, 2008, 5 were in the area of nutrition, micro-nutrient supplements, micro-nutrient fortification, biofortification, de-worming and other nutrient programmes at school and community level8. These are also needed to achieve the millennium development goals.

Iron Deficiency Anaemia (IDA)

Moderate and severe IDA adversely affects immunity (resistance to fight infections), cognitive and motor development, physical performance (and hence productivity) and reproductive health: (premature birth, low birth weight and perinatal mortality)[9,10]. It is estimated that anaemia is the direct cause of maternal deaths in 20 per cent and contributory cause in another 20 per cent. Apart from dietary deficiency, helminthic infections, inhibitors of iron absorption, in the diet and repeated pregnancies (in women) also contribute.

Iodine Deficiency Disorder (IDD)

Goitre is the clinical manifestation of iodine deficiency disorder. The functional consequences are: permanent brain damage, (cretinism, - mental retardation, and deaf mutism), reproductive failure, and decreased child survival. Milder deficiency also adversely affects mental development[9,10].

Vitamin A Deficiency (VAD)

The earliest ocular manifestation of vitamin A deficiency (VAD) is night blindness, and Bitot spots on the white of the eye. Severe vitamin A deficiency leads to keratomalacia (ulceration and sloughing of the cornea) and total blindness. Though kerotamalacia is no longer a public health problem, night blindness is prevalent particularly in pregnant mothers and subclinical deficiency (low Serum levels of Vitamin A), is still encountered. In addition to the ocular manifestations, vitamin A deficiency has been shown to cause growth retardation, decreased resistance to infections, and even death[9,10]. Opinion regarding efficacy of vitamin A supplementation in reducing mortality even in populations with sub-clinical vitamin A deficiency is however divided[11,12,13].

B-Complex Deficiencies

Though there is marked dietary, biochemical and clinical evidence of riboflavin (vitamin B2) deficiency (metabolically a very important vitamin), it has not received adequate attention because its deficiency is neither a killer nor a crippler. Impaired psychomotor performance in school children and adults and impaired reproduction in animals associated with riboflavin deficiency has been reported[9,10,14]. There is evidence of dietary and biochemical folic acid deficiency in India[9,10]. It can cause megaloblastic anaemia due to impaired red cell maturation. Folic acid deficiency has also been implicated in congenital malformation (neural tube defects), Folic acid supplementation in early pregnancy or even pre-pregnant state has been shown to prevent it. Folic acid deficiency leads to raised levels of serum homocysteine – an independent risk factor for cardiovascular disease (CVD)[9,10,15,16]. Fragmentary evidence suggests that Indians do tend to have high levels of homocysteine which responds to treatment with folic acid16. Till recently, vitamin B12 deficiency was not considered to be a problem in India since its daily requirement is only 1 microgram. However, reports of vitamin B12 deficiency in developing countries like India and its link with homocysteinaemia, besides megaloblastic anaemia, have started appearing[17]. Both folic acid and B12, besides vitamin B6 and B2 are required for homocysteine metabolism.

In view of the rising incidence of CVD in India, B- complex vitamin deficiency needs to be taken more seriously and its link with homocysteinaemia and CVD needs to be investigated. Research is also needed to examine the role and dosage of folic acid for prevention of neutral tube defects- which are not uncommon in India. A balance of folic acid with vitamin B12 has to be ensured.

Vitamin D Deficiency

Main function of vitamin D is in bone calcification by facilitating calcium absorption and maintaining blood calcium levels. Since generation of vitamin D in the skin from its precursor 7-dehydrocholesterol is through exposure of skin to sunlight, adequacy of vitamin D in a tropical country like India was assumed. However, recent studies suggest existence of vitamin D deficiency in all age groups in India. As mentioned earlier, osteoporosis associated with calcium and vitamin D deficiency is common in post-menopausal women. Low levels of vitamin D are also associated with chronic diseases like certain malignancies, and chronic inflammatory and autoimmune diseases like type 1 diabetes, and impaired resistance to infections[5,9,10]. However, according to a recent review "Health-benefits often reported in the media—were from studies that provided often mixed and inconclusive results and could not be considered reliable"[5].

Research is needed to determine the dietary requirement of vitamin D to ensure adequate vitamin D status.

Vitamin C Deficiency

Vitamin C is a powerful antioxidant. Dietary vitamin C deficiency does exist, but severe clinical manifestation (scurvy) has become rare. Vitamin C is an iron absorption promoter and hence its deficiency can contribute to IDA. Antioxidants delay degenerative diseases[9,10].

Zinc Deficiency

Zinc is essential for growth and development. Zinc supplementation has been reported to help linear growth, reduce severity and duration, of diarrhoeas, and respiratory infections and reduce child mortality[10].

The magnitude and consequences of zinc deficiency in India need to be determined. For the purpose suitable indicators have to be developed to assess zinc deficiency.

Strategies for Increasing Access to Micronutrients – Current Response and Research Opportunities

Basically there are four types of approaches to augment MN intake. 1) Pharmaceutical supplements, 2) Food fortification 3) Biofortification and 4) Food fortification (dietary diversification).

Micronutrient Supplementation

National Nutritional Anaemia Control Programme (NACP)

In this programme supplements containing 100 mg of elemental iron + 500 ìg folic acid are given to pregnant women for 100 days during pregnancy; 20 mg elemental iron and 100 ìg folic acid are given daily to preschool children for 100 days in the year.9,10. Recently adolescent girls have also been included as part of the life cycle approach with same dose as pregnant women, and weekly once administration throughout the year. Unfortunately despite scientific basis for the programme, ironfolic acid supplementation has failed to have an impact on the incidence or severity of anaemia, allegedly due to 1) lack of awareness regarding its importance and consequently poor compliance and 2) poor outreach in a vast country like India. NFHS 3 survey shows less than 20 per cent full compliance in pregnant and lactating women.4 3) Uniformly giving one tablet of IFA (which is meant for preventing anaemia in non-anaemic women) regardless of the severity of anaemia. Screening of all pregnant women for anaemia and treatment of anaemic women besides the prophylactic treatment with iron, folic acid has been recommended to reduce the prevalence and severity of anaemia in pregnancy..

Suggestions for Research to Improve NACP

Public-private-NGO partnership may have a role in improving the outreach. Absorption of non-haeme Iron from the diet is only 3-5 per cent. The challenge is to translate filling of iron stores in to improvement in haemoglobin.

Research is needed to find out socio-cultural, behavioural factors and administrative bottle neck to improve the efficiency of NACP.

All pregnant women should get Hb estimation done using reliable method and anaemic women treated with appropriate route and dose of iron, folic acid.

Children coming to hospital for any illness and undernourished children should be screened and those found anaemic should be appropriately treated.

Prevalence of iron, folic acid and B12 deficiency in non anaemic and anaemic pregnant women in different regions of the country should be assessed to find out if the current practice of prescribing iron and folic acid without B12 is appropriate.

Efficacy of present regimen of giving uniformly one tablet of iron –folic acid needs re-evaluation and replaced with treatment after screening.

Since nutrients, besides iron and folic acid, are also involved in haemoglobin synthesis/formation of red blood cells/absorption of iron, inclusion of MN like vitamins B12, C, B2, and zinc may improve the efficacy of the oral supplements.

Clinical research in a hospital setting with appropriately worked out dose and schedules for multi-vitamin supplements Vs IFA supplements should be conducted..

Massive Dose Vitamin A Supplementation to Prevent Nutritional Blindness

In this programme children between 6-60 months are given 200,000 IU of vitamin A, every six months as prophylactic dose. The rationale is: vitamin A being fat soluble, is stored in the liver and a massive dose would ensure adequate storage to last for at least 6 months. This programme also suffers from the infirmity of the other programmes *viz*., poor outreach, inadequate and irregular supplies. As mentioned earlier the severity of vitamin A deficiency has reduced despite inefficient operation of this programme and several eminent nutrition scientists have raised doubts about its continuation[11,12,13]. However, as of now, this programme should continue in areas where vitamin A deficiency is a public health problem (incidence of Bitot spots more than 0.5 per cent).

According to the Annual Report of Micronutrients Initiatives India, an International Non Government Organization, out of 32 Million US Dollars available in the Annual Budget 2009-2010, more than 20 Million US Dollars were spent on Vitamin A Procurement and Interventions. A meager sum of 2.5 Million US Dollars was spent on Iron interventions (Umesh Kapil, personal communication). Children with VAD should be identified and should be administered VA. The present Universal VAS approach needs to be reviewed.

There is need for operation/translational research to improve efficiency of the programme

Food Fortification

Iodised Salt

Food fortification is a powerful method of reaching out a deficient nutrient to populations, provided the vehicle used for fortification is consumed by the poorest of the poor. In India salt is the only such vehicle and it has been effectively used for reaching out iodine. This is one successful programme in the country[3,9], but its efficiency has to be improved in terms of stability of iodine in the salt, pricing, and outreach. Private companies tend to seek easy urban markets even if the problem is more acute elsewhere.

Iron Fortified Iodised Salt (IFIS, also called Double Fortified Salt-DFS)

This technology was developed by the National Institute of Nutrition, Hyderabad to address the dual problem of iron and iodine deficiency.9,10,18. Government order for its production has been released. Its efficacy has been tested in small scale studies.

Systematic programme for scaling up is needed to examine the effectiveness of DFS in preventing iron and iodine deficiency and replacing iodised salt with IFIS. Evaluation of DFS should be done by an independent agency rather than ICMR.

Iron Fortified Wheat Flour (Atta) and Rice

Since staple grains are consumed in substantial quantity, their fortification makes sense. In some countries wheat flour is fortified with iron and other micronutrients. Doubts have been raised about bio-availability of iron from wheat 'atta' because of high phytate (inhibitor of absorption) content. The inhibitory effect of phytate may be bypassed by some potential compounds like Na-Fe-EDTA and or enzyme phytase. The higher cost of this salt may be off-set by better bioavailablity and hence lesser dose of fortification.

For more than a half the population in India, rice is the staple. Fortification of rice has been tried by mixing fortified extruded grains from rice flour with rice (Ultra rice). More research is needed to make this technology cost-effective and acceptable.19

Programmatic studies are needed to examine the effectiveness of fortifying cereals with iron.

Fortification of Cereal Products with Folic Acid

In many countries, cereal products are fortified with folic acid to reduce the incidence of neural tube defects. Folic acid fortification, perhaps along with vitamin B12 may also reduce serum homocysteine levels.

In view of the rising incidence of CVD in India, this strategy needs to be researched.

Fortification of Oil with Vitamins A and D

Gujarat has taken the initiative. *Impact studies are needed. Fortified oil should be packed in dark bottles to cut off UV radiation, and prevent oxidation of vitamin A* Vehicles like sugar or soya sauce used in some countries are not suitable for India.

Fortification of Food for Supplementary Feeding

India has large feeding programmes like supplementary feeding of preschool children through ICDS and school children through Mid Day Meal. Opinion of nutritionists is divided about fortifying these foods with MNs – sprinklers (micronutrient powder- MNP), spreads.

Well planned field research to examine efficacy, feasibility and cost of fortification vs. enrichment with MN rich foods, of meals for supplementary feeding need to be undertaken. Impact of Golden rice rich in provitamin A, ultra rice rich in iron and other MN and red palm oil in feeding programmes needs to be researched for feasibility, acceptability and cost effectiveness.

Food fortification is suitable only for prevention of MN deficiencies and not for treating severe forms of the disease. Food fortification programme for country should consider the Recommended Dietary Allowances (RDA) of different nutrients for Indians, the types of foods to be fortified, nutrients to be added and the percentage of the RDA to be supplied through fortification etc. to avoid excess intakes by the rich and deficient intakes by the poor. Fortified foods should be affordable by the poorest of the poor, and reach the unreachable in a vast country like India.

Bio-fortification

Enriching germplasm with MNs through conventional breeding methods, molecular breeding and genetic engineering is a promising method for increasing dietary access to MNs. Bio-fortification is a sustainable intervention being a seed-based technology. There is no recurring cost, once the varieties are developed and adopted. It can benefit the farmer as well as the consumer if the cost of seed is kept low and not exploited by seed companies. Bio-fortified plants grow better. Potential nutritional impact of iron biofortificaion in India seems to be encouraging.[20] "The Harvest Plus: bio-fortification challenge programme is an interdisciplinary, global alliance of research and implementing institutions[21,22]. India is part of this. It includes: α–carotene (pro-vitamin A)- rich sweet potato, and cassava, zinc and iron rich rice, wheat, maize, pearl millet, and beans. DBT network project on biofortification of rice, wheat and maize is currently being implemented by ICAR Institutions and state agriculture universities and National Institute of Nutrition. Golden Rice rich in pro-vitamin A; high- iron rice (high ferritin gene from mangrove); are examples of transgenic technologies."

For conventional breeding or molecular breeding to improve the MN content of foods, sufficient within species diversity with appropriate gene pool would be necessary. Also this approach is slow. GM route with genes from other plant species, preferably edible varieties, is faster and with suitable safe guards regarding safety to health, protection of biodiversity, and cost; is a useful option for a country like India While the conventional strategies of food fortification and plant breeding to improve nutritional quality should be pursued, the GM technology despite some concerns needs to be vigorously investigated. This technology has received support from many science academies of the world including Indian academies.

Food–Food Fortification – Dietary Diversification

The problem of MN deficiency can to a great extent be addressed by encouraging dietary diversification and household access to MN-rich foods. Overall emphasis of agriculture, horticulture and livestock breeding has been to ensure calories, income and export. MNs are not on the antenna of agriculture planning. Decentralised planning to ensure district, village and household level adequacy of MN- rich foods is required. India being a vast country, long distance transport of perishable foods like vegetables and animal products may not be feasible. Promotion of homestead gardens even in urban areas, with emphasis on MN- dense varieties, back- yard poultry, dairy, fish ponds etc is needed.

Operation research to study the impact of such a strategy needs high priority. Farmbased approaches are low-cost, and sustainable if the community is empowered.

While it may be difficult to meet the requirement of some nutrients like iron and riboflavin through low-cost diets, there is ample supply of â carotene (pro-vitamin A) in nature in green leafy vegetables and yellow orange fruits and vegetables. This wealth needs to be exploited through education and advocacy. Low bio-availability of beta carotene is an issue. Food habits may be hard to change.

Research is needed to establish the factor for conversion of â *carotene to vitamin A and improve its bioavailability. Red palm oil is an important source of* â *caroteneand can be used in supplementary feeding programmes if supply and acceptability can be assured.*

Absorption and utilisation of micronutrients is influenced by competition between nutrients, (particularly minerals and trace elements), and other biotic (microorganisms, helminthic) and a-biotic (chemical agents like phytates, tannins, xenobiotics etc).

These influences have to be researched to improve utilisation of dietary MNs. Treatment with phytases and tannases needs to be considered. Bioavailability studies to examine the interaction within nutrients, and between nutrients and promoters, inhibitors and xenobiotics (biotic and abiotic), have to be planned.

Enhanced Production to Consumption of Millets and Pulses

MN rich coarse grains and millets are being forgotten. They are referred to as orphan grains, but with the threat of global warming, they may well be the grains of the future. There is 2-3 fold gap between optimal productivity of millets and farm-level productivity. Apart from being high in proteins, pulses are also rich in MN. Their production and consumption is on the decline.

Productivity of these grains has to be increased through research and robust extension effort. Agriculture research needs to give high priority to millets and pulses – a production to consumption strategy. Some of the World Bank funded National Agriculture Innovation Projects (NAIP) are attempting to do that.

Plant Foods as Speciality Foods for Protection against Chronic Diseases

This is a vast field for research since plant foods are rich in health giving phytochemicals-nutraceuticals. Some nutrients like vitamin C, E, Se also have antioxidant properties.

Economic Logic

World Bank has compared the benefits of MNs (iron) supplementation vs. Food fortification in terms of 1) cost per life saved, 2) productivity gained- a measure of efficient use of resources and defined as least cost discounted method of reducing clinical deficiency in the population and 3) social benefit cost- DALY or healthy life years saved. For saving life at least cost, targeted iron supplementation to pregnant women is more effective than iron fortification of flour or other staple. However, for enhanced productivity delivered by a programme and social benefit cost, iron fortification is most cost effective (For details see, Hunt[23]).

Similar analysis is needed in the Indian context. Farm-based interventions are believed to have high benefit-cost ratio in terms of economic cost. They also help the farmer and empower the community.

Food Analysis

India is rich in biodiversity with lots of local foods with potential benefits in terms of MN as well as health-promoting phytochemicals. Food analysis labs equipped to do co-ordinated MN analysis with proper quality control should be set up in agriculture universities and home science colleges which have the expertise and *country-wide research programmes to analyse local foods and identify MN rich foods should be started so that their production and consumption can be enhanced.*

Food Storage and Processing

Almost 30 per cent of farm food perishes due to lack of storage facilities and value addition. *Development of suitable storage structures not only for grains but also for vegetables, fruits and animal products and promoting those which are already available is urgently needed. There are indigenous time-tested traditional technologies for storage. Those have to be re-discovered and promoted.*

Value addition should not take away food from the poor. *Both, high-end and low-cost, processed foods need to be developed with public private partnership, with the triple objectives of reducing wastage, generating employment and enhancing nutrition value and security.*

Bioavailability and Stability of Micronutrients in Processed Foods

While minerals are reasonably stable during processing, vitamins, particularly vitamins A and C are sensitive to light and heat. Microencapsulation is done to increase the stability, but that increases the cost. Low-cost technologies to increase the bioavailability and stability of MNs in processed foods are needed to make processed foods accessible to all sections of the society.

Nutrition Security and Climate Change

Rise in temperature due to climate change is associated with increase in atmospheric carbon dioxide. While the latter helps to increase plant growth if there is enough supply of other nutrients in the soil, in tropical climate, further temperature rise has an adverse effect. It is suspected that apart from plant growth, climate change may also affect the nutrient composition (protein and minerals) in foods. No information is available on vitamin content of foods and climate change[24].

Impact of climate change on productivity and nutrient composition of major food items in India needs to be researched.

Monitoring, Surveillance and Management Information System for Early Detection

India needs a robust monitoring and surveillance programme in nutrition for early identification of the problems and capacity building to tackle them in a decentralized way. For every frank case of deficiency, there are dozens of others who suffer from subclinical deficiencies which contribute substantially to human suffering, medical expenses and economic cost.

Sensitive biochemical and functional tests to detect such deficiencies, using new knowledge of molecular and cellular biology and state of the art instruments need to be developed.

Nutrition and Health Education

Behavioural Change Communication (BCC) programmes to targeted audience can improve both nutrition and health status. Nutrition education for professionals-(agriculture, medical, social scientists and others), needs to be strengthened.

Innovative strategies need to be developed and tested not only to improve knowledge and attitudes but practices as well. Behavioural modification modules are needed. Nutrition education policy is nonexistent. Awareness among school children, teachers, consumers and women has to be enhanced. Women empowerment is essential to improve diets and health. Nutrition component of medical and agriculture curricula needs to be strengthened.

Conclusion

The current mindset of looking at food security only in terms of energy security has to change. Pumping cereals alone to quench hunger will not ensure nutrition and health. The goal should be to ensure a balanced diet adequate in macro- and micronutrients. Laboratory, clinical, and community (operations) -based research is needed to ensure MN security. An optimum mix of food-food fortification, (dietary diversification), biofortification, and early detection and effective treatment of clinical deficiencies needs to be worked out. Extension methodology has to be robust. Media support for creating awareness and compliance is important. The most important low-cost intervention is following WHO guidelines for breast feeding- early initiation, exclusive breast feeding for 6 months and complimentary feeding with continuation of breast feeding up to 1 year. This requires media blitz to educate the community.

Summary Table of Recommendations for Research and Action

The following table attempts to prioritize the suggestions made as priority 1+ (very high), which has evidence-based strength and need immediate action and scale up. Priority 1, (high) has some evidence, but needs more research. Priority 2 (moderately high) needs more evidence and research inputs. Research areas have been grouped under: 1.Translational and operation research other than agriculture, 2. Laboratory, clinical and controlled field research (other than agriculture),

3. Agriculture. The term agriculture includes horticulture, livestock and fisheries. Nutrition being a complex and multidimensional subject, inputs are needed from various areas and hence the wish list is long.

Sl.No.	Research Area	Type of Research	Agency	Priority
Translational, Operation Research other than Agriculture				
1.	Improve the efficiency of Micronutrient supplementation programmes-NACP, Massive dose vitamin A supplementation, iodised salt distribution	Multi-dimensional investigations to find out administrative and outreach bottlenecks and socio cultural, behavioural factors	Health ministry should contract it out to a good management research group	I+
		Efficacy of present regimen of giving uniformly one tablet of iron-folic acid needs reevaluation and replaced with treatment of anaemia with higher dose after screening.	ICMR	
2.	Replace iodised salt with iron fortified, iodised salt (double fortified salt)	Study the effectiveness, and acceptance in a scaled-up programme	ICMR, Ministry of Health, Ministry of Civil supplies, Private sector, Home Science Colleges	I+
3.	Nutrition monitoring, surveillance and management information system (MIS)	Put in place effective system for early detection of nutritional problems in the community	Ministry of Health in collaboration with ICMR, Dept. of Women and Child Development and Department of Space, Home Science Colleges	I+
4.	Infant and child feeding-promotion of breast and complimentary feeding according to WHO norms	Large scale awareness programme through media blitz	Ministry of Health, HRD, I and B, Space, Home Science Colleges etc.	I+
5.	Fortification of cereals-wheat flour and rice with iron	Investigate the effectiveness, outreach, and cost benefit ratio.	ICMR, CSIR, DST, DBT, Ministry of Health, Ministry of Food processing industries, international agencies	I
6.	Nutrition education, extension and training	1. Strengthen nutrition component of medical and agriculture curricula. 2. Nutrition education at all levels	Ministries/Depts. of Health, Women and Child Development, Agriculture, Human Resource development, I and B, space	I
7.	Fortification of oil with vitamins A and D-Gujarat experiment	Investigate the effectiveness/cost benefit ratio, stability of the fortificants. Learn from Gujarat	Gujarat state government, ICMR, Ministry of Health	II

Contd...

Contd...

Sl.No.	Research Area	Type of Research	Agency	Priority

Laboratory, Clinical and Controlled Field Research other than Agriculture

Sl.No.	Research Area	Type of Research	Agency	Priority
1.	Multi nutrient pill containing besides iron and folic acid, zinc, and vitamins B12, B2 and C for anaemia prophylaxis	Controlled laboratory and clinical trials to examine the efficacy, and cost benefit	ICMR	I+
2.	Bioavailability studies to examine interaction within nutrients and between nutrients and promoters and inhibitors-biotic and abiotic	Select research studies at institutional level	ICMR, ICAR, CSIR, DST, DBT	I
3.	Bioavailability of beta-carotene from plant foods-Conversion factor for beta carotene to vitamin A in Indian context. Effect of vitamin A deficiency	Controlled laboratory and clinical studies	ICMR, CSIR, DBT	I
4.	Role of B-vitamins deficiencies in aetiology of raised levels of homo-cysteine–anindependent risk factor for CVD. Genetic predisposition	Controlled laboratory and clinical studies	ICMR, CSIR, DBT	I
5.	Development of non-invasive methods for assessing MN status. *e.g.* Dry blood spot method	Laboratory studies in select well equipped institutions.	ICMR, ICAR, CSIR, DST, DBT	I
6.	Ready to cook, MNfortified foods suitable for institutional meals-ICDS, MDM, and homes. Comparison with food-food fortification with vegetables, fruits and products like red palm oil	Field studies to compare chemical fortification with food-food fortification	ICMR, ICAR, CSIR, DST, DBT, Ministry of Food Processing industries, NGO, Home Science Colleges	II
7.	Assessment of the magnitude of deficiency and requirement of less-recognised MN like zinc, vitamin B12, an vitamin D by developing appropriate laboratory tests	Laboratory studies in select well equipped institutions.	ICMR, CSIR, DST, DBT	II
8.	Functional significance of subclinical MN deficiencies	Laboratory/clinical studies in select well equipped institutions.	ICMR, CSIR, DST, DBT	II
9.	Plant foods containing nutraceuticals as speciality foods	Laboratory and clinical studies in select institutions.	ICMR, CSIR, DBT, DST	II

Contd...

Contd...

Sl.No.	Research Area	Type of Research	Agency	Priority
10.	Bioavailability and stability of MN in processed foods	Laboratory and field studies	CSIR, DST, DBT Ministry of Food Processing Industries	II

Agriculture, (Including horticulture livestock research)

Sl.No.	Research Area	Type of Research	Agency	Priority
1.	Increased production to consumption of pulses, coarse grains and millets	Agriculture extension work to bridge the productivity gap where products exist.	ICAR, Central and State Agriculture Universities, Home Science Colleges, ICMR	I+
2.	Food-food fortification-dietary diversification through decentralized agriculture planning, including homestead production.	Field trials to examine the impact of village-level and household production of vegetables, fruits, dairy, and poultry, fish using improved varieties, on householdfood security.	ICAR, Central and State Agriculture Universities, Home Science Colleges, ICMR, NGOs	I+
3.	Biofortification using conventional and molecular breeding	Research and development. Scaled up field trials with products available.	DBT, ICAR, Central and State Agriculture Universities	I+
4.	Genetic engineering to develop micronutrient - enriched foodgrains, vegetables and fruits	Research for health and environment safety etc. Develop new products.	ICAR, Central and State Agriculture Universities	I
5.	Golden rice to improve vitamin A nutrition	Validation studies	ICMR, ICAR, DBT	I
6.	Development of storage and packaging methods to preserve the MN content of foods, and prevent infestation.	R&D work, identification of time tested traditional methods. Promotion of methods already developed	CAR, Home Science Colleges	I
				I
7.	Analysis of local foods for micronutrients and health promoting phytochemicals. (These can also be resource for molecular breeding studies)	A large scale multi-centric effort using quality control at reference laboratories	Select institutions from ICAR, Agriculture Universities, Home Science Colleges with NIN and CFTRI as reference laboratories	II
8.	Impact of climate change on productivity and micro-nutrient content of staple foods	Controlled simulation studies in agriculture institutions and select laboratories	ICAR, DBT, Agriculture Universities	II

References

1. UNICEF-State of World's Children, Report, 2010, special edition

2. National Nutrition Monitoring Bureau (NNMB), Prevalence of micronutrient deficiencies; Technical report No.22, National Institute of Nutrition, ICMR, 2003

3. National Nutrition Monitoring Bureau (NNMB), Diet and nutrition status of populations and prevalence of hypertension among adults in rural areas;. Technical Report No 24. National Institute of Nutrition, ICMR, 2006

4. National Family Health Survey (NFHS-3), International Institute of population Sciences, Mumbai, India, 2005-2006; Jan.2011, **96 (1)**.

5. Ross C, Manson JE, Abrams SA et al. The 2011 Report on dietary reference intakes for calcium and vitamin D from the Institute of Medicine: What clinicians need to know; *J Clin Endocrinol Metab*, **96:** 53-58, 2011

6. Research and development in millets. Present status and future strategies. National seminar on millets-2010; Directorate of sorghum research, Hyderabad, November, 2010.

7. Coalition for sustainable nutrition security for India, Leadership agenda, 2010; http://www.nutritioncoalition.in/publication.asp

8. Copenhagen Consensus, 2008. http://www.copenhagenconsensus.com/Home.aspx.

9. Textbook of human nutrition, 3rd ed. (2010). Editors: Bamji MS, Krishnaswamy, K and Brahmam GNV

10. Nutrient requirement and recommended dietary allowances for Indians. A report of the expert group of the Indian Council of Medical Research, 2010; http://www.icmr.nic.in/pricepubl/content/11.htm

11. Latham M. The great vitamin A fiasco; *World Nutr*, **1:**12–45, 2010.

12. Gopalan C. Massive dose vitamin A prophylaxis should now be scrapped;. *World Nutr.*, **1:** 79–85. 2010.

13. Kapil U and Sachdev HPS. Universal Vitamin A Supplementation Programme in India: The need for a re-look; *The National Medical Journal of India.* **23:** 193-195, 2010.

14. Bamji MS and Lakshmi AV, Less recognised micronutrient deficiencies in India; *NFI Bulletin*, **19:** 5- 8, 1998.

15. Lakshmi AV and Bamji MS. Aetiology and health implications of hyperhomocysteinaemia; *PINSA* **B64:** 235-248, 1998.

16. Joshi S. Demystifying relevance of homocysteine hypothesius in native Asian Indian population. Editorial. *J. Assoc. physicians, India*; **54:** 763-764, 2006.

17. Stabler S. and Allen RH. Vitamin B12 deficiency as a worldwide problem; *Annual Review of Nutrition*, **24:** 299-326, 2004.

18. Ranganathan S and Sesikaran B. Development of the double fortified salt from the National Institute of Nutrition; Comprehensive Rev Food Sci, *Food Safety.***7:**390-396,2008.

19. Transforming Development through Science, Technology and Innovation. Ultra Rice Challenge; http://www.usaid.gov/scitech/ur.html. (Accessed January 3, 2011).

20. Stein A, Meenakshi J, Qaim, JV, et.al. Potential impacts of iron biofortification in India; *Social Science and Medicine*, **66:** 1797-1808, 2008.

21. Harvest plus in Asia. Breeding crops for better nutrition. harvestplus@cgiar.org. www.harvestplus.org. Accessed Jan 3, 2011.

22. "Nutrition Security for India– issues and the way forward" INSA Position paper, www.insa.ac.in.

23. Hunt JM, Forging effective strategies to combat iron deficiency; *J. Nutr.* **132:** 794S–801S, 2002.

24. Manoj-Kumar and Patra AK. Nutritional security: a missing link in climate change debates; *Current Science*, **99:** 287-289, 2010.

Policy Paper No. 30: Organic Farming: Approaches and Possibilities in the Context of Indian Agriculture

Issued by National Academy of Agricultural Sciences (NAAS), New Delhi

During the era of Green Revolution, introduction of high-yielding varieties, extension of irrigated areas, use of high analysis NPK fertilisers and increase in cropping intensity, propelled India towards self-sufficiency in food production. In the process, relative contribution of organic manures as a source of plant nutrients vis-a-vis chemical fertilizers declined substantially. With increase in cost of production inputs, inorganic fertilizers became increasingly more expensive. Another issue of great concern was the sustainability of soil productivity as land began to be intensively tilled to produce higher yields under multiple and intensive cropping systems. Water logging and secondary salinisation have been the banes associated with excess and irrational irrigation. Groundwater table declined sharply as more and more deep bore wells were drilled. Recharging of groundwater has also been reduced due to severe deforestation. Indiscriminate use of chemical pesticides to control various insect pests and diseases over the years, has destroyed many naturally occurring effective biological control agents. An increase in resistance of insect pests to chemical pesticides has also been noticed. Health hazards associated with intensive modern agriculture, such as pesticides residues in food products and groundwater contamination are matter of concern. The occurrence of multi-nutrient deficiencies and overall decline in the productive capacity of the soil due to non-judicious fertiliser use, have been widely reported. Such concerns and problems posed by modern-day agriculture gave birth to new concepts in farming, such as organic farming, natural farming, biodynamic agriculture, do-nothing agriculture, eco-farming, etc. The essential features of such farming practices imply, *i.e.,* back to nature.

Organic Farming and Country's Food Security

The primary concern of all organized communities and civilized societies is to meet the food requirements of its people. The cultivated area required to maintain the present level of food grain production in India without using the fertilisers, reaches more than the total geographical area of the country. At present, there is a gap of nearly 10 million tones between annual addition and removal of nutrients by crops which are met by mining nutrients from soil. A negative balance of about 8 million tones of NPK is foreseen in 2020, even if we continue to use chemical fertilizers, maintaining present growth rates of production and consumption. The most optimistic estimates at present, show that only about 25-30 per cent nutrient needs of Indian agriculture can be met by utilizing various organic sources. It is proved beyond doubt that on long-term basis, conjoint application of inorganic fertilizers along with various organic sources is capable of sustaining higher crop productivity, improving soil quality and soil productivity. The organic sources should be used in integration with chemical fertilizers to narrow down the gap between addition and removal of nutrients by crops as well as to sustain soil quality and to achieve higher crop productivity. The food security demand of the country requires that inorganic fertilizers be used in balanced doses.

Issues

Organic farming has the twin objectives of the system being sustainable and environmentally benign. In order to achieve these two goals, it has developed some rules and standards which must be strictly adhered to. There is a very little scope for change and flexibility. Organic farming thus, does not require best use of options available but the best use of approved options. These options are usually more complex and less effective than the conventional ones. The philosophy of the proponents of organic farming presents considerable difficulty to the scientists as the organic movement generally resists the comparisons between the two systems.

The organic farming movement presents a challenge to the scientists who cannot and would not want to abandon a scientific approach. The present lower productivity of the system is a result of constraints that its practitioners have put upon themselves. Difficulty faced is with evangelising statements, that the products of the organic farming system - food and fibre - are in some way better than those from conventional agriculture. The Indian farmer should get the advantage of emerging global market on organic farming which is at present around 26 billion US dollars, and is expected to grow to 102 billion US dollars in 2010. Currently, 130 countries are producing certified organic products. A vast scope for promotion of organic farming in the export market, without compromising with the national food security exists in the country, as farming by tribals and under rainfed conditions is generally organic, since very little chemical inputs are used.

Organic farming as a concept/philosophy is well tested in some of the western countries, though the same is not unknown to most of the nations. But in the Indian context, it needs to be looked into more critically, seeking answers to the following questions: What level of crop productivity is acceptable? Is it suitable for a country like India with such a large population to feed or can it fit in the niche areas? Are available organic sources of plant nutrients sufficient for organic farming in the form it is advocated? Are organic farming technologies sustainable in the long run?

Various aspects relating to organic farming were addressed and discussed in a workshop organised by the NAAS. Following issues were taken up:

☆ Redefining the concept of organic farming in Indian context.

☆ Opportunities and constraints in organic agriculture.

☆ Organic agriculture perspective and future challenges.

☆ Soil quality, biodiversity and fertility in organic farming system.

☆ Standards in organic farming and quality markers for organic produce.

☆ Setting up of certification and accreditation agencies for export.

☆ Setting up of testing units for soil and organic produce for chemical residues and quality control.

☆ Market development and assessment of organic produce for export.

☆ Evaluation of organic/biodynamic farming *vis-a-vis* integrated nutrient and pest management.

☆ Documentation and confirmation of existing Indigenous Technological Knowledge (ITK) on biodynamic farming.

☆ Development of bio-control measures for various plant diseases and pests.

☆ Research agenda for comparing organic and conventional agriculture.

☆ Holistic approaches in organic farming research and development.

Recommendations

Major recommendations that emerged from the workshop are listed below:

1. Organic Farming: Prospects and Limitations

☆ For India with its ever increasing population, the sustainable agriculture has to be based on site-specific balanced and adequate fertilization, and an integrated plant nutrient supply system (IPNS) involving organics, inorganics and biofertilizers. What the country needs today, is the conjunctive use of organic and inorganic sources of plant nutrients for sustainable productivity.

☆ Organic farming is a market demand-driven agriculture, aimed to cater to the foreign export and affluent section of the society in the country. However, in order to make a dent in the export market, we need to develop high-tech organic technology with strict quality control, meeting international quality standards prescribed for organic produce.

☆ Niches of the organic farming need to be identified. However, the real niches will be determined by the market infrastructure and the international links. The practice should be considered for lesser endowed regions of the country rather than in resource endowed regions which serve as the backbone of the country's food security.

☆ The availability of organic manures in adequate amounts and at costs affordable by the farmers is a major problem. The increased mechanization has further reduced the availability of manures with the farmers and this problem will become more acute in future. In such circumstances, post-harvest residues should be utilized to the fullest extent. However, to accomplish this objective, feasible technologies are needed for *in situ* recycling/rapid composting of on-farm residues and wastes, in addition to extension efforts to change the mindset of the farmers. Possibilities of using non-traditional organic sources e.g., slaughter house waste, should be exploited to partly supplement, plant nutrient needs of the organic farming systems.

2. Organic Farming: Its Relevance to Indian Agriculture

☆ In Indian context, organic farming has to be practiced without synthetic pesticides, but complete exclusion of fertilizers may not be advisable under all situations. A holistic approach involving integrated nutrient management (INM), integrated pest management (IPM), enhanced input-use efficiency, and adoption of region-specific promising cropping systems would be the best organic farming strategy for India.

☆ As organic farming is attracting worldwide attention, and there is a potential for export of organic agricultural produce, this opportunity has to be tapped with adequate safeguards so that the interest of small and marginal farmers is not harmed.

☆ Organic farming may be practiced in crops, commodities and regions where the country has comparative advantage. To begin with, the practice of organic farming should be for low volume high-value crops, like spices, medicinal plants etc., beside fruits and vegetables, for which R and D support is required.

☆ Organic farming should not be confined to the age old practice of using cattle dung, and other inputs of organic/biological origin, but an emphasis needs to be laid on the soil and crop management practices that enhance the population and efficiency of below-ground soil biodiversity to improve nutrient availability. Performance of cultural techniques for weed control and that of biopesticides for pest management need to be evaluated under field conditions, preferably under cultivators' management conditions. Besides the identification of regions suitable for the adoption of organic farming, the crops and their products should also be identified which are amenable for production through organic ways and have the potential to fetch a premium price in the international organic market.

3. Organic Farmers and NGOs

☆ Region-specific resource inventory, including animal wealth, farm residues/by products and their competitive uses, nonconventional nutrient sources of organic/biological origin etc., has to be prepared, for development of rational technology packages of organic farming.

☆ A strong technological back-up by scientific community should be provided in order to verify, confirm and further refine some selected ITKs like, *Agnihotra, Panchgavya,* pertaining to organic farming.

☆ Crop-specific and farming situation-specific package of practices for organic cultivation should be developed and after thorough on-farm validation, recommended for adoption. Such proven technology packages need to be documented in regional languages.

☆ Entrepreneurial potential with respect to production of organic inputs, processing and marketing of organic food should be fully exploited.

4. Trade and Certification Issues

☆ Certification of organic produce is an important issue that is central to organic production itself, as the prime goal of organic farming is and will be to fetch premium price in domestic and international market for the producers.

☆ For desired economic gains out of organic farming, the upcoming global market on organic produce has to be exploited, for which strict phyto-sanitary measures have to be followed.

☆ A strong research back up has to be put in place to develop and improve national standards for organic farming. The policy documents brought out so far by APEDA and DAC, including the report of Task Force on Organic Farming etc., should be considered in developing such scientifically sound standards, mainly for the crops and commodities that have export potential. The national standards should be the same for domestic market and for export.

☆ Weak links in the certification systems have to be identified and researchable areas flagged.

☆ Whereas organic certification will continue to be a process certification, strong research set-up/laboratories are required to monitor the quality of organic produce so as to prevent the sale of substandard material in the name of organic produce, and to save the interests of producers and consumers.

☆ Organic produce fetches premium price owing to better quality, and for credibility in the market it ought to be quality goods. Therefore, there should be adequate provision for their grading, packaging, storage and transportation. Marketing outlets on the lines of milk unions, may be established.

☆ There is a need to organise Producers' Cooperative Marketing Societies and establish credible marketing channels for steady flow of organic foods and materials in accordance with the demand, as also to safeguard the interest of small farmers opting for organic agriculture.

5. The Quality Aspects of Organic Food

☆ There is an urgent need to compare the quality of organically produced food with conventionally produced food. There appears to be a widespread perception amongst consumers, that organically produced foods are of superior nutritional quality. However, to prove or disapprove this contention, very limited research has been conducted, and whatever meagre scientific data is available, is often out-dated or based on inadequate study designs, lacking proper controls. In view of this, the following points merit consideration.

☆ No clear-cut evidence is available to support consumer perceptions regarding potential health benefits of organic foods. An in-depth research on quality aspects is required to arrive at any valid acceptable conclusion.

☆ Valid nutritional quality comparisons between organic and conventional food requires that plants be cultivated in similar soils, under identical climatic conditions, be sampled at the same time, pretreated similarly, and analysed by validated methods.

☆ Well-designed controlled studies in animal models and human subjects are needed.

☆ It is also necessary to undertake well-controlled studies to evaluate sensory properties, shelf life, and nutrient load of organic produce *vis-a-vis* produce from the conventional farming techniques.

☆ More importantly, organically and conventionally produced food should also be analysed for pesticide residues and microbiological safety *i.e.,* presence of pathogenic organisms which could pose health hazards.

6. Research Agenda

☆ Develop package of practices for integrated pest management for organic farming in different agro-climatic regions for specific crops involving components like bio-inoculants, pheromones etc., among others.

☆ The issue of microbial contamination of food arising from the use of manures has to be addressed, and measures suggested to mitigate it.

☆ It is frequently documented that fertilizers and pesticides applied at recommended rates have had no adverse effect on soil biological activity, and that integrated farming systems are best for nutrient management, yield sustainability and soil biodiversity conservation. There is thus, a need to develop modern organic farming system integrating the best available options.

☆ There is a need to establish referral laboratories for analysis of pesticides, heavy metals and mycotoxins in the produce with appropriate accreditation to help organic farming movement. The maximum residue limits in organic food must be set in accordance with the CODEX standards.

☆ The Green Revolution technologies have been alleged to have caused depletion of soil organic carbon. The critical values of soil organic matter that can support the sustainable crop production under organic management have to be worked out.

☆ The patterns of rate and amount of nutrients released from various organic sources and their goodness of fit with the nutrient requirement of the crops at different growth stages need to be worked out. This information could be used in evolving appropriate nutrient management schedule, so as to ensure optimal nutrient supply to the crop at active physiological stages having peak nutrient demands.

☆ Allelopathic effects of various plant species need to be tapped, particularly for weed and pest management.

Policy Paper No. 7: Diversification of Agriculture for Human Nutrition

Issued by National Academy of Agricultural Sciences (NAAS), New Delhi

The current agricultural production scenario reflects some disturbing tendencies from the point of view of human nutrition, particularly in rural areas where production and consumption are directly linked, and for the poor everywhere. Decline in areas of coarse cereals and pulses and other so-called 'low-value' crops which provide access to better nutrition for the poor, illustrates this concern. On the other hand, substitution of these crops by those with higher productivity has improved calorie availability and incomes of farmers who can increasingly afford better nutrition. Over the last 10-15 years, there has been a remarkable spurt in production of horticultural crops, livestock and fisheries, driven largely by buoyant domestic demand. Average per capita consumption of these commodities has increased.

Data on nutritional status in the country in relation to norms developed by nutrition scientists, on the other hand, reveal the appalling status of nutrition, in general and of vulnerable sections of the population in particular. The National Academy of Agricultural Sciences organised a symposium, to deliberate on these issues. Some salient observations, policy issues and recommendations emerging out of the deliberations are summarised below.

Food and Nutrition Security

Thanks to the Green Revolution, per capita availability of foodgrains has increased. With 200 million tonnes of foodgrains, calorie and protein deficiencies have largely been overcome. The country has achieved basic food security in terms of foodgrains. The averages, however, conceal some important differences. For example, distribution of food is highly unequal and the poor, who constitute nearly one-third of our population, continue to be food-deprived. Interregional and intrahousehold deficiencies are also significant.

Also, almost all the increase in foodgrain production has come from wheat and rice. The production of coarse grains (millets) and pulses has declined or remained stagnant, implying a decline in per capita availability of these commodities over time. Coarse grains have been important in the diets of the poor. These have relatively higher nutritive value in terms of proteins, vitamins and minerals compared to rice. Declines here, thus, affect food security of the poor qualitatively. Similar are the implications relating to pulses which are the major source of protein in vegetarian diets. Both pulses and coarse grains are important crops for dryland and fragile environments where poverty levels are high. Legumes have been the traditional restorer of soil fertility and declining area poses a threat to sustainability. Finally, the spectacular increase in foodgrain production and a reasonable overall agricultural performance has failed to eradicate malnutrition.

Nutritional security, thus, goes beyond food security as we understand the term. The latter means ensuring adequate availability of foodgrains to provide calorie and, may be, protein needs of the people, the former implies adequate supply of micronutrients such as vitamins and minerals as well. As we move up the development ladder, this becomes more relevant indicator of food security. To ensure nutritional security, increased availability of diverse types of food, such as millets, pulses, fruits and vegetables, foods of animal origin (milk, eggs, meat, fish), besides cereals, is essential. While fruits and vegetables are rich sources of micronutrients, animal-based foods abound in quality proteins as well. Vegetables and fruits also contain some health-giving phytochemicals which are powerful antioxidants and detoxifying agents which protect against degenerative diseases. Marine fish is a rich source of long chain n-3 fatty acids which have important physiological role.

Diversification Scenario

Cropping patterns have traditionally been dominated by food needs. Commercial crops were confined to some regions and on relatively larger farms. The system we inherited at the time of independence became unsustainable as rapid population growth outstripped our capacity to produce food. Fruits, vegetables, milk, meat, fish, became luxury foods, whose demand was confined to the very small rich class in rural and urban areas. A food insecure nation, despite devoting bulk of its agricultural resources to food production, became chronic importer of food. Even extension of cultivation to marginal and sub-marginal lands did not help.

The Green Revolution transformed this scene. In less than a decade we were able to achieve reasonable food security. High growth in productivity of cereals spurred agricultural growth and incomes. Rising incomes prompted shifts in consumption patterns and demand for non-cereal food became buoyant. By mid-eighties expansion of area under cereals ceased. Producers too began to look for alternatives and the process of diversification set in. It became the *mantra* for agricultural development in the nineties. The following paragraphs document these developments briefly:

- ☆ In the *horticulture* sector, there has been 2-3 fold increase in production of fruits and vegetables in the last 10-15 years. Their area has gone up, particularly in rainfed, hilly, and coastal regions prompting fears of diversion of land from foodgrains. This has so far not materialised. Among fruits, banana, papaya, citrus and mango have contributed to the significant increase. All these are nutritionally rich fruits and their increased production has resulted in decline in real prices and penetration in rural markets.

- ☆ *Livestock* production has been increasing at 10-15 per cent over the last decade or two. Milk, eggs and meat production have registered spectacular growth. Despite these increases, the annual per capita availability is much below the amounts recommended by ICMR. In terms of these, the present availability is 86.4 per cent for milk, 16.7 per cent for eggs and 29.6 per cent for meat.

- ☆ *Fish* is one of the best sources of protein and fisheries sector has grown substantially. Marine products' industry is the largest foreign exchange earner. Scientific acquaculture and deep sea fishing have contributed to the growth of these sectors in the nineties. There are, however, reports that these developments have been unsustainable and have not contributed to livelihood security of small fishermen in coastal areas. The sector faces infrastructural constraints which inhibit exploitation of the potential.

- ☆ *Genetic engineering* is a powerful tool for improving the yields and quality of both plant and animal foods. Both macro- and micro-nutrient content can be enhanced with the now available and evolving biotech tools. Some promising breakthroughs in improving the protein and micronutrient content of crops, such as maize, rice, rapeseed, tomato, etc., have been achieved. However, safety aspects of GMO foods have become an area of concern and debate, globally.

Issues

Changes in Food/Nutritional Security Status

There is agreement that increased per capita availability of cereals and a number of other foods has contributed to greater food security at the national level. At the household level, consumption expenditure surveys by the National Sample Survey Organisation reveal declining consumption of cereals and increases in livestock and horticultural products, sugar and edible oils implying an average improvement in nutritional status. These, as well as, results of the surveys of the National Nutrition Monitoring Board, bear out gains in calorie and protein content of diets, essentially from improvements in non-grain food consumption, driven by income growth and poverty reduction. These surveys, however, also reveal the inequities across regions, farms of different sizes, sociocultural groups, as well as in intrafamily distribution of food. To illustrate, nearly 26 per cent of the rural farming households, mainly sub-marginal and marginal farmers, were nutritionally deprived.

Nutritionists evaluate nutritional security in terms of adequacy of macro- and micronutrient with reference to clinical nutritional norms. NNMB surveys indicate that, in general, the consumption of cereals and millets in rural and tribal groups was comparable to balanced diets, but the intakes were low in urban slums. The consumption of micronutrientrich foods, such as green leafy vegetables, other vegetables and fruits falls below recommended levels, particularly among the poorer population groups. These studies also show sub-standard consumption of green leafy vegetables, milk, fats and oils, sugar and jaggery in the diets of children. The deficit in mean energy intakes among children of preschool and school age was about 25 per cent, but intake of iron, vitamin A and riboflavin was higher across the board. In case of rural pregnant and nursing women, vitamins A, C, and B complex and calcium deficits were higher than those of energy and protein. Both these kinds of surveys, suggest the need to look beyond total production, availability, and average profiles.

Economists view food consumption as driven by self-provisioning, food habits and changes therein, education, incomes, prices, and availability. These and other interacting variables determine the food choices of households as consumers. Empirical studies on recent Indian data indicate that, on the average, calorie consumption has ceased to be income-responsive, implying a switch to non-calorie food with further income growth. At lower end of the income distribution, however, overcoming calorie deficiency remains a priority. All such analyses point to the need for proper targeting of food intervention programmes. Universal public distribution kind of programmes have outlived their utility.

Diversification

Driven by self-sufficiency motive, cropping patterns in India were locked to foodgrain production for a long time–a few regions devoting some resources for commercial crops like cotton, sugarcane, groundnut, etc. or plantation crops. As foodgrain production saturated markets, a trend towards diversification set in. Changes in pattern of domestic demand and, to some extent, export demand in the wake of trade liberalisation, resulted in changes in resource use and increasing diversification of enterprises. States which diversified the crop sector in a big way, have attained relatively higher growth in the net state domestic product of agricultural sector during the past two decades.

The factors that led to diversification of agriculture have varied, over time. During the first 15 years following the onset of Green Revolution, irrigation played the most important role, predominance of small holdings discouraged it. Abundant and cheap supply of electricity also fostered specialisation. Since early eighties, credit availability emerged as a significant determinant of diversification. Smaller farms continued to face rigidity in cropping patterns because of binding food production constraint. They diverted their attention to livestock enterprises. At the end of the millennium, there was consensus that diversification to higher value enterprises like, vegetables, fruits, other specialty crops, livestock products, fisheries, value-added agricultural products etc., was the new pathway for income growth in agricultural and rural sector. This would also help in bridging the quality gaps in terms of nutrition. Most of these enterprises were traditionally supplementary in nature and had become the domain of women. This bias adds to the strength of this paradigm.

Concerns and Constraints

☆ There is some apprehension that increasing diversification of land to non-food crops may affect basic food security adversely. Stagnation or decline in area under foodgrains would undermine our self-sufficiency in food. Some projections suggest massive imports of foodgrains, driven by rising food needs of a growing population and for animal consumption. A country of our size cannot reliably depend on the world market to meet these needs.

☆ This perception is countered by the scientific community. They contend that productivity levels are very low for most of the crops and animals. Exploitation of this yield gap through research can ease the pressure on land resources substantially. It was also pointed out that animal production systems in India are based on byproduct and waste utilisation. Increased efficiency of this system would moderate demand for foodgrains for animal consumption.

☆ In fact, there have been instances of ecological degradation in the wake of growth in such enterprises. Brackishwater aquaculture, massive use of pesticides on fruits and vegetables, biodiversity erosion, salinisation and waterlogging in Punjab-Haryana and parts of Rajasthan, are examples sometimes alluded to, as consequences of diversification.

☆ These instances need to be carefully diagnosed because analysts also attribute these negative externalities to deficiencies in policies relating to pricing, investments, etc. The real challenge would be to achieve such intensification in an ecologically benign manner, through appropriate policies and technologies.

☆ High post-harvest losses significantly undermine the prospects of diversification, particularly through high-value perishables, like horticultural and livestock produce. Losses through inefficient handling, transport, storage etc., rise exponentially as scale of output expands. There are increasing instances of total loss in the fields as farmers cannot even cover harvest costs due to crash in prices at harvest time. We need to strengthen processes, institutions and R and D efforts to tackle these constraints. The priority accorded to agroprocessing industries is a welcome policy initiative in this direction.

☆ Also important are infrastructural constraints. Most of the commodities in the new diversification basket were traditionally confined to local markets in unprocessed form and in small volumes. Facilities like cool chains, refrigerated transport, modern abattoirs, processing plants and other infrastructure were not relevant. Now the market is global and weak support in infrastructure restricts growth prospects. Similar constraints emerge in institutional infrastructure. Extension, market, credit, information—all have focused on food and commercial enterprises of major importance. In the present context, these create bottlenecks.

☆ There is evidence to show that because of binding food needs, small farmers are able to diversify only to a limited extent. This implies that as this process accelerates, such farms will lose ground further in terms of income growth. There are special programmes like IRDP which provide support for diversification in respect of small and poor farmers, but more needs to be done. Increase in productivity of their basic staples would provide a significant leeway for incorporation of other enterprises. Small producers also face scale related constraints–the size of their output is small and incomeenhancing options like access to superior markets, value addition, storage, etc. become non-viable. New institutional arrangements are necessary if this handicap is to be overcome.

☆ There is also an apprehension that shift from traditional foods many of which are rich in vitamins and micronutrients, would result in decline in nutritional quality of diets. Substitution of millets by wheat and rice, for example, does not augur well for poor consumers who cannot afford supplementary alternatives. Fall in consumption of pulses is also a cause for concern. This nutritionally regressive substitution in production patterns arises from technology and price changes which affect profitability and incomes of producers. Lack of nutritional awareness leads to undervaluation of nutritive food by consumers and in the marketplace. Nutrition scientists highlight such dimensions of commercially-driven diversification and plead for investment in nutrition education.

Conclusions and Recommendations

Shifting agricultural resources to higher-valued options is the new strategy for agricultural development. Buoyancy in domestic demand for such commodities has generated congenial incentive environment for such transition and the process has begun. Good export prospects reinforce this trend. Non-conventional crops like aromatic and medicinal plants, floriculture, etc., figure importantly in this strategy but the major impetus comes from horticulture, livestock, dairy poultry, fisheries, etc., which have traditionally been minor constituents of Indian diets. The nutritional implications are obvious. Growth in incomes has spurred demand for these commodities even as food grain consumption stabilises, and producers have responded to such market signals.

General inadequacy of Indian diets in terms of micronutrients and vitamins is well established. For the poor, access to even macronutrients is constrained. As overall wellbeing improves with future growth in incomes, special attention will have to be paid to nutritional aspects. Salient recommendations relating to diversification and nutrition emanating from the symposium are summarised below.

1. Policy Imperatives

☆ Thrust on raising productivity of foodgrains must remain a central feature of agricultural policy. Only through this route can the twin objectives of self-reliance and rapid rural income growth be realised. It will enable unlocking of resources which would otherwise remain tied to less remunerative enterprises.

☆ Diversification of production base of Indian agriculture requires massive investments in rural and other infrastructure. Apart from roads, electricity, irrigation, greater emphasis on storage, specialised handling and transport, assembling, wholesale and retail markets, effective market intelligence, etc. will be needed.

☆ The emerging economic regime requires dismantling all distortions in input-output pricing. These impart incorrect price signals and farmers are distracted from efficient production patterns. Massive irrigation subsidy and its effect on cropping patterns is an illustration. The structure of tariffs is also a case in point.

☆ A reorientation of the institutional support for agriculture will be necessary for exploitation of new opportunities by small producers who constitute more than 80 per cent of the farming households. Input-output marketing, value addition and processing, credit, insurance, R and D, extension, etc. need to shift from foodgrain and large-farm-based approaches to a more holistic paradigm. Leasing and tenancy reforms will be necessary.

☆ Agroprocessing investments must move to the countryside where production is concentrated. While technology and quality considerations may necessitate foreign investments in this sector, mechanisms will have to be developed to ensure effective small farm participation.

☆ A task of this magnitude and complexity will necessitate a dominant role for the private sector. The public sector will need to withdraw from some areas and strengthen others, like R and D, information, natural resource management, regulatory processes and so on. A set of policies to provide incentives to the private sector will be necessary.

☆ A large population will continue to be economically and nutritionally deprived. People below the poverty line, women and children, particularly in rural areas, and urban slums, will need strong safety nets. Weaknesses in existing programmes have been well identified and are being addressed. These need to be pursued more vigorously.

☆ These challenges are beyond the competence and resources of governments. It will be necessary to involve people in planning and executing decentralised initiatives. Nongovernmental organisations, self-help groups, cooperatives, *panchayats* will need to play a greater role and these must be strengthened.

2. Nutrition

☆ Nutrition education must be made part of regular curricula in schools. Sustained drives using mass media, particularly in rural areas and urban slums are necessary to create greater awareness.

☆ Programmes like homestead gardening, urban gardening, household preservation and enrichment of food, etc. must be actively supported. Health and hygiene, sanitation, etc. make significant contributions to nutritional well-being and should be accorded greater priority.

☆ Food enrichment, fortification strategies need to be supported. Assessment and incorporation of indigenous ingredients offer considerable opportunities and should be exploited.

☆ National nutrition monitoring effort must be further strengthened and focused on target themes and populations.

3. R&D

☆ Continuous increase in productivity of agricultural enterprises—crops, animals, fish, is essential. This would ease the subsistence pressure on natural resources, relating them to commercial enterprises. The research system must continue to accord high priority to food crops, particularly those which are of importance to the rural poor and tribal farmers.

☆ Advances in modern sciences, particularly biotechnology, offer exciting opportunities for incorporation of marketable and nutritional qualities in food crops of various kinds. Even as research on genetically modified organisms is accelerated, proper testing and safeguard procedures need to be put in place. The point to note is that this area of research can tackle several constraints inhibiting yield, quality, and nutrition.

☆ More resources should be allocated for nutrition research. There are basic as well as applied research issues relating to indigenous food, nutraceuticals, formulations, food safety, standards, etc. which need to investigated. There is enormous variability in food habits, tastes and preference products etc. across the country and these must be captured and analysed.

☆ Unmindful pursuit of market opportunities often exacerbates pressure on natural resources and ecology. Safeguarding future production potential and ecological balance should be high priority for research.

☆ Wide diversity in growing conditions implies a wide range of options for diversification and income growth. This is a big strength for Indian agriculture, but this necessitates decentralised research approaches which maximise comparative advantage of different regions. A careful regional prioritisation of research is called for.

☆ Preventing post-harvest losses has emerged as a critical element, and so has value addition and processing. Known technological options in these areas are highly capital intensive and not really appropriate for small scale operations which are characteristic of the Indian rural scene. The research system faces this unique challenge of developing efficient small scale technologies which will benefit small scale rural producers and entrepreneurs.

☆ Finally, success of the diversification strategy would demand research on a number of socioeconomic parameters like, market structure, conduct and performance, input-output demand, comparative cost and returns, price analysis, organizing producers' and entrepreneurs' private sector role, etc. While the research system is gearing up to meet production research challenges, this area must also receive attention.

Contents

List of Contributors

Aastik Jha
ICAR-Indian Institute of Vegetable Research, Varanasi – 221 305, U.P.

Anant, Bahadur
ICAR-Indian Institute of Vegetable Research, Varanasi – 221 305, U.P.

Babu, K. Nirmal
Indian Institute of Spices Research, P O Marikunnu, Calicut – 673 012
e-mail: nirmalbabu30@hotmail.com, nirmalbabu@spices.res.in

Bakshi, Prashant
Division of Fruit Science, SKUAST-Jammu Campus, Chatha, Jammu – 180 009, J&K
e-mail: bakshi_parshant@rediffmail.com

Basak, B.B.
Directorate of Medicinal and Aromatic Plants Research (DMAPR), Boriavi, Anand – 387 310, Gujarat

Bharati, M.
Untereggweg 7, 4147 Aesh, Basel, Switzerland
e-mail: bharati_bth@yahoo.com

Bonny, Binoo
Department of Agricultural Extension, Kerala Agricultural University, P O Mannuthy – 680 651, Kerala
e-mail: binoobonny@gmail.com

Geetha, P.
Radio Tracer Laboratory, Kerala Agricultural University, P O KAU Thrissur – 680 656, Kerala

Indira, P.
Department of Olericulture, Kerala Agricultural University, P O KAU Thrissur – 680 656
e-mail: indusowhrudham@yahoo.co.in

Irenaes, T.K.S.
Department of Fruits and Orchard Management, Faculty of Horticulture, Bidhan Chandra Krishi Vidyalaya, Mohanpur, Nadia – 741 252, West Bengal

Koley, T.K.
ICAR-Indian Institute of Vegetable Research, Varanasi – 221 305, U.P.

Kour, Ganganpreet
Division of Fruit Science, SKUAST-Jammu Campus, Chatha, Jammu – 180 009, J&K

Kumar, Suresh
Radio Tracer Laboratory, Kerala Agricultural University, P O KAU Thrissur – 680 656, Kerala
e-mail: sureshkumar.paikad@gmail.com

Mahtab S. Bamji
211, Sri Datta Sai Apartments, RTC Cross Road, Musheerabad, Hyderabad – 500 020
e-mail: msbamji@gmail.com

Maity, A.
ICAR-National Research Centre for Pomegranate, Solapur – 413 006, Maharashtra

Masood, F.A.
Department of Food Science and Technology, Kashmir University, Hazratbal, Srinagar – 192 001, J&K
e-mail: masoodi_fa@yahoo.co.in

Mitra, Sisir K.
Department of Fruits and Orchard Management, Faculty of Horticulture, Bidhan Chandra Krishi Vidyalaya, Mohanpur, Nadia – 741 252, West Bengal
e-mail: ssirm55@gmail.com

Muthamilarasan, Mehanathan
National Institute of Plant Genome Research, Aruna Asafali Marg, New Delhi – 110 067

Nirmala Devi, S.

Department of Olericulture, Kerala Agricultural University, P O KAU Thrissur – 680 656, Kerala
e-mail: drsnirmala@yahoo.com

Pal, R.K.

ICAR-National Research Centre for Pomegranate, Solapur – 413 006, Maharashtra
e-mail: krishnapal@gmail.com

Peter, K.V.

Director, World Noni Research Foundation, 12, Rajiv Gandhi Salai, Sreenivasa Nagar, P O Perungudi, Chennai – 96, T.N.
e-mail: kvptr@yahoo.com

Prasad, Manoj

National Institute of Plant Genome Research, Aruna Asafali Marg, New Delhi – 110 067
e-mail: manoj_prasad@nipgr.ac.in

Pradeepkumar, T.

Department of Olericulture, Kerala Agricultural University, P O KAU Thrissur – 680 656, Kerala
e-mail: pradeepkau@gmail.com

Prema, A.

Department of Agricultural Extension, Kerala Agricultural University, P O Mannuthy – 680 651, Kerala

Raigond, Pinky

ICAR-Central Potato Research Institute, Shimla – 171 001, H.P.

Rakshit, Rajiv

Bihar Agricultural University, Sabour, Bagalpur – 813 210, Bihar
e-mail: biraj.ssac@gmail.com

Rani, Pathipat Usha

Indian Institute of Chemical Technology, Hyderabad – 500 007
e-mail: usharani65@yahoo.com

Sadhankumar, P.G.

Department of Olericulture, Kerala Agricultural University, P O KAU Thrissur – 680 656, Kerala
e-mail: psadhankumar@yahoo.co.in

Saxena, K.B.

ICRISAT, Hyderabad
e-mail: kbsaxena1949@gmail.com

Singh, Amit Kumar
ICAR-Indian Institute of Vegetable Research, Varanasi – 221 305, U.P.
e-mail: singhab98@gmail.com

Singh, B.P.
ICAR-Central Potato Research Institute, Shimla – 171 001, H.P.
e-mail: directorcpri@gmail.com

Singh, Brajesh
ICAR-Central Potato Research Institute, Shimla – 171 001, H.P.

Singh, B.
ICAR-Indian Institute of Vegetable Research, Varanasi – 221 305, U.P.
e-mail: bsinghiivr@gmail.com

Suma, S.
College of Horticulture, Kerala Agricultural University, P O KAU
Thrissur – 680 656, Kerala

Tarafdar, Jagdish
ICAR-Central Arid Zone Research Institute, Jodhpur – 342 003, Rajasthan
e-mail: jctarafdar@yahoo.com

Introduction

The Tamil saint poet Thiruvalluvar wrote in the classic Thirukkural "What else is more painful than poverty? Poverty alone as painful as poverty". He further writes "Farmers are the linchpin of the whole world as they feed even those not in the field". Gandhiji liked to be addressed as a humble farmer. He wrote in the visitors book at Buckingham Palace against the column 'occupation' A Farmer. He encouraged the inmates of Sabarmathi Ashram to grow fruits and vegetables to meet their nutritional requirements. The infamous Bengal Famine during 1942-43 devastated millions of people. The resurgence of India from the status of 'ship to mouth economy to farm to ship' with a buffer stock of 82 million tones of food grains demonstrates the symphony among policy support, science and technology and hard working farmers. The three "A"-availability, access and absorption-are the key words of food and nutrition security. Purchasing power is key to access the food which unfortunately is not within the hands of 360 million people in India. Malnutrition-under and over-is rampant and the most worrying is undernutrition and hidden hunger. After more than 70 years of independence, India continues to battle with pre-transition diseases like infections and under nutrition. Over 50 per cent of pre-school children and 30 per cent adults are undernourished as judged by anthropometric indices and over 70 per cent of women and children suffer from iron deficiency anaemia. Every third child is borne with low birth weight, and is condemned to poor mental and physical development and immunity unless rehabilitated within the first year of life. Intrauterine malnutrition epigenetically predisposes to cardiovascular diseases in later life. Almost 60 per cent of deaths and 42 per cent DALYs (disability adjusted for year) lost are due to communicable diseases, perinatal and maternal conditions and nutritional deficiencies. Post-transition life-style related diseases like obesity and chronic degenerative diseases are increasing, with India becoming world capital of diabetes. Over 10 per cent of Indians are obese, the incidence being almost 20 per cent in urban areas. Apart from human sufferings due to morbidity and mortality, malnutrition is severely denting India's productivity and development and adding to medical expenditure.

Nutrition security implies physical, economic and social access to balanced diet, clean drinking water, safe environment and health care-preventive and curative-for every citizen. Education and awareness are needed to utilize these services. Malnutrition has a complex aetiology and its prevention requires awareness and access at affordable price to all the above-unreached and undernourished-. Diet surveys show that Indian diets are qualitatively more deficient in vitamins and minerals(hidden hunger) than proteins due to low intake of income elastic foods like fruits, vegetables and foods of animal origin. Nutritious millets are disappearing. Within the family, diet of pre-school children is particularly inadequate due to ignorance and time constraints on mothers rather than affordability (Mahtab Bamji, 2009). More than 70 per cent of preschool children consume<50 of recommended amount (RDA) of iron, vitamin A and some B vitamins, particularly riboflavin and folic acid.

India is committed to achieve millennium development goal (MDG) of UN by 2015 especially reducing the percentage of people below poverty level to half (13 per cent). The National Food Security Bill-2013 enacted by Indian Parliament makes access to food as a fundamental right especially to people below poverty level. The bill "provides for food and nutritional security in human life cycle approach, by ensuring access to adequate quantity of quality food at affordable prices to people to live a life with dignity and for matters connected therewith or incidental thereto". Indian horticulture is in sound footing, the sub-continent hosting two hotspots of biodiversity, suitable soils and climate for an array of fruits, vegetables, tuber crops, spices and plantation crops and above all indigenous traditional knowledge supplemented with world's largest scientific man power support. India is also the second largest domestic market making horticulture crops production attuned to a demand-supply chain. Threats to Indian horticulture is the high harvest and post harvest losses (30-40 per cent) and poor infrastructure like near absence of cold storage chains and packing. The recent interest in Protected Cultivation of fruits and vegetables under fertigation is creating newer opportunities. Urban and peri-urban horticulture are providing new opportunities, challenges and threats. Demands for fruits and vegetables have gone up in urban markets and production nearby areas is turning conventional horticulture into a commercial venture.

India inhabiting 1200 million people has around 240 million homesteads. With urbanization, industrialization and migration area under traditional agriculture is going down. Homesteads are evolving to production centres for self consumption and marketing in near by markets. Concepts of co-operative farming and farmers co-operatives are adding to family income and family nutrition. The state of Kerala alone has 6 million homesteads and efforts are made to transform them to homestead vegetable/fruit/spice gardens. Named kitchen garden, nutrition gardens and home garden, they are saving space, water, nutrients, energy and in addition provide remuneration to family members. Nutrition garden therapy is highly recommended to people suffering from depression and also to differentially disabled. The Japanese system of "double digging" where one meter depth pits with four meter long and two meter width is enough to produce fresh vegetables to a family of 5 members. Recently vertical farming, terrace gardening, container gardening, hydroponics, aeroponics and vegetable forcing are providing concepts and opportunities to produce green and fresh vegetables.

Four policy papers one on 'Innovation in family farming' by Food and Agricultural Organization of United Nations, second on 'Micro-nutrient security for India: Priorities for research and action' by Indian National Sciences Academy, New Delhi, third on 'Organic farming: Approaches and Possibilities in the context of Indian Agriculture-Policy paper 30' and fourth on 'Diversification of Agriculture for Nutritional Security-Policy paper 7" by the National Academy of Agricultural Sciences, New Delhi are given as preambles to the book. The book HORTICULTURE FOR NUTRITION SECURITY carries 21 chapters dealing with crops, farming systems, soil fertility, green technologies for plant protection, emerging technologies like nano technology, hybrids using male sterility and the last common man's greens and potatoes. The year 2014 being the year of family farming, the book is designed to explain the strategy, policy initiatives, targets and perceived action points.

K.V. Peter

2015, Horticulture for Nutrition Security

Editor: **Prof. K.V. Peter**

Published by: **DAYA PUBLISHING HOUSE, NEW DELHI**

Chapter 1

Aetiology and Consequences of Malnutrition and Way Forward

Mahtab S. Bamji

The term malnutrition implies both under-nutrition and over-nutrition. After over 60 years of independence, India- a country in developmental transition continues to battle with the pre-transition diseases like infections and under-nutrition. Over 50 per cent of preschool children and 30 per cent adults are undernourished as judged by anthropometric indices and over 70 per cent of women and children suffer from iron deficiency anaemia. Every third child is born with low birth weight, and is condemned to poor mental and physical development and immunity unless rehabilitated within the first year of life. Intra-uterine malnutrition epigenetically predisposes to cardiovascular diseases in later life. Almost 60 per cent of deaths due to major infections and diseases are caused by superimposed malnutrition. In India, 36 per cent deaths and 42 per cent DALYs lost are due to communicable diseases, perinatal and maternal conditions and nutritional deficiencies. In the mean time post-transition life-style related diseases like obesity and chronic degenerative diseases are increasing, with India becoming world capital of diabetes. Over 10 per cent Indians are overweight or obese, the incidence being almost 20 per cent in urban areas. Apart from human suffering caused due to morbidity and mortality, malnutrition, is severely denting India's productivity and development, and adding to medical expenditure.

Nutrition Security implies physical, economic and social access to balanced diet, clean drinking water, safe environment, and health care (preventive and curative) for every individual. Education and awareness are needed to utilise these services. Thus malnutrition has a complex aetiology and its prevention requires **A**wareness, and **A**ccess at **A**ffordable price to all the above. Women's health, nutrition, education and decision making through empowerment are important for nation's nutrition security but remain neglected due to cultural and allegedly scriptural biases.

Countrywide diet surveys show that Indian diets are qualitatively more deficient in vitamins and minerals (hidden hunger) than proteins due to low intake of income-elastic foods like vegetables, fruits, pulses and foods of animal origin. Nutritious millets are disappearing. Within the family diet of preschool children are particularly inadequate, due to ignorance and time constraint on mothers rather than affordability. More than 70 per cent preschool children consume < 50 of recommended amount (RDA) of iron, vitamin A, and some B vitamins particularly riboflavin and folic acid.

Within India states like Kerala and Tamil Nadu have relatively better nutrition parameters than states with higher calorie intake (Madhya Pradesh) or economic growth (Gujarat, Maharashtra) suggesting that the situation is more complex than mere access to food (calories) or income, important as they are. Time trends suggest that over the years intake of all the food groups and nutrients has declined, but the magnitude of malnutrition, has not worsened, in fact there is marginal improvement and severe clinical forms are rare, except anaemia, whose incidence and severity have not changed- in fact marginally increased.

Nutrition-infection is a vicious cycle. Malnutrition reduces immunity and infections and disease reduce appetite, impair absorption and lead to catabolic losses of precious nutrients. Thus apart from physical and economic access to food, access to clean environment and drinking water are areas of great concern. Increasing incidence of obesity and chronic diseases is due to more sedentary lifestyles, shift to less fibre, high fat refined carbohydrate diets, stress and addictions. Crowded urban areas leave little space for physical activity like walking or play even for children.

Neither government nor scientists can be faulted for being silent spectators. Efforts have been made, but something is missing and situation continues to be grim. Food grain (wheat and rice) production went up markedly and kept ahead of population till mid nineties, but is tending to plateau. Unfortunately pulse production has stagnated and per capita availability has declined. There is erosion of millets production and consumption. Milk, fruit and vegetable production has increased markedly with India holding 1ˢᵗ and 2ⁿᵈ positions respectively, but that is not reflected in the diet of the poor due to poor purchasing power, and lack of awareness about their nutritional importance among the producers. Loss of almost 30 per cent of farm produce is occurring due to inadequate post harvest storage facilities, and food processing for value addition. New technologies for bio-fortification of crops have been developed, but languish due to uninformed opposition and inability to put in place convincing safety guidelines and measures. Several programmes, missions and acts including a National Nutrition Policy (1993), National Nutrition Plan of Action (1995) and National Nutrition Mission (2001), have been formulated with scientific and technological underpinning. But they have failed to achieve nutrition goals. Some of the reasons are:

1. Nutrition is a poor cousin even in health and agriculture planning and execution.

2. Nutrition improvement is not a stated goal with measurable parameters for monitoring, in missions like National Food Security Mission, National Horticulture Mission and even the recent National Rural Health Mission, leave aside others aimed at income, sanitation and drinking water.

3. Top-down approach without preparing the community and making them partners in planning and execution.

4. Poor targeting, accountability, and governance.

5. Inadequate importance to nutrition in school, college and even professional (health, agriculture, social science) education.

6. Neglect of female health, education and empowerment.

7. Vertical programmes with poor convergence and synergy between functioning of ministries and departments.

The Way Forward

Concentrate on proven interventions which have reduced the scale of malnutrition in less endowed countries. Some of these and other implementable suggestions are:

1. Proper breast feeding and complimentary feeding practices, as prescribed by WHO/UNICEF and support systems to enable infant care.

2. Nutrition management during illness, including diarrhoeas.

3. Early detection and effective home-based management of mild and moderate under nutrition and referral and therapeutic feeding for rehabilitation of severe under nutrition.

4. Full immunisation.

5. Women's education, health and empowerment- a life cycle approach.

6. Access to clean environment, drinking water, and food safety

7. Increased food production using conventional and new technologies, nutritionally oriented cropping pattern, decentralised planning for food production including homestead production of income- elastic protective foods and advocacy for dietary diversification.

8. Distribution of salt fortified with adequate iodine and ensuring its consumption in all areas particularly endemic areas for iodine deficiency. Now that salt double fortified with iron and iodine with proven efficacy is available, it should replace iodised salt.

9. Effective distribution of iron folic acid tablets for pregnant and lactating women, children and adolescent girls and de-worming.

10. Bi-annual supplementation of massive dose vitamin A in areas, where vitamin A deficiency is a public health problem. Emphasis should be on

promotion of nutritionally well endowed vegetables and fruits for food-food fortification. There is enough pro-vitamin A in dark green leafy vegetables, leafy portion of some vegetables like cauliflower, radish etc. and yellow orange fruits and vegetables, and they should be promoted.

11. Popularisation of the Food guidelines for Indians through media and educational blitz.

12. Universalisation of public distribution system and broadening the basket with inclusion of millets, pulse and blended oils

13. Integrated post-harvest management including establishment of silos in every block should receive high priority to prevent wastage and generate employment.

14. Town planning should ensure lung space and place for walking and exercise. All schools should have play ground and physical training.

15. Nutrition should be clearly stated as an important input and output parameter for judging development and should not be treated as trickle down beneficiary of economic and industrial development. It should not get subsumed under curative or preventive health care in general, where emphasis tends to be on chronic diseases and immunisation- important as they are. Without Nutrition, neither communicable nor non-communicable diseases can be prevented and hence it should have an important status as an independent entity. Malnutrition is the worst form of non-communicable disease.

16. Leadership and efficient governance is required at all levels to ensure synergy through convergence between Programmes/Missions/Acts which impact nutrition directly or indirectly (income, sanitation, drinking water, feeding programmes etc.) run by different departments/ ministries like health, women and child development, agriculture, civil supplies, etc.

17. Planning and execution should be done with community participation and involvement of trained nutrition leaders from the community. Anganwadi workers, along with ASHA workers in collaboration with KVK functionaries should be trained as first-contact health and nutrition activists ('hunger fighters').

18. Greater dialogue and interaction between nutrition scientists and scientists belonging to agriculture, food technology, medicine, public health, and basic sciences as well as social scientists is needed.

19. National Nutrition Monitoring Bureau (NNMB) which now operates in 9 states should cover all the states and have wider coverage, with additional component of Nutrition surveillance.

This long wish list cannot be curtailed if the dream of Nutrition Security has to become reality.

Apart from human suffering, developing nations like India suffer substantial economic loss due to malnutrition. World Bank and other economists have emphasised the importance of nutrition for national development as follows:

"Malnutrition is costing poor countries up to 3 percent of their yearly GDP. Malnourished children are at risk of losing more than 10 percent of their lifetime earnings potential. Malnutrition may increase the risks of HIV infection, while reducing the numbers of children and mothers who survive malaria. Developing nations should reposition nutrition as central to development."

According to a group of economists including three Nobel laureates (the Copenhagen Consensus 2004) "investment in nutrition is one of the `best buys' that developing countries can make for economic growth".

Source: Current Science C V Raman Avenue P B No.8001 Bangalore-560080.

2015, Horticulture for Nutrition Security
Editor: **Prof. K.V. Peter**
Published by: **DAYA PUBLISHING HOUSE, NEW DELHI**

Pages ***7–18***

Chapter 2

Economics of Family Farming

K.V. Peter, Binoo P. Bonny and A. Prema

Despite tremendous gains in global food production, there is unusual increase in the number of undernourished in recent years with figures crossing 1 billion mark in 2009-10 with marginal reductions there after (FAO, 2008). More importantly, the majority of the people who cannot meet their nutritional needs depend on agriculture for livelihood. The FAO (2010) estimates of the global trends of poverty and under nourishment as reported in Table 2.1 show that the number of undernourished people in 2007 increased by 75 million, over and above FAO's estimate of 848 million

Table 2.1: Trends in Global Under Nutrition (1990-92 base year)

Sl.No.	Country/Region	1990-92 Million (per cent)	2003-2005 Million (per cent)
1.	World	842	848
2.	Developing countries	823	832
3.	India	207 (26)	231 (28)
4.	China	178 (21)	123 (14)
5.	Asia Pacific (excluding India and China)	(22)	
6.	East and North Africa	19 (2)	33(4)
7.	Developed countries	19 (2)	16 (2)
8.	Latin America and Carribia	53 (6)	45 (5)
9.	Sub Saharan Africa	169 (20)	212 (25)

Source: FAO.

undernourished in 2003-05. The Asia-Pacific region with 578 million undernourished and Sub-Saharan Africa with 239 million hungry people account for 62 and 26 percent respectively of the world's food insecure population. Much of this increase is attributed to the continued and drastic increase in prices of staple cereals, pulses and oilseeds crops (Figure 2.1) since 2004 leading to an estimated 100 million more people pushed into poverty. Unlike earlier food price shocks which were mainly driven by short term food shortages, the 2007-08 price spikes came in a year when the world reaped a record grain crop with more than 100 million metric tons of extra grain compared to the previous year. In this the relation between price volatility and hunger has been explained in terms of widening demand-supply gap attributed to population growth, energy cost leading to conversion of farm lands into biofuel crops and climate variability. This brought the number of undernourished people worldwide to 925 million in 2010 (FAO, 2012) and the global food demand is expected to be doubled by 2050.

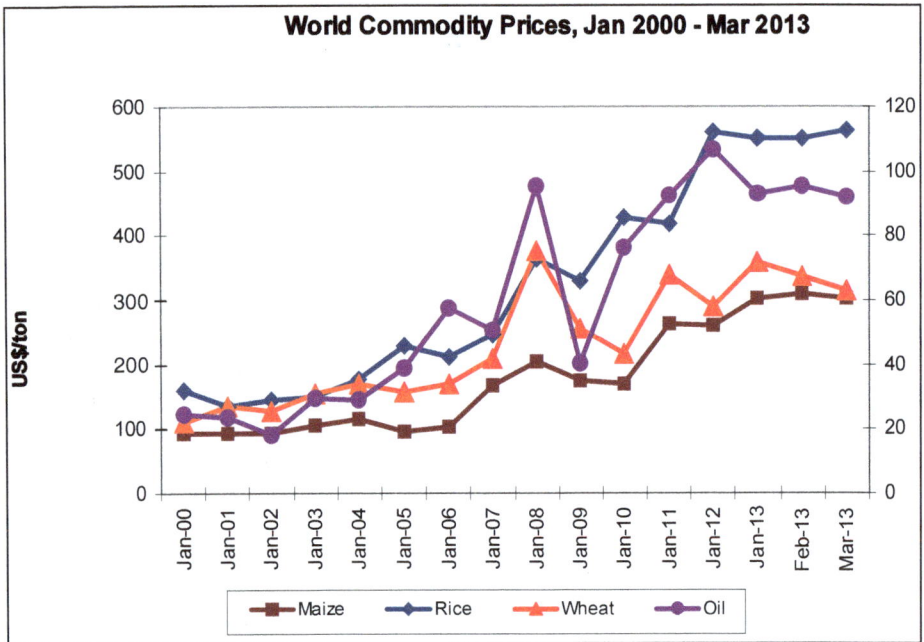

Figure 2.1: World Commodity Prices.

The first World Food Summit held in 1974 was aimed at eradicating world hunger within a decade. Since then 50 years have passed and 4 more Food Summits were held at short intervals in 1996, 2000, 2002 and 2006 which resolved to reduce hunger to half by 2015.Unfortunately the hunger and food security statistics show that even this target is not likely to be met due to conglomeration of reasons both at lopsided policy level and occurrence of natural calamities. Therefore, the onus to eliminate hunger continues to rest with agriculture and its capacity to address the following inherited and future challenges.

☆ Increase food production by 70 per cent to feed a growing population of 9.00 billion by 2050

☆ Meet the dietary requirement changes linked with rising income in developing countries

☆ Sustainable management of over exploited and depleting natural resources through investments and incentives

☆ Compacting the impact of increasing food prices and energy crisis

☆ Initiatives for climate resilient agriculture

☆ Family farming initiatives

Indian Scenario

Journey of Indian agriculture from the years of acute food crisis to food self sufficiency and food security entitlements has been tremendous. From the years of Bengal rice famine of 1943 that killed over 3 million children, women and men from hunger and starvation to the National Food Security Bill of 2013, it has been posed with the challenges of population, poverty and climate change. Enormity of the problems faced by Indian agriculture is evident from the fact that despite the national gains in food self sufficiency (51million tones in 1951 to 258 million tons in 2013) and security through the heralded Green revolution, India still has a population of 250 million food insecure inhabitants which roughly estimate to one fourth of world's food insecure people (Table 2.2). The Global Hunger Index (GHI), a more comprehensive tool used to measure and track global hunger, ranks India in the category of alarming poverty along with Bangladesh as evident from Table 2.3 (IFPRI, 2008). More disturbing are the rates of malnutrition among the most vulnerable sections of Indian population like women and children (Table 2.4).

Table 2.2: Number and Percentage of Under Nourished People in India since the Base Year 1990-92

Year	Total Population (Million)	Under-nourishment	
		Number (Million)	Percentage
1990-92	863	215	25
1995-97	949	202	21
2001-03	1050	212	20
2005-07	1116	221	20
2009-10	1168	250	21

Green Revolution

Despite rampant hunger and poverty, the food production trends show an impressive record of taking the country out of serious food shortages. The annals of Green revolution that demonstrated favorable interplay of infrastructure, technology, extension, risk taking farming community and policy support backed by strong political will provided India the much needed impetus. Indian farmers achieved as

Table 2.3: Contributions of the Three Components of Global Hunger Index (GHI) and the Underlying Data for Calculating the 1990 and 2010 GHI

Country	Proportion of Under Nourished in the population (per cent)		Prevalence of Underweight in Children Under 5 Years (per cent)		Under Five Mortality Rate (per cent)		GHI	
							With Data from 1988-92	With Data from 2003-08
	1990-92	2004-06	1990-92	2003-08	1990	2008	1990	2010
India	24.0	22.0	59.5	43.5	11.6	6.9	31.7	24.1
Bangladesh	36.0	26.0	56.5	41.3	14.9	5.4	35.8	24.2
China	15.0	10.0	15.3	6.0	4.6	2.1	11.6	6.0
Pakistan	22.0	23.0	39.0	25.3	13.0	8.9	24.7	19.1
World	–	–	–	–	–	–	19.8	15.1

Source: Global Hunger Index, IFPRI, 2010.

Table 2.4: Levels of Malnutrition among Vulnerable Sections of the Population (per cent) in India

Indicators	NFHS-2 1998-99	NHFS-3 2005-06
Children under 3 years who are stunted	45.50	38.40
Children under 3 years who are wasted: weight for height	15.50	19.10
Children under 3 years who are underweight: weight for age-less than 2 S.D	47.0	45.9
Anaemia among children aged 6 to 59 months	74.0	70.0
Women in 15 to 24 years in BMI 18.5	NA	44.1
Women in age 25 to 49 year BMI 18.5	NA	30.7
Women in age 15 to 49 year BMI 18.5	36.2	35.6

Source: Ministry of Women and Child Development, Government of India.

much progress in wheat production in four years (1964–68) of green revolution, as during the preceding 4000 years (Table 2.5). The main source of long-run growth was technological augmentation of yields per unit of cropped area. This resulted in tripling of food grain yields, and food grain production increased from 51 million tones in 1950–51 to 217 million tones in 2006–07. Production of oilseeds, sugarcane and cotton have also increased more than four-fold over the period, reaching 24 million tones and 355 million tones and 23 million bales, respectively, in 2006–07. But, although GDP from agriculture has more than quadrupled, from Rs. 1083740 million in 1950–51 to Rs. 4859370 million in 2006–07 (both at 1999–2000 price), the increase per worker has been rather modest (Figure 2.2). Moreover, the percentage contribution of agriculture to the overall GDP also decreased from around 50 percent in 1950-51 to 30 percent in 1990-91 to 15.7 percent in 2008-09. In the last two decades, agricultural growth rate (3 percent) has been around half of the overall growth rate of 6-7 percent

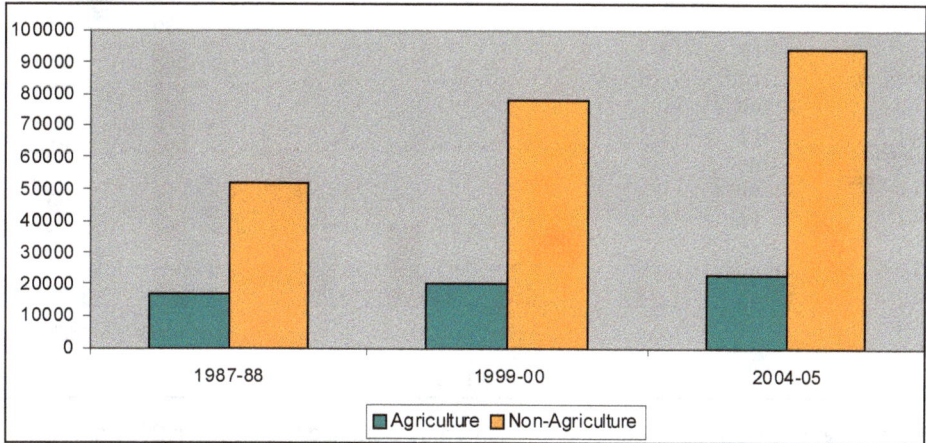

Figure 2.2: Per Worker GDP in Agriculture and Non-agriculture Sectors, Rs. at 1999-00 Prices.

(NAAS, 2009). But there has been only slight fall in the population dependent on agriculture for employment and livelihood with 61 percent in 1990-91 to 52 percent in 2008-09. The average farm size also recorded significant fall from 2.3 ha in 1970-71 to 1.32 ha in 2000-01 with an increase in the number of small farm holdings from 70 million to 121 million during the period due to arable land conversion and fragmentation. At this trend it is projected to be 0.68 ha in 2020 and 0.32 ha by 2030.

Table 2.5: Achievement in Wheat Production over the Years

Sl.No.	Period	Production (million tonnes)
1.	Earliest evidence of cultivation	Mohanjo-daro excavations 2300 BC
2.	Production in 1947-48	06
3.	Production in 1963-64	10
4.	Production in 1967-68	17
5.	Production in 2012	92

Fast growing population is one of the major factors that negated the Indian achievements in food production. Population of 520 million in 1947 has crossed the billion mark to reach 1120 million in 2013. In meeting the predicted food requirement of 345 million tones in 2030 the food production needs to be increased at the annual rate of 5.5 million tones. Moreover, there is a predicted increase of 100 percent in the demand for high value food products of horticulture, dairy, livestock and fish during the years 2000-2030 (Figure 2.3). This is attributed to the increased rural urban migration and increase in quality of life. This necessitates high investments in infrastructure and cool chain development for its handling, value addition, processing and marketing. The required growth over the base year production of 2006-07 to achieve domestic demand by 2020 is presented in Table 2.6. But the production environment remains hampered by the continuously shrinking and deteriorating

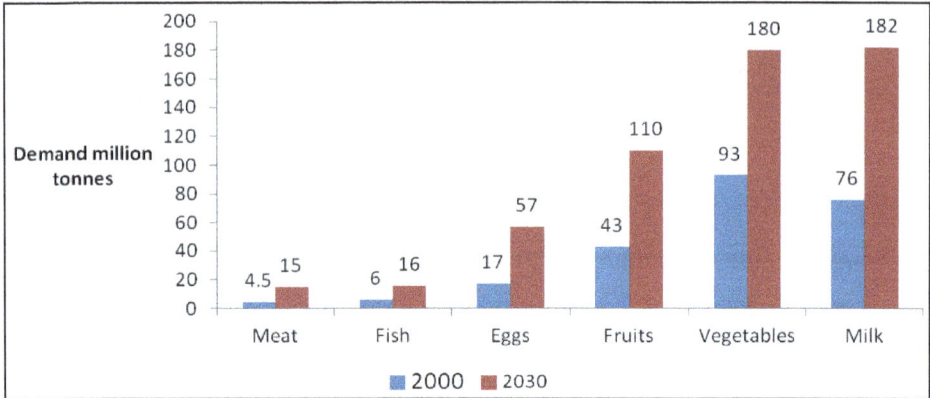

Figure 2.3: Comparison of Demand for High Value Commodities between 2000 and 2030.

natural resources and impact of climate variability. Around 120.72 million ha land is degraded out of agriculture due to soil erosion and 8.4 million ha is affected by soil salinity and water logging. Green house gases from agriculture have been estimated to contribute 30 percent of global emissions causing climate change. The total amount of GHGs emitted in India is estimated at 1228 million tones, which accounted for only 3 per cent of the total global emissions, and of which 63 per cent was emitted as CO_2, 33 per cent as CH_4, and the rest 4 per cent as N_2O. The GHG emissions in 1990, 1994 and 2000 increased from 988 to 1228 to 1484 million tones respectively and the compounded annual growth rate of these emissions between 1990 and 2000 has been 4.2 per cent which is low compared to many developed countries.

Table 2.6: Required Growth Over the Base Year Production of 2006-07 to Achieve Domestic Demand by 2020

Commodity	Domestic Production 2006-07 (million tons)	Projected Demand (2020-21)	Growth Rate during 1998-99 to 2006-07	Required Growth Rate Over 2006-07 to meet the Demand (per cent)
Cereals	201.9	262.0	0.62	1.9
Pulses	14.2	19.1	0.47	2.1
Food grains	216.1	281.1	0.61	1.9
Oilseeds	23.6	53.7	1.96	6.0
Vegetables	111.8	127.2	3.68	0.9
Fruits	57.7	86.2	3.06	2.9
Sugarcane	315.5	345.3	−0.60	0.6
Milk	100.9	141.5	3.65	2.4
Fish	6.9	11.2	2.89	3.5
Eggs (billion)	50.7	81.4	6.60	3.4

Thus agriculture in India is unfolding a new agrarian model where domestic product is low, average size of holding is contracting/ fragmenting and number of operational holding is increasing. Reduction in farm size without any alternative income augmenting opportunity is resulting in fall in farm income leading to agrarian distress. It is in this context the augmentation of food production through small farm holdings and family farming assumes significance. Crop diversification and establishment of Nutri-Farms in small holdings by integrating principles of farm ecology, nutrition and farm economics are proposed to defend the gains of food self sufficiency.

International Year of Family Farming – 2014

Assuming the significance of small holdings and family farms in future food production, General Assembly of the United Nations in its 66th session adopted 2014 as the "International Year of Family Farming" (IYFF). It aims at repositioning family farming at the centre of agricultural, environmental and social policies in the national agenda by identifying gaps and opportunities to promote a shift towards a more equal and balanced development. It helps to raise the profile of family farming and smallholder farming by focusing world attention on its significant role in the fight for eradication of hunger and poverty, providing food security and nutrition, improving livelihoods, managing natural resources, protecting the environment, and achieving sustainable development, particularly in rural areas.

Family farming includes all family-based agricultural activities, and is linked to several areas of rural development. It is defined as a means of organizing agricultural, forestry, fisheries, pastoral and aquaculture production which is managed and operated by a family and predominantly reliant on family labor. Family and small-scale farming are inextricably linked to world food security as it forms the predominant type of agriculture in both developing and developed countries. It involves around 2 billion people, almost one-third of humanity. Around 70 per cent of the world's food is produced on 404 million small-scale farms of less than 2 hectares operated by 1.5 billion men and women farmers. It is estimated that 40-60 percent of rural income in developing countries is generated in these small family farms. In addition to the essential role they play in food production, they have significant contribution in sustaining rural economies and stewardship of biodiversity. Yet most of these are very poor and operate in disadvantaged conditions with insufficient access to resources and support.

Expensive and short-term solutions proposed by conventional agriculture do not hold good for the social and economic problems of smallholder farmers. Many farmers are forced to purchase expensive inputs that often push them into a cycle of debt. Beyond the financial implications, these inputs also pollute the environment, destroy biodiversity, and degrade the nutritional quality of our food. Therefore, organic agriculture and other agro-ecological models are suggested as the most appropriate way to achieve ecological, agronomic and socio-economic intensification of family farming and smallholder agriculture. It also provides healthy and nutritious food. Organic farmers ensure that traditions are kept alive and facilitate market place success by sharing experiences and developing new farming methods. Together, organic smallholders can strengthen social structures, develop innovative networks and promote entrepreneurship. The impact of which includes more job opportunities, a stronger local value chain and improved rural development.

Indian Context

At national level, there are a number of factors that are key for a successful development of family farming, such as: agro-ecological conditions and territorial characteristics; policy environment; access to markets; access to land and natural resources; access to technology and extension services; access to finance; demographic, economic and socio-cultural conditions; availability of specialized education among others. Family farming has important socio-economic, environmental and cultural role as 81 percent of total agricultural holdings belong to the category with an average holding size of 1.33 ha. Statistics on the farming population of the country (Table 2.7) indicated 49 percent of the farms are in the family farm category. Contribution of these farms to the national food security is enormous as 60 per cent of the production come from these family farms. These family farms grow traditional crops mostly following organic and natural farming practices. These practices help in the maintenance of traditional flavor of commodities, soil fertility, environment and reduction in the Green House Gases (GHGs) and effects of climate change.

Table 2.7: Statistics of Farming Population Over Total Population of India

	Total Population (Million)	Farming Population (Million)	Small and Family Farmers (Million)
Male	530	318	249
Female	495	297	249
Total	**1025**	**615**	**498**

However, income of Family Farm farmers are not in tune with increase in the prices of agricultural inputs. Inadequate access to natural and basic resources for production, markets, value addition and processing are responsible for the uneconomic functioning of these farms. This has been instrumental in perpetuating poverty, hunger and malnutrition in the country and keeping these farm families in a highly disadvantaged socio-economic situation. Many food aid programs like midday school meals for school children, subsidized food grains through public distribution system and the attempts to enact the legendary Food Security Enactment of 2013 which ensure legal right to food for all are operational in the country to combat problems of hunger and malnutrition. In fact, the Food Security Bill addresses the problems of calorie deprivation, protein hunger and micronutrient malnutrition. It ensures legal entitlements to food through Life Cycle Approach with special attention to the first 1000 days in a child's life, and considers women as head of the household with regard to food entitlements. It provides the right to enlarged food basket with 35 kg per family per month of wheat, rice or climate smart nutri-cereals. These enabling provisions on food availability and food absorption have been ensured through national grid of grain storages and Public Distribution System reforms. However, these can only be considered as short term solutions against hunger and the governments needs to focus more on sustainable strategies in agricultural production which target the poorest households in meeting the hunger reduction targets of the Millennium Development Goals (MDG) and World Food Summits. Homestead farms and nutritional gardens provide great scope in overcoming widespread malnutrition prevailing in our country by enabling farm families to have balanced diet. The programme for organizing nutri-farms will involve introducing in the cropping system, biofortified crops which are rich in iron, zinc, vitamin A, vitamin B_{12} and other micro-nutrients as well as those which are rich in protein like quality protein maize. The programme will be taken up in the districts which are suffering from a high burden of hidden hunger caused by micro-nutrient deficiencies.

Homesteads in Family Farming

Small holder homestead farming, at this juncture has proved to be a time tested method for ensuring food sufficiency. Home gardening or homestead farming is an oldest land use activity next only to shifting cultivation that has evolved through generations of gradual intensification of cropping in response to increasing human pressure and the resulting shortage of arable lands. The origin of home gardening could be traced to South East Asia and East Africa. In Java(Indonesia) and Kerala (India), home gardening has been a way of life and is still critical to local subsistence economy and food security (Nair and Kumar, 2006). A homestead is an operational farm unit in which a number of crops (including tree crops), vegetables, fruits,spices, ornamentals, tubers, medicinal plants are grown along with livestock, poultry, and/ or fish production mainly to satisfy the farmers basic need (Tejwani, 1994). A home garden could also be viewed as an integrated system which comprises different components in its small area: the family house, a living area, a kitchen garden, a mixed garden, a fish pond, stores, an animal house and, of course people. The system with a mix of vegetables, flowers and medicinal plants as intercrops in fruits and plantations involving spices was able to sustain the biodiversity of vegetation around the household with a change in the local micro-climate having long-term benefits on health of family members.

However, the traditional home gardens were managed unscientifically without any planning and with very low resource use efficiency. Increasing population and low per capita availability of arable lands have necessitated the need to have better managed home gardens. Evidences from various parts have showed that homestead farming and interventions in home gardens could play a significant role in improving food security especially for the resource poor rural households in developing country. Ali *et al.* (2008) reported that in Bangladesh, implementation of a project aiming to provide farmers a subsidiary source of income through potential homestead farming system, to increase and sustain productivity of existing homestead farms and to encourage maximum utilization of available family labor, through vegetable could meet the daily vegetable requirement of small farmers by supplying 200-220 g/head/ day against the bench marklevel of 40-50 g/head/day. Also vegetable intake largely provided required amount of vitamins A, C, calcium and iron. Thus food security was increased; nutritional deficiency was reduced along with women empowerment.

The home gardens of Kerala evolved in response to the pressure of a shrinking land resource base coupled with a high population density, which necessitated a conscious attempt on the part of farmers to achieve their goals by living within biophysical, ecological, and economic constraints is another classical example of family farm. As a traditional farming system, home garden agriculture is essentially subsistence-oriented, with need based production and labor provided mostly by the family itself. High-density farming involving several species of seasonal, annual, and perennial crops thus evolved to meet their demands and to achieve highly efficient use of resources. To supplement the subsistence level on-farm income from smallholdings, household members are being driven to off-farm employment and the high opportunity cost of family labor has led to the development of part time farming also (Salam *et al.,* 2000). A study conducted to determine the socio-economic factors and constraints that affect farming in 400 homesteads of southern Kerala by John and Nair (1999) revealed that 17.5 per cent and 30.25 per cent of the homesteads raised cattle and poultry, respectively, as a complementary enterprise. The system in general was reported profitable, resulting in a net profit of Rs. 280532/year with a benefit:cost ratio of 2.35. Nearly 78 percent of the farmers reported labor scarcity, while 97.75 per cent said that the high labor cost resulted in increased cultivation cost.

A study conducted by ICAR in 20 homesteads in Goa indicated that nearly two-thirds of the homestead-grown vegetables are consumed in the household with only one-fifth of the vegetable produce being sold in the market while a part of it is exchanged among the neighboring households. Further, it was also observed that most of the vegetable production in the homesteads was concentrated during November to March period coinciding with the post-monsoon period. Nutritionally, the homestead met most of the mineral and vitamins requirement of the household family members although the carbohydrates and protein requirements were met partly only by the homestead production. The project also helped to develop model homestead units for different holding sizes of the household based on the family requirement and the marketing potential keeping in view the resource situations including part or fulltime availability of the family members to work in their gardens. The Homestead Food Production (HFP) program, introduced in Bangladesh by Helen Keller International nearly two decades ago, to promote an integrated package of home gardening, small livestock production and nutrition education with the aim of increasing household production, availability, and consumption of micro-nutrient-rich foods and improving the health and nutritional status of women and children is now operating in several countries of Asia and Sub-Saharan Africa. Evidence shows that HFP in Bangladesh has improved food security for nearly 5 million vulnerable people in diverse agro ecological zones. This has been achieved through increased production and consumption of micro-nutrient-rich foods; increased income from gardens and less expenditures on micro-nutrient-rich foods; women's empowerment; enhanced partner capacity; and community development (Iannotti *et al;* 2009). Asaduzzaman *et al.* (2011) quantified the costs/benefits of traditional and developed homestead vegetable production systems, and suggested that developed gardening has better performances in terms of calorie intake and economic performances over traditional but the optimal calorie intake with least-cost technology could be a feasible livelihood strategy for resource poor people. Developed garden contributed 10 percent to achieve threshold poverty level whereas traditional contributes only 2 percent. On the basis of input costs other than family labor, BCR was higher in developed (3.24) than traditional (2.63) but when labor cost was included in the analysis, traditional gardening was still feasible (BCR 1.25) but developed garden (BCR 0.60) appeared as non-profitable.The study concluded that developed gardening is profitable compared to traditional if the opportunity cost of family labor is ignored as an item of cost of production.

Family farming saves space, water, energy and makes use of the principles of protected organic cultivation primarily to meet the house hold nutritional security and the surplus marketed in nearby markets or directly sold to customers. The fresh vegetables and fruits supply needed minerals, vitamins, fibres, antioxidants and pigments. It is also a therapy to drive off loneliness and consequent depression. Wastes are transformed to wealth through composting and bio-degradation.

References

1. Ali M. Yusuf, M. MustaqueAhmed and Badirul Islam, M. (2008) "Homestead Vegetable Gardening: Meeting The Need Of Year Round Vegetable Requirement of Farm Family" Paper presented in the National workshop on Multiple cropping held at Bangladesh Agricultural Research Council, Farmgate, Dhaka-1215, Bangladesh on 23-24 April, 2008.

2. Asaduzzaman Md, Naseem A and Singla,R.(2011). Benefit-Cost Assessment of Different Homestead Vegetable Gardening on Improving Household Food and Nutrition Security in Rural Bangladesh, Paper presented at the Agricultural and Applied Economics Association's 2011 AAEA and NAREA Joint Annual Meeting, Pittsburgh, Pennsylvania, July 24-26, 2011.

3. FAO (2008). The State of Food Insecurity in the World 2008. Economic and Social Development Department, Food and Agriculture Organization of the UN, Rome.

4. FAO (2010). The State of Food Insecurity in the World 2010. Economic and Social Development Department, Food and Agriculture Organization of the UN. Rome.

5. FAO, (2012). Food Insecurity in the World, Food And Agriculture Organization Of The United Nations, Rome, 2012.

6. Iannotti ora, Kenda Cunningham and Marie Ruel (2009). Improving Diet Quality and Micronutrient Nutrition Homestead Food Production in Bangladesh IPRI Discussion Paper 00928 (www.ifpri.org/millionsfed).

7. IFPRI. (2009). Global Hunger Index. Washington, DC, IFPRI.

8. John, J. and Nair, M. A. (1999). Socio-economic characteristics of homestead farming in south Kerala. *Journal of Tropical Agriculture,* Vol. 37 No. 1/2 pp. 107-109.

9. NAAS. (2009). State of Indian Agriculture- A NAAS Publication. National Academy of Agricultural Sciences, New Delhi.

10. Nair, P.K.R and Kumar, B.M.(2006). Introduction (in) *Tropical Home gardens*, Springer, Netherlands, p. 1-10.

11. Salam, M.A., Noguchi T., and Koika, M. (2000). Understanding why farmers plant trees in the homestead agro-forestry in Bangladesh. *Agroforestry Systems*. 50. 77-93.

12. Singh, R. B.(2011). *The Hungry Child Cannot Wait* (Presidential Address), Foundation Day and 18th General Body meeting, June 5, 2011, National Academy of Agricultural Sciences, New Delhi.

13. Tejwani, K. G. (1994). Agroforestry in India, Oxford and IBH Publishing Co. Pvt. Ltd. New Delhi.

2015, Horticulture for Nutrition Security
Editor: **Prof. K.V. Peter**
Published by: **DAYA PUBLISHING HOUSE, NEW DELHI**

Pages **19–35**

Chapter 3

Nutrition Garden in the Context of Nutritional Security

P. Indira and K.V. Peter

Malnutrition and undernutrition are widely prevalent in Asia and Africa due to consumption of inadequate and faulty diets. Fruits and vegetables are good sources of vitamins, minerals, proteins, carbohydrates, fibres, antioxidants, lipids pigments and phytochemicals and thus they have important role in our daily diet. Among the rural community, their consumption is very low due to lack of purchasing power, ignorance and other factors including unavailability. Cultivation of these crops by gardening in a systematic manner in small pieces of land available in households is known as "Nutrition Garden". The nutrition garden ensures access to healthy diet with adequate macro and micro-nutrients at doorstep.

Nutritional security has become the political focus and concern all over the world. Without nutritional security, health care, wellness and zero hunger can not be achieved-millenium development goals-. Although the South-East Asian region is blessed with natural resources- fertile land, water and human resources-, the people still suffer from under-nutrition and non-communicable chronic diseases (NCD) which are more prevalent among low income group. In Asia and Pacific alone 642 million people are undernourished, out of 1017 million hungry in the world (Singh, 2012).

It is widely recognized that intervention on food security should also take into account concern underlying nutritional security so as to ensure food and nutritional security to millions of people in Asia where traditional food basket is cereal dominated (Dasgupta, 2012). Lack of availability of different ingredients of food to the needy is one of the major causes of malnutrition, other factors being low purchasing power, ignorance, large family size, lack of sanitation, hygiene and inability to absorb. The intake of protective foods like pulses, vegetables, milk and fruits are very low which leads to many nutritional deficiency disorders. According to Hungama (2012) - Hunger and Malnutrition, 2012 Report- malnutrition in Indian children continues to be of higher level with 42.3 per cent being under weight, 58.8 per cent stunted and 11 per cent wasted (Indumathi *et al.*, 2012).

Common nutritional problems of human beings are protein energy malnutrition (PEM), micro nutrient deficiencies like vitamin A deficiency (VAD), Iron deficiency anaemia (IDA), Iodine deficiency disorder (IDD) and vitamin B complex deficiencies (NIN, 2011). To tackle these problems recommendations of National Institute of Nutrition are:

1. Eat variety of foods to ensure balanced diet.
2. Eat protective foods rich in vitamins and minerals.
3. Eat folate rich foods for haemoglobin synthesis.
4. Eat plenty of fruits and vegetables which are rich sources of micro and macro-nutrients.

Fruits, vegetables, spices, medicinal and aromatic plants are boon to mankind in several ways.

According to Prof. M.S. Swaminathan *there is a horticultural remedy for every nutritional malady*. The National health survey-2002 indicated that 65 per cent of women and children in India are anaemic and 50 per cent of world's anaemic are in India as per WHO report. India inspite of its horticultural strength is unable to mitigate or reduce undernutrition. Horticultural therapy in the form of raising terrace garden, roof garden and protected cultivation is getting attention to manage time, space and waste biocycling and as a treatment against lonliness and depression.

The expert committee of Indian Council of Medical Research (ICMR) recommends that every individual should consume at least 300 g vegetables and 100 g fresh fruits/day (green leaf vegetables – 50 g, other vegetables – 200 g, roots and tubers – 50 g). Pregnant women should consume 100 g leaf vegetables/day.

Importance of Vegetables in Human Health and Nutrition

Vegetables are rich sources of nutritional bioactive compounds. They are important sources of protective nutrients like vitamins, minerals, antioxidants, folic acid and dietary fibres. The diversified and highly nutritive vegetables are affordable and cost effective solution to hidden hunger and malnutrition. A person on average needs a daily diet which should provide 2800 calories, 55 g protein, 450 mg Ca, 20 mg Fe, 3000 mg β – carotene, 50 mg vitamin C, 100 mg folic acid, 1.0 mg vitamin B, 1.4 mg thiamine, 1.5 mg riboflavin, 19 mg niacin and 5 mg vitamin D. Vegetables are good source of all these nutrients (Tables 3.1 and 3.2) (Sharma, 2009).

Table 3.1: Nutritional Composition of Vegetables per 100 g Edible Portion

Vegetable Crops	Energy (Kcal)	Moisture (g)	Protein (g)	Fat (g)	Carbohydrates (g)	Fibre (g)
Amaranth	45.0	85.7	4.0	0.5	6.1	1.0
Asparagus	26.0	91.7	2.5	0.2	5.0	0.7
Basella	32.0	90.8	2.8	0.4	4.2	–
Bittergourd	25.0	92.4	1.6	0.2	4.2	1.7
Bottle gourd	12.0	96.1	0.2	0.1	2.5	1.5
Brinjal	24.0	92.7	1.4	0.3	4.0	–
Broad bean	48.0	85.4	4.5	0.1	7.3	–
Broccoli	37.0	89.9	3.3	0.2	5.5	2.6
Brussel's sprout	45.0	85.2	4.9	0.4	8.3	1.5
Bengal gram leaves	97.0	73.4	7.0	1.4	14.1	–
Cabbage	24.0	92.4	1.3	0.2	5.4	1.5
Capsicum	29.0	92.5	1.2	0.2	4.0	2.5
Chilli	29.0	82.6	2.9	0.6	6.1	6.7
Carrot	42.0	88.6	1.1	0.2	9.1	1.0
Coriander leaves	44.0	66.3	3.3	0.6	6.3	–
Cassava	157.0	59.4	0.7	0.2	38.1	–
Cauliflower	27.0	91.0	2.7	0.2	5.2	0.9
Celery	17.0	94.1	0.9	0.1	3.9	1.4
Cucumber	18.0	96.3	0.4	0.1	2.5	0.6
Drumstick leaves	92.0	76.0	6.7	1.7	12.7	0.9
Fenugreek leaves	49.0	86.1	4.4	0.9	6.0	1.1
French bean	30.0	62.0	6.3	0.1	29.8	4.0
Garlic	32.0	90.1	1.9	0.2	7.1	0.8
Kale	53.0	82.7	6.0	0.8	9.0	1.5
Knolkhol	29.0	90.3	2.0	0.1	6.6	1.1
Leek	77.0	78.9	1.8	0.1	17.2	1.3
Lettuce	14.0	95.1	1.2	0.2	2.5	0.6
Mint	48.0	84.9	4.8	0.6	5.8	2.0
Musk melon	17.0	95.2	0.3	0.2	3.5	0.6
Mustard leaves	34.0	89.8	4.0	0.6.	3.2	1.5
Okra	35.0	89.6	1.9	0.2	6.4	1.2
Onion	50.0	86.6	1.2	0.1	11.1	0.5
Pea	84.0	78.0	6.3	0.4	14.4	4.0
Potato	97.0	74.7	1.6	0.1	22.6	1.6
Parsley	16.0	90.0	2.2	0.3	1.3	–
Pointed gourd	20.0	92.0	2.0	0.3	2.2	3.0

Contd...

Table 3.1–*Contd...*

Vegetable Crops	Energy (Kcal)	Moisture (g)	Protein (g)	Fat (g)	Carbohydrates (g)	Fibre (g)
Pumpkin	25.0	92.6	1.4	0.1	4.6	0.5
Radish	17.0	94.4	0.7	0.1	3.4	0.7
Sponge gourd	18.0	93.2	1.2	0.2	2.9	0.5
Spinach	26.0	90.7	3.2	0.3	4.3	–
Spinach beet	46.0	86.4	3.4	0.8	6.5	0.7
Sweet potato	124.0	68.5	1.8	0.7	28.0	1.0
Tomato	22.0	93.5	1.1	0.2	4.7	0.7
Turnip green	67.0	8.9	4.0	1.5	9.4	–
Watermelon	26.0	92.6	0.5	0.2	6.4	0.2
Nadroo (Lotus root)	53.0	85.9	1.7	0.1	11.3	–
Yam	102.0	74.0	1.5	0.2	24.0	–
Taro corm (Arvi)	97.0	73.1	3.0	0.1	22.1	–
Giant Taro (Kachloo)	71.0	81.2	0.6	0.1	17.0	–
Cowpea leaves	51.0	84.6	4.3	0.2	8.0	–

Table 3.2: Vitamins in Vegetables per 100 g Edible Portion

Crops	Vitamin A (IU)	Thiamine (mg)	Riboflavin (mg)	Niacin (mg)	Vitamin C (mg)
Amaranth	9,108	0.03	0.30	1.2	99.9
Asparagus	900	0.18	0.20	1.5	33.0
Basella	12,276	0.03	0.16	0.5	87.0
Bittergourd	208	0.07	0.09	0.5	88.0
Bottle gourd	traces	0.03	0.01	0.2	6.0
Brinjal	122	0.04	0.11	0.9	12.0
Broad bean	15	0.08	–	0.8	12.0
Broccoli	2,500	0.10	0.23	0.9	113.0
Brussel's sprout	550	0.10	0.16	0.9	102
Cabbage	130	0.05	0.05	0.03	47
Capsicum	900	0.06	0.06	0.5	175
Chilli (green)	454	0.06	0.03	0.6	111
Carrot	11,000	0.06	0.05	0.6	8
Coriander leaves	11,168	0.50	0.06	–	135
Cassava	700	0.05	0.10	0.3	25
Cauliflower	60	0.11	0.10	0.7	78
Celery	240	0.03	0.03	0.3	9

Contd...

Table 3.2–*Contd...*

Crops	Vitamin A (IU)	Thiamin (mg)	Riboflavin (mg)	Niacin (mg)	Vitamin C (mg)
Cucumber	0.00	0.03	0.0	0.2	7
Drumstick leaves	11,187	0.06	0.05	0.8	220
Fenugreek leaves	3,861	0.04	0.31	0.8	52
French bean	10	0.06	0.23	0.4	13
Garlic	600	0.08	0.11	0.5	19
Kale	10,000	0.16	0.26	2.1	186
Knolkhol	20	0.06	0.04	0.3	66
Leek	30	0.23	0.06	0.5	25
Lettuce	900	0.06	0.06	0.3	8
Mint	2,700	0.05	0.20	0.4	750
Musk melon	279	0.11	0.08	0.3	26
Mustard leaves	4,200	0.03	–	–	33
Okra	172	0.07	0.10	0.6	13
Onion	35	0.08	0.01	0.4	11
Pea	640	0.35	0.14	2.4	27
Potato	32	0.10	0.01	1.2	17
Parsley	5,200	0.08	0.11	0.7	90
Pointed gourd	252	0.05	0.06	0.5	29
Pumpkin	2180	0.06	0.04	0.05	2
Radish	5	0.06	0.04	0.05	15
Bengal gram leaves	1564	0.09	0.10	–	61
Sponge gourd	369	0.02	0.06	0.4	0
Spinach	8100	0.10	0.20	0.6	51
Spinach beet	5862	0.26	0.56	3.3	70
Sweet potato	8800	0.10	0.06	0.6	21
Tomato	900	0.06	0.04	0.7	23
Turnip green	15691	0.31	0.57	5.4	180
Taro corm (Arvi)	166	0.09	0.03	–	0
Giant Taro (Kachloo)	–	0.31	0.57	5.4	180
Cowpea pods	941	0.07	0.09	0.9	13
Yam	–	0.1	0.01	0.8	15

Dietary Fibre

Dietary fibres are mainly plant cell walls containing cellulose, pectins, xyloglucans, xylans, mannans, and free arabinans. There are both soluble (plant gums, pectins and mucilaginous materials) and insoluble fibres (lignin, cellulose and part of hemicellulose). Fibres have high water binding capacity and produce viscosity. They possess cholestrolemic effect, bind bile acids and promote faecal excretion. They delay absorption of glucose and fats after meal, increase faecal bulk and speed up the passage through digestive tract, thus preventing the risk of constipation, haemorrhoids, colon cancer and diverticulosis.

Leaf vegetables like celery, cabbage, spinach, lettuce and amaranth are rich sources of dietary fibre. Dry chilli, peas, french bean, cluster bean, and okra are also good sources of fibre.

Proteins

Vegetables are quite low in protein content when compared to dry pulses. Protein rich vegetables are peas, lima bean, drumstick leaves, french bean and celery. Potato, cauliflower, okra, cowpea, beet leaf (palak), fenugreek (methi) leaves and onion are also sources of protein.

Vegetables and Antioxidants

There are numerous compounds in vegetables which function as antioxidants. Primary among them are â-carotene, vitamin 'C', vitamin 'E', selenium and flavonoids. There is evidence that antioxidants like carotene, ascorbic acid (vitamin C), β-tocopherols (vitamin E), flavonoids and selenium are significantly associated with reduced cancer risks (Singh, 2012). Vegetables are rich sources of antioxidant nutrients *viz.*, vitamin C, vitamin E and the carotenoids. Certain flavonoids like quercetin, kaempferol, myricetin and luteolin are also present in vegetables.

Carotenoids

They are the pigments in fruits and vegetables which protect from damage to lipids, blood and other fluids. Among 600 and more naturally occurring carotenoids, majority have antioxidant properties (Table 3.3). Beta carotene is present in dark green leaf vegetables, carrot and pumpkin. It provides protection from coronary heart diseases and lung cancer. Lycopene is a carotenoid present in tomato and watermelon which prevents oxidation of LDL cholesterol and reduces damages to arteries. It will also reduce the risk of bladder cancer and pancreatic cancer.

Ascorbic Acid

Ascorbic acid or vitamin C is a water soluble antioxidant which is easily oxidized to form semi-dehydro ascorbic acid that is relatively stable. Being an electron donor, ascorbic acid serves as a reducing agent for many reactive oxidant species. It reduces tocopherol radicals to their active form. Ascorbic acid is seen abundantly in green leaf vegetables like drumstick leaves, coriander leaves, turnip green, sweet pepper, chilli, Brussels sprouts, broccoli and cabbage. Bittergourd, cauliflower, amaranth and beet leaf are also good sources of vitamin C.

Vitamine E

It is the major lipid soluble antioxidant responsible for protecting the polyunsaturated fatty acids in membranes against lipid peroxidation by free radicals and singlet oxygen species. Cruciferous vegetables have high level of antioxidants

Table 2.3: Qualitative and Quantitative Carotenoid Distribution in Vegetables

Vegetables-Carotenoids (mg/g fresh wt)	Total	β-carotene	Major Carotenoids
Asparagus (*Asparagus officinalis* L.)	8.5	4.3–7.0	α-carotene. Lutein, Violaxanthin, Neoxanthin
Bitter gourd (*Momordica charantia* L.)	5.3	2.3	α-carotene, β-carotene. Zeinoxanthin. Lutein
French bean (*Phaseolus vulgaris* L.)	17.1	2–4	α-carotene, β-carotene. Lutein 5,6- epoxide. Neoxanthin, Violaxanthin
Broccoli (*Brassica oleracea* var. *italica* Plenck.)	42.4	4.8	α-carotene. Lutein, Isolutein, Luteoxanthin, Violaxanthin, Neoxanthin, Chrysanthemaxanthin
Cabbage (*Brassica oleracea* var. *capitata* L.)	8.9	0.8	β-carotene. Lutein, α-carotene 5,6- epoxide, Neoxanthin, Violaxanthin, Chrysanthemaxanthin
Carrots (*Daucus carota* L.)	54–124	76.0	α-carotene, β-carotene, J-carotene, â-Zeacarotene, ρ-carotene, Neurosporene
Cauliflower (*Brassica oleracea* var. *capitata* L.)	0.44	0.11	β-carotene, Lutein, Violaxanthin, Neoxanthin
Cucumber (*Cucumis sativus* L.)	17.2	2.20	α-carotene, β-carotene and Cryptoxantin
Pepper (Green) (*Capsicum annuum* L.)	10.0	6.8	Capsanthin, Capsorubin, Cryptocapsin, β-carotene
Pepper (Red)	127–284	1.27–2.84	β-carotene, Violaxanthin, Neoxanthin
Lettuce (*Lactuca sativa* L.)	68.0	10.8–24.5	β-carotene, Lutein, Violaxanthin, Neoxanthin
Spinach (*Spinacia oleraeea* L.)	69.0	40.0	β-carotene. Lutein epoxide, Violaxanthin, Lutein, Antheraxanthin, Neoxanthin
Tomato (*Lycoopersicon esculentum* L.)	70–190	7.8	Lycopene, β-carotene, Phytoene.

Source. Pigments in vegetables: Cholorophylls and Carotenoids, Jeana Gross (1991). Van Nostrand Reinhold, New York.

viz. carotenoid, tocopherol and ascorbate. Kale has the higher level followed by broccoli and Brussels sprouts.

Flavanoids

They are one of the most widely occurring groups of secondary metabolites or phytochemicals in plants. They induct enzymes which detoxify carcinogens. The flavonoids are a large family of lower molecular weight poly-phenolic compounds found in plant tissues, which include the flavones, flavonols, flavonones, anthocyanin anthocyanidins, catechins and isoflavonoids. Other phenolics like p-coumaric, caffeic, ferrulic and chlorogenic acids are reported in tomato. They are mostly present in the skin. Among cucurbits, flavonoids are present in genera *Cucumis, Lagenaria, Citrullus* and *Luffa.* Quercetin and kaemferol are reported in french bean (Singh, 2012). Cheratin is a flavonoid isolated from bitter gourd having hypoglycaemic effect. Onion and garlic contain several sulphur containing compounds like allicin, garlicin etc. which reduce blood cholesterol and help to prevent coronary heart diseases and heart attacks.

Folates

Folates or folic acids are found in abundance in spinach followed by other green leaf vegetables and beans. It is one among the two vitamins associated with magaloblastic anaemia often seen in children and pregnant women. Diets high in folic acid reduce the risk of colon cancer and cardio vascular diseases. The actual requirement of folic acid ranges from 50-100 µg but for pregnant women the recommendation is 150-300 µg/day.

Natural Pigments

This group includes anthocyanin, betalains, carotenoids, chlorophylls etc. They are exploited as the source of neutraceuticals used against many human ailments. These pigments are useful to maintain human health, to protect from chronic diseases or to restore wellness by repairing tissues.

Anthocyanin pigment gives purple colour to brinjal, amaranth, cowpea, dolichos bean etc. It is an important and widely distributed group of water soluble natural pigments which can prevent cardio vascular diseases, neurological diseases and cancer. But the amount of anthocyanin being negligible in vegetables, transgenic tomatoes, with purple colour were developed in U.S.A. by transferring genes from snapdragon flowers. Similarly purple cauliflowers, purple capsicums etc. are also catching the market.

Nutraceuticals from Coloured Vegetables

"Eating a rainbow" of vegetables means including as much coloured vegetables into our daily diet. As the colours are an indication of vitamins and other nutrients, the coloured vegetables enhance the body's ability to prevent and fight diseases. The multitude of phytochemicals present in vegetables acts as antioxidant, antiallergic, anticarcinogenic, antiinflammatory, antiviral and antiproliferative. Pigments from paprika are used as natural colourants in medicines and food items. Some of the neutraceutically rich vegetable crops are red and black carrot, beet root, tomato, chilli and broccoli. Based on colour vegetables are grouped (Table 3.4), (Singh, 2012).

Table 3.4: Groups of Colored Vegetable Crops

Green	Artichoke, asparagus, amaranth, broccoli, Brussels sprout, celery, squash, Chinese cabbage, cucumber, endive, egg plant, beans green, cabbage green, onion green, pepper green, leek, lettuce, okra, peas, spinach, snap pea, watercress
White	Cauliflower, garlic, Jerusalem artichoke, kholrabi, egg plant, onion, parsnip, potato, shallot, turnip, white corn, white radish.
Red	Beet, radish, red capsicum, red pepper, red onion, red potato, tomato, watermelon, red amaranth
Yellow/ Orange	Pumpkin and squashes, carrot, sweet corn, sweet potato, beet yellow, yellow capsicum, potato yellow, yellow tomato, watermelon yellow.
Blue/ Purple	Egg plant, purple potato, purple cabbage, black carrot, purple cowpea, purple dolichos bean, purple capsicum and chilli

Therapeutic Values of Vegetables

Vegetables are an integral part of Indian system of medicine. The whole plant or plant parts of vegetables are used as crude drugs or drug formulations in naturopathy, ayurveda and homeopathy. Vegetables like brinjal, ashgourd, snakegourd, drumstick, curry leaf, onion and garlic are used in ayurveda. Watermelon seeds, seeds of summer squash, ivy gourd etc. are used in homoeopathic medicines.

'Bittergourd plus' and 'Karela Jamun Juice' are two important products from bittergourd used for curing diabetes and hypertension. Bittergourd plus is a health supplement which increases glucose metabolism and maintains a healthy blood sugar level. Ivygourd extract capsule is used against diabetes. Moringa body butter from moringa seeds and moringa leaf powder capsule are catching the market recently.

Vegetables for Protection against Cancer

According to recent estimates, diet is responsible for 20-30 per cent of all cancers occuring in economically developed countries. Frequent consumption especially green and yellow vegetables is associated with decreased susceptibility to some forms of cancer, although the mechanisms for their protective action have not been fully understood.

A diet rich in cruciferous vegetables are associated with inhibition of chemically induced carcinogenesis in laboratory animals and humans. These vegetables are rich in sulfur containing glucosides called glucosinolates. About 100 different forms of glucosinolates were identified in the cruciferous family. Broccoli, Brussels sprouts and Chinese cabbage are the main sources of glucosinolates in human diet. The dominant glucosinolates in cabbage, Brussels sprouts, cauliflower and kale are sinigrin and glucobrassicin (Singh, 2012).

Toxic Metabolics and Antinutritional Compounds

Though vegetables are rich in many nutrients and vitamins, there are many toxic substances and anti-nutritional compounds in them, which restrict their use in animal nutrition. A few among them are cucurbitacins, glycoalkaloids, glucosinolates, lathyrogens, saponins, trypsin inhibitors, tannins, phenols and non-amino organic acids. They are capable of inducing adverse effects as neurological disorders, kidney stone, high blood pressure and gastric disorders.

Anti-nutrients like phytate, tannins, oxalates, nitrates and glucosinolates have toxic effect. Phytate and tannins interfere with iron bioavailability, oxalates present in leaf vegetables and legumes interfere with calcium absorption and nitrate accumulation leading to serious deleterious effects. In the gastro intestinal tract, nitrate is (NO_3) is reduced to nitrite (NO_2) which is absorbed in the blood stream and bind with haemoglobin.

Importance of Nutrition Garden

Promotion of local plants is an appropriate strategy for increasing vegetable consumption in a particular region. Many local plants have antioxidative, antimutagenicity and anti-inflammatory properties (Chavasith, 2012). Nutrition awareness programmes stress the need for inclusion of locally available fruits and vegetables like papaya, mango, guava and leaf vegetables in their daily diet. Hence every housewife or every citizen has a vital role in converting his surrounding vacant land into alive kitchen garden, where location specific seasonal vegetables and fruits are grown.

The main purpose of a nutrition garden is to provide the family daily with fresh vegetables rich in nutrients and energy. A scientifically laid out nutrition garden helps to meet the entire requirements of fruits and vegetables for a family all the year round (Sheela *et al.,* 1998). Establishment of nutrition garden is advocated as a means of preventing malnutrition among 250 tribal families and five tribal villages of Udaipur district, Rajasthan. They found that it is a low cost sustainable approach for reducing malnutrition, increasing awareness of vegetable production, increasing working hours and achieving food, nutrition and economic security for tribal families (Nandal and Mathur, 2012).

The fruits and vegetables are consumed by purchasing them from the market but for each small and marginal family it is not possible to include them in daily life. According to Prathiba and Rani (2012), a healthy vegetarian person should consume 125 g leafy vegetables, 100 g root vegetables, 75 g other vegetables and 85 g fruits apart from 475 g cereals and 85 g pulses in his daily diet. In order to ensure a healthy diet, fruits and vegetables are to be grown systematically in a small piece of land available in a home which is known as nutrition garden. This is important in rural areas where people have limited income and poor access to markets. Location specific programmes like promotion of nutrition gardens will play a major role in solving malnutrition problem. The concept of nutrition garden aims at continuous supply of vegetables to cater the daily needs of the family from the available area utilizing household wastes including water other organic matter (Indumathi *et al.,* 2012).

According to Shukla *et al.* (2012) the development and maintenance of nutrition garden is a collective effort of family members led by a woman or housewife. They conducted a study in two villages of Udham Singh Nagar district of Uttarakhand to assess the impact of nutrition garden on the nutrient intake of families. The result showed that the mean nutrient uptake in both villages improved after establishing nutrition garden and it is an effective measure to ensure household food security, health and nutrition.

Advantages of Nutrition Gardening

1. Nutrition gardening is the best means of recreation and exercise.
2. Helps in lowering down the family budget for vegetable.
3. Ideal medium for training children through gardens, in beauty and order.
4. Secures enough vegetables within the means of all classes at very cheap rate.
5. The cost of raising vegetables utilizing family labour is affordable and it gives a special appeal to palate.
6. Provides fresh vegetables free from infection from unsanitary markets.

Design and Layout of a Nutrition Garden

A model nutrition garden sustains fruits and vegetables either separately or in combination. An area of 20 m² (1/2 cent) is sufficient to provide required quantity of fruits and vegetables to one adult. A nutrition garden with an area of 100 m² will be sufficient for a family of 5 members (Figure 3.1).

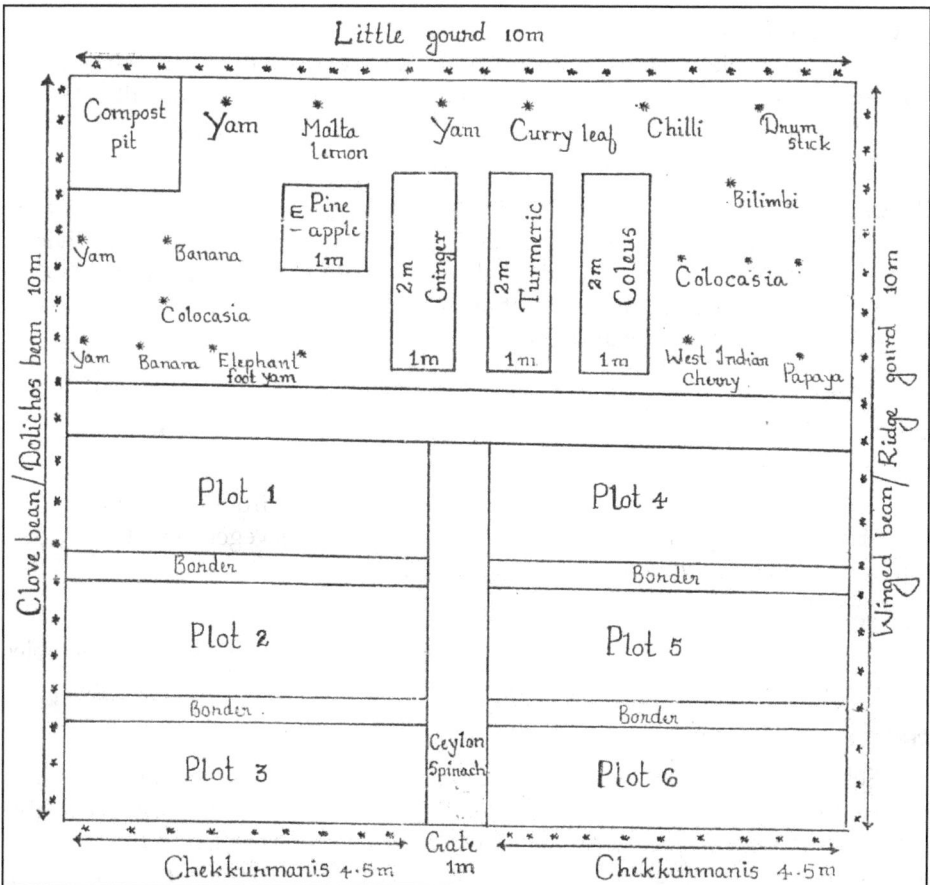

Figure 3.1: Layout of a Nutrition Garden (100 m²).

Location

Select an open area receiving plenty of sunshine. As far as possible it should be near the house and well protected from stray cattle. It should be near a source of irrigation. The garden should be located on the eastern or southern side of the house to have maximum sunlight.

Plan or Layout

The main objective in layout is the most economic utilization of space. Cropping intensity should be maximum in a nutrition garden. Fences, borders and interspaces of perennial crops are utilized for vegetable cultivation. A rectangular garden is preferable than a square garden. Planning is very essential for the successful management of a nutrition garden. The plan will vary in size and shape according to the availability of space, interest of family members, soil, climate, selection of vegetables etc. It is always better to have a small, well kept garden than a poorly managed large one.

Components of a Nutrition Garden

1. Perennial Crops

Perennial fruits and vegetables are located at one end of the garden so that they will not interfere with the intercultural operations of other crops and will not shade them. Usually grafts or dwarf varieties of fruit plants are preferred in a nutrition garden as they occupy only lesser space.

2. Annual Plants Plots

After the allotment of land for perennial crops, the remaining area may be divided into equal plots for raising annual vegetables. For continuous and steady availability of vegetables round the year, crop rotation has to be followed in each plot, which will reduce the pest and disease incidence also.

3. Borders

On the borders of garden plots dwarf and bushy vegetables are grown.

4. Fences

Fence can be made up of barbed wire to protect the crop from stray animals. Live fence is made up of plants like chekkurmanis or other tailing vegetables like coccinea, winged bean and dolichos bean.

5. Bee Hive

A bee hive is essential to ensure pollination in cross pollinated vegetables especially cucurbits. The farmer will get honey as an additional income.

Selection of Crops

Select locally adaptable crops and varieties as far as possible. Preference should be given for raising pest and disease resistant varieties. F_1 hybrid vegetables give early and high yield. Seeds or planting materials should be collected from reliable source like State Agricultural Universities, State department of Agriculture/ Horticulture, National seeds corporation and registered seed companies.

Fruits in a Nutrition Garden

Layers, grafts and cuttings of fruit plants are used as planting materials. Dwarf varieties of banana or culinary banana are usually grown in a nutrition garden. Guava, pineapple, papaya, carambola, west Indian cherry etc. are some other fruit plants grown in a nutrition garden. The interspaces of these crops are utilized for growing ginger, turmeric, amorphophallus etc. Ratooning is practiced in banana and pineapple. After harvest of bunches, one or two suckers are retained and the ratoon crop raised from them.

Cultivation Practices

Nursery raising, land preparation, manure application, sowing/planting, fertilizer application, irrigation, earthing up, weeding etc. are the major operations in vegetables.

Nursery Raising

Solanaceous vegetables and cole crops like cabbage and cauliflower are usually transplanted to the main field. Seeds of these vegetables are sown in open nursery beds or in plug trays. Since the open field nursery beds are facing various problems, vegetable seedlings are mostly raised in hi-tech nurseries in plugtrays. Plug trays of various size are available in the market with reasonable cost. They are filled with coconut pith or potting mixture. F_1 hybrid seeds being very costly, only one seed/pit is sown in these trays. Tray sowing ensures 100 per cent seed germination when compared to nursery beds in open field. After 20-25 days, vigorous seedlings are ready for transplanting. Care should be taken to irrigate the nursery every day, but there should not be waterlogging. Damping off is a serious disease in the nursery which can be controlled by seed treatment with captan @3g/kg of seed. The seedlings are to be hardened 3-4 days before planting by withholding irrigation and it will reduce the transplanting shock.

Land Preparation

The land should be prepared thoroughly by proper digging or ploughing. Planting can be done in flat beds, ridges, furrows, pits and mounts, according to location and season. Burning of dry leaves or stubbles in pits or furrows is a good practice which will eliminate soil borne pests and improve soil fertility.

Manures and Fertilizers

Organically grown vegetables are catching the market now-a-days. People are more health conscious and they are forced to grow as much vegetables in their home garden as possible. Most of the vegetables purchased from markets are contaminated with plant protection chemicals. Hence organically grown vegetables and fruits are preferred especially when they are consumed in raw form. Farm yard manure, poultry manure, goat manure, wood ash, vermicompost, neem cake etc. are a few organic manures used in vegetables. A compost pit at one corner of nutrition garden helps to dump crop residues and other wastes of a kitchen garden which can be converted into good organic manure. Organic manures are applied @ 100 kg/40 m^2 at the time of land preparation. In organic farming, vermicompost or other manures are top dressed at two or three stages.

Chemical fertilizers are an integral part of scientific cultivation. Fertilizer requirement varies with soil type, location, season, crop and variety. NPK fertilizers are given as basal dose and also in split doses. Micronutrients like Zn, Mo and Boron are given at lower doses as foliar sprays. Fertigation is recently adopted by many farmers to increase the yield of vegetables.

Irrigation

Irrigate the crop daily during summer. But it should not be excess. Waste water can be used for irrigation, which is free of any toxic chemicals or detergents. Irrigation interval can be reduced by mulching the plant basins with dried leaves, straw or polythene mulches. Mulching will not only conserve moisture but also check weed growth.

Intercultural Operations

Weeding and earthing up are the intercultural operations in vegetables. Weeding is done before fertilizer application which ensures more nutrient absorption. Earthing up is done along with fertilizer application. Shallow intercultural operations like hoeing is done to improve the soil aeration. For trailing vegetables, iron frames or permanent pandals are to be erected. Though their initial cost is high, they can be used for a longer period.

Plant Protection

Pests and diseases are the major constraints in vegetable production. Systematic and timely control measures are necessary to protect the crop from the pathogen. Emphasis should be given to follow biological, cultural and mechanical method for controlling pest and diseases. As far as possible use home made formulations which is free from phytotoxicity

Growing resistant varieties is the most economic method of plant protection. Seed treatment with fungicides minimizes nursery diseases. Addition of lime @ 5 kg/ cent at the time of land preparation reduces soil acidity and incidence of bacterial wilt diseases. Application of neem, glyricedia leaves, sesbania leaves and paddy straw reduce incidence of bacterial wilt in vegetables. In corporation of neem cake in soil reduces nematode population. Home made formulations like tobacco decoction, neem kernel extract, neem oil, Bordeaux mixture etc. are safe and effective against pests and diseases in a nutrition garden.

In the mechanical method of control, catch and kill the pests like borers and beetles in the early morning hours. Raising seedlings in polyhouse or net house reduces the pest incidence. In cucurbits, covering the fruits with poly covers or paper covers at early stages give protection from fruit flies. Use of pheromone traps or bait traps are also effective in cucurbits. Cultural methods are very important in plant protection. Use of disease free planting materials, field sanitation, clean cultivation, barrier crops and controlled irrigation are some of the cultural methods to reduce biotic stresses.

Harvest and Postharvest Handling

Vegetables and fruits are to be harvested at their optimum stage of maturity for better quality and yield. Harvesting should preferably be done at cooler part of the day and while harvesting care should be taken to avoid mechanical injury to the plant. After harvest direct contact with sunlight, heat or rain should be avoided. Avoid damage or bruise to vegetables, wash root crops in running water and do not heap fruits and vegetables after harvest.

Vegetables are consumed either raw and fresh or in cooked form. Excess vegetables are processed as sauce, squashes, pickles or chips. While cooking care should be taken to minimize the loss of nutrients and some tips are given below.

1. Vegetables are to be used immediately after harvest otherwise their water and vitamins will be lost.

2. Wash vegetables under running water before cutting or peeling to remove dirt, dust, insects, etc. Soak vegetables like cauliflower in salt water to reduce the pest load. Do not wash vegetables after peeling, cutting or slicing, since this would result in the loss of water soluble nutrients.

3. Trimming of vegetables should be reduced or avoided whenever possible.

4. Peeling is done as thin as possible since many nutrients are found just below the skin.

5. Do not cut vegetables into too large or too small pieces. Cut vegetables are cooked immediately.

6. Vegetables are to be cooked for as short time as possible. Boil water first and then add vegetables.

7. Never use excess water to cook. Steaming is better to conserve nutrients than boiling.

8. Pressure cooking is desirable as it reduces cooking time and conserve nutrients.

9. Don't overcook vegetables, never use soda while cooking since it destroys vitamins.

10. Avoid frying of vegetables as it destroys vitamin. Use prepared dishes immediately and avoid reheating the dishes.

Loss of nutrients at different stages like storage, transport, cooking and curing should be prevented by developing appropriate practices. By discarding edible and highly nutritional parts of plants such as leaves, stem, peel and seed due to prevailing eating preferences, lot of nutrients are lost. Cooking practices often destroy nutrition and unhealthy eating habits result in suboptimal nutritional benefits. In addition to food and water quality personal and kitchen hygiene are also important for proper nutrient utilization (Kumar, 2012). Hence, every housewife or woman who is the backbone of not only their home but also the society is to be properly trained in various aspects of kitchen gardening starting from cultivation up to consumption.

Conclusion

Rome Declaration on food security in 1996 emphasises access to safe and nutritious food which is an indication of global concern about nutritional security. Though India is the second largest producer of fruits and vegetables, their consumption is meagre especially among the rural population. Now-a-days people are more health conscious and good food shall be our medicine. Increased consumption of fruits and vegetables is one of the easiest and cheapest ways of enhancing health. Asia and Africa are blessed with diverse climatic condition and an array of fruits and vegetables are cultivated in different parts of the continents. Many underutilized fruits and vegetables, which are rich sources of phytochemicals are to be cultivated in backyard nutrition gardens. There is an increasing demand for indigenous, location specific underutilized vegetables and fruits throughout the world. Backyard nutrition gardening is a low cost sustainable approach for mitigating malnutrition especially in rural households.

References

1. Chavasith, V. (2012). Importance of vegetables to achieve food and nutrition security in South-east Asia. Proc. Regional Symposium on high value vegetables in South-east Asia: Production, Supply and Demand (SEAVEG 2012). Chiang Mai, Thailand 24-26 January p.33.

2. Dasgupta, S. (2012). Vegetable production and productivity gains are key to increasing food and nutrition security in Asia. Proc. Regional Symposium on high value vegetables in South-east Asia Production, Supply and Demand (SEAVEG 2012). Chiang Mai, Thailand 24-26 January p.34.

3. Indumathi, K., Shanmugam, P.S. and Tamilselvan, N. (2012). Nutrition garden as a valuable intervention to fight malnutrition in rural India. Abstracts. Global Conference on Horticulture for food, nutrition and livelihood options, Bhubaneswar, Odisha, India 28-31 May p.19.

4. Kumar, S. (2012). Horticulture as an instrument of frugal health and nutrition security. Proc. Global Workshop on Topical Issues (In) Global Conference on Horticulture for food, nutrition and livelihood options, Bhubaneswar, Odisha, India 28-31 May p.13.

5. Nandal, V. and Mathur, I.J. (2012). Backyard nutrition gardening: an effective solution for achieving food and nutrition security of tribal families in Sirohi (Rajasthan).Abstracts Global Conference on Horticulture for food, nutrition and livelihood options, Bhubaneswar, Odisha, India 28-31 May p.25.

6. NIN (National Institute of Nutrition) (2011). Dietary Guidelines for Indians – A Manual, National Institute of Nutrition, Hyderabad, India,p.127.

7. Prathiba and Rani, S. (2012). Nutrition gardening – A venture to mitigate malnutrition. Abstracts. Global Conference on Horticulture for food, nutrition and livelihood options, Bhubaneswar, Odisha, India 28-31 May p.11-12.

8. Sharma,J.P. (2009). Principles of Vegetable Breeding, Kalyani Publishers, New Delhi.p.288-312.

9.   Sheela, K.B., Peter, K.V. and Krishnakumari, K.(1998). Nutrition garden, Directorate of Extension, Kerala Agricultural University, Vellanikkara, Thrissur, p. 86.

10. Shukla, P., Rajkumari and Limbu, R. (2012). Nutrition garden as a tool towards combating nutrient deficiencies. Abstracts. Global Conference on Horticulture for food, nutrition and livelihood options, Bhubaneswar, Odisha, India 28-31 May p.16-17.

11. Singh, B. (2012). Vegetables for health and nutrition *"Shodh Chintan"*. Global Conference on Horticulture for food, nutrition and livelihood options, Bhubaneswar, Odisha, India 28-31 May p.102-108.

12. Singh, H.P. (2012). Trend of horticultural research particularly vegetables in India and its regional prospects. Proc. Regional Symposium on high value vegetables in South-east Asia: Production, Supply and Demand (SEAVEG 2012). Chiang Mai, Thailand 24-26 January p.34.

2015, Horticulture for Nutrition Security
Editor: **Prof. K.V. Peter**
Published by: **DAYA PUBLISHING HOUSE, NEW DELHI**

Pages **37–41**

Chapter 4

Good Agricultural Practices and Organic Farming (GAP and OF)

*K.V. Peter, S.Nirmala Devi, P.G. Sadhankumar
and T. Pradeepkumar*

As defined by the Food and Agriculture Organization (FAO) of the United Nations(UN), "Good Agricultural Practices(GAPs) are a collection of principles to apply for on-farm production and post-production processes, resulting in safe and healthy food and non-food agricultural products, while taking into account economical, social and environmental sustainability". GAPs may be applied to a wide range of farming systems through sustainable agricultural methods, such as integrated pest management (IPM), integrated fertilizer management (IFM) and conservation agriculture (CA).

GAPs related to soil, water and cropping system depend on principles of economic and sufficient production of safe and nutritious food, enhancement and sustainability of natural resources and maintenance of viable farming enterprises which contribute to sustainable livelihoods and meeting cultural and social demands of society.

Soil and Water

Reducing soil erosion through hedging and ditching, adoption of cultural practices which maintain soil structure and application of adequate doses of manures and fertilizers to restore soil organic matter content are some of the good agricultural practices related to soil. Scheduling irrigation depending on plant needs, soil water reserve status, *in situ* water harvesting by digging catch pits, crescent bunds across slope, restoration of wetlands, selection of suitable crops and avoiding run-off are GAPs related to water.

Cropping System

In agriculture, the major crops are cereals and millets, pulses, oil seeds, industrial crops,vegetables, fruits, plantations, spices,tubers and medicinal plants. Cereals, millets and pulses serve as source of energy and protein for human beings. Vegetables and fruits serve as sources of minerals and vitamins. Being a wellness crop, vegetables provide nutrition for the poor and health for all. Vegetables score more as source of income to farmers when compared to other crops like cereals, pulses and oilseeds. The Integrated Plant Nutrient and Water Management (IPNWM) and Integrated Pest and Disease Management (IPDM) are good agricultural practices in crop production.

Irrigation and fertilizer application are standard management practices for crop production in developed countries as well as in developing countries where water and fertilizer are not limiting. Most farmers apply water and fertilizer at maximum levels, decreasing the economic efficiency of these resources. Excessive fertilizer application also leads to environmental pollution and degradation due to leaching of nutrients to waterways and underground water. Integrated Plant Nutrient and Water Management (IPNWM) promotes efficient use of water and fertilizers in crop production. Compared with furrow irrigation, drip irrigation uses less water, improves yield, promotes efficient use of fertilizers and reduces the risk of groundwater contamination. Water use in drip irrigation was 45-77 per cent less than in furrow irrigation and yields achieved under drip irrigation were almost similar or better than yields under furrow irrigation. Furthermore, nutrient uptake was more efficient in drip-irrigated vegetables (Palada and Wu, 2005).

The GAPs include cultivation of varieties with durable and multiple resistance, use of grafted seedlings which can resist soil borne pathogens, adoption of cultural practices like rotation with non-host crops, removal and destruction of infested plant parts and flooding. Prolonged exposure to pesticides has been associated with several chronic and acute health effects like cardiopulmonary disorders, neurological and hematological symptoms and skin diseases. The health costs incurred by farmers exposed to pesticides are 61 per cent higher than those of farmers who are not exposed. IPDM focuses on management of important insect pests and diseases through biological control (use of predators and parasitoids, sex pheromones, biopesticides, etc.).Growing vegetable crops under protective structures covered with nylon netting, usually 16-mesh, prevents entry of insects and allows free flow of air.The porous nature of the net helps it to withstand strong winds without much damage. Keeping the soil moisture high throughout the growing season could reduce thrips damage in onion. Flooding the soil for up to 48 hours causes heavy mortality of soil-inhabiting immature stages of common armyworm, *Spodoptera litura* in cruciferous vegetable production system (AVRDC 2002). Overhead sprinkler irrigation would significantly reduce the diamondback moth (DBM), *Plutella xylostella* L., infestation. The aphid population was also less in the sprinkler irrigation. Sprinkler irrigation presumably drowns and washes away the DBM larvae and aphids feeding on the leaf surface, and the aphids.

The natural enemies like parasitoids, predators and pathogens can be used to control the insect pests in vegetable and fruit production The insect pathogenic bacteria, *Bacillus thuringiensis* entomo-pathogenic fungi like *Beauveria bassiana* and *Metarhizium anisopliae,* nuclear polyhedrosis virus (NPV), egg parasitoids like *Trichogramma chilonis,* larval parasitoids like *Diadegma semidausum* and *Cotesia plulellae* and the pupal parasitoid, *Diadromus collaris* are effectively be used in the management of several insects.

Organic farming evolved on the basic theoretical expositions of Rodale in the United States, Lady Balfour in England and Sir Albert Howard in India in the 1940s, has progressed to cover about 23 million hectares of land all over the world. Howard's magnum opus, 'An Agricultural Testament' has a special significance in India as it is based on an analysis of the environment friendly farming practiced here for centuries. However, India lags behind a majority of agriculture based countries in the world in the practice of organic farming.

In India, fertilizers are used only in 30 per cent of total cultivable area where irrigation facilities are available and in the remaining 70 per cent of arable land, which is mainly rain-fed, negligible amount is being used. Farmers in these areas often use organic manure that is readily available either in their own farm or in their locality as a source of nutrients. The North Eastern Region of India provides considerable opportunity for organic farming due to very low use of chemical inputs. Application of microbes in agriculture is one of the best options for organic agriculture. The report of the Task Force on Organic Farming (TFOF)appointed by the Government of India also observed that in vast areas of the country, where limited amount of chemicals is used and have low productivity, could be exploited as potential areas for organic agriculture (Anon, 2001). Organic farming can be more profitable in fruit crops, vegetable crops, tea, coffee, cashew, pepper, ginger, nutmeg, turmeric, cardamom, clove and vanilla where instead of quantity, quality is more important and also in soils having high fixation capacity of the nutrients like the calcareous, acidic and alkali soils.

Organic farming targeted to produce nutritive, healthy and safe food focuses on components like sustainable soil management, disease-resistant cultivars and biological pest and disease control methods. Among the different soil components, soil biological component is considered as a good indicator for soil health. With the development of better monitoring and identification methods for soil microorganisms, recent studies are focusing on the relationship between microbial dynamics and disease management (Garbeva *et al.*, 2004) and productivity (Webburn *et al.*, 2004). It is an important research component for promoting sustainable production in the developing countries.

In addition to agricultural farming, animal husbandry, poultry, fisheries, etc. should be practiced in organic farming system for rejuvenating degraded soil and ensure sustainability of crop production. It relies on crop use of resistant varieties, rotations, crop residues, animal manures, legumes, green manures, off farm organic wastes to maintain soil productivity and to supply plant nutrients and aspects of biological methods to control insect pests,diseases and weeds. Thus, the organic farming implies recycling of waste and residue to the native soil itself, replenishing the nutrients depleted from the soil during the crop growth, encouraging the growth of microorganisms which could regulate phased release of stored nutrients in the soil to the crop growth in right proportion, maintaining soil health by balancing the soil moisture and soil aeration and ensuring soil fertility by firmly binding the nutrient elements in the complex organic molecules.

A farmer can switch over to the system of organic farming from the conventional system based on a conversion plan. The time between the start of organic management and certification is called conversion period. Generally, conversion period is two years for annual crops and three years for perennial crops. Species and varieties cultivated should be adapted to soil and climatic conditions and resistant to pests and diseases. Seeds/planting materials should be procured from organic sources. If not available, chemically untreated seeds/planting materials can be used. Soil fertility should be maintained/enhanced through raising green manure crops and leguminous crops. The residues of plants after harvest should be incorporated into soil as far as possible. Bio-degradable materials of microbial, plant or animal origin shall be applied as manures. (eg. compost, vermin- compost, farm yard manure, sheep penning etc.).Natural enemies shall be encouraged and protected. Slash weeding is to be done between plants. Weeds under base of the plants shall be cleaned and put as mulch around the plant base. The products that are permitted for control of pest and diseases are neem oil, neem seed kernel extract, chromatic traps, mechanical traps, pheromone traps, plant based repellants, cow urine spray, garlic extract, chilli extract and release of parasite predators of insect pests

Certification of organic farms is required to satisfy the consumers that the produce is totally organic. Certification agency conducts inspection to verify that minimum requirements prescribed for organic agriculture are fully met and certificates issued

Conclusion

Though organic farming is one of the best approaches to get sustainability in crop production, the progress of organic agriculture in India is very slow which is only 41,000 ha (0.03 per cent of the cultivated area). The production of organic farms came to about 14,000 tones and 85 per cent of it was exported. Domestic consumption is marginal and is concentrated in the metropolitan cities in the country. There are certain constraints in its full fledge adoption. They are yield reduction in initial years of conversion, problem in availability, transportation, and application of biological materials to meet the nutrient demand of the crops, lack of package of practices involving organic farming practices along with cost benefit ratio of different crops, lack of adequate research and development backup as well as training in organic farming in India.

References

1.   Anonymous (2001). Report of Task Force on Organic Farming, Department of Agriculture and Cooperation, Ministry of Agriculture, Government of India, 2001, p. 76.

2.   AVRDC.(2002).AVRDC report 2001.Shanhua,Tainan,Taiwan: AVRDC. p.29.

3. Khurana,P.S.M and Singh,D.K(2012).Horticultural renaissance in India: A Compendium of selected lectures of Dr. K.L.Chadha.Published by IARI,New Delhi.767-779.

4. Garbeva, P. J.A.vanVeen and J.D.vanElsas.(2004).Microbial diversity in soil: Selection of microbial populations by plant and soil type and implications for disease suppressiveness. *Ann. Rev.Phytopathol.*42: 243-270.

5. Palada,M.C.and D.L.Wu.(2005).Influence of rainshelter and irrigation method on yield, water and fertilizer use efficiency of chilli pepper. *HortScience*40: 1143 (abstract).

2015, Horticulture for Nutrition Security
Editor: **Prof. K.V. Peter**
Published by: **DAYA PUBLISHING HOUSE, NEW DELHI**

Pages 43–65

Chapter 5

Organic Spices

B. Suma and T. Pradeepkumar

India, the land of spices, is a traditional producer, consumer and exporter of spices. The Western Ghats region is the centre of origin of many spices particularly, black pepper, cardamom and many zingiberaceous spices. Spices are high value and low volume commodities of commerce in the world market. All over the world, the fast growing food industry depends largely on spices as taste and flavor makers. Health conscious consumers in developed countries prefer natural colors and flavors of plant origin to low cost synthetic ones. Thus, spices are the basic building blocks of flavor in food industry.

Global awareness of health and environmental issues is spreading fast in recent years, especially in developed countries. Sustainability in production has become the primary concern in agriculture development. The organic method of farming is the best option to ensure that the air, water and soil around us remain unpolluted, leaving the environment safe for present and future generations.

In many countries, exploitative agriculture using industrial inputs is the norm since 1960s, to cater to an increasing population and to combat occurrence of famine and natural calamities. Such a system of farming causes imbalances in the constituents of biosphere, bio-forces, bio-forms and bio-sources. As a result, health of 'Mother Earth' is deteriorating. Organic agriculture aims to tackle the above concern and also aims at protecting the environment from continuous decline.

The concept of organic farming is based on holistic approach, where nature is perceived to be more than just an individual element. In this farming system, there is dynamic interaction among soil, humus, plant, animal, eco-system and environment.

Hence organic farming differs from industrial agriculture as in the latter, biological systems are replaced by technical production systems with liberal use of chemicals.

Organic farming improves structure and fertility of soil through balanced choice of crops and implementation of diversified crop rotation systems. Biological processes are strengthened without re-coursing to chemical remedies like synthetic fertilizers and pesticides. In this farming system, control of pests, diseases and weeds is primarily preventative, and if required, adopting organic products, which will not adversely affect the environment. Genetically modified organisms are not normally acceptable because of manipulations made in their natural set up. Organic matter of various kinds, nitrogen fixing plants, pests and disease resistant varieties, soil improvement practices like mulching and fallowing, crop rotation, multiple cropping, mixed farming, etc., are freely adopted. In brief, organic farming merges traditional and respectable views on nature with modern insights.

Bio-dynamic agriculture is yet another approach to organic farming. It is based on anthropsophy on the ideas formulated by the Austrian expert, Rudolf Steiner, in 1924. In this system, maintenance and furtherance of life processes on Earth are achieved by harnessing cosmic energy and various influences of sun, stars, moon and other planets. Bio-dynamic agriculture, most often, combines animal husbandry and crop production and use of compost and bio-dynamic preparations to revitalize soil and plants and subsequently animals and human beings.

Organic Spices - Indian Capability

Indian spices are the most sought-after globally, given their exquisite aroma, texture and taste. India has the largest domestic market for spices in the world. There are about 109 spices listed by International Standard Organization (ISO) and India grows about 60 of these spices. The varying climatic conditions in India provide ample scope for the cultivation of a variety of spices. Almost all Indian states produce spices, with a total area under spice cultivation to be 3.21 million hectares. India commands a formidable position in world spice trade. A total of 508555 tones of spices and spice products valued US$1396.51 million were exported during April-November 2013. The spice export basket consists of whole spices, organic, spice mixes, spice blends, freeze dried, curry powders/mixtures, oleoresins, extracts, essential oils,de-hydrated, spice in brine and other value added products.

Organic Cultivation

In recent years, organic agriculture is gaining considerable importance. Many farmers today show interest in organic cultivation. Several of them have begun switching to this traditional method of cultivation as a means to produce safe foodstuffs and preserve the environment. The concept of sustainable farming has caught on in India.

World Demand for Organic Foods

The world demand for organically produced foods is growing rapidly in developed countries like Europe, USA, Japan and Australia. The current estimated share of organic foods in these countries is approximately 1 to 1.5 per cent. Worldwide, food trends are changing with a marked health orientation. Since organic foods are free from chemical contaminants, demand for these products should steadily increase in the new millennium. According to the ITC, UNCTAD/GAT, more than 130 countries produce certified organic foods, 100 of them are from Asia and Africa.

Internationally, there is definite shift towards traditional/ethnic medicines. Since spices form part of many of these medicines, demand for organically produced spices should grow. Organic cultivation is nothing new to India. The country has always been practicing the traditional ways of using indigenous technologies and inputs mostly in line with modern organic farming principles. The per capita consumption of fertilizers and pesticides in India is far below that of developed countries, which means, it is very easy for Indian farmers to embrace organic spice farming in its true sense.

Spices Board of India has taken a major initiative in promoting the production and export of organic spices in a big way. Spices Board supports production, processing, certification and marketing of organic spices. Assistance is provided for organic cultivation of paprika, chillies, ginger, turmeric, kokkam, cumin, fennel, fenugreek, coriander and dill in Maharashtra, Odisha, Chattisgarh, Uttarakhand, Bihar, Gujarat and Rajasthan. The Board has also programmes to encourage production of organic pepper, herbal spices, Lakadong turmeric and ginger in the North Eastern States. Research is on to develop suitable package of practices for organic cultivation of chilli and cardamom. In certain spice crops, where organic production practices have not fully developed, Board encourages Integrated Pest Management (IPM) Programmes. The National Programme for Organic Production (NPOP) generally conforms to guidelines of IFOAM, EU, NOP and JAS in major aspects.

Export of organic products is allowed only if they are produced, processed or packed under the approved production programme and certified by an accredited certifying agency. Spices Board is designated as one of the agencies empowered to accredit certification agencies.

The Indian Initiative

Export of organic spices from India started in right earnest. The country at present exports around 50 tones of different varieties of organic spices. Exports will get a significant boost in coming years as more farmers switch to organic methods.

Spices Board of India prepared a document on production of organic spices. It features organic concepts, principles, basic standards, production guidelines, documentation, inspection and certification. The document is published after approval by National Standards Committee constituted by the members of IFOAM in India.

Research programmes on organic cultivation of important spices are commenced. The work is carried out at the Spices Board's Indian Cardamom Research Institute at ldukki District in Kerala. Besides organizing demonstrations to educate and motivate prospective organic spice growers, the Board is simultaneously involved in training programmes to existing spice growers on organic principles and practices.

Organic Farming – Guidelines

Organic farming is a crop production method respecting the rules of nature. Organic farming is targeted to produce nutritive, healthy and pollution free food. It maximizes use of on farm resources and minimizes the use of off-farm resources. It is a farming system that seeks to avoid use of chemical fertilizers and pesticides. It is not profit oriented but social profit oriented. Commitment to nature protection is a pre-requisite for practicing organic farming. In organic farming, entire system - plant, animal, soil, water and micro-organism- are to be protected.

Scope

Organic farming helps in rejuvenating degraded soil and ensures sustainability of crop production. Common man and farmers are aware of hazards from use of chemicals and pesticides. It is a common practice that farmers maintain part of their rice fields without pesticide application for their own consumption. When vegetables are grown in the Kitchen garden, no chemical fertilizers or pesticides are used since the house wife knows that the vegetables are meant for their own consumption. Now, the consumers prefer to consume natural/ethnic foods, particularly organic foods across the world. Moreover, they are ready to pay a premium price for such foods. The demand for organic agricultural products is increasing day by day.

Minimum Requirements

In organic farming system, certain minimum requirements are to be met to fulfill its objectives. Steps should be taken to maintain bio-diversity, *viz.* swamps, gross lands, forests, etc.

1. **Mixed farming:** Animal husbandry, poultry, fisheries, etc. should be practiced in addition to agricultural farming. Shifting cultivation is not allowed. Then only the farm is certified as organic.

2. **Conversion:** When a farmer switches over to the system of organic farming from the conventional system of farming, it is known as conversion. The time between the start of organic management and certification is called conversion period. The farmers should have a conversion plan prepared, if the entire field is not converted into organic at a time. In that case, it is necessary to maintain organic and non-organic fields separately. In the long run, entire farm including livestock should be converted into organic. The conversion period is decided based on past use of land and ecological situation. Generally, conversion period is two years for annual crops and three years for perennial crops. However, the conversion period can be relaxed based on verification by certification agency if the requirements are fully met.

3. **Cropping Pattern:** Crop rotation should be followed if annual crops are grown. Intercropping should be practiced when perennial crops are grown. Crop rotation should cover green manure and fodder crops. In case of perennial crops, cover crops like Kolinji (*Tephraria purpurea*) should be grown to protect the soil. Monocropping should be avoided.

4. **Planting:** Species and varieties cultivated should be adapted to soil and climatic condition and resistant to pests and diseases. Seeds/planting materials should be procured from organic sources. If not available, chemically untreated seeds/planting materials can be used.Even one time use of genetically engineered seeds or planting materials such as tissue culture, pollen culture, transgenic plants is not allowed.

5. **Manure Policy:** Soil fertility should be maintained/enhanced through raising green manure crops and leguminous crops. The residues of plants after harvest should be incorporated into soil as far as possible. Bio-degradable materials of microbial, plant or animal origin shall be applied as manures. (compost, vermi-compost, farm yard manure, sheep penning etc.) Use of synthetic/chemical fertilizers is not permitted. The mineral based materials like rock phosphate, gypsum and lime, can be applied in limited quantities when there is absolute necessity.

Following products are permitted for use in manuring/soil conditioning in organic fields:

☆ Farm yard manure, slurry, green manures, crop residues, straw and other mulches from own farm.

☆ Saw dust, wood shaving from untreated wood, calcium chloride, lime stone, gypsum and chalk, rock magnesium chloride, sodium chloride, Bacterial preparations (Bio-fertilizers) like azospirillum and rhizobium, Bio-dynamic preparations, extracts of neem cake, vermicompost etc.

☆ Following products shall be used when they are absolutely needed and taking into consideration factors like contaminations, depletion of natural resources, and nutritional imbalances. If proposing for certification, the certification agency may be consulted before using these inputs such as farm yard manure, slurry, urine and straw from other farms, blood meal, bone meal, fish meal without preservatives, minerals like basic slag and sulphate of potash, wood ash from untreated wood and vermi compost from other farms.

6. **Pests, Diseases and Weed management:** Use of synthetic/chemical pesticides, fungicides and weedicides are prohibited. Natural enemies shall be encouraged and protected. (raising trees in the farm attracts birds which kill pests of crops and nest construction). Products collected from the local farm, animals, plants and micro-organisms and prepared at the farm are allowed for control of pests and diseases. (Neem seed kernel extract, cow urine spray). Use of genetically engineered organisms and products are prohibited for controlling pests and diseases. Similarly, use of synthetic growth regulators is not permitted.

Slash weeding is to be done between plants. Weeds under base of the plants shall be cleaned and put as mulch around the plant base. The weeded materials should be applied as mulch in the ground itself.

The products which are permitted for control of pests and diseases are:

Neem oil and other neem preparations like Neem seed kernel extract, Chromatic traps, Mechanical traps, Pheromone traps, Plant based repellants, and Soft soap clay.

The following products shall be used when they are absolutely necessary and taking environmental impact into consideration. The certification agency shall be consulted before using these inputs such as copper salts *e.g.* Bordeaux mixture, Plant and animal preparations, Cow urine spray, Garlic extract, Chilli extract, Light mineral oils like Kerosene, release of parasites and predators of insect pests *e.g.* Trichogramma.

7. **Soil and water conservation:** Measures like stone pitching/contour wall construction are to be taken up to prevent soil erosion. In saline soils, saline resistant varieties may be grown. Judicious irrigation is to be practiced. Mulching is required. Pollution of surface and ground water shall be prevented. Clearing of primary forest is prohibited. Cleaning of land through straw burning should be restricted to minimum.

8. **Contamination control:** It is necessary to take up following measures to reduce contamination from outside and within farm.

   a) If neighboring fields are non-organic, a buffer zone should be maintained. The height of buffer crop shall be twice the height of organic crop and width of buffer shall be 8-16m. (When chilli is grown as the main organic crop, castor or Agathi (*Sesbania*) can be grown as buffer crop. The crops from buffer zone should be sold as non-organic).

   b) If the farm is under conversion, equipments used in conventional areas shall be well cleaned before using in organic areas.

   c) Products based on polythene, polypropeline and other polycarbarnates are allowed to cover protected structures, insect netting, nursery and drying subject to the condition that these materials shall be removed from field after use and they shall not-burnt or put in the soil. Use of poly-chloricle based products like PVC pipe is prohibited.

9. **Processing:** Processing technologies like solar drying, freeze drying and hot air chambers are permitted. Irradiation of agricultural produce is not permitted. No synthetic additives/dyes are to be added during processing. Lot number should include the crop, country; field no., date of harvest (in Julian Calendar) and production year.

10. **Labeling:** The label should convey clear and accurate information on organic status of products. (conversion in progress or organic). The labels for organic and conversion in product should carry, quantity of product, name and address of producer, name of certification agency, certification and lot number.

Lot number is helpful in tracing back the product particularly the field number in which it is grown in case of contamination. Calendar ranges 1 to 365 or 366, starting 1st January, as I and December 31st as 365/366.

Crop	OC (organic chillies)
Country	I (India)
Field No.	05
Date of harvest	32 (1 st Feb)
Year	1999
Lot number	OC 1 0532 1999

11. **Packaging:** For packing, recycling and reusable materials like clean jute bags shall be used. Use of bio-degradable materials shall also be encouraged. Unnecessary packaging material should be avoided. Organic and non-organic products shall not be stored and transported together except when labeled.

12. **Social Justice:** Social right and justice are integral part of organic agriculture. The laws relating to labor welfare and rights of children should be honored. All employees and their families should have access to potable water, food, housing, education, transportation and health services. All employees should have equal wages when doing same job. They must have equal opportunities irrespective of color, creed and gender. Social security which includes maternity, sickness and retirement benefits, should be met. Labor conditions regarding noise, dust, light and exposure to chemicals should be within acceptable limits and they should have adequate protection. The rights of indigenous people should be respected.

13. **Documentation:** Documentation of farm activities is a must for acquiring certification especially when both conventional and organic crops are raised. The following documents/records are to be maintained.

a) Field map
b) Field history sheet
c) Activity register
d) Input record
e) Output record
f) Harvest record
g) Storage record
h) Sales record
i) Pest control records
j) Movement record
k) Equipments cleaning records
l) Labeling records.

Certification Process

Certification of organic farms is required to satisfy the consumers that the produce is totally organic. Certification agency conducts inspection that minimum requirements prescribed for organic agriculture are fully met and certificates issued

The producer makes contact with certifying agency. Certification agency provides information on standards, fees, application, inspection, certification and appeal procedures. The producer then submits application along with field history, farm map and record keeping system. Then the contract, indicating scope, obligation, inspection and certification, sanction and appeals, duration and fee structure, is executed.

Then the Inspector of agency comes and carries out inspection. The Inspector gives inspection report with his recommendation to the agency, the agency issues approval or denial of certificate. Certificate is given for current year's harvest only and hence annual certification is required.

Production of Organic Pepper

General Guidelines

When pepper is grown in mixed cultivation system, it is essential that all the crops in the field are maintained following organic methods of production. It is also advisable that the entire farm is converted to organic. Livestock should be reared according to organic principles. All crop residues and farm wastes should be recycled through composting so that soil fertility is restored and maintained at a very high level. Soil and nutrient loss through soil wash, run off and percolating water should be minimized through proper agronomic practices. Weeding is to be limited to slashing. Slashed materials should be used for mulching the plant base. The plantation should have a green cover with leguminous crops. They could be cut and mulched during summer to prevent moisture loss. Shade and support trees, leguminous and green manuring shrubs to provide bio-mass and neem and other sources for plant protection agents should find a place with in or on the borders of such organic farm.

Guidelines for Organic Pepper Production

An isolation belt of at least 25 m wide should be maintained around the organic plantation. The produce from this isolation belt shall not be treated as organic. Precaution should be taken to avoid the entry of run off water and drift from the neighboring farms. While converting an existing plantation to organic cultivation, a minimum three years conversion period is a must. The conversion period can be relaxed for the organic farm raised on a land where chemicals were not used previously. In all these cases, proper records must be maintained for submission to the certifying agency.

Sources of Planting Materials for Organic Production of Pepper

Initially, cuttings from conventional plantations can be used in the absence of purely organic sources. The production of rooted cutting should be done according to organic principles. In subsequent years, the runner shoots or aerial shoots collected from elite mother vines grown organically only shall be used for generation of planting materials.

Adoption of following practices in the nursery is recommended. The soil used for potting mixture should be solarised prior to use. Solarised soil should be fortified with cultures of VAM and *Trichoderma* (250 g mass multiplied in 25 kg compost).

The vines in the rapid multiplication units may be sprayed with vermi wash (50 ml/unit) for enhancing growth.

Disease Management in Nursery

The two important nursery diseases *viz.*, leaf rot caused by *Rhizoctonia solani* and basal wilt caused by *Sclerotium rolfsi* can be minimum, if solarised soil inoculated with VAM and *Trichoderma* are used. If isolated incidence of these diseases is noticed, timely adoption of phytosanitary measures and spot application of Bordeaux mixture (1 per cent) may be done. In areas, where nematode problems occur, addition of crushed neem seed or neem cake is recommended.

Prevention of Run Off and Drifts from other Farms

In sloppy lands, adequate soil and water conservation measures are necessary at the time of preparing the land for planting to avoid entry of run off water and drift from neighbouring farms.

Use Variety of Standards

As many varieties of standards as possible should be used. Use of *Erythrina* sp. should be minimized as the plant harbors root knot nematode.

Nutrients

Application of 2 kg compost of rotten cow-dung mixed with 125 g rock phosphate at the time of planting rooted cuttings should be done as a basal dose. Micronutrient mixture from IISR can be applied on need basis (Sheeja *et al.*, 2014).

Cultural Operations

As the cuttings grow the shoots are to be tied to the standard and the young vines should be protected from sun during summer by providing artificial shade. Regulation of shade of the standard by lopping, mulching, green leaf manuring, weeding and cover cropping are recommended to improve the soil nutrient status.

Manuring

Compost or farmyard manure may be applied from second year onwards@ 4kg/ vine which can be gradually increased to10 kg/vine as it grows during May-June. This can be partially or completely substituted by vermicompost in which case, the quantity needed will be half. If found necessary, based on soil test, application of rock phosphate, bone meal, lime and dolomite may be carried out. Wood ash may be used in potash deficient areas. Compost made from green loppings, crop residues, grasses, cow dung and poultry droppings, fortified wood ash and/or rock phosphate should be used regularly instead of farmyard manure alone. Such compost can be further enriched with non-edible oil cakes and right microbial cultures prior to withdrawal from compost pit and before applying to field. Crushed neem seed @ 2 kg/vine/year may be applied in areas infested with nematodes. Use of bio-fertilizers can also be resorted to in a restricted manner.

Pest, Disease and Weed Management

Various pest and diseases of pepper can be managed by the adoption of following methods such as timely surveillance of the garden; adoption of proper plant and field sanitation, appropriate cultural practices, control of pest by mechanical means, need based prophylactic application of BM, application of soil conditioners such as neem cake and application of bio agents such as *Trichoderma*. Bioactive compounds from actinomycetes are effective against phytopathogens of pepper and ginger. The potential isolates were characterized morphologically and molecularly and identified as *Streptomyces* sp. and *Kitasatospora* sp., besides they were also found as growth promoters in blackpepper (Bhai, 2014).

Processing Methods

Processing methods of black or white pepper should be based on mechanized, physical and biological processes. Care should be taken to maintain the quality throughout processing. DNA barcoding is now used to detect the plant based adulterants in traded spices such as black pepper powder. The barcoding loci *psba – trnH* and *rbcl* are useful in this regard along with findings supplemented from HPLC (Sasikumar *et al.,* 2014).

Packaging

The dried produce is to be packed in eco friendly packing materials, stored and transported separately to protect it from co mingling with non organic products.

Ginger

Ginger (*Zingiber officinale*) is one of the important spices grown in India. Ginger of commerce is the dried rhizome. It is marketed in different forms raw ginger, dry ginger, bleached dry ginger, ginger powder, ginger oil, ginger oleoresin, gingerale, ginger candy, ginger beer, brined ginger, ginger wine, ginger squash and ginger flakes.

Climate and Soil

Ginger is cultivated in almost all states in India. Kerala is the major ginger growing state. Other major ginger growing states are Odisha, Meghalaya, Himachal Pradesh and Karnataka. Ginger grows in warm and humid climate. It is mainly cultivated in the tropics from sea level to an altitude of above 1500m. amsl and it can be grown both under rainfed and irrigated conditions. For successful cultivation of the crop, a moderate rainfall at sowing time till the rhizomes sprout, fairly heavy and well distributed showers during growing period and dry weather for about a month before harvesting are necessary. Ginger thrives the best in well drained soils like sandy or clay loam, red loam or lateritic loam. A friable loam rich in humus is ideal. However, being an exhausting crop, it may not be desirable to grow ginger in the same site year after year.

Maintenance of Buffer Zone

To cultivate ginger organically, a buffer zone of 8 to 16m is to be left all around from conventional farm, depending upon location of farm. The produce from this buffer zone belt shall not be treated as organic. Being an annual crop, the conversion period required will be two years. Ginger can be cultivated organically as an inter or mixed crop provided all other crops are grown following organic methods. It is desirable to include a leguminous crop in rotation with ginger. Ginger-banana- legume or ginger-vegetable-legume can be adopted.

Sources of Planting Material

Carefully preserved seed rhizomes free from pests and diseases which are collected from organically cultivated farms can be used for planting. To begin with, seed material from high yielding local varieties may be used in the absence of organically produced seed materials. Seed rhizomes should not be treated with any chemicals. The seed rate varies from region to region and with method of cultivation adopted. The seed rate varies from 1500- 2500 kg/hectare.

Preparation of Land and Planting

While preparing the land, minimum tillage operations may be adopted. Beds of 15 cm height, 1m width and of convenient length may be prepared giving at least 50 cm spacing between beds. Solarisation of the beds is beneficial in checking the multiplication of pests and disease causing organisms. Solarisation is a technique by which polythene sheets are covered over moist beds of field, reaching all the sides and exposing to sun for a period of 20-30 days. The polythene sheets used for soil solarisation should be kept away safely after the work is completed.

At the time of planting, apply 25g powdered neem cake and mix well with the soil in each pit taken at a spacing of 20-25 cm within and between rows. Seed rhizomes may be put in shallow pits and mixed with well rotten cattle manure or compost mixed with *Trichoderma*, an antagonistic (parasitic) fungi (10g compost inoculated with *Trichoderma*). The best time for planting in West Coast is first fortnight of May with receipt of monsoon. Under irrigated conditions, it can be planted well in advance during middle of February or early March.

Cultural Practices

Mulching ginger beds with green leaves is an essential operation to enhance germination of seed rhizomes and to prevent washing off of soil due to heavy rain. This also helps to add organic matter to soil and conserve moisture during the later part of cropping seasons. The first mulching is to be done with green leaves @ 10-12 t/ha at the time of planting. It is to be repeated @ 5 t/ha at 40th and 90th day after planting. Use of *Lantana camara* and *Vitex negundo* as mulch may reduce infestation of shoot borer. Cow dung slurry or liquid manure may be poured on bed after each mulching to enhance microbial activity and nutrient availability. Weeding may be carried out depending on intensity of weed growth. Such materials may be used for mulching. Proper drainage channels are to be provided in the inter rows to drain off stagnant water.

Manuring

Application of well rotten cow dung or compost @ 5-6 t/ha may be made as basal dose while planting the rhizomes in pits. In addition, application of neem cake @ 2 t/ha is also desirable

Plant Protection

Pests

Shoot borer is the major pest infesting ginger. Regular field surveillance and adoption of phytosanitary measures are necessary for pest management. It appears during July -October. Spot out shoots infested by borer and cut open the shoot and pick out caterpillar and destroy. Spray neem oil (0.5 per cent) at fortnightly intervals, if necessary. Light traps will be useful in attracting and collecting the adult moths.

Diseases

Soft rot or rhizome rot is a major disease of ginger. While selecting the area for ginger cultivation, care should be taken to see that the area is well drained as water stagnation pre- disposes plants to infection. Hence provide adequate drainage. Select seed rhizomes from disease free areas since this disease is seed borne. Solarisation of soil done at time of bed preparation reduces the fungus inoculum. However, if the disease is noticed, the affected clumps are to be removed carefully along with the soil surrounding the rhizome to reduce the spread. *Trichoderma* may be applied at the time of planting and subsequently, if necessary. Restricted use of Bordeaux mixture (1 per cent) in disease prone areas as spot application may be made to control it.

Harvesting and Postharvest Operations

The crop is ready to harvest in about eight to ten months depending upon maturity of variety. When fully mature, leaves turn yellow and start drying up gradually. Clumps are lifted carefully with a spade or digging fork and rhizomes are separated from dried leaves, roots and adhering soil. The average yield of fresh ginger/hectare varies with varieties ranging from 15 to 25 tones. For making vegetable ginger, harvesting is done from 6th month onwards. The rhizomes are thoroughly washed in water twice or thrice after harvest and sun-dried for a day. For preparing dry ginger, the produce is kept soaked in water overnight. Rhizomes are then rubbed well to clean them. After cleaning, rhizomes are removed from water and outer skin removed with a bamboo splinter or wooden knife having pointed ends. Iron knife is not recommended, as colour will be faded. In order to get rid of last bit of the skin or dirt, dry rhizomes are rubbed together. Peeled rhizomes are washed and dried in sun uniformly for one week. Rhizomes are to be dried to a moisture level of 11 per cent and they are stored properly to avoid infestation by storage pests. Storage of dry ginger for longer periods is not desirable. The yield of dry ginger is 16-25 per cent of the fresh ginger depending upon the variety and location where the crop is grown.

Preservation of Seed Rhizomes

The rhizomes to be used as seed material should be preserved carefully. The indigenous practices like spreading layers of leaves of *Glycosmis pentaphylla* being followed by farmers can very well be adopted for this purpose. To get good germination, seed rhizomes are to be stored properly in pits under shade. For seed materials, big and healthy rhizomes from disease-free plants are selected immediately after harvest. For this purpose, healthy and disease-free clumps are marked in the field when the crop is 6-8 months old and still green. Seed rhizomes are stored in pits of convenient size, made inside the shed to protect from sun and rain. Walls of pits may be coated with cow dung paste. Seed rhizomes are stored in these pits in layers along with well-dried sand/saw dust (*i.e.* put one layer of seed rhizomes, then put 2 cm thick layer of sand/saw dust). Sufficient gap is to be left at top of pits for adequate aeration. Seed rhizomes in pits need inspection once in twenty days to remove shrivelled and disease affected rhizomes. Seed rhizomes can also be stored in pits dug in the ground under shade of a tree provided there is no chance for water to enter pits. In some areas, rhizomes are loosely heaped over a layer of sand or paddy husk and covered with dry leaves in a thatched shed.

Turmeric

Turmeric (*Curcuma longa*), the ancient and sacred spice of India is a major spice produced and exported from India. Turmeric is used as condiment, dye and cosmetic in addition to its use in religious ceremonies. India is the leading producer, consumer and exporter of turmeric in the world.

Climate and Soil

Turmeric can be grown in diverse tropical conditions from sea level to 1500m amsl at temperature of 20-30°C with a rainfall of 1500 mm or more/annum or under irrigated conditions. It is grown on different types of soils from light black, loam and red soils to clay loams. It thrives the best in a well-drained sandy or clayey loam.

Maintenance of Buffer Zone

To cultivate turmeric organically, a buffer zone of 8 to 16m shall be maintained if the neighbouring farms are non-organic. The produce from this zone shall not be treated as organic. Turmeric being an annual crop, the conversion period required will be two years. Turmeric can be cultivated organically as an intercrop with other crops provided organic methods of cultivation are followed for all the companion crops.

Sources of Planting Material

Carefully preserved seed rhizomes free from pests and diseases collected from organically cultivated farms, should be used for planting. To begin with, seed material from high yielding local varieties may be used in the absence of organically produced seeds. A seed rate of 2500 kg rhizomes is required for planting one hectare.

Preparation of Land and Planting

While preparing land, minimum tillage operations may be adopted. Beds of 15 cm height, 1 m width and of convenient length may be prepared giving at least 50 cm spacing between beds. Solarisation of such beds is beneficial in checking the multiplication of pests and disease causing organisms. The polythene sheets used for soil solarisation should be kept away safely after the work is completed.

At the time of planting, apply 25 g powdered neem cake and mix well with the soil in each pit taken at a spacing of 20-25 cm within and between rows. Seed rhizomes may be put in shallow pits and covered with well rotten cattle manure or compost mixed with *Trichoderma* (10 g compost inoculated with *Trichoderma*). Turmeric can be planted during April-May with receipt of pre monsoon showers.

Cultural Practices

Mulching turmeric beds with green leaves is an essential operation to enhance germination of seed rhizomes and to prevent washing off of soil due to heavy rain.

This also helps to add organic matter to soil and conserve moisture during the later part of cropping season. Judicious mix of leguminous leaves with high nitrogen content and leaves rich in phosphorous like *Acalypha* weed and leaves rich in potassium like *Calatropis* can be used according to availability. The first mulching is done at planting with green leaves @ 10-12 tones/ha. It is to be repeated again @ 5 tones/ha at 50th day after planting. Cow dung slurry may be poured on bed after each mulching to enhance microbial activity and nutrient availability. Weeding may be carried out depending on intensity of weed growth. Such materials may be used for mulching. Proper drainage channels are to be provided in inter rows to drain off stagnant water.

Manuring

Application of well rotten cow dung or own compost from own farm @5-6 tones/ha may be made as basal dose while planting rhizomes in pits. In addition, application of neem cake @2 tones/ha is also desirable.

Plant Protection

Pests

Regular field surveillance and adoption of phytosanitary measures are required for pest management. If shoot borer incidence is noticed, such shoots may be cut open and larvae picked out and destroyed. Spray neem oil (0.5 per cent) at fortnightly intervals, if necessary.

Diseases

No major disease is noticed in the crop. Leaf spot and leaf blotch can be controlled by restricted use of Bordeaux mixture (1 per cent). Application of *Trichoderma* at the time of planting checks incidence of rhizome rot.

Harvesting and Postharvest Operations

Turmeric is to be harvested at correct maturity. Depending upon variety, crop becomes ready for harvest in 7-9 months, medium varieties in 8-9 months and late varieties after 9 months.

Usually the land is ploughed and rhizomes are gathered by hand picking or clumps are carefully lifted with a spade. Harvested rhizomes are cleaned of mud and other extraneous matter adhering to them. Average yield/hectare comes to 20-25 tones of green turmeric.

Fingers are separated from mother rhizomes. Mother rhizomes are usually kept as seed material. The fresh turmeric is cured for obtaining dry turmeric. Curing involves boiling of rhizomes in fresh water and drying in sun.

No chemical should be used for processing. The cleaned rhizomes are boiled in copper or galvanized iron or earthen vessels, with water just enough to soak them. Boil till the fingers/mother rhizomes become soft. The cooked turmeric is taken out of pan by lifting troughs and draining the water into pan itself. The same hot water in the pan can be used for boiling next set of raw turmeric which is already filled in troughs. The cooking of turmeric is to be done within 2-3 days after harvest.

Rhizomes may also be cooked using baskets with perforated bottom and sides. The mother rhizomes and fingers are cured separately.

The cooked fingers/mother rhizomes are dried in sun by spreading in 5-7 cm thick layers on bamboo mats or cement floor. A thinner layer is not desirable as colour of dried product may be adversely affected. During night time, the material should be heaped or covered. It takes 10-15 days for rhizomes to become completely dry. Artificial drying using cross-flow hot air at a maximum temperature of 60°C gives satisfactory product. In sliced turmeric, artificial drying has clear advantages in giving brighter coloured product than sun drying which tends to suffer due to surface bleaching. Recovery of dry product varies from 20-30 per cent depending upon variety and location where the crop is grown.

Dried turmeric has poor appearance and rough dull colour outside the surface with scales and root bits. Smoothening and polishing outer surface by manual or mechanical rubbing improve appearance.

Manual polishing consists of rubbing dried turmeric fingers on a hard surface. The improved method is by using hand-operated barrel or drum mounted on a central axis, sides of which are made of expanded metal mesh. When the drum filled with turmeric is rotated, polishing is effected by abrasion of surface against mesh by mutual rubbing against each other as they roll inside the drum. The turmeric is also polished in power-operated drums. The yield of polished turmeric from raw material varies 15-25 per cent.

Colour of turmeric always attracts buyers. To impart attractive yellow colour, turmeric suspension in water is added to polishing drum in the last 10 minutes. When rhizomes are uniformly coated with suspension, they may be dried in sun.

Rhizomes for seed purpose are generally stored after heaping under shade of tree or in well ventilated shade and covered with turmeric leaves. Sometimes, the heap is plastered with earth mixed with cow dung. The seed rhizomes can also be stored in pits with sawdust. The pits can be covered with wooden planks with one or two holes for aeration.

Small Cardamom

Small cardamom the "Queen of spices" is indigenous to evergreen forests of Western Ghats in South India. It is found to grow within an altitude ranging between 600 and 1200m amsl with an annual rainfall between 1500 and 4000 mm and temperature ranges from 10 to 35°C.

Choice of Crops and Varieties

Varieties naturally resistant to pest and disease adapted to the soil and climatic conditions should be selected. When certified organic planting materials are not available, conventional planting materials shall be used initially.

Sources of Planting Material

Seeds are to be collected from elite plantation even if they are not grown organically, seedlings should be raised using organic standards. The plantation should follow organic method of production at least one year prior to collection of rhizomes. Tissue culture plantlets should not be used as planting material. Avoid acid treatment of seed and treatment of seeds with *Trichoderma* culture (50 ml spore suspension for 100g of seed) is desirable as a prophylactic measure for managing nursery rot diseases. At the time of preparation of beds, incorporation of VAM multiplied in recommended organic medium is done. For raising polybag seedlings potting mixture @ of 3:1:1 soil, cowdung and sand can be used. To this cow dung fortified with VAM and *Trichoderma* can also be added. Regular surveillance, adopting phytosanitary measures, restricted application of BM (1 per cent) may be done to control the disease at the initial stage itself.

Cultural Practices

Clean weeding is to be limited to plant bases (50 cm) and inter rows are maintained by slash weeding. The weeded materials should be used for mulching. Trashed materials and fallen leaves may also be used for mulching. Trashing the dry leaves and leaf sheaths and removal of yielded old suckers may be carried out once in a year after completion of final harvest, which can be used for composting. The inter rows should not be dug at any cost. Clean water free from insecticides, fungicides, other chemicals and fertilizer leaches should only be used for irrigation under organic system of cultivation. This implies that watersheds for irrigation sources should also be maintained following organic methods of production. In areas, where adequate soil conservation measures and mulching are practiced, there will not be any necessity for earthing up.

To facilitate penetration of sufficient light, restricted lopping of shade tree branches may be done. However, even under such situations, no tree top shall be cut. In areas, which are overexposed, planting of shade trees is an essential operation and while doing so, maximum bio-diversity suited to local situation may be considered. Trees having desirable characters like defoliation during rainy season, self-pruning habit, flowering during summer and medicinal values may be considered. If such trees belong to leguminous species, they are preferred. Restricted loppings and leaf litter may be used for green leaf manuring or composting. Preservation of bee fauna is an integral part of organic cultivation. Integration of apiculture will not only ensure bio-diversity, but also help in increasing production through assured pollination.

Buffer Zone

An isolation distance of at least 25m width is to be left from all around the conventional plantation. The produce from this zone shall not be treated as organic. In sloppy lands, adequate soil and water conservation measures are necessary at the time of preparing the land for planting to avoid entry of runoff water and drift from neighbouring farms.

Nutrient Management

Application of organic manures like neem cake@ one kg or poultry manure/ farm yard manure/compost/vermin or compost@ two kg/plant may be done May-June. Well rotted cattle manure or FYM, sheep manure, bone meal, fish meal and caster cake enhances yield. Applications of cow dung slurry and slurry of ground nut oil cake and neem cake are widely practiced for better growth. Applications of bio-agents like *Trichoderma* or *Pseudomonas* help to check disease.

Pests

Removal of drooping leaves, dry leaf sheath, old panicles and other dry plant parts are to be done as part of sanitation. Mechanical collection and destruction of egg mass, larvae and adult beetles are effective method for reducing pest damage. Injection of *Bacillus thuringiensis* @ of 0.5 mi/10 ml of water is effective against stem borer. Under organic methods of cultivation outbreak of white flies Dia leurodes cardamom is seldom observed. However in the event of such outbreaks collection of adults using yellow sticky traps and control of nymphs Application of neem cake @ 1kg/plant or bio-agents like *Metarrhizium* is effective for control of nematodes and root grubs.

Diseases

Incorporation of *Trichoderma* multiplied in suitable organic medium @ of 1kg/clump prior to the onset of monsoon is a prophylactic treatment for clump rot disease.

Processing Methods

Processing methods be based on mechanized, physical and biological processes. Care should be taken to maintain the quality throughout processing.

Packaging

The dried produce is to be packed in eco friendly packing materials, stored and transported separately to protect it from co mingling with non organic products.

Large Cardamom

Large cardamom (*Amomum subulatum* Roxburg) a member of Zingiberaceae family is one of the main cash crops cultivated in Sub Himalayan state of Sikkim and Darjeeling region at altitude ranging from 1000 to 2000 m above msl with an annual rain fall of 3000 to 3500 mm. For successful cultivation the mean minimum and maximum temperature requirements are 6°C and 25°C during warm month. It performs well under the shade of forest trees with medium dense canopy.

Choice of Crops and Varieties

Naturally resistant to pest and disease adapted to the soil and climatic conditions should be selected. When certified organic planting materials are not available, conventional planting materials shall be used initially.

Cultivars

There are mainly five popular cultivars *viz.* Ramsey, Sawney, Golsey, Ramla and Varlanga.

Sources of Planting Material

Seeds are to be collected from elite plantation even if they are not grown organically, seedlings should be raised using organic standards. The plantation should follow organic method of production at least one year prior to collection of rhizomes. Tissue culture plantlets should not be used as planting material.

Cultural Practices

They are largely eco friendly and are grown without use of chemical fertilizers. Controlling weed growth in plantations is important for maximum utilization of available soil moisture and nutrients by the crop. Weeding may be done as and when necessary either by hand weeding or sickle weeding depending upon intensity of weed growth. Around the plant base, weeds can be pulled out and weeds in inter space need only slashed with sickle. Along with weeding, dry shoots and other thrash materials are also cut and removed preferably in September-October and the trashed materials can be used as mulch around the plant base to a thickness of about five cm. Intensive operations which loosen and expose soil increase soil erosion and therefore, only minimum tillage operations should be practiced. As far as possible, contour terraces should be made only at planting points well before taking up planting operations. For sustainable and better yield, plants may be watered during dry months.

Nutrient Management

For sustained production, soil fertility should be maintained at a fairly high level. Well rotten cattle manure, compost or organic products or non edible oil cakes should be applied @ 2 kg/plant. If all the crop residues are recycled in the plantation itself, it is possible to maintain fertility of soil. Mulching base of plants after application of second dose of manuring helps better intake of nutrients by plants.

Buffer Zone

An isolation distance of at least 25m width is to be left from all around the conventional plantation. The produce from this zone shall not be treated as organic. In sloppy lands, adequate soil and water conservation measures are necessary at the time of preparing the land for planting to avoid entry of run-off water and drift from neighbouring farms.

Vanilla

Vanilla prefers loamy laterite soil with plenty of organic matter with proper perfect drainage. Organic matter and decomposed mulch, are main sources of nutrients for plants. A thick layer of organic debris also helps to retain enough moisture and gives a loose soil structure for roots to spread. Hence, it is very important that easily decomposable organic matter is applied around plant base at least three to four times a year. Timely irrigation should be provided in initial two to three years of growth. Weeds growing at base of support tree around the plant within a radius of 50 cm should be pulled out and left there itself to decompose. Any operation done in plantation should not disturb roots, which are mainly confined to mulch and surface layer of soil. Any new shoot of vanilla should be allowed to attach to support tree with their own roots. Vines should grow upwards on support tree. By bending/nipping off tip of vines, one can produce more number of shoots or branches in a particular plant.

Support tree such as Glyricidia, Jatropha, Plumeria, Casuarina should be allowed to form branches to different directions to have an umbrella shaped appearance about 150 cm above ground. The support trees should be pruned to allow 50 per cent shade. If the trees do not shed leaves, they may be pruned before onset of heavy rains to allow more sunlight. Another light pruning may be done just before beans are harvested to facilitate their maturity and initiate flower bud formation for next crop.

Vines are allowed to grow up to a height of 1.20 to 1.50 meters and allowed to hang down the branches. Such vines should be brought back to ground and a portion of it, is placed under mulch. Later it is allowed to grow up on same support again. Thus the vines should be lopped up and down adjusting height in such a way that it is not more than the shoulder height of workers. This makes operations such as pollination and harvest easy and will make the crop more productive. Trailing is an important cultural operation as it helps accumulation of carbohydrate and other flower forming compounds at leaf axils, which facilitate early induction of flowering.

Vanilla, being an orchid, prefers organic manures. Decomposed organic materials may be applied two to three times a year. Well rotten cowdung, compost or vermicompost @4 to 5 kg/plant/year may be applied. Application of *Azospirillum, Phosphobacteria* @ 25-50 g/vine is useful. Use of *Trichoderma* and *Pseudomonas* helps to prevent/suppress fungal diseases.

Cloves and Nutmeg

Cloves and nutmeg are tree spices grown under appropriate climatic situations for maximum yields. Clove is planted in medium elevated places without shade and near river basin with abundant water facilities for irrigation to yield well. Slash weeding may be resorted with intensified mulching practices to arrest weed growth and favour retention of moisture. Apply well rotten cattle manure or compost @ 15 kg/plant/year during May/June in initial years. The quantity may be increased gradually so that a well grown tree of 15 years and above gets 40 to 50 kg of organic manure. Inter cropping is also advisable in cloves and nutmeg plantations. Clove and nutmeg are crops very much easier for organic management.

Clove

Seedlings are planted during monsoon in pits of 1 m³ size filled with soil + FYM. Mulching during initial years is done and adopting cover crops Calopogonium and Peuraria are beneficial. Banana, albizzia or jack may be planted along boarders. Compost of FYM at four kg/plant may be applied during initial years at onset of monsoon. For bearing trees, 50 kg compost, half dose at on set of monsoon and the leaf as top dressing in September may be beneficial. Basins are mulched with dry leaves, 50 g Azospirillum and Phosphobacteria.

Nutmeg

Well drained, deep rich alluvial loamy and laterite soil is the best suited. Semi shaded conditions in coconut and arecanut gardens are desirable. If shade trees are not available, fast growing Erythrina, Glyricidia or banana are planted prior to planting seedlings. For permanent shading, jack and mango are suitable.

Manuring

Every year, 15 kg compost (Cattle manure) in split doses of two at outset of monsoon and during weeding are applied. For bearing trees 50 kg compost is beneficial. Prawn dust and ground cake are also suitable. Bio agents Azospirillium and Phosphobacteria at 50 g each are also beneificial. Mulch with dry leaves during summer to conserve moisture. No water logged condition is allowed.

Good Agricultural Practices

This Code of Good Agricultural Practices (CGAP) is a general guide to production and handling of spices. The objectives are to ensure that the spices are (a) of the highest quality, (b) cultivated or processed hygienically minimizing microbial contamination and (c) produced with care minimizing physical and chemical contamination. There are no effective and legally permitted methods for reducing microbial load of dried spices. Therefore, these guidelines are intended to help minimize such contamination during growing, harvesting, drying, packing and storage stages. Farmers, processors and distributors are encouraged to devise practical measures for their workers to implement the code. It is intended that buying company should circulate these GAP guidelines to producers and processors. Application of the GAP guidelines can be advised when preparing agreements with growers/processors of organic spices.

Spices are exposed to microbial and other contamination from a wide variety of sources in the field. Such contaminations cannot be removed effectively by washing and peeling techniques which are applicable to many other crops, nor can they be reduced significantly by low temperature, drying, necessary for conservation of colour and flavour characteristics of spices. This code of practices set out here are guidelines to minimize such contaminations at primary producer/processor's level.

Cultivation

Spices should not be cultivated in soils contaminated with sewage sludge, heavy metals, pesticides and other industrial chemicals. The soil should be well drained and irrigation (if necessary) should be regular and uniform to avoid water logging and high humidity microclimates which promote mould growth. The source of irrigation water should be free from human as well as animal faecal contamination. Organic manures (no human faecal material) should be well composted before use and generally applied after the final harvest and before next planting for seasonal crops. No cattle should be allowed to move freely in the cultivated area. Plants should be spaced to minimize weed growth. Slash weeding should be regular and dead weeds with other plant debris should be removed from the cultivated area and destroyed, if they promote fungal infection and pest damage.

Organic pesticides and herbicides should be approved by the customer whenever possible and applied by the trained personnel. Application should be carried out at pre-harvest intervals specified by customer.

Harvesting

Harvesting should not be carried out in wet (dew or rain) or high humidity conditions, *i.e.*, wherever possible harvesting should be carried out in dry, low humidity conditions. Harvesting equipment should be clean and well maintained.

All containers used for primary collection of produce must be kept free from previously accumulated plant materials, and when not in use, they must be kept in a dry place, free from vermin and inaccessible to farm and domestic animals.

Damaged and spoiled produce should be sorted and discarded. Harvested produce should be collected in dry sacks, baskets, trailers or hoppers. It must not be collected on ground. Mechanical damage, high compaction and storage which promote composting should be avoided for which plastic sacks should not be used during harvesting. Sacks must not be overfilled and stacking should be done avoiding compaction. The time between harvest and transport of the produce to the drying site should be kept as short as reasonably practicable. The harvested produce should be protected from pests and domestic as well as farm animals.

Drying

The produce should be unloaded and unpacked as soon as possible on arrival at drying facilities. It must not be allowed to stand for extended periods in direct sunlight and must be protected from rain. Buildings used for drying the produce should be well ventilated and never used for keeping livestock. The building should be constructed to protect the produce from birds, insects, domestic and farm animals. Drying racks, if any, should be kept clean and regularly maintained. Drying on the floor and in direct sunlight are not adopted.

Dried produce should be inspected and sieved or winnowed to remove discoloured, mouldy and damaged material, soil, stones and other foreign bodies. Sieves should be kept clean and maintained well.

Clearly marked waste bins should be provided, emptied daily and cleaned. Dried as well as drying produces should be protected from pests and domestic and farm animals. Dried produce should be packed as soon as possible for protection and also to lessen opportunity of pest infestation.

Packing

After removal of damaged materials and foreign bodies, the sound dried produce should be packed in clean dry sacks, bags or boxes, preferably new. Packaging materials should be stored in a clean dry place free from pests and inaccessible to animals. Re-usable packaging materials like jute sacks and plastic bags should be well cleaned and dried before re-use. The packed produce should be stored in a dry place away from wall and off the ground and be protected from pests, domestic and farm animals. Whenever possible, the packaging materials used should be agreed between supplier and buyer.

Storage and Transport

Packed dried produce should be stored in a dry, well ventilated building with minimal variation in diurnal temperature avoiding high heat and with good air ventilation. Shutter and door openings should be protected by wire screens to keep out pests farm and domestic animals.

Packed and dried produces should be stored in buildings with concrete floors, on pallets, away from wall and well separated from all other produces. For bulk deliveries, use of vented containers for transport and storage in temporary warehousing is highly recommended to minimize contamination risks. Alternatively, suitably vented transport vehicles and temporary storage facilities are recommended. Whenever possible, conditions for transport and temporary storage should be agreed between supplier and buyer.

Equipments

Equipments used for cultivation and handling of produces should be well cleaned to minimize contamination. Dry cleaning is recommended. Where the use of water is unavoidable, equipments should be dried as quickly as possible after washing.

All equipments should be installed to allow easy access and should be well maintained and cleaned regularly.

Personnel and Facilities

Working practices should comply with General Principles of Food Hygiene of Codex Alimentarius Commission. Personnel handling of food material should maintain a high degree of personal hygiene and be provided with suitable changing facilities and toilets with hand washing facilities. Personnel should not be permitted to work in the material handling area if they are suffering from, or carriers of diseases likely to be transmitted through food, including diarrhoea. Personnel with open wounds, sores and skin infections should be transferred away from handling areas until completely recovered.

Training

Training of personnel, whether engaged in crop production, handling or processing is highly recommended. This can be achieved by using experts from local agricultural institutes or certifying agencies.

Market Potential for Organic Spices

With the trend towards pre-processed foods (convenience foods), the demand for organic spices increases likewise. Processed food can only be labelled as organic, if 95 per cent of all ingredients originate from organic agriculture. The remaining 5 per cent can be products, which are listed in a special annexure of EU-regulation. Spices and culinary herbs are not listed and must, therefore, always be of organic origin.

Assuming that within the next 10 years, organics will have a market share of 10 per cent in Europe, US and Japan. The current production is still small and demand much higher.

References

1. Bhai, S.R. (2014). Actinomycetes – a new potential biocontrol agent for blackpepper pathogens. *Indian J. Arecanut, Spices and Medicinal Plants.* 16 (4): 22–26.

2. Sasikumar,B. Sheeja,T.E. Swetha, V.P. and Parvathy, V.A. (2014). DNA barcoding debars adulterants in traded spices. *Spice India,* 27(12): 23–25.

3. Sheeja T. E., Jayashree, E., Srinivasan, V. and Anandaraj, M. (2014). Promoting entrepreneurship in spices sector. *Spice India,* 27(12): 20–23.

4. Spices Board, 1996 – (2010). Various miscellaneous publications in Spice India. *Indian Spices and Spice Bulletin.*

5. Spices Board (2001). Guidelines for production of organic spices in India. Cochin, p. 125.

2015, Horticulture for Nutrition Security
Editor: **Prof. K.V. Peter**
Published by: **DAYA PUBLISHING HOUSE, NEW DELHI**

*Pages **67–124***

Chapter 6

Vegetables and our Health

B. Singh, T.K. Koley and Aastik Jha

Malnutrition is the worst in Sub-saharan Africa and Souh Asia, but over the next 40 years, food production in these regions will need to increase more than double to meet the need of population growth. Under nutrition is a serious problem in under developing and developing countries, whereas forty percent of the world's malnourished children and thiry five percent of the developing world's low birth weight infants live in India. These micronutrient deficiencies often go unnoticed despite their insidious effects on the immune system. Consumption of less than 200g of vegetables per capita per day in many countries today is common. This meagre is a matter of concern amount, often conjugation with poverty and poor medical services, each associated with unacceptable levels of mortality and malnutrition in pre-school children and other valuable groups is a matter of concern. An increase in the avalaibilty, affordability and consumption of nutrient-dense vegetables is one way, malnutrition be substantially reversed. Yet nutritional security appears to be of less value than food security by key descision markers, and vegetable crops thus receive inadequate research investment. Vegetables are our most important sources of micronutrients, vitamins, antioxidants and vital sources of protein, but production in many countries is below the minimum to provide a basic balanced diet for all. Due to this vegetables are being considered as protective food to combat many chronic diseases like cancer, diabetes, blood sugar etc.

In the present circumstances, human civilization is facing a major threat of nutritional insecurity. Majority of the population is suffering from various lifestyle diseases linked with deficiency of specific nutritional compounds. Thus the major challenge to the scientists is to provide nutritional security to the growing population. During last a few decades hectic search is going to find out natural source of bioactive nutritional compounds. Some of the natural sources for bioactive compounds are fruits, vegetables, pulse, cereals, fish etc. Among them vegetables are unique in many aspects. They are the unique source of various bioactive compounds like vitamins, trace elements, antioxidants, dietary fibers and Poly-Unsaturated Fatty Acid (PUFA) which not only improve the state of health and well being, they also reduce the risk of various degenerative diseases like cancer, cardiovascular disease, mascular degeneration and ageing. Besides, they are the part of our normal food pattern and consumed in various occasions as fresh, cooked, dried, juice and other processed form. They are also relatively cheap and easily accessible throughout the year. Thus, in the present context of nutritional insecurity, vegetables could play a major role in alleviation of nutrition related lifestyle disease.

Nutritional security denotes the consumption and physiological use of adequate quantities of safe and nutritious food by every member and encompasses the process of equitable distribution among members of household and communities (Nandi and Bhattacharjee, 2002). This would entail the need to ensure a varied food intake, comprising all the essential vitamins and minerals through a diversified diet. In this context, production and consumption of vegetables that are the valuable sources of micronutrients and the highly beneficial phytonutrients hold the key in assuring nutritional security. Vegetables also provide nutritionally less defined, yet important component of the diet like dietary fibre. Dietary fibre has also attracted global attention due to their role in protection against certain types of cancer, regulation of transit, lowering of blood cholesterol, etc. (Smith *et al.,* 1998). Nutritional studies are now concentrating on examining food for their protective and disease preventing potentials (Nicoli *et al.,* 1999). Recently phytochemicals in vegetables have attracted a great deal of attention mainly due to diseases caused as a result of oxidative stress. Oxidative stress, which releases free oxygen radicles in the body, has been implicated in a number of disorders including cardiovascular malfunction, cataracts, cancers, rheumatism and many other autoimmune diseases besides ageing. These phytochemicals (mainly carotenoids, tocopherol, ascorbic acid, flavonoids, *etc.*) act as antioxidants, scavenge free radicles and act as saviours of the cell.

Vegetable production has made spectacular progress in recent years and touched a new height. Due to increasing awareness about nutritional security, vegetable production is getting continuous momentum in our country. Indias high growth has had little impact on food security and the nutrition levels of its population. In spite of the progress, per capita consumption of vegetable in India is only about 229 g/day/person, which is far below the minimum dietary requirement of 300g/day/person. Forty per cent of the world's malnourished children are in India and 60 per cent of Indian women are anaemic.

Vegetables possess high medicinal and nutritive values; there exists an enormous potentiality in vegetable technologies in India to address the micronutrient malnutrition, often called "hidden hunger". Vegetable being a rich and cheap source of vitamin and mineral, occupies an important place in the food dishes of Indian consumers, a majority of the population of the country are vegetarian. More than 70 types of vegetable are grown, higher emphasis given to popular vegetables like tomato, brinjal, chilli, okra, pea, cowpea, french bean, cauliflower, cabbage, onion, garlic, cucurbits, carrot, radish and leafy vegetables.

Bioactive Nutritional Compounds

Knowledge regarding health benefits of vegetables is not new. From the time immemorial many vegetables are used to recover or reduce many physiological malfunctions. During the last a few decades as a result of rigorous scientific investigation many bioactive nutritional compounds have been identified, directly or indirectly linked with reduction of degenerative diseases and improvement of state of health and well being. The major bioactive compounds in vegetables are vitamins, trace element, glucosinolates, phenolics, carotenoids, Poly-Unsaturated Fatty Acid (PUFA), dietary fiber, prebiotics etc.

Vitamins

The most important bioactive compounds in vegetables are vitamins. They are the organic compounds required in minute quantity for maintaining basic physiological functions. Besides, some of them possess potent antioxidant activity and play a critical role in human disease prevention. Vegetables are good sources of water soluble as well as lipid soluble vitamins. Although vegetables are sources of ascorbic acid but they are fair supplier of vitamin B-complex. Green leafy vegetables like spinach, amaranths, fenugreek, drumstick etc. are rich sources of vitamin B - Complex. Presently, vitamin B- Complex is drawing much attention due to its effect on improving cognitive performance and mental health in elderly people (Penninx *et al.*, 2000; Duthie *et al.*, 2002). In the B complex group, vitamin B-complex folates (Folic acid and tetrahydrofolate) are very important. Folates are involved as cofactor in carbon transfer reactions in DNA biosynthesis. In vegetables major sources of folates are yard long bean, green leafy vegetables, broccoli and brussels sprout (Scott *et al.*, 2000). Deficiency of folates leads to neural tube defect *i.e.* spina bifida in foetus and neurotoxicity, cognitive defect and colon cancer risk. Most of the green vegetables are good sources of lipid soluble vitamins. Carrot, pumpkin, pepper, green peas and beans are good sources of provitamin A. Green leafy vegetables are good sources of vitamins E and K. These lipid soluble vitamins play major role in improvement in immune response, bone health and age related degenerative diseases.

Table 6.1: Folic Acid Content of some Common Vegetables

Vegetables	Free Folic Acid (µg/100g)	Vegetables	Free Folic Acid (µg/100g)
Cabbage	13.3	French bean	15.5
Curry leaves	23.5	Okra	25.3
Spinach	51.0	Pumpkin	3.0
Carrot	5.0	Cluster bean	50.0
Colocasia	16.0	Snake gourd	7.5
Brinjal	5.0	Chillies green	6.0
Cucumber	12.6	Tomato ripe	14.0

Minerals and Trace Element

Wide ranges of minerals and trace elements are present in vegetables. Leafy vegetables and crucifers are rich sources of minerals and trace elements. Calcium, phosphorus and iron are important minerals with major role in bone health and prevention of anaemia. The important trace elements in vegetables are Zn, Cu, Mn, Se and play an important role in immune functions, body defense against oxidative stress. Interest in selenium as a bioactive elements has increased dramatically in the last three decades as a result of several studies which demonstrated increased risk of cancer with low selenium intake. Two important vegetables are rich sources of selenium, broccoli and garlic when grown on high selenium rich soil (Finley and Penland, 1998). Chromium is another trace element which may be effective in optimizing insulin metabolism and lowering plasma cholesterol levels.

Glucosinolate

Glucosinolates are the major organosulfur compounds found in cruciferous vegetables like cauliflower, cabbage, broccoli, knolkhol, brussels sprout, kale, radish, turnip etc. Epidemiological studies revealed that a diet rich in cruciferous vegetable can reduce several types of cancer like lung cancer and colon cancer. Cancer inhibition property of glucosinolate is due to its effect on Nrf2, polymorphism, anti-inflammatory, inhibition of histone deacetylase activity and influence on estrogen metabolism (Manchali *et al.,* 2011). Nutraceuticals in cruciferous vegetables help in maintenance of good heart health mainly through their ability to reduce low density lipoprotein (LDL), combat free radicals and up-regulate glutathione-S transferase activity (Manchalia *et al.,* 2011). Intact Glucosinolates have no biological activity against cancer, however the breakdown products have been shown to stimulate mixed-function oxidases involved in detoxification of carcinogens. This break down takes place through the action of endogenous myrosinase enzyme released by disruption of the plant cell through harvesting, processing and mastication and by the microflora present in stomach (Campbell *et al.,* 1995).

Carotenoids

Another major group of bioactive compounds in vegetables are carotenoids. Among different carotenoids lycopene and lutein have significant role in human health protection.Lycopene is unsaturated carotenoid with lack of β-ionone ring structure responsible for vitamin A activity. Tomato and water melon are two important sources of lycopene. Besides these two, carrot and *Momordica chochinchinensis* are also potential sources of lycopene. Cherry type tomato contains higher amount of lycopene than normal tomato (George *et al.,* 2004). Processed tomato product contains 2-40 times higher lycopene. The consumption of tomato and tomato products containing lycopene has been shown to associate with lower risk of cancer and cardiovascular diseases (Minorsky, 2002). Disease preventing mechanism of lycopene is due to its strong free radical quenching activity. Scientific investigation revealed that high amount of lycopene intake is associated with reduced risk of prostate cancer. Tang *et al.* (2005) reported increased concentration of lycopene reduced the viability of prostate cancer cell pc-3. Lutein is nonprovitamin A Xanthophylls which is transported by lipoprotein and selectively accumulated in macular region of retina. In that region, it absorbs blue light, quench photochemically induced singlet oxygen and prevent our eye from photo oxidation. Thus lutein protects the lens and macula from development of cataract and age related macular degeneration (Olmedilla *et al.,* 2001). Principle lutein rich vegetables are Kale, Parsley, Spinach, Broccoli, Collard and other green leafy vegetables.

Table 6.2: Qualitative and Quantitative Carotenoid Distribution in Vegetables

Vegetables	Total Carotenoids (mg/g fresh wt)	β-carotene (mg/g fresh wt)	Major Carotenoids
Asparagus	8.5	4.3-7.0	β-carotene. Lutein, Violaxanthin, Neoxanthin
Bitter gourd	5.3	2.3	α-carotene, β-carotene, Zeinoxanthin, Lutein
French bean	17.1	2-4	α-carotene, β-carotene, Lutein, Lutein 5, 6 epoxide. Neoxanthin, Violaxanthin
Broccoli	42.4	4.8	β-carotene, Lutein, Isolutein, Luteoxanthin, Violaxanthin, Neoxanthin, Chrysanthemaxanthin
Cabbage	8.9	0.8	β-carotene, Lutein, β-carotene 5,6-epoxide, Neoxanthin, Violaxanthin, Chrysanthemaxanthin
Carrot	54-124	76.0	α-carotene, β-carotene, J-carotene, β-Zeacarotene, r-carotene, Neurosporene
Cauliflower	0.44	0.11	β-carotene, Lutein, Violaxanthin, Neoxanthin
Cucumber	17.2	2.20	α-carotene, β-carotene, and Cryptoxanthin
Pepper (Green)	10.0	6.8	Capsanthin, Capsorubin, Cryptocapsin, β carotene
Pepper (Red)	127-284	1.27-2.84	β-cryptoxanthin, Violaxanthin, Neoxanthin
Lettuce	68.0	10.8-24.5	β-carotene, Lutein, Violaxanthin, Neoxanthin
Spinach	69.0	40.0	β-carotene, Lutein epoxide, Violaxanthin, Lutein, Antheraxanthin, Neoxanthin
Tomato	70-190	7.8	Lycopene, γ-carotene, Phytoene,

Phenolics

Among the major antioxidants in vegetables phenolics in particular have received renewed focus (Neuhauser, 2004; Chun *et al.,* 2005). This is due to strong antioxidant properties exhibited by chelating metal ions, inhibiting lipid oxidation, inhibiting radical forming enzyme or quenching free radicals. Moderate intake of phenolics rich beverages reduces the risk of cardiovascular, cerebrovascular and peripheral vascular diseases. Phenolic compounds appear to interfere with the molecular processes underlying the initiation, progression and rupture of atherosclerotic plaques (Szmitko and Verma, 2005). A more recent area is the emerging role of phenolic compounds in the treatment of neurodegenerative diseases, like Alzheimer's and Parkinson's found to have clear association with oxidative stress. Increasing number of studies demonstrate the efficacy of phenolic antioxidants to reduce or to block neuron death. In addition to intrinsic antioxidant properties, phenolics also seem to exert beneficial effects through different pathways like signalling cascades, anti-apoptotic processes or the synthesis/degradation of the amyloid β peptide (Ramassamy, 2006). Several reports suggest involvement of natural phenolics in the prevention of skin damage (Svobodova *et al.,* 2003; Hsu, 2005). In-depth studies have dealt with phytoestrogen properties of different sources of polyphenols: isoflavonoids, stilbenes, coumestrans and lignans (Cornwell *et al.,* 2004). The strongest body of evidence supports the view that soy isoflavonoids may be effective in preventing osteoporosis. In this context, phenolic and flavonoids assume the status of "Guardian of Health" and are being commonly referred as "Star Nutrient of the Millennium" (Trewavas and Stewart, 2003).

These are classified into four major groups, anthocyanin, flavonoids, flavones and isoflavone (Table 6.3). Flavonoids are the largest group of phenolics in vegetables.

Table 6.3: Common Sources of Polyphenolic Compound in Vegetables

Phenolic Group	Principle Sources
Anthocyanin	Red cabbage, purple broccoli, brinjal, rhubarb, radish, black carrot, onion
Flavonoids	Onion, lettuce, endive, horse radish, tomato, beans.
Flavones	Celery, tomato, brinjal, garlic, onion.
Isoflavone	Soybean, pea, broccoli, asparagus, alfalfa, okra.

Poly-unsaturated Fatty Acid (PUFA)

Poly unsaturated fatty acid is one of the important bioactive compounds which have profound influence on human health. There are two series of polyunsaturated fatty acids that are deemed essential, the ω-6 and ω-3 series. Animal meat and vegetable oil are good sources of ω-6 fatty acids. Although vegetables are not good sources of ω-6 fatty acid, they are excellent sources of ω-3 fatty acid in particular alpha linolenic acid (ALA) which have major role in prevention of coronary artery disease, hypertension, diabetes, and other inflammatory and autoimmune diseases. Green leafy vegetables, broccoli, kale, vegetable soybeans and radish sprouts are good sources of ALA (Simopoulos, 2002). Other important ω-3 fatty acids are docosahexanoic acid and eicosapentanoic acid which are generally not detected in vegetables. However, they can be synthesized inside the human body from sufficient intake of ALA. Besides quality of fatty acids, balance between ω-6 and ω-3 is very important. Studies indicate that a high intake of omega-6 fatty acids shifts the physiological state to one that is prothrombotic and proaggregatory, characterized by increases in blood viscosity, vasospasm, and vasoconstriction and decreases in bleeding time, whereas omega-3 fatty acids have anti-inflammatory, antithrombotic, anti-arrhythmic, hypolipidemic, and vasodilatory properties. As green leafy vegetables are good sources of ALA, regular intake of them lead to balance the ω-6 and ω-3.

Dietary Fiber

Dietary fiber is the edible plant or animal material not hydrolyzed by the endogenous enzymes of the human digestive tract. Based upon fermentation, dietary fibers are classified into two major groups: partially fermented and low fermented. Vegetables which good source of well fermented dietary fiber are chicory, onion, jerujalem artichoke and guar. Corn and legumes are good sources of partially or low fermented dietary fiber. Consumption of dietary fiber is associated with reduction of cholesterol, increasing stool bulk, attenuating glycemie and insulin response and improving laxation (Schneeman, 1999) thus reducing risk of cardiovascular disease, certain type of diabetes, colon cancer etc.

Table 6.4: Fibre Content of some Common Vegetables

Vegetables	Crude Fibre (g/100g)	Vegetables	Crude Fibre (g/100g)
Cabbage	4.6	French bean	1.8
Curry leaves	6.4	Okra	1.2
Spinach	6.0	Pumpkin	0.7
Carrot	1.2	Cluster bean	3.2
Colocasia	1.0	Snake gourd	0.8
Brinjal	1.3	Chillies green	6.8
Cucumber	0.4	Tomato ripe	0.7

Thiosulphides

Thiosulphides are organo-sulfur compounds mainly found in onion, garlic, shallot, chive, leek etc. Alliin and Methiin, are major thiosulphides in intact tissues of *Allium* species. Thiosulphides are highly unstable. When the tissues are cut, chewed, cytosolic allin is rapidly lysed by vascular enzyme alliinlyase or allinase into highly unstable diallylthiosulphinate (Allicin) which is again converted into alkyl alkane thiosulphinates. These compounds are related to reducing cancer mainly stomach cancer and cardiovuscual disease. Thiosulphides reduce cholesterol and other fatty acid synthesis, prevent lipid peroxidation of LDL, enhance fibrinolysis and improve fluidity of erythrocyte, increase antioxidant status and inhibit angiotension converting enzyme (Rahman and Lowe, 2006). Thiosulphides stimulate the immune system, by activating T cell proliferation and reducing blood glucose level in diabetes by stimulating insulin secretion by the pancreases (Augusti and Sheela, 1996).

Saponine

The major saponins are found in vegetables, which have hypoglycemic properties or other action of potential benefit against diabetes mellitus (Shrivastava *et al.*, 1993). Bitter gourd is the excellent sources of cheratins. Other hypoglycemic compounds in bitter gourd which have same property are Insulin like peptide and Alkaloids. These compounds prevent glucose absorption from intestine, regenerate insulin producing cell and potentiate the effect of insulin and thus prevent diabetes.

Phytosterol and Stanol

Phytosterol and stanol are alcoholic derivatives of cyclopentanoperhy-drophenanthrene an essential constituent of cell membrane. Over forty plant sterols identified, β-sitosterol, compesterol, stigmasterols are the most abundant. Among vegetables soybean, moringa and sweet corn are good sources of phytosterol and stanol. They reduce blood cholesterol by inhibiting cholesterol absorption and thus prevent cardiovascular diseases (Makhal *et al.*, 2006).

Vegetables and their Health Benefit

Solanaceous Vegetables

Solanaceae is one of the major families of plant supplying vegetables, fruit and staple food crops in the world. Tomato, capsicum and brinjal are the major solanaceous vegetables grown worldwide. Fruits of solanaceous vegetables provide important bioactive compounds to human diet. They provide maximum part of dietary nutrition as they are consumed frequently.

Tomato

Tomato (*Solanum lycopersicum*) is the most important vegetable among solanaceous vegetables cultivated through the world for its ripe fruit. It is consumed in various ways like salad, cooked or in different processed forms. It plays a crucial role in human health due to its popularity, availability and high per capita consumption. Tomato is a rich source of potassium, folate and vitamins A, C, and E. Tomato products contain similar amounts of potassium and folate compared with other popular vegetables, but tomato products have a superior source of alpha-tocopherol and vitamin C which are potential dietary antioxidants. In addition to basic nutritional compounds like vitamins and minerals, tomatoes also contain valuable bioactive compounds, including carotenoids and phenolic compounds. The major carotenoids in tomatoes are lycopene, β-carotene, α-carotene, lutein, zeaxanthin and unique lycopene metabolite (Li *et al.,* 2012). These carotenoids are getting much attention due to their multiple health beneficial properties. Lycopene is the predominant lipid soluble compound and constitutes more than 80 per cent of total tomato carotenoids in fully red ripe fruits. Although devoid of provitamin-A activity, it is a well known biomolecule of interest in epidemiological studies (Rao and Agarwal, 2000). It showed strong antioxidant activity in both *in vitro* and *in vivo*, having quenching rate constant with singlet oxygen almost twice that of β-carotene. Lycopene reduces several cancer types and the risk of coronary heart disease (Rao *et al.,* 2006). Serum lipid peroxidation and LDL (Low-Density Lipoproteins) oxidation were significantly decreased by lycopene. Lutein, another carotenoids in tomato is associated with reduced risk of age-related macular degeneration, cataracts and atherosclerosis (Bone *et al.,* 2000). Phenolics including flavonoids are important hydrophilic phytochemicals present in tomato. Recently phenolics received wide attention due to its strong biological activities. In tomato hydroxycinnamic acids and their derivatives are the most abundant family of the total phenolics. Within this family, the major compounds were caffeoyl-hexose I, coumaric acid hexose I, chlorogenic acid and 5-caffeoylquinic acid. Other important groups of phenolics in tomato are phenylacetic acids and its derivatives, hydroxybenzoic acids and derivatives. Flavonoids are a diverse group of phenolic secondary metabolites. Many of the compounds belonging to this group are potent antioxidants *in vitro* and epidemiological studies suggest a direct correlation between high flavonoid intake and decreased risk of cardiovascular disease, cancer and other age related diseases. The dihydrochalconephloretin 3', 5'-di-C-beta-glucopyranoside and the flavonolquercetin 3-*O*-(2''-O-beta-apiofuranosyl-6''-*O*-alpha-rhamnopyranosyl-beta-glucopyranoside) are the major flavonoids in tomato. Besides chalconaringenin, kaempferol 3-rutinoside and quercetin 3-rutinoside (rutin) are also present in good quantity in tomato.

Brinjal

Brinjal (*Solanum melongena* L.) is another common and popular solanaceous vegetable crop grown in the subtropics and tropics of the world (Sarker *et al.,* 2006) with a worldwide production of around 32 million tonnes (FAO, STAT, 2009). It is a perennial crop but grown commercially as an annual. The unripe fruit of eggplant is primarily used as a cooking vegetable for the various dishes all over the world. Traditionally brinjal is used for treatment of several diseases. Various parts of the plant are useful in the treatment of inflammatory conditions, cardiac debility, neuralgia, ulcers of nose, cholera, bronchitis and asthma. It is analgesic and hypolipidemic (Sudheesh *et al.,* 1997). Studies show that brinjal extracts suppress the development of blood vessels required for tumor growth and metastasis (Matsubara *et al.,* 2005), and inhibit inflammation that can lead to atherosclerosis (Han *et al.,* 2003). Anthocyanin from brinjal could be potent natural pigment that prevents chemical induced mutation (Azevedo *et al.,* 2007). Phenolic extract from brinjal had high alpha-glucosidase inhibitory activity and in specific cases moderate to high angiotensin I-converting enzyme inhibitory activity (Kwon *et al.,* 2008). Brinjal contains ascorbic acid and phenolics, both of which are powerful antioxidants (Vinson *et al.,* 1998). The main phenolic compound in brinjal includes N-caffeoylputrescine, 5-caffeoylquinic acid, and 3-acetyl-5-caffeoylquinic acid. In addition, trace quantities of flavonols, namely, quercetin-3-glucoside, quercetin-3-rhamnoside, and myricetin-3-galactoside are also observed in its pulp (Singh *et al.,* 2009). It is also a natural source of vitamin A. From the 120 vegetable species evaluated for antioxidant activity using four different assays (2,2-azinobis-[3-ethylbenzthiazoline-6-sulphonic acid]), 2, 2-diphenyl-1-picrylhydrazyl radical, inhibition of lipid peroxidation, and Superoxide scavenging), it ranked among the top 10 species for superoxide scavenging activity (Yang, 2006). Extracts from fruit peel have been demonstrated to possess high capacity in scavenging of superoxide free radicals and inhibition of hydroxyl radical generation by chelating ferrous iron (Noda *et al.,* 2000). Nasunin, an anthocyanin isolated from the peel of purple fruit, is one phenolic compound implicated in both inhibition of hydroxyl radical generation and superoxide scaveging activity (Noda *et al.,* 2000). Flavonoids isolated from its fruit showed potent antioxidant activity (Sadilova *et al.,* 2006) against chromosomal aberrations induced by Doxorubicin. Nutritional component and functional activity of vegetable is function of genotype. Natural out crossing and mutation could lead to change in genotype resulting diversity in nutritional value as well as functional properties. Wide ranges of diversity in terms of colour, size and shape exist in brinjal genotypes. Among different types of brinjal studied, purple colour small size fruit demonstrated better antioxidant activities, higher phenolic and anthocyanin content than the other type (Nisha *et al.,* 2009). Among the five brinjal cultivar studied by Akanitapichat *et al.* (2010) total phenolic content in methanol extracts ranged from 739.36 ± 1.59 to 1116.13 ± 7.30 gallic acid equivalents mg/100 g extract and total flavonoid content from 1991.29 ± 6.32 to 3954.20 ± 6.06 catechin equivalents mg/100 g extract. Land races had higher content of phenolics than commercial cultivars. Small fruited cultivars tend to have higher antioxidant activity

Chilli

This immense horticultural and biological diversity has helped to make *C. annuum* globally important as a fresh and cooked vegetable (*e.g.* for salads, warm dishes, pickled) and a source of food ingredients for sauces and powders and as a colourant, used in cosmetics as well, (Andrews, 1995; Bosland, 1994; Bosland and Votava, 2000). Moreover, the species and used for medicinal purposes and provide the ingredient for a non-lethal deterrent or repellent to some human and animal behaviour (Krishna De, 2003; Cichewicz and Thorpe, 1996; Reilly *et al.,* 2001). Chilli peppers are also cultivated for ornamental purpose especially for their brightly glossy fruits with a wide range of colours. Chilli pepper comprises numerous chemicals including steam-volatile oil, fatty oils, capsaicinoids, carotenoids, vitamins, protein, fiber and mineral elements (Bosland and Votava, 2000; Krishna De, 2003). The ripe fruits are especially rich in vitamin-C (Marin *et al.,* 2004). Two chemical groups of the greatest interest are the capsaicinoids and the carotenoids. The capsaicinoids are alkaloids that give hot chilli peppers, their characteristic pungency. The rich supply of carotenoids contributes to chilli peppers' nutritional value and colour (Hornero-Mendez *et al.,* 2002; Perez-Galvez *et al.,* 2003).

The fruits of most *Capsicum* species contain significant amount of vitamins B, C, E and provitamin A (carotene) when in fresh state. The large type of *C. annuum* is among the richest known source of vitamin C, which may be present up to 340 mg/ 100 g in some varieties. Fruits of *Capsicum* species have relatively low volatile oil content, ranging from 0.1 to 2.6 percent in paprika. The characteristic aroma and flavour of the fresh fruit are imparted by the volatile oil. The composition of the volatile oil of fresh green bell peppers has been examined. Twenty four components in this oil were positively identified by Buttery *et al.* (1969) using gas chromatography. One of the major components, 2-methoxy-isobutyl pyrazine, possessed an aroma characteristic of the fresh fruit and dominated the organoleptic profile of paprika. The colour of the paprika powder is the principal criterion for assessing its quality. The pigment content of paprika powder ranges from 0.1 to 0.8 percent. The major colouring pigments in paprika are capsanthin and capsorubin, comprising 60 percent of the total carotenoids. Other pigments are β-carotene, zeaxanthin, violaxanthin, neoxanthin and lutein (Anu and Peter, 2000). In international trade the colour intensity of paprika pods is generally expressed as ASTA units unlike oleoresin which is described in colour value. Nearly 40 percent of paprika oleoresin is used to blend in chicken feed.

Capsicum

Capsicum is another solanaceous species grown throughout the world for vegetables and spices purpose. Among more than 30 species, five main domesticated species commercially cultivated are: *Capsicum annuum* (hot pepper, bell pepper, paprika) *C. frutescens* (tabasco pepper), *C. chinense* (naga), *C. pubescens* (South American rocoto peppers) and *C. baccatum* (wax pepper). Among five domesticated species, *C. annuum* is the largest group of capsicum grown worldwide. Capsicums play an important role in human health as antioxidants and immune enhancers, helping in the prevention of cancers, cardiovascular diseases, age related macular degeneration, cataracts, diseases related to low immune function, and other degenerative diseases (Perera and Yen, 2007). Strong functional properties of capsicum are attributed to its various bioactive compounds. The major bioactive compounds in Capsicum are capsaicin carotenoids, phenolics and vitamin E. The main carotenoids in Capsicum are β-carotene, α-carotene, β-cryptoxanthin, zeaxanthin, luteine, capsanthin, capsorubin, and cryptocapsin (Collera-Zuniga *et al.,* 2005).Among them only β-carotene and β-cryptoxanthin have vitamin A activity (Minguez-Mosquera and Hornero-Mendez, 1994). Spicy character of capsicum is attributed to its capsaicin, a phenolic compound closely related to vanillin, which gives pungency to capsicums and shows a significant antioxidative effect (Adegoke *et al.,* 1976).However, other carotenoids are potent antioxidants and they potentially reduce the risk of age related life style diseases. In addition, capsicum is rich in polyphenols, particularly the flavonoids, quercetin, myrcetin and luteolin (Lee *et al.,* 1995). Quercetin was not detected in green chili. Bird chili contained the highest level of luteolins (1035.0 mg/kg) among all the samples tested. Therefore, consumption of bird chilli may reduce risk of tumorigenesis because luteolin was a potent inhibitor to enzyme lipoxygenase and prostaglandin synthetase (Larson, 1988).

Cucurbits

Cucurbitaceous vegetable is major vegetable family cultivated in various parts of the world since time immemorial. They are consumed in many forms like salad, cooked, dehydrated and osmotically dehydrated. Among cucurbitaceous crops bitter gourd, pumpkin, cucumber, bottle gourd, watermelons are grown worldwide. Besides, various minor cucurbits like luffa, snake gourd, pointed gourd and chayote are grown in various parts of the world for local consumption. Cucurbitaceous vegetables provide important bioactive compounds to human diet. They are the sources of carotenoids, phenolic, triterpenoids etc.

Bitter Gourd

Bitter gourd (*Momordica charantia* L.) is a tropical and subtropical annual cucurbitaceous vegetable, widely grown in Asia, Africa, and the Caribbean for its edible fruit. Being a vine crop, it grows and yields well on trellis and bower. Unripe fruits are commonly consumed as vegetables. Traditionally bitter gourd fruits are used as anthelmintic, antiemetic, carminative, purgative and for the treatment of anaemia, jaundice, malaria, cholera, etc. (Ross, 1999). Fruits possess wide range of pharmacological activities such as gastroprotective and ulcer healing activities, hypoglycaemic, antidiabetic, antifungal, inhibition of p-glycoproteins, antihyperlipidemic and hepatoprotective activity (Kushawa *et al.,* 2005). The roots are used in head trouble, urinary calculi and as an errhine in jaundice. The leaves are aphrodisiac and antihelmintic and need against fever, asthma, bronchitis, high cough and piles. Studies on animal model revealed that leaves of related species *Momordica dioica* have potent hepatoprotective action against carbon tetrachloride induced hepatic damage in rats (Jain *et al.,* 2008). *Momordica cymbalaria* is another related species of bitter gourd found in Deccan, Mysore and Konkan regions of India and its tuber is used as an abortificient (Nadkarni, 1994). Fruit is routinely used as a vegetable and also for the treatment of diabetes mellitus by local people. Rao *et al.* (1999) reported that fruit powder of *M. cymbalaria* possessed antidiabetic and hypolipidemic effects in alloxan-induced diabetic rats.

Bitter gourd is a rich source of phenolic phytochemical. The predominant phenolic compounds are gallic acid, followed by caffeic acid and catechin (Kubola and Siriamornpun, 2008). Besides phenolics, many cucurbitane type triterpenoids have been isolated from fruits, seeds, leaves and vines and stems (Tan *et al.*, 2008). A new cucurbitane-type triterpene glycoside taiwacin A (23, 24, 25, 26, 27-pentanorcucurbitane), taiwacin B and a known cucurbitane-type triterpene glycoside, and a known steroid glycoside, were isolated from stems and fruits of bitter gourd (Lin *et al.*, 2011). Diverse phenolics and terpenoids in fruits, leaves and stem are responsible for strong antioxidant activity of bitter gourd. The leaf extract showed the highest value of antioxidant activity, based on DPPH radical-scavenging activity and ferric reducing power, while the green fruit extract showed the highest value of antioxidant activity, based on hydroxyl radical-scavenging activity, b-carotene–linoleate bleaching assay and total antioxidant capacity. Wild bitter gourd (*Momordica charantia* Linn. var. *abbreviata* Ser.) possesses potent DPPH radical scavenging activity, even better than vitamin E (Wu and Ng, 2008).

Pumpkin

Pumpkin (*Cucurbita moschata*) and its related species (*C. pepo; C. maxima*) are important vegetables in cucurbits cultivated for its immature as well as mature fruits. Open flower is also consumed in many parts of the world. Its unique texture and colour enable making of various form of cooked and processed products. It is a rich source of vitamins and minerals. Besides it is a good source of various bioactive compounds. It is high in carotenoids, especially β-carotene and lutein, both of which are important antioxidants. Other carotenoids are α-Carotene and minor carotenoides æ-carotene, zeaxanthin, violaxanthin, β-carotene 5,6-epoxide, β-cryptoxanthin, taraxanthin, luteoxanthin, auroxanthin, phytofluene, neurosporene and neoxanthin (Toshiro *et al.*, 1986; González *et al.*, 2001; Rodriguez-Amaya *et al.*, 2008). Besides, *Cucurbita sp* contain protein-bound polysaccharide with strong hypoglycemic effect (Li *et al.*, 2005). Oil from un-germinated pumpkin seeds and proteins from germinated pumpkin seeds also possess hypoglycemic activity (Li *et al.*, 2001). Pumpkin polysaccharide and seed oil showed antihypercholesterolemia activity.

Ash Gourd

Ash gourd (*Benincasa hispida* (Thunb.) Cogn) is an important vegetable of cucurbitaceous family known for its multiple health benefits. Fruit is a rich source of vitamins and minerals like ascorbic acid, riboflavin, thiamin, niacin, potassium, sodium, calcium, iron etc. It also contains essential amino acids, mono as well as polysaccharides and various volatile compounds. The fruit traditionally is used for treatment of diabetes, urinary infection, chronic inflammatory disorder, epilepsy and other nervous disorders, peptic ulcer etc. It is used as an ingredient for preparation of medicine 'Kusmanda Lehyam' (Kumar and Vimalavathini, 2004). Therapeutic properties of ash gourd are attributed to its bioactive compounds with strong functional properties. Fruits demonstrated strong antioxidant activity *in vitro* as well as *in vivo* condition (Huang *et al.,* 2004; Roy *et al.,* 2007). It also exhibited angiotension-converting enzyme (ACE) inhibitor property (Huang *et al.,* 2004). Other functional properties demonstrated by fruits include anti-compulsive (Girdhar *et al.,* 2010), anti-ulcer (Rachchh and Jain, 2009), anti-inflammatory (Gill *et al.,* 2010), anti-obesity (Kumar and Vimalavathini, 2004), anti-diarrheal (Mathad *et al.,* 2005) etc. The major bioactive compounds related to its strong functional properties are plant phenolics like astilbin, catechin, naringenin, flavones iso-vitexin, terpenoids etc.

Chayote

Chayote (*Sechium edule* (Jacq.) Swartz) is an underutilised vegetable mostly cultivated in tropical and subtropical parts of the world for its immature fruits, seed, leaves as well as tuberous root. Young stems, roots and seeds have high carbohydrate content. Fruits are good sources of micro as well as macronutrients. The fruit and seeds are rich in several important amino acids. A protein isolated from seed aqueous extracts of *S. edule*, called Sechiumin, possesses ribosomal inactivation properties and is a potential chemotherapeutic agent (Wu *et al.,* 1998). Edible part of this plant exerts different biological properties like diuretic, antihypertensive, cardiovascular and anti-inflammatory (Bueno *et al.,* 1970; Losoya, 1980; Salama *et al.,* 1986; Gordon *et al.,* 2000; Ribeiro *et al.,* 1988). Water extracts of *S. edule* showed antimutagenic activity. The heat-stable antimutagenic substances might be phenolic compounds (Yen *et al.,* 2001). Ethanolic extracts of *S. edule* showed potent antibacterial activity (Ordonez *et al.,* 2003) and antioxidant activity (Ordonez *et al.,* 2006).

Bottle Gourd

Bottle gourd (*Lagenaria siceraria*) is a commonly used cucurbitaceous vegetable in India. Both fruits and leafy twigs are used as vegetable throughout the world. It is a good source or minerals. The edible portion of bottle gourd fruit contains 96.1 per cent moisture, 0.2 per cent protein, 2.5 per cent carbohydrates, 0.1 per cent fat, 0.6 per cent fibre, 0.5 per cent mineral matter, 0.044 mg thiamine, 0.023 mg riboflavin, 0.33 mg niacin and 12 mg ascorbic acid per 100g edible portion. Bottle gourd fruits are consumed in various form. As a vegetable, it is the most easily digestible, even by patients. The pulp is good for overcoming constipation, cough and night blindness and as antidote against certain poisons. A decoction made from the leaf is a very good medicine for curing jaundice.

Muskmelon

Muskmelon (*Cucumis melo*) is an important desert vegtable grown in many countries of the World. Immature fruits are often used as a cooked vegetable but the mature fruits are eaten raw as dessert, and occasionally used in preserves. Seeds that have the cooling and diuretic effect are eaten after roasting, and oil extracted from seeds is used for cooking purposes. The fruit pulp is diuretic and used in chronic eczema. The plant extract has the blood sugar lowering principle, and roots have the emetic and purgative properties. Its ripe fruits are very much useful in human kidney disease. The fruits are extensively used as dessert fruits and are highly esteemed in summer months, sometimes eaten as vegetable when unripe. Melons of drier regions are sweeter and tastier than those of moisture situation (Chakravarty, 1982). The seeds are diuretic, cooling, nutritive and beneficial to the enlargement of prostate gland. The pulp is diuretic and beneficial to chronic or acute eczema. The ripe fruit is cooling, flattening, tonic, laxative, aphrodisiac nd cures biliousness (Vashista, 1974). It can be a great substitute for high calorie snacks and can aid in losing weight healthily. Since it is rich in potassium, it can help control blood pressure and can prevent the risk of strokes. Potassium in the fruit can also reduce the problem of developing kidney stones. The folic acid in muskmelons helps create healthy fetuses (in pregnant women) and can even prevent cervical cancer and osteoporosis. It also serves as a mild antidepressant. Muskmelon comprises of a significant amount of dietary fiber, making it good for those suffering from constipation. Since the fruit is not high in sugar or calories, it serves as a good snack for all those who are trying to lose weight and maintain a healthy body. Muskmelon is a good source of vitamin C, which is an anti-oxidant. This prevents heart diseases and even cancer. It has beta-carotene too. Researches reveal the fact that the combination of beta-carotene and vitamin-C can prevent many chronic conditions. In addition to health benefits, muskmelon takes care of your skin too. It contains Vitamin A, which is useful in maintaining healthy skin. Regular consumption of muskmelon juice can help to treat lack of appetite, acidity, ulcer and urinary tract infections. It can reduce the heat in the body largely and therefore can prevent heat related disorders. It can relieve tiredness. It has effective laxative properties and so can help curing insomnia.

Cucumber

Cucumber (*Cucumis sativus*) is grown extensively in India. It is rich in vitamins B and C as well as minerals such as calcium, phosphorous, iron and potassium. The fruit is eaten raw as salad. There are reports, which show that cucumbers prevent constipation, jaundice and indigestion. The typical flavour of cucumber is attributed to two compounds 2, 6 nonadienal and 2, 6 nonadienol. The pleasant aroma of cucumber is derived mainly from the 2, 6 nonadienal and 2 hexenal, the more astringent taste being contributed by 2 noenal. Some volatile compounds are also identified in cucumber inonanol, trans-2-nonen-l-ol, cis-3-nones and C10-15 saturated straight chain aldehydes. Cucumbers arc characterized by the presence or bitter principles called cucurbitacins, which are tetra cyclic triterpenes. They occur in nature as free glycosides or in bound form. A high concentration or bitter principle is found in fruits. Group of compounds found in cucumber is called cucurbitacins-cucurbitacins A, B, C, D and E have all been identified within fresh cucumber. Cancer cell development and cancer cell survival can be blocked by activity of these cucurbitacins. Second groups of cucumber phytonutrients are lignans. The lignans pinoresinol, lariciresinol, and secoisolariciresinol have all been identified within cucumber. The role of these plant lignans in cancer protection involves the role of bacteria in our digestive tract. These reduce the risk of estrogen-related cancers, including cancers of the breast, ovary, uterus, and prostate and associated with intake of dietary lignans from plant foods like cucumber.

Watermelon

Watermelon (*Citrullus lanatus* (Thunb.) Mansfeld) is an important tropical cucurbit mostly consumed as desert rather than in cooked form. Its edible placenta is rich source of minerals like potassium, iron and vitamins like vitamin A, thiamine and riboflavin. Besides, it is a good source of dietary antioxidants like phenolics, flavonoids, ascorbic acids, lycopene, β-Carotene etc. Watermelon contains moderate but significant quantities of phenolics (Perkins-Veazie *et al.,* 2002; Brat *et al.,* 2006). Watermelon fruits are identified as a good source of vitamin C, mainly in the reduced form of ascorbic acid (Vanderslice *et al.,* 1990). The lycopene content of watermelon is higher than many other fruits and vegetables. Consumption of watermelon juice increases plasma concentrations of lycopene and β-carotene in humans. Lipophilic as well as hydrophilic extract of watermelon showed good antioxidant activity. Watermelon is a rich source of citrulline, an amino acid that can be metabolized to arginine, a conditionally essential amino acid for humans. Arginine is the nitrogenous substrate used in the synthesis of nitric oxide which plays an essential role in cardiovascular and immune functions. Consumption of watermelon juice increases plasma concentration of arginine in human adult (Collins *et al.,* 2007). Dietary supplementation with watermelon pomace juice containing citrulline increases arginine availability, reduces serum concentrations of cardiovascular risk factors, improves glycaemic control, and ameliorates vascular dysfunction in obese animals with type-II diabetes (Wu *et al.,* 2007).

Luffa

Luffa (*Luffa cylindrica)* is another underutilized cucurbits cultivated from ancient time. Young fruits of *Luffa cylindrica* and *L. acutangula* are used as vegetables. Fruits are rich sources of minerals like calcium, magnesium, potassium, sodium and vitamins like thiamin, riboflavin, niacin and ascorbic acid (Bangash *et al.,* 2011). It contains moderate amount of dietary fiber including soluble as well as insoluble (Khanum *et al.,* 2000). Fruits of *Luffa sp* showed potential free radical scavenging activity and antioxidant activity (Shekhawat *et al.,* 2010; Raghu *et al.,* 2011). Scientific investigation reveals that high antioxidant potential of fruits of *Luffa sp* due to presence of phenolics compounds such as cinnamic acid derivates (1- *O*-feruloyl-*â*-D-glucose, 1-*O*-*p*-coumaroyl-*â*-D-glucose, *p*-coumaric acid, and 1-*O*-caffeoyl-*â*-D-glucose) and the flavonoids glycosides (diosmetin-7-*O*-*â*-D-glucuronide methyl ester, apigenin- 7-*O*-*â*-D-glucuronide methyl ester, and luteolin 7-*O*-*â*- D-glucuronide methyl ester) (Du *et al.,* 2006). Ethanolic extract of seed of *Luffa acutangula* showed potential antioxidants, anti-inflammatory and analgesic activity (Gill *et al.,* 2011). In recent times research finding focused ridge gourd as potential antitumor food due to its significant antiproliferative and antiangiogenic activity (Reddy *et al.,* 2009). Presence of ribosome inhibition protein *viz.* luffin a and luffin b in seed revealed the potentiality of luffa as anti-HIV agent. Ribosome inactivating proteins are group of protein that are able to inactivate eukaryotic protein synthesis by attacking the 28s ribosomal RNA. Scientific investigation revealed that luffin possesses strong anti-immunodeficiency virus deficiency activity (Au *et al.,* 2000).

Pointed Gourd

Pointed gourd (*Trichosanthes dioica* Roxb.) is one of the most nutritive cucurbit vegetables; it has higher nutrition content than other cucurbits. Fruits are richer sources of minerals like phosphorus, calcium, magnesium, sodium and vitamins like ascorbic acid and vitamin A. Since time immemorial, various parts of pointed gourd used as anthelmintic, antipyretic, antidiuretic, appetizing, digestives and expectorant (Sharma and Pant, 1988). Fruits and leaves have blood lowering properties in experimental animal and mild diabetic human subjects (Rai *et al.,* 2008a; Rai *et al.,* 2008b). Major bioactive compound present in pointed gourd is phenolics and triperpenoids. Bioactive lectin has been isolated from its seed having sugar binding capacity. The seeds of *T. dioica* contain a large amount of peptides and have the unique property of being resistant to the action of silver nitrate, a sensitive reagent commonly used to stain proteins. Being very rich in protein and vitamin A, it has certain medicinal properties, and many reports are available regarding its role in lowering of blood sugar and serum triglycerides. The fruits are easily digestible and diuretic in nature. They are also known to have antiulcerous effects. It grows as a vegetable all over India. It is prescribed to improve appetite and digestion. The decoction of TD is useful as a valuable alternative tonic, and as a febrifuge, which is given for boils and other skin diseases. The juice of the leaf is applied to patches of alopecia areata. The root is used as a hydragogue cathartic tonic and febrifuge. The fruits are used as a remedy for spermatorrhoea, and the juice of unripe fruits and also tender shoots, are used for cooling and as a laxative. The fruits and seeds have some prospects in the control of some cancer- like conditions and haemagglutinating activities.

Snake Gourd

Snake gourd (*Trichosanthes anguina*) is another relative of pointed gourd cultivated for its immature fruits and rarely for its ripe fruits. It is a good source of calcium, potassium, iron, substantial amount of carotene, little thiamine, riboflavin and niacin. As calorific value of fruit is very less it is very much suitable for diabetic patient. Juice of the fresh leaves traditionally used to treat several heart related disorder, jaundice etc. The edible part of the immature fruit is 86 – 98 per cent per 100g fresh fruit, which contains water 94g, protein (0.6g), fat (0.3g), carbohydrate (4g), fibre (0.8g), Ca (26mg), Fe (0.3mg), P (20mg), Vitamin B_1 (0.02mg), Vitamin B_2 0.03ng, Niacin 0.3mg, Vitamin C (12mg), energy value 70kJ (Siemonsma and Piluck, 1993). The fruits become inedible upon ripening, they taste bitter and developed hardened fibro vascular bundles. Fruits of the wild forms are very bitter and inedible. They are used in traditional medicine as a purgative. The use of the pulp of ripe fruits as a substitute for tomato paste is the major use of snake gourd known.The moisture content of the snake gourd seed is 3.13 per cent which is low when compared with values reported for melon seed (5.6 per cent) and water melon (5.7 per cent) (Ige *et al.,* 1984; Osagie and Odutuga, 1986). The ash content of the seed is quite high (2.93 per cent) and this may make it unsuitable for animal feed production. Snake gourd seeds are rich in potassium (121.6mg/100g) and very poor in copper and manganese content (0.31 mg/100g). The protein concentrate of snake gourd flour is 78 per cent.

Round Melon

Round melon or squash melon or Indian squash (*Citrullus lanatus* var. *fistulosus*) is commonly known as tinda in India. They are equally esteemed for vegetable purpose as well as for medicinal use and animal fodder. They are apple-sized, round, light green fruit, which are eaten cooked, pickled, or candied. The seeds are roasted and eaten as a snack. Tinda is an immature fruit used in rayata or vegetable curries. The seeds of tinda are roasted and consumed in the same way as watermelon or egusi seeds. In India, tinda is used as fodder and in medicine. The entire immature fruit is used as a cooked vegetable. In India, the fruits are also pickled and candied. The 100 g edible part of tinda contains water (93.5g), protein (1.4g), fat (0.2g), carbohydrate (3.6g), fiber (1.6g), calcium (25mg), iron (0.9mg), potassium (24mg), carotene (13µg), thiamin (0.04mg), riboflavin (0.08mg), niacin (0.3mg) and ascorbic acid (18mg). The fruits are also helpful in elimination of dry cough and improving blood circulation.

Ivy Gourd

The fruits of ivy gourd (*Coccinia indica*) are commonly eaten in Indian cuisine. People of Indonesia and other south east Asian countries also consume the fruit and leaves. In Thailand it is one of the ingredients of the Kaeng khae curry. Cultivation of ivy gourd in home gardens has been encouraged due to its being a good source of several micronutrients, including vitamins A and C. In India it is eaten as a curry, by deep-frying along with or without chilli and garlic; stuffing it with masala and sauteing it, or boiling it first in a cooker and then frying it. It is also used in sambar. Ivy gourd is rich in beta-carotene.In traditional medicine, fruits are used to treat leprosy, fever, asthma, bronchitis and jaundice. The fruit possesses mast cell stabilizing, anti-anaphylactic and anti-histaminic potential. The leaves are a rich source of protein (3.3 - 4.9 g), minerals and vitamins, in particular vitamin A (8000-18000 IU.) (Boonkerd *et al.,* 1993). The plant possesses antioxidant property and administration of its leaf extract in streptozotocin diabetic rats caused a significant increase in plasma vitamin C and reduced glutothione (Venkateswaran and Pari, 2003). The roots and leaves of ivy gourd are used in Ayurvedic and folk medicines to treat the diabetes mellitus, skin eruption, tongue sores and earache. The extract of the root shows a hypo glycaemic effect in fasted albino rats (Vaibhav and Gupta, 1995; Vaibhav and Gupta, 1996). Ivy gourd extracts and other forms of the plant can be purchased online and in health food stores. It is claimed that these products help regulate blood sugar levels. There is research to support that compounds in the plant inhibit the enzyme glucose-6-phosphatase. Glucose-6-phosphatase is one of the key liver enzymes involved in regulating sugar metabolism. Therefore, ivy gourd is sometimes recommended for diabetic patients. Although these claims have not been supported, there currently is a fair amount of research focused on the medicinal properties of this plant focusing on its use as an antioxidant, anti-hypoglycemic agent, immune system modulator, etc. Some countries in Asia like Thailand prepare traditional tonic like drinks for medicinal purposes.

Legume Vegetables

Leguminous vegetable is another major vegetable family cultivated in various parts of the world. They are consumed either as vegetable or as pulse. Among leguminous vegetables, yard long bean, french bean and vegetable pea are consumed in various parts of world. Leguminous vegetables provide important bioactive compound to human diet. They are the sources of folic acid, carotenoids, phenolic, bioactive peptide etc.

Garden Pea

Garden peas, botanically classified as a fruit is used as a vegetable, for cooking purposes, since ages. They can be described as the small, spherical seeds or pods of the legume *Pisumsativum*. In India, it is mainly grown as winter crop for its green pods and for mature dry seed in the plains of northern region. It is a good source of protein (7.2g), minerals (0.8 g) and vitamins (Vitamin A 139 IU, Vitamin C 9 mg, thiamine 0.25 mg). Being low in calories, green peas are good for those who are trying to lose weight. These are rich in dietary fiber, making them good for those suffering from constipation (Choudhary, 1967). Studies indicated that green peas might prove beneficial for those suffering from the problem of high cholesterol. The high amount of iron and vitamin C in green peas help strengthen the immune system. The lutein present helps reduce the risk of age-related macular degeneration and cataracts. Green peas slow down the appearance of glucose in the blood and thus, help keep the energy levels steady (Duke, 1981). They also aid energy production, nerve function and carbohydrate metabolism. Green peas provide the body with those nutrients that are important for maintaining bone health. The folic acid and vitamin B_6 in green peas are good for promoting the cardiovascular health of an individual. Anti-nutritional factors, although present in pea, are relatively minor and do not adversely affect crop use. The nutritional value and agronomic benefits of pea have contributed to its sustained production and use in cropping systems worldwide.

Yardlong Bean

Yardlong bean [*Vigna unguiculata* L. (Walp.) *subsp.sesquipedalis*] is an important leguminous vegetable mostly cultivated in China, Southeast Asia, the Caribbean, Central and West Africa for its immature pods.The pods are rich in calcium, phosphorus, sodium, and potassium and fair amounts of vitamin A, thiamine and ascorbic acid (Piluek, 1994). It is a rich source of another very important water soluble vitamin-folic acid, essential for prevention of neural tube defect in new born baby (Devi *et al.,* 2008). Immature pods showed potential free radical scavenging activity and antioxidant activity measured by various methods (Yang *et al.,* 2005; Wen *et al.,* 2010). Antioxidant potential of yard long beans is attributed due to presence of natural antioxidants like ascorbic acid, beta carotene and phenolics (Wen *et al.,* 2010). Due to proven potential, physiological health benefit of anthocyanin and purple podded variety of yard long bean is now in focal point of nutraceutical research. Ha *et al.* (2010) identified five anthocyanins, *viz.* delphinidin-3-*O*-glucoside, cyanidin-3-*O*-sambubioside, cyanidin-3-*O*glucoside, pelargonidin-3-*O*-glucoside, and peonidin-3-*O*-glucoside from immature purple pod of yard-long beans. A potent antimicrobial peptide, 'sesquin' with a molecular mass around 7 kDa and an N-terminal sequence highly homologous to defensin was isolated from yard long beans. This peptide exerted antimicrobial activity against fungus and bacteria and antiproliferative activity towards breast cancer (MCF-7) cells and leukemia M1 cells. It also exhibited some inhibitory activity towards human immunodeficiency virus-type 1 reverse transcriptase (Wong and Ng, 2005).

French Bean

French bean (*Phaseolus vulgaris*) is another important leguminous vegetable used for its immature green pod. Dried seeds are used as pulse popularly known as *rajma*. Immature green pods are good sources of calcium, phosphorus, thiamin, riboflavin, folic acid, ascorbic acid and relatively low amount of α-tocopherol. It contains moderate amount of dietary fiber. Immature pod showed moderate antioxidant activity measured by various in *vitro* methods (Kaur and Kapoor, 2002; Isabale *et al.*, 2010; Wen *et al.*, 2010). Antioxidants activity is attributed due its water soluble and lipid soluble antioxidant nutraceuticals. Major water soluble antioxidants in french bean are ascorbic acid and phenolics including flavonoids. Miean and Mohamed (2001) identified three flavonoids *i.e.* quercetin (114.5 mg/kg dw), myrecetin (45 mg/kg dw) and luteolin (11 mg/kg dw). Hempel and Böhm (1996) identified another flavonoids kaempferol (5.6 to 14.8 µg/g fw) in both yellow and green colour french bean. Currently, kaempferol is in interest because of its antioxidant (Vinson *et al.*, 1995), antitumor, anti-inflammatory and antiulcer activity (Goet *et al.*, 1988), and its inhibitory activity of HIV protease (Brinkworth*et al.*, 1992). Major lipid soluble antioxidants including carotenoids (Neoxanthin, Violaxanthin, Lutein, Zeaxanthin, β-carotene) and tocopherol are identified by Isabale *et al.* (2010). Recently a dimeric 64-kDa hemagglutinin was isolated from the seed of French bean which have antifungal and anti-HIV-1 reverse transcriptase activities. It also inhibited proliferation of hepatoma HepG2 cells and breast cancer MCF-7 cells (Lam and Ng, 2010).

Winged Bean

Winged bean (*Psophocarpus tetragonalobus* L.D.C.) is an important legume vegetable and its every part *viz.* immature pods, seeds, flowers, shoots and leaves and root are edible. The immature pods are used in curries, soups, salads and in several other preparations. The flavour of winged bean pods is similar to that of green beans. The average protein content is 2.4 g per 100 g of edible portion. It is also rich in minerals and vitamins. Besides immature pods, the dry seeds of winged bean are the most nutritious. Seeds are rich in protein content (30-42 per cent) and possess favourable amino acid composition. Its seeds can be steamed, boiled, fried, roasted, fermented or made into milk, tofu (bean curd) or tenpeh. The amino acid composition is also similar to that in soybean like most legume seeds. The winged bean is relatively deficient in sulphur containing amino acids (Cerny and Addy, 1973). Winged bean is fairly rich in available lysine and it may be supplemented cereal diets which are deficient in lysine content. Consumption of winged bean is advocated to counter the problem of protein energy malnutrition. The leaves of winged bean are also consumed as a green vegetable in some parts of Africa. The leaves have relatively low lysine content but an uncommonly high content of tryptophan, a nutritionally essential amino acid. Its incorporation in the cereals which are deficient in tryptophan like corn is highly beneficial. About 15 percent protein could be recovered from the leaves of the winged bean cultivars (NAP, 1981). The vitamin A content in fresh and tender leaves is measured around 20,000 I.U. Winged bean develops edible tuberous roots which are commonly used in the Papua New Guinea and Myanmar. Roots are boiled, steamed, fried or baked. The brown skin peels off readily after boiling, leaving a white or cream coloured flesh that is firm and moist, with a distinct nutty, creamy flavour. Roots are rich in protein content (8 - 20 percent on dry weight basis) compared 1-5 percent in potato (Sri Kantha *et al.,* 1978). It possesses several medicinal properties and consumption of winged bean stimulates mother milk production (Anonymous, 1991).

Sword Bean

Sword bean/jack bean (*Canavalia* spp.) being a legume crop is used for forage and as a green manure. Pink seeds are sometimes employed in traditional Chinese medicine. Maiti and Mishra (2000) reported anti -venom property of sword bean. Sword bean has received much attention of biochemists for its various antinutritional factors like canavalin, concanavalin A and B, urease and canavanine. Urease activity in leaf of sword bean was reported by Watanabe *et al.* (1983). In mature seeds of sword bean, Rodrigues and Torne (1992) analysed 3.06 per cent canavanine. Fujihara (1986) reported that polyam-ines sym. homospermidine and canav-almine appear in the immature seed and their concentration increase during seed development. At the same time, the content of spermidine decreases while that of the arginine analogue canavalmine greatly increases, especially in the cotyledons. Toxicity seems largely due to concanavalin A, which binds to mucosal cells of the intestine, thus reducing the body's ability to absorb nutrients from the intestine (Kooi, 1993). However, Laurena *et al.* (1994) reported low concentrations of condensed tannins and protein perceptible polyphenol, total cyanide and trypsin inhibitor activity and these substances do not constitute nutritional problems. Further, phytate phosphorus like antinutritional factors is in tolerable limit.

Velvet Bean

Velvet bean (*Mucuna pruriens*) is a crop of multiple use *viz.* foodstuffs, medicinal, green manure, cover crop and as a fodder (Fujii, 1990). Its tender pods are used for vegetable purpose. Being a leguminous crop, its seed is a good source of protein. Analysis of seeds gives moisture 9.1 per cent, protein 25.03, ether ester 2.96, fiber 6.75 and mineral matter 3.95, calcium 0.16, phosphorus 0.47, and iron 0.02 per cent. Seed oil is rich in palmitic, oleic and lindeic acid (Gupta *et al.,* 1995). Aguiyi *et al.* (1997) reported 23 ± 0.65 per cent oil in its species *Mucuna sloanei* and it had light yellow colour with high saponification and iodine values. Its oil can be used in preparation of resin, paints, polish, wood tarnish and skin cream. It could also be a constituent of liquid soap. About 5 per cent of the weight of the bean is a psychoactive substance called "L dopa". Dopa is still a commonly prescribed treatment for Parkinson's disease, though it has side effects such as uncontrolled muscle twitches and extreme cases even psychotic disorders including schizophrenia. Velvet bean has long been valued in medicine. It is used to promote blood circulation, to relieve stasis and to treat irregular menstruation and joint pains or numbness (Ding *et al.,* 1991). Seeds of

Macuna pruriens are used in traditional medicine in West Africa to treat a variety of disorders including snake bites. The pharmacological effects of aqueous extract of seeds were investigated against the skeletal muscle and against smooth muscle of the gastrointestinal tract. Seeds of velvet beans are widely used for treating male sexual dysfunction in Tibbe unani (Unani medicine). The effect of powdered seeds on the mating behaviour, libido and potency of sexually normal male rats was investigated. The drug produced a striking and sustained increase of sexual activity (Amin *et al.*, 1996). Besides seeds and arial parts; root also has medicinal properties being used as a stimulant tonic and diuretic. Iauk *et al.* (1993) reported analgesic effects and anti inflammatory activities with the most pronounced effects of leaf extracts. Both extracts also exhibited antipyretic activities.

Cowpea

Cowpea (*Vigna unguiculata*) is grown for its long green pods as vegetable, seeds as pulse and foliage as vegetable and as fodder. It has been difficult to find comprehensive information on the composition of cow peas. However, it is known that they are a good source of protein, complex carbohydrates and fibre (Phillips *et al.*, 2003; Amjad *et al.*, 2006; Olivera-Castillo *et al.*, 2007) and like other legumes are likely also to provide vitamins such as folate, thiamine and riboflavin (Phillips *et al.*, 2003) and minerals potassium, magnesium, phosphorus, calcium, copper, iron and zinc (Amjad *et al.*, 2006). Wu *et al.* (2004) established that they contain levels of phenolic compounds that are moderately high in terms of other vegetables (6.47 mg GAE/g), though toward the lower end of the scale in comparison with other dry pulses. Tannins (particularly in the seed coat), and phenolic acids (protocatechuic, p-hydroxybenzoic, caffeic, p-coumaric, ferulic, 2,4-dimethoxybenzoic, cinnamic, gallic, vanillic, p-hydroxybenzoic and protocatechuic) have also been identified (Cai *et al.*, 2003; Duenas *et al.*, 2005; Siddhuraju and Becker 2007). Myricetin and quercetin glycosides flavonols were also reported by Duenas *et al.* (2005). Apart from being rich in proteins, these legumes are also packed in essential minerals and vitamins that are vital for good health. Regular consumption of cowpeas tones the spleen, thereby enhancing the production of cells that improve immune responses of our body. Consuming these legumes can induce the process of urination and also provide relief from leucorrhea. Cowpeas are also good for the stomach and the pancreas.

Dolichos Bean

Dolichos bean (*Lablab purpureus*) or hyacinth bean (Sem) is grown throughout the country. Its immature green pods are used as vegetables and dry seeds as pulse. Like other legumes, hyacinth beans are a good source of protein (3.8 g), minerals and vitamins. Amino acid composition of hyacinth beans is comparable to other legume proteins. Methionine is the most limiting amino acid in hyacinth beans. Like other legumes, hyacinth bean seeds contain trypsin inhibitor, lectin, phytic acid and polyphenol. The toxicity of lectin can be eliminated by autoclaving (heating) the meal. Apart from their nutritional value the plant is also used in traditional systems of medicine. In China the boiled ripe seeds are thought to be a tonic, and good to get rid of flatulence. The beans are also used as an aphrodisiac, for fevers, stomach problems, and as an antispasmodic. When the beans are used regularly in a diet, they prevent a build-up of cholesterol, Vadde *et al.* (2007).

Cole Crops

Among different groups of vegetables, members of cruciferous family are cultivated and widely consumed universally as a part of daily diet since time of immemorial. The major vegetable includes broccoli, cauliflower, kale, Brussels sprouts, and cabbage that are used either fresh (salads), steamed or cooked. Besides nutritional components, these vegetables are also rich in health beneficial secondary metabolites, which include sulfur containing glucosinolates and S-methyl cysteine sulfoxide, flavonoids, anthocyanins, coumarins, carotenoids, antioxidant enzymes, terpenes and other minor compounds. Wide ranges of bioactive compounds are responsible for strong health beneficial properties of cole crops.

Broccoli

Broccoli (*Brassica oleracia* var *italica*) is the most important cole crop recently received wide attention due to its multiple health benefit property. It is a good source of vitamins, minerals and dietary fiber. It accumulates selenium, an antioxidant trace element when grown in selenium enriched soil (Banuelos and Meek, 1989). It ranked among the highest source of folic acid, contributing about 70-90 µg/100g (Scott *et al.*, 2000). Besides, it contains wide range of health promoting compounds which are responsible for its strong functional properties. Broccoli exhibits strong antioxidant activity under *in vitro* condition measured by different methods (Kaur and Kapoor, 2002). Antioxidant properties of broccoli are attributed due to presence of wide ranges of water soluble and lipid soluble antioxidants. Lutein, β-carotene and α-tocopherol are the important lipid soluble antioxidants in broccoli. Among water soluble antioxidants, ascorbic acid and polyphenols including flavonoids are important. The predominant hydroxycinnamoyl acids were identified as 1-sinapoyl-2-feruloylgentiobiose, 1,2-diferuloylgentiobiose, 1,2,20-trisinapoylgentiobiose, and neochlorogenic acid (Vallejo *et al.*, 2003). In addition 1,20-disinapoyl-2-feruloylgentiobiose and 1,2-disinapoylgentiobiose,1-sinapoyl-2,2-diferuloyl gentiobiose, isomeric form of 1,2,20-trisinapoylgentiobiose, and chlorogenic acids are found in broccoli (Price *et al.*, 1997; Vallejo *et al.*, 2003). Total amount of feruloylsinapoyl esters of gentiobiose and caffeic acid derivatives in 14 cultivars of broccoli varied from 0 to 8.25 mg/100 g and from 0 to 3.82 mg/100 g, respectively. The two main flavonol glycosides present in broccoli florets are 3-*O*-sophoroside-7-*O*glucoside of quercetin and kaempferol and other minor glucosides isoquercitrin, kaempferol 3-*O*-glucoside and kaempferol 3-*O*-diglucoside (Price *et al.*, 1998; Vallejo *et al.*, 2004). Another important bioactive compound in broccoli is glucosinolates. With the presence of mirosinase enzyme, these glucosinolates break into various bioactive isothiocianates. Main glucosinolates break down products are sulphoraphane, indole-3-carbinol, benzyl isothiocyanate and phentylisothiocyanate, may be responsible for disease prevention properties of broccoli especially different types of cancer. *In vitro* and *in vivo* studies reported that isothiocyanates affect many steps of cancer development, including modulation of phases I and II detoxification enzymes. They function as a direct antioxidant or as an indirect antioxidant by phase II enzyme induction, modulating cell signalling, induction of apoptosis, control of the cell cycle and reduction of *Helicobacter* infections (Sarkar *et al.,* 2003, Visanji *et al.,* 2004).

Cauliflower

Cauliflower (*Brassica oleracia* var *botrytis*) is another cole crop having many functional properties with ranges of nutraceutical compounds. Hypertrophied inflorescence known as 'curd' is commonly used as vegetables. Curd is a good source of vitamins and minerals. Besides, it is a good source of dietary fiber. Many functional properties investigated by various researchers are attributed to various bioactive compounds. The major carotenoids present in cauliflower are lutein, neoxanthin, viola-xanthin and β-carotene. It is also a good source of tocopherols (Isabelle *et al.,* 2010). Other important bioactive compounds present in cauliflower are glucosinolate, ascorbic acid and phenolics including flavonoids. Major glucosinolates present in cauliflower are glucoalyssin, glucoiberverin, glucobrassicin, neoglucobrassicin, 4-hydroxy glucobrassicin, 4-methoxyglucobrassicin etc. (Gratacós-Cubarsí *et al.,* 2010). Cabello-Hurtado *et al.* (2012) assessed *in vitro* antioxidant activity of cauliflower glucosinolates and their derivatives. Glucosinolates showed significant antioxidant activity measured by oxygen radical absorbance capacity assay and superoxide radical scavenging activity assays. Among five glucosinolates *i.e.* gluconapin, glucoiberin, progoitrin, sinigrin and glucoraphanin, glucobrassicin showed the highest antioxidant capacity. Mixtures of glucosinolates did not show either synergy or antagonism in general. Interestingly enzymatic hydrolysis-derived products of glucosinolates were far more active than the native glucosinolates. In cauliflower, predominant hydroxycinnamic acid conjugates have been identified as 1,2-disinapoyl diglucoside, 1-sinapoyl-2-feruloyl-diglucoside, 1,2,2'-trisinapoyl diglucoside and 1,2'-disinapoyl-feruloyl-diglucoside. The major flavonol glycosides in cauliflower are Quercetin-3-diglucoside-7-glucoside, Kaempferol-3-diglucoside-7-glucoside and Kaempferol -3-diglucoside-7-diglucoside (Gratacós-Cubarsí *et al.,* 2010). Anthocyanins are important group of flavonoids identified in some varieties of violet pigmented cauliflowers. The violet curded landraces of cauliflower are commonly found in agricultural agri-systems of east Sicily. Violet cauliflower extracts show significant antioxidant properties, which are the scavenging activity of the very reactive hydroxyl radical (Pizzocaro *et al.,* 2000). The predominant anthocyanins present in purple cauliflower are *p*-coumaryl and feruloyl esterified forms of cyanidin-3-sophoroside-5-glucoside (Lo Scalzo *et al.,* 2008).

Cabbage

Cabbage (*Brassica oleracea* var *capitata*) is an important cole crop cultivated for its apical bud from time immemorial. It is consumed in many forms *viz.* cooked, salad, fermented and other processed forms. It is rich in various minerals and vitamins. Sauerkraut is a fermented cabbage product rich in ascorbic acid recommended for people suffering from scurvy. Many bioactive compounds are identified like glucosinolates, phenolics and carotenoids which are responsible for health beneficial properties of cabbage. Nielsen *et al.* (1993) showed that cabbage contains a mixture of more than 20 compounds of which seven are identified as 3-*O*-sophoroside-7-*O*-

glucosides of kaempferol and quercetin with and without further acylation with hydroxycinnamic acids. In addition, unmodified kaempferol tetraglucosides or their derivatives acylated with either sinapic, ferulic or caffeic acid were found in cabbage leaves (Nielsen *et al.,* 1998). Red cabbage, another type of cabbage presently receives wide attention due to its health potential and commercial application. Red pigmentation of red cabbage is caused by anthocyanins, which belong to flavonoids. Red cabbage contains more than 15 different anthocyanins which are acylglycosides of cyanidin (Dyrby *et al.,* 2001; Mazza and Miniati, 1993). Total anthocyanin content in red cabbage is 25 mg/100 g (Wang *et al.,* 1997) or 44.4–51.2 mg/100 g for anthocyanidins released after acidhydrolysis (Franke *et al.,* 2004). In Japan red cabbage is a source of red food colorants and the preparation of these pigments is described in several patents (Bridle and Timberlake, 1997).

Knol-khol

Knol-khol or kohlrabi (*Brassica oleracea* var. *gongylodes*) is a cool season, stout, round shaped tuberous modified stem (knob) vegetable of the Brassicaceae family grown for its knob as well for its turnip-flavored succulent leaves (top greens). The succulent kohlrabi knobs are rich in mineral content (Ca, K, Mn, Fe, P and Cu), carotenoids, antioxidants, vitamins and dietary fiber; while it has low calorie and fat and almost zero in cholesterol content (West and Poortvliet 1993, Chakrabarti 2003, Kala and Jamuna 2006, and Ahmed and Beigh 2009). Potassium (K) is an important component of cell and body fluids that helps controlling heart rate and blood pressure by countering effects of Sodium (Na). It is a rich source of ascorbic acid (60-75 mg/100 g); a water-soluble vitamin possessing antioxidant properties. It helps the body maintain healthy connective tissue, teeth and gum and additionally antioxidant property helps the human body protect from diseases and cancers by scavenging harmful free radicals from the body. Like other Brassica vegetables, knoll-khol also contains health-promoting phytochemicals like aisothiocyanates, sulforaphane and indole-3-carbinol which are supposed to protect against prostate and colon cancers (Fan *et al.,* 2009). It also contains good amount of many B-complex vitamins, namely niacin, pyridoxine, thiamin and pantothenic acid which act as co-factors to enzymes during various metabolisms inside the body. The top greens are also very nutritious and abundant in carotenes, vitamin A, vitamin K, minerals and vitamin B-complex.

Brussels Sprouts

Brussels sprouts (*Brassica oleracea* var, *gemmifera*) are one of the many variations of cabbage. Instead of single head at the top of the stem, a large bud or miniature head is borne in the axil of each leaf so that little heads are distributed all along a tall stem crowned with a cluster of loose leaves. The little solid heads or "sprouts" are tender and delicious. Brussels sprouts are rich in protein (4.4 per cent). vitamin-A (520 I.U.) and ascorbic acid (72 mg) and contain appreciable amounts of riboflavin, niacin, calcium and iron. Brussels sprouts are also rich source of glucosinolates. This vegetable is rich in vitamin K, used by body for blood clotting so some heart patients taking anticoagulants can avoid its excess intake. In short, all cruciferous vegetables contain glucosinolates and have great health benefits for this reason.Glucosinolates are important phytonutrients for our health because they are the chemical starting points for a variety of cancer-protective substances".Brussels sprouts are very beneficial to our lives. The cancer protection from Brussels sprouts is largely related to four specific glucsinolates found in this cruciferous vegetable: glucoraphanin, glucobrassicin, sinigrin and gluconasturtiian.

Kale

Kale (*Brassica oleracea* var. *acephala*), characterized by headless leafy greens, belongs to the family Brassicaceae. Among cole crops and brassicas, it is one of the richest source of protein, dietary fiber, minerals, vitamins (Vitamin B complex, vitamin K, tocopherol, ascorbic acid, B complex) and health benefiting polyphenolic flavonoid compounds like α-carotene, β-bcarotene, lutein and zea-xanthin (Kurilich *et al.,* 1999, Kopsell *et al.,* 2004, Singh *et al.,* 2004, Alibas 2009, Sikora 2008 Korus 2010, Heimler 2006, Miller-Cebert *et al.,* 2009a, b, Acikgoz and Deveci 2011). The polyphenolic flavonoids have strong antioxidant and anticancerous activities. The carotenes namely α-carotene and β-carotene are converted to vitamin A in the body. The leaves possess low fat, low calorie and no cholesterol. Kale, like other member of the Brassica family, contains health promoting phyto-chemicals, sulforaphane and indole-3-carbinol which appear to protect against prostate and colon cancers. Di-indolyl-methane, a metabolite of indole-3-carbinol is an effective immune modulator, antibacterial and antiviral agent through its action of potentiating "Interferon-Gamma" receptors (Fan *et al.,* 2009).

Root Crops

Root crop is another major vegetable group cultivated in various parts of the world. They are consumed in many forms like salad, cooked, canned, dehydrated as well as fermented. Among root crops radish, carrot and beet root are consumed in various parts of world due to their high nutritional value. Root vegetable provide important bioactive compounds to human diet. They are the sources of carotenoids, phenolic, dietary fiber, folic acid etc.

Carrot

Carrot (*Daucus carota*) is an important root vegetable having an important role in human nutrition. Its tender root is consumed either in the form of salad, cooked or processed forms. It is a rich sources of vitamins, minerals and dietary fibers. Besides, it is a rich source of both hydrophilic as well as lipophilic antioxidants. The major lipophilic antioxidants are beta carotene, alpha carotene, lutein nd lycopene. Yellow carrot accumulates xanthophylls lutein, an essential bioactive pigment protecting eye from age related macular degeneration and cataract. Red carrot contains abundant lycopene in addition to beta carotene and alpha carotene. Lycopene is a functional ingredient which prevents prostate cancer and scavenges free radicals from body. Studies revealed that lycopenes in red carrot are about 44 per cent more bioavailable than tomato paste (Horvitz *et al.,* 2004). Major hydrophilic antioxidants are phenolics including flavonoids and anthocyanin. Carrots contained mainly hydroxycinnamic acids and derivatives like chlorogenic acid, caffeic acid, 3'-caffeoylquinic acid, 4'p-coumaroylquinic acid, 3',4'-dicaffeoylquinic acid, 3',5'-dicaffeoylquinic acid and some unidentified hydroxycinnamic derivatives. Among them, chlorogenic acid

(5'-caffeoylquinic acid) is a major hydroxycinnamic acid in carrot tissues, representing 42.2 per cent to 61.8 per cent of total phenolic compounds detected in different carrot tissues (Zhang and Hamauzu, 2004). The major flavonoids identified in carrot are quercetin, luteolin and kaempferol (Miean and Mohamed, 2001).Recently, among various bioactive compounds anthocyanin received wide attention due to their multiple health benefits. Stintzing *et al.* (2002) identified four major anthocyanins in purple carrot extract and found 41 per cent of anthocyanins to be acylated, namely cyanidin 3 sinapoly-xylosyl-glucosyl-galactoside (27.5 per cent) and cyanidin 3-sinapoyl-xylosyl-glucosyl-galactoside (13.5 per cent).

Beet

Beet (*Beta vulgaris*) is an important vegetable belonging to chenopodiaceae family. Roots are good sources of calcium, phosphorus, iron, thiamin, riboflavin, folic acid and ascorbic acid. The main pigments in beet are betalains. Betalains are water-soluble nitrogenous pigments. They can be divided into two major structural groups, the red to red-violet betacyanins and the yellow betaxanthins. Betacyanins can be further classified by their chemical structures into four kinds: betanin-type, Amaranthusin-type, gomphrenin-type and bougainvillein-type (Strack *et al.,* 1993). Common beets usually contain both red betacyanins (consisting of 75–95 per cent betanin) and yellow betaxanthins (95 per cent vul-gaxanthin I), in various ratios depending on cultivars (Francis, 1999; Piattelli, 1981). Betalains are stable between pH 3.5 to 7.0 which cover nearly all foods with maximum colour stability at pH 5.5. Betanin is susceptible to light and temperature damage which limit the use to fresh foods, foods packed under modified atmospheric condition or food will not undergo heat treatment. It is mainly used in frozen products (ice cream and yoghurt). Dried betanin is more stable and used as colour additives in instant food and powdered soft drink. Besides vegetables use it is also cultivated for sugar production thus used in jelly, candy, fillings. Root powder is very popular colouring agent for soap and cosmetics. However, colour properties, betalains from beet roots possessed high antiradical effect and antioxidant activity, representing a new class of dietary cationized antioxidant (Pedreno and Escribano, 2000; Kanner *et al.,* 2001). Besides betalains it also contains phenolics including flavonoids which are potential antioxidants (Jiratanan and Liu, 2004). It also contains soluble fiber which helps in reducing blood cholesterol level.

Radish

Radish (*Raphanus sativus*) is a good source of vitamin-C containing 15.40 mg per 100g of edible portion and supplies a variety of minerals. It has 0.6 per cent protein. 0.3 per cent fat, 0.6 per cent minerals, 0.6 per cent fibre and 6.8 per cent other carbohydrates, Pink-skinned radish is generally richer in ascorbic acid than the white-skinned one. Fertilizers also influenced vitamin-C content of radish roots. Application of phosphorous significantly increased the ascorbic acid content. Radish contains glucose as the major sugar and smaller quantities of fructose and sucrose. Pectin and pentosan are also reported present. The characteristic pungent flavour of radish is due to presence of volatile isothiocyanates (trans-4- methyl-thiobutenyl isothiocyanate) and the colour of pink cultivars is due to the presence of anthocyanin in pigments. Radish is used in liver and gall bladder troubles. In homeopathy, they are used for neuralgic headaches, sleeplessness and chronic diarrhoea. Roots, leaves, flowers and pods are active against gram-positive bacteria. The roots are useful in urinary complaints, piles and in gastrodynia.

Turnip

Turnip (*Brassica rapa* var. *rapa*), one of the cool season vegetables is grown for swollen tap root and green leaves (turnip top). Although its bulbous root is widely popular, yet the turnip top is more nutritious, several times richer in vitamins, minerals and antioxidants content (Persson *et al.,* 2001, Fernadese *et al.,* 2007, Saeed *et al.,* 2012). The antioxidants help the body scavenge harmful free radicals, prevent from cancers, inflammation, and help boost immunity. The root is high only in vitamin C content; while green leaves are a good source of dietary fiber, vitamin A, folic acid, niacin, riboflavin, riboflavin, pyridoxine, pantothenic acid, thiamin, vitamin C, vitamin K, iron, manganese and calcium. Turnip greens are high in carotenoid, xanthin and lutein. It is a very low calorie root vegetable, contains only 25-30 calories/100 g. Turnip is characterized by a particular bitter and pungent taste, which has been related to the content of some glucosinolate degradation products (Carlson *et al.,* 1987). It controls and reduces the risk of cancers of the lung, bladder, pancreas and the stomach. It also decreases the chances of diabetes, hypertension and obesity. Turnip greens are good for rheumatoid arthritis patients, since the greens contain beta-carotene. Beta-carotene helps in the formation of vitamin A in the body, which is important for the proper working of the immune system. It also helps to produce as well as maintain healthy membranes, especially the membrane that lines the joints. Turnip greens are packed with vitamin E, vitamin C and beta-carotene. Vitamin C helps to reduce the levels of free radicals in the body and maintains good immune function. Studies show that the vitamin is also good for colon function and can reduce colon tumors. Vitamin E is also good for colon health and studies have shown that it can prevent colon cancer.

Leaf Vegetables

Leaf vegetable is another major vegetable group cultivated in various parts of the world. They are the sources of antioxidants, dietary fiber, folic acid etc. Realising the importance of leaf vegetables, ICMR recommended per day per capita consumption of 125g leaf vegetables. Among leaf vegetables, fenugreek is consumed in various parts of our country.

Fenugreek

Fenugreek (*Trigonella foenumgraecum*) commonly known as methi is cultivated throughout India and other parts of the world for leaf vegetable, condiment and medicinal purposes. Traditionally seeds are used as antidiabetic agent. Fenugreek leaves are rich in minerals, protein, vitamins-A and C. Fresh tender leaves, pods and stems are consumed as vegetable and seeds of fenugreek are mainly used as spice. The leaves are very rich in protein containing about 18.6 to 40.9 per cent at different stages of growth, on dry weight basis and 100 g of fresh methi leaves contain 3900 I.U. of vitamin A and 140 mg of vitamin C. The *in vitro* available iron in fenugreek leaves is only about 4.6 per cent of the total iron. Vitamin B, (0.8 mg/100g), vitamin K (240 ppm), α and β tocopherols (0.87 and 0.37 mg/g dry weight respectively) are reported to be present. Flavonoids identified in the leaves of fenugreek are Kaempferol and 3' 4' – dio Me-Quercetin. The fenugreek leaves lose 7.4 to 10.8 per cent of vitamin-C after boiling in water, steaming and frying. It also has high medicinal and industrial importance. It prevents constipation, removes indigestion, stimulates spleen and liver, and is appetizing and diuretic. The soluble dietary fibre *viz.*, gum, pectin and other mucilaginous substances present in fenugreek seeds reduce post- parandial levels of glucose in blood.

Asparagus

Asparagus (*Asparagus officinalis*) is a delicate, nutritious and appetizing vegetable grown in temperate as well as tropical regions. Asparagus is a good source of vitamin-A. Green asparagus (100 g edible portion) contains 93 g of water, 2.2g of protein, 21 mg of calcium, 700 IU of vitamin A, 30 mg of ascorbic acid, 0.2 mg of thiamine, 0.16 mg of riboflavin and 1.0 mg niacin. It is a good source of vitamin B_6, calcium, magnesium and zinc, and a very good source of dietary fiber, protein, vitamin A, vitamin C, vitamin E, vitamin K, thiamin, riboflavin, rutin, niacin, folic acid, iron, phosphorus, potassium, copper, manganese and selenium,(USDA, data base) as well as chromium, a trace mineral that enhances ability of insulin to transport glucose from bloodstream into cells. The amino acid asparagine gets its name from asparagus, as the asparagus plant is relatively rich in this compound. Asparagus root possesses aphrodisiac, demulcent, general tonic, diuretic, anti-inflammatory, antiseptic, anti-oxidant and antispasmodic properties. Regular use of asparagus root treats infertility, impotence, leucorrhea, menopause syndromes, hyperacidity, and certain infectious diseases like herpes and syphilis. It is also useful in treatment of epilepsy, kidney disorders, chronic fevers, excessive heat, stomach ulcers and liver cancer, increases milk secretion in nursing mothers and regulates sexual behaviors. *Asparagus racemosus* cleanses, nourishes, and strengthens the female reproductive organs and so, it is traditionally used for PMS, amenorrhea, dysmenorrhea, menopause and pelvic inflammatory disease (PID) like endometriosis. *Asparagus racemosus* is considered as the most potent female health tonic.

Lettuce

Lettuce (*Lactuca sativa*) accupies the largest area of the salad crop world wide. It is a pleasure food with low nutrient density. It has a crisp texture, a large surface to volume ratio and serves as source of bulk for diet conscious consumes. It is one of the few vegetable crops used extensively as a fresh raw product. Lettuce is generally not canned, dried or frozen and rarely cooked. The loose lettuce and Romaine types are more nutritious than the head type lettuce, mainly because of its high vitamins A and C values. It is also a good source of calcium and phosphorus. Lettuce is also known to be sedative, diuretic and expectorant. Lettuce extracts are sometimes used in skin creams and lotions for treating sunburn and rough skin. It was once thought to be useful in relieving liver issues. Some American settlers claimed that smallpox could be prevented through the ingestion of lettuce, Watts (2007), and an Iranian belief suggested consumption of the seeds when afflicted with typhoid. Duke *et al.* (2007) in folk medicine has also claimed it as a treatment for pain, rheumatism, tension and nervousness, coughs and insanity.

Amaranth

Amaranth (*Amaranthus* spp.) or Chinese spinach, is a common leaf vegetable throughout the tropics and in many warm temperate regions. Cooked Amaranths leaves are a good source of vitamin A, vitamin C, and folate; they are also a complementing source of other vitamins like thiamine, niacin, and riboflavin, plus some dietary minerals including calcium, iron, potassium, zinc, copper, and manganese. Cooked Amaranthus grains are a complementing source of thiamine, niacin, riboflavin, and folate, and dietary minerals including calcium, iron, magnesium, phosphorus, zinc, copper, and manganese - comparable to common grains such as wheat germ, oats and others. (USDA data base).

Amaranth seeds contain lysine, an essential amino acid, limited in other grains or plant sources. Most fruits and vegetables do not contain a complete set of amino acids, and thus different sources of protein must be used. Amaranth too is limited in some essential amino acids, such as leucine and threonine. Ricardo, *et al.* (1989). Amaranth seeds are promising complement to common grains such as wheat germ, oats and corn because these common grains are abundant sources of essential amino acids found to be limited in Amaranth (Pisarikova, *et al.,* 2006). Amaranth has medicinal values, which can reduce or combat common diseases such as diabetes, hypertension, liver disease, TB, HIV/AIDS, wound non-healing, skin diseases among others. Amaranth seeds and leaf biomass are rich in soluble and insoluble diet fibers important in prevention of coronary heart diseases of the colon. The compounds in Amaranth can enhance human growth and development, improve general health, and strengthen immune responses to combat diseases. In situations where dietary choices are limited or when immune systems are compromised, Amaranth consumption may make the difference between normal health and life-threatening diseases.

Spinach

Spinach (*Spinacia oleracea*) is an excellent source of antioxidant nutrients including vitamin C, vitamin E, beta-carotene and manganese as well as a very good source of zinc and selenium. It is no wonder that spinach lower risks of numerous health problems related to oxidative stress. Our blood vessels, for example, are especially susceptible to damage from oxidative stress, and intake of spinach has been associated with decreased risk of several blood vessel-related problems, including atherosclerosis and high blood pressure. Interestingly, the blood pressure benefits of spinach may be related not only to its antioxidants, but also to some of its special peptides. Peptides are small pieces of protein, and researchers have discovered several peptides in spinach that can help lower blood pressure by inhibiting an enzyme. The strong natural mineral and vitamin content in spinach can be an effective source of healing for a multitude of health ailments.

Curry Leaf

Curry leaf [*Murraya koenigii* (L.) Spreng.] grown throughout Central and South

East Asia, is used medicinally as tonic, stomachic, carminative, anti-dysenteric, stimulant, febrifuge and anti-periodic, as an anti-vomiting agent and to treat bite by poisonous animals (Reisch *et al.,* 1994). The leaves are used as a spice in different curries and impart a very good flavour to the preparations. The leaves are a fair source of vitamin A and folic acid. They are also rich sources of calcium, but due to presence of oxalic acid in high concentration (total oxalates 1.35 per cent, soluble oxalates 1.15 per cent), its nutritional availability is affected (Philip *et al.,* 1981). The free amino acids in the leaves are asparagine, glycine, serine, aspartic acid, glutamic acid, threonine, alanine, proline, tyrosine, tryptophan, tri-amino butyric acid, phenyl alanine, leucine, isoleucine and traces of ornithine, lysine, arginine and histidine (WOI, 1972). Curry leaf contains more than 20 carbazole alkaloids and among them important are murrayazididine, mukonine isomurrayazoline, mukonicine and murrayacinine (pentacyclic carbo-zole alkoloid) and mukoic acid (carbozole carboxylic acid) (Chakraborty *et al.,* 1978; Chakraborty *et al.,* 1973; Chakraborty *et al.,* 1974; Chaudhary and Chakraborty, 1971; Bhattacharya *et al.,* 1982, Mukherjee *et al.,* 1983). Root bark contains Mahanimboline-carbazole alkaloid (Adesina *et al.,* 1988; Roy *et al.,* 1979). Maha-nimbinol (C_{23} - carbazole alkaloid) found in stem is a key precursor in the biosynthesis of other carbozole alkaloids (Rao *et al.,* 1980). A minor alkaloid, maha-nine was isolated from the leaves (Rahman *et al.,* 1988). Fresh leaves on steam distillation yield volatile oil (curry leaf oil), which may find use as a fixative for heavy type of soap perfume (WOI, 1972). Terpenes are the main component of volatile essential oil of leaves; and the major constituents of terpenes are caryophy-llene, gurjunene, elemene, phelan-ddlrene and thujene (Macleod and Pieris, 1982). Essential oil of leaves possesses anti-bacterial and antifungal properties (Gautam and Purohit, 1974; Deshmukh *et al.,* 1986). Seeds of curry leaf contain 8-geranyloxypsoralene, imperatorin, heraclenin and isoxalin. Besides these, seed also contains mahanimbine, isomahanine, mahanine, girinimbine and koenimbine (Reisch *et al.,* 1994). Koenoline of roots exhibits cytotoxic activity against the KB cell-structure test system (Fiebig *et al.,* 1985). Bhattacharya *et al.* (1989) isolated 2 coumarins *viz.,* 3 (1', 1'-Dimothyl allyl) xanthyletin and scopoletin from the bark of curry leaf.

Drumstick

Drumstick (*Moringa* spp.) is the most versatile tree with rich horticultural potential. Its vitamin rich, mineral packed nutritious pods are valued for culinary preparations. Along with pods and leaves, all parts of the tree are considered medicinal and are used in the treatment of ascites, snakebites and as a cardiac stimulant. Roots are used as an ointment of scurvy, catarrh, wounds and as an emetic, seed oil as edible oil; for lightening and in cosmetics, and the stem gum exudates in calico printing and medicines. The wood has suitable characteristics for pulp, paper and cellophane and textile production (Nautiyal and Venkataraman, 1987). In Guatemala, drumstick is used to treat many disorders, in particular infectious diseases of the skin, digestive system and respiratory tracts. The antimicrobial activities of the leaves, roots, bark and seeds have been investigated *in vitro* against bacteria, yeast, dermophytes and helminthes pathogenic to man. Vitamin C content of pods and leaves of 11 clones of *Moringa oleifera* and 2 clones of *M. concanensis* was in the range of 55-143 mg/100 g (Verma *et al.,* 1976). Leaves have high protein value as well as low fibre content which made it suitable for the extraction of leaf protein for use as low cost source of proteins (Awasthi and Tandon, 1988). Its tender green leaves are good source of neutral detergent fibre (NDF) and acid detergent fibre (ADF). Leaves are rich source of major and trace elements. Calcium, phosphorus and zinc vary from 0.9 to 2.9, 0.4 to 1.2 per cent and 17.5 to 46.2 mg/kg, respectively. It contains most flatus factors (sucrose + raffinose + stachyose) 5.6 per cent (Gupta *et al.,* 1989). Drumstick leaves are rich in sulphur (175.35 mg/100 g) and other micronutrients like iron, boron and zinc. Boron content in drumstick is around 41.63 ppm (Shingade and Chavan, 1996). Soluble oxalate content (as percentage of total oxalates) was 28, Ca: oxalate ratio was 3.6 (Meena *et al.,* 1987). Saluja *et al.* (1978) isolated 4-hydroxymellein from drumstick stems *viz.* Vanillin, ~-sistostenone, octacosanoic acid and ~-sistosterol. Seeds of *Moringa oleifera* and *M. aptera* have oil content of about 47 per cent (Ibrahim *et al.,* 1974). Seeds contain fatty acid which is rich in oleic acid (Verma *et al.,* 1976). Amino acid extraction and HPLC analysis performed on homogenates of the edible parts of drumstick showed glutamic acid 200g/100 g sample (Mallorca *et al.,* 1992).In Northern India, its flowers are used in preparation of curry and pickles. The tree from which they isolated this substance is *"Moringa pterygosperma,"* (now regarded as an archaic designation for *M. oleifera*). Although others were to show that pterygospermin and extracts of the Moringa plants from which it was isolated were antibacterial against a variety of microbes, the identity of Pterygospermin has since been challenged (Eilert *et al.,* 1981) as an artifact of isolation or structural determination.

The widespread combination of diuretic along with lipid and blood pressure lowering constituents make this plant highly useful in cardiovascular disorders. Moringa leaf juice is known to have a stabilizing effect on blood pressure. "Thiocarbamate glycosides" and isolated from moringa leaves, which are found responsible for the blood pressure lowering effect (Faizi *et al.,* 1994, 1995).

Most of these compounds, bearing thiocarbamate, carbamate or nitrile groups, are fully cetylated glycosides, which are very rare in nature (Faizi *et al.,* 1995). Bioassay guided fractionation of the active ethanol extract of moringa leaves led to the isolation of our pure compounds, niazinin A, niazinin B, niazimicin and niazinin, which showed a blood pressure lowering effect.

Another study on the ethanol and aqueous extracts of whole moringa pods and its parts *i.e.,* coat, pulp and seed revealed that the blood pressure lowering effect of seed was more pronounced with comparable results in both ethanol and water extracts indicating that the activity is widely distributed (Faizi *et al.,* 1998). Activity-directed fractionation of the ethanol extract of pods of *M. oleifera* has led to the isolation of thiocarbamate and isothiocyanate glycosides, which are known to be the hypotensive principles (Faizi *et al.,* 1995). Methyl phydroxybenzoate and ~ sitosterol investigated in the pods of *M oleifera* have also shown promising hypotensive activity (Faizi *et al.,* 1998).

Moringa roots, leaves, flowers, gum and the aqueous infusion of seeds possess diuretic activity (Morton, 1991; Caceres *et al.,*1992) and such diuretic components are likely to play a complementary role in the overall blood pressure lowering effect of this plant. The crude extract of moringa leaves has a significant cholesterol lowering action in the serum of high fat diet, which might be attributed to presence of a bioactive phytoconstituent, ~-sitosterol (Ghasi *et al.*, 2000)

Anti-spasmodic, Anti-ulcer and Hepatoprotective Activities

M. oleifera roots possess anti-spasmodic activity (Caceres *et al.*, 1992). Moringa leaves were extensively studied pharmacologically and was found that the ethanol extract and its constituents exhibit anti-spasmodic effects (Dangi *et al.*, 2002).

The anti-spasmodic activity of the ethanol extract of *M. oleifera* leaves is attributed to the presence of 4-(L-rhamnosyloxy) benzyl]- o-methyl thiocarbamate, which forms the basis for its traditional use in diarrhoea. Moreover, spasmolytic activity exhibited by different constituents provides pharmacological basis for the traditional uses of this plant in intestinal motility disorder. The methanol fraction of *M. oleifera* leaf extract showed anti-ulcerogenic and hepatoprotective effects (Pal *et al.*, 1995). Aqueous leaf extracts also showed anti-ulcer effect indicating that the anti- ulcer component is widely distributed in this plant.

Moringa roots also possess hepatoprotective activity Ruckmani *et al.*, 1998). The aqueous and alcohol extracts from moringa flowers have a significant hepatoprotective effect, which may be due to presence of quercetin, a well known flavonoid with hepatoprotective activity.

The bark extracts possess antifungal activity (Bhatnagar *et al.*, 1961), while the juice from the stem bark showed antibacterial effect against *Staphylococcus aureus*. The fresh leaf juice inhibits the growth of microorganisms (*Pseudomonas aeruginosa*, *Staphylococcus aureus*) and pathogenic to human (Caceres *et al.*, 1992).

Other Diverse Pharmaceutical Activities

M. oleifera exhibits other diverse activities. Aqueous leaf extracts regulate "thyroid hormone" and can be used to treat hyperthyroidism and exhibit an antioxidant effect (Pal *et al.*, 1995; Tahiliani and Kar, 2000). Moringa leaves are effective for regulation of thyroid hormone status (Tahiliani and Kar, 2000). Extract of *M. oleifera* leaves conferred significant radiation protection to the bone marrow chromosomes in mice (Rao *et al.*, 2001). A recent report showed that *M. oleifera* leaf may be applicable as a prophylactic or therapeutic anti-HSV (Herpes simplex virus type-I) medicine and may be effective against the acyclovir-resistant variant (Lipipun *et al.*, 2003).

The flowers and leaves also possess high medicinal value with anthelmintic activity (Bhattacharya *et al.*, 1982). An infusion of leaf juice was shown to reduce glucose levels in rabbits (Makonnen *et al.*, 1997). *M. oleifera* is coming to the forefront as a result of scientific evidence that moringa is an important source of naturally occurring phytochemicals and this provides a basis for future viable developments.

Bulb Crop

Onion

Onions (*Allium cepa*) are rich source of amino acids and glutamyl peptide, anthocyanins, flavonols and phenolics. Non-structural carbohydrates consisting of free sugars, trisaccharides and fructans contribute the major portion the dry weight of onions. High dry matter onion cultivars have reduced glucose and fructose contents and much higher fructan levels than varieties with low dry- matter contents. Onion is rich in sulfur containing compounds. The enzyme aliinase hydrolyses s- alkenyl eystein sulfoxides and produces pyruvate, ammonia and many volatile sulfur compounds are associated with flavour and odour of onion. Onions contain primarily the s-(l-propenyl), propyl and to a lesser degree, methyl alliins. The typical flavour of onion is due to presence of propyl and l-propenyl containing allicins and di-and tri-sulfides. Onion contains an acrid volatile oil (0.05 per cent) with a pungent smell. The oil is rich in sulfer and contains a variety of aliphatic disulfide including allyl or propenyl propyl disulfide, dipropyl disulfide, methyl propyl disulfide and their trisulfides. However, the chief component is propenyl propyl disulfide, an isomer of allyl propyl disulfide. The precursors of onion oil are the cysteine sulfoxide derivative of amino acids known as alliins. Onion contains an enzyme called alliinase, which converts alliins to disulfide oxides. These oxides are allicin type compounds.

Onion bulb and leaves are essential ingredients of vegetarian and non-vegetarian diets having medicinal and nutritive value. The World Health Organization (WHO) supports the use of onions for the treatment of poor appetite and to prevent atherosclerosis. Onion extracts are recognized by WHO for providing relief in the treatment of coughs and colds, asthma and bronchitis. Wide range of claims are made for the effectiveness of onions against conditions ranging from common cold to heart disease, diabetes, osteoporosis and other diseases. They contain chemical compounds with anti-inflammatory, anticholesterol, anticancer, and antioxidant properties, such as quercetin. Preliminary studies show increased consumption of onions, reduce the risk of head and neck cancers.

The raw bulb is eaten to improve eyesight; taken orally for gastronomic purposes, arnenorrhea, menstrual and uterine pains. These hot bulbs are applied externally to treat furuncles. Fresh bulb if taken orally along with leaf exrtract of *Adhatoda vasica* in honey, helps to cure tuberculosis. The fresh bulb, eaten raw with salt helps in relieving stomachache. If taken orally it acts as a sedative, blood purifier, expectorant, carminative tonic, antipyretic, hypotensive and diuretic. Hot water extract of the fresh bulb if taken orally, acts as an aphrodisiac for both men and women. It is used to regulate blood pressure (hypertension), treat inflammation, diabetes, urinary problems, dysentery, fever, dropsy, colic, renal and biliary calculi, catarrh, chronic bronchitis, scurvy, body heat, epilepsy, hysterical fits, nosebleed, jaundice, unclear vision, spleen enlargement, rheumatic pain and strangury and to induce miscarriage and dieresis. Extract is also used externally for acne treatment. The dried/roasted bulb is used either orally as a contraceptive, antiphlogistic, or intra-vaginally to induce menses. It is applied externally, as an emmenagogue, in the form of a pessary in Unani medicine. Dried bulb is also used to treat infections. Hot water extract of the dried bulb is taken orally for diabetes, dropsy colic, catarrh, chronic bronchitis, scurvy, epileptic and hysterical fits, epistasis, jaundice, enlarged spleen, rheumatic pain and strangury. It is applied externally for wounds, ulcers, bruises, sores, skin diseases, irritations, inflammation, eruptions, erysipelas and burns (Nadkarni, 1927). Fresh bulb juice is used externally as an anti-inflammatory agent on insect bites and for bronchitis, applied opthalmically to improve eyesight; orally with sugar is given to children for worms, mixed with the juice of *Achyranthues bidentata* leaves taken orally every 2 hours for cholera and aurally for earache (juice warmed with coconut oil is dropped in the ear). Butanol extract of the bulb is taken orally for asthma, and also used as an expectorant and diuretic.

Wine extract of the fresh bulb is taken orally for renal function and urinary disease; externally it is used for boils and whitlow. Butanol extract of the dried bulb is taken orally to treat high blood pressure. Onion in the diet may play a part in preventing heart diseases and other ailments and the medicinal uses of onions have been reviewed by Hanley and Fenwick (1985) and Augusti (1990). The hypoglycemic *i.e.* blood sugar lowering, effects of onion extract have long been recognized. Petroleum ether soluble fraction of the diethyl ether onion extract contained the hyperglycemic fractions, whereas, insoluble component contained the hypoglycemic fraction. In both the fractions the disulfides are the major sulphur compounds as they inactivate insulin and the hypoglycemic action may be attributed to their unsaturated disulfides and related components. Another anti- hypoglycemic agent in onions has been identified, *viz.* diphenylamine. This is the first non-sulphur compound in onion claimed to have hypoglycemic action. The lipid lowering effects of onion extracts are also reported. The hypolipidemic effects of onion are attributed to their sulphur containing compounds. Onions contain anthocyanins and the flavonoids quercetin and kaempferol. Quercetin is the major flavonoid of interest in onions. Mechanisms of action include free radical scavenging, chelation of transition metal ions, and inhibition of oxidases such as lipoxygenase. The homogenate fresh onion and hot water extract of fresh aerial parts of *Allium cepa* exhibit significant inhibition of lipid peroxidation. The antioxidative effects of consumption of onions are associated with a reduced risk of neurodegenerative disorders, many forms of cancer, cataract formation, ulcer development and prevention of cardiovascular diseases by inhibition of lipid peroxidation and lowering of low density lipoprotein (LDL) cholesterol levels. Another antioxidant effect of onions and their extracts includes the reduction of rancidity in cooked meat.

Garlic (*Allium sativum* L.)

It is mostly used for culinary purposes and is one of the most popular and widely used flavourings. People all over the world use it as a condiment for different food items. In India and other countries, it is used in several food preparations like chutneys, pickles, curry powders, curried vegetables, meat preparations, tomato ketchup etc. The garlic powder has also earned significant popularity in the recent times. Garlic cloves contain an enzyme alliinase, which is released when they are crushed. Allinase acts on SACS and produce allicin, ammonia and pyruvic acid. A powerful antibiotic and antifungal compound (phytoncide). However, it is of limited use for oral consumption due to poor bioavailability. It also contains alliin, ajoene, enzymes, vitamin B, minerals and flavonoids. Garlic contains about 62.8 per cent water, 6.3 per cent protein, 0.1 per cent fat, 29 per cent carbohydrate including 3. 9 per cent sucrose, traces of Ca, Fe, Zn and phosphate salts, and small amount of vitamins thiamin, riboflavin, niacin and ascorbic acid. Garlic can yield 0.06 to 0.1 per cent essential oil made up of mainly diallyl disulfide and small amounts of allylpropyl disulfide. This oil is formed from the decomposition product of alliin called S-allyl cysteine sulfoxide (SACS). Garlic is used in making remedies for various ailments and physiological disorders and is one of the oldest medicines in the world. In Ayurveda, garlic is one of the most effective antimicrobial herbs, as it has anti-bacterial, anti-fungal, anti-viral, anthelmintic and antiseptic properties. It has healing capacity and effectiveness against cholera, as well. It has useful anti-bacterial action against *Eberthella typhosa, Escherichia coli,* acid fast *Bacilli, Aerobacter aerogenes, Staphylococcus aureus* etc.

According to Ayurveda, Garlic is useful for increasing sexual energy and in combating impotence. It can help kill parasites like hookworms and pinworms, as well. Several benefits of garlic are described in Unani medicine. According to Unani medicine, garlic is used as carminative and can also act as a gastric stimulant. It aids in digestion and absorption of food and is also given in flatulence. In modern Allopathic treatment also, garlic is used in a number of patented medicines and other preparations. The residue of garlic, obtained by alcoholic extraction and distillation, contains a bacteriostatic and bactericidal substance identified as 'allyl disulfide oxide'. It is used traditionally for ages to treat a wide array of diseases, namely, respiratory infections, ulcers, diarrhea and skin infections (Fenwick and Hanley, 1985). Reuter *et al.* (1996) reported garlic as a plant with antibiotic, anticancer, antioxidant, immunomodulatory, anti-inflammatory, hypoglycemic and cardiovascular protecting effects. Garlic is very rich in aromatic oils, which enhance digestion and positively influence respiratory system being inhaled into air sacs and lungs of birds. It was found that garlic has strong antioxidative effects (Gardzielewska *et al.,* 2003). Garlic extract and/or garlic components prevent chemically induced tumors or acute toxic effects of chemicals. The chemo-preventive potential of garlic is attributed to presence of several bioactive organosulfur compounds. Theses compounds might act as antioxidants (Fanelli *et al.,* 1998; Siegers *et al.,* 1999). The antioxidative stress properties of garlic might result from the contributions of its sulfur component in different steps and not necessarily from the contribution of only one of them (Fanelli *et al.,* 1998). Garlic also has strong antimicrobial action (Iwalokun *et al.,* 2004; Gbenga *et al.,* 2009). Allicin and its derivatives have larvicidal and bacteriostatic action against both Gram positive or Gram negative organisms as well as fungi such as *Candida albicans* and viruses including influenza viruses (Chang and Cheong, 2008). *Allium sativum* taken at a low dose may have therapeutic potentials against gastric ulcers associated with *H. pylori* infection (Adeniyi *et al.,* 2006). Garlic extracts have significant inhibitory effects against microorganisms associated with dental caries, Masaadeh *et al.* (2006).

Garlic is used as an antidote to snake and scorpion bites and is also a very good medicine for running cold and saliva formation. As herbal medicine, it is also used to treat diseases like chronic bronchitis, respiratory catarrh, whooping cough, bronchitic asthma, influenza and other health problems. It can also fight infection, reduce cholesterol, protect against heart diseases and stroke, control diabetes and prevent cancer. It prevents blood clots and destroys plaque preventing atherosclerosis and reduce the chances of stroke and heart attacks. Its extract is also used in homeopathy medicines. The inhalation of garlic oil or garlic juice is commonly recommended in pulmonary tuberculosis, rheumatism and impotence. The Garlic juice is extensively used for treating various ailments of stomach, and also used as a rubefacient in skin diseases. It is used as eardrop in ear aches, as well. The garlic juice can also be used against duodenal ulcers, after diluting with water. Garlic is used for killing bacteria which cause tuberculosis and it reduces blood sugar level. It improves the immune functions, fights against chronic diarrhea etc. and it also heals open pores, activates and stimulates blood circulation, and improves hair growth. The bioactive substances like alliin, allicin, gamma-glutamylcysteine, thiosulfinates etc. in garlic can help in fighting against bacterial, parasitic and fungal infections. Garlic can be effective in treatment for congestion in the respiratory system, if taken along with honey.

Other Vegetables

Okra

Okra or lady's finger (*Abelmoschus esculantus*) is a popular vegetable. Its tender fruit contains Vitamins A, B and C (13 mg). It is rich in iodine, calcium, potassium and other mineral matters. It has 1.9 per cent protein. 6.4 per cent carbohydrate. 0.2 per cent fat and 1.2 per cent fibre. Sucrose is present in the developing and dry seeds of okra at all stages of development. The raffinose family of oligosaccharides is present in mature and dry seeds, while free glucose and fructose were detected at most stages. Dry seeds contained raffinose and stachyose sugars of oligosaccharides family. Mucilage in okra fruits is acidic polysaccharides. Hydrolysis of the mucilage gave polysaccharide composed of galaciuronic and glucuronic acids and minor contents of galactose, rhamnose, glucose and arabinose. The maximum protein content was 2.08 per cent in pods and 2.09 per cent in seeds. Mature dry okra seeds contain 20.58 per cent protein. Okra seeds contain 14-19 per cent oil having good proportion of linoleic acid. Okra fruit contains high amount of calcium, phosphorus, sodium, sulphur and nitrogen in the developing seeds, embryo, seed coat and fruit wall. Embryo was consistently rich in phosphorus and sulfur. Young leaves which are cooked as 'spinach' by the Africans (Busson) also act as a diuretic, and abortifacient as well as having gastric ulcer and wound healing properties (Weniger and Robineau, 1988). The fruit mucilage is of interest as it could replace blood plasma (Benjamin *et al.*, 1951), reduce fluid friction in turbulent flow (Castro and Neuwirth, 1971), and stabilize foams (Woolfe *et al.*, 1977) and suspensions (Wahi *et al.*, 1985). It also has medicinal properties as an emollient, laxative and expectorant (Muresan and Popescu, 1993). Okra seeds are a source of oil, protein and a coffee substitute (Martin, 1982), while the seed powder is used as a substitute for aluminium salts in water purification (Vaidya and Nanoti, 1989).

Toxic Metabolites and Anti-nutritional Compounds in Vegetables

A major factor, which is restricting the used of vegetables in animal nutrition, is the presence of a diverse array of toxic substances and anti-nutritional compounds *viz.* alkenyl benzenes, s-alk (en) yl cysteine sulphoxides, biogenic amines, cinnamic acids, cyanogenic glycosides, cucurbitacins, flavonoids, glycoalkaloids, glucosinolates, lectins, hydrazines, isoflavones, lathyrogens, lignans, raffinose family of oligosaccharides, oxalate, pyrrolizidine alkaloids, furanocoumarins, quinolizidine alkaloids, sesquiterpene lactones, saponins, trypsin inhibitors, xanthine alkaloids, mineral toxins, resinoids, tannins, phenols, non-amino organic acids, alcohols and terpenoids restriet used in nutrition (Table 6.5). These anti-nutritional compounds are capable of inducing adverse effects ranging from neurological disorders, kidney stones, elevated blood pressure, gastric disorders and even death. Significant advances have been made in recent years to establish the nature of these compounds and to assess their toxic effects in animals and human beings.

Table 6.5: Anti-nutritional Compounds/Toxicants of Vegetables

Vegetables	Toxic Compounds	Adverse Effects
Carrot	Carota-toxin (Polyacetylenic alcohol)	Neurotoxic symptoms
Lettuce	Nitrates, Alkaloids	Methemoglobinaemia
Brassica (Cruciferous vegetables)	Glucosinolates, Choline-esterase inhibitor, S-methyl cysteine sulfoxides	Goiter, Digestive disorders
Beets, Spinach	Oxalates, Nitrates, Phytate, Tannins, Saponins, Nitrosamine	Methemoglobinaemia reduces bio-availability of certain minerals such as calcium, iron, and zinc. Careinogenic
Sweet potato	Ipomeamarone	Enzyme inhibitors
Watermelon	Serotonin	Elevates blood pressure
Pumpkin and squashes	Choline-esterase inhibitor	Neurotoxic
Legumes (Vegetables)	Lectins, Cyanogenic glucosides, Haemagglutinins, Trypsin, Amylase, Glucose-6-P-dehydrogenase inhibitor, Compounds having anti-vitamin properties (Vitamin A, E and D)	Allergens
Asparagus	Saponins, Choline-esterase inhibotor	Neurotoxic
Solanaceous vegetables	Alkaloids	Birth defect, protease inhibitor
Potato	Solanine and chaconine	Invertase inhibitor
Tomato	Tomatine	Gastric discomfort
Pungent pepper (chillies)	Capaicin	Skin irritation, gastric disorders
Parsley, Celery	Psoralens, Terpenoids, Alkaloids, Choline-esterase inhibitor	Dernatitis

Phytate

Phytate is hexaphosphate of inositol. These phytates bind iron, zinc, calcium and magnesium. In presence of Ca^{2+} and $Mg2+$ it forms insoluble complexes with iron and thus reduces iron bioavailability. On germination of the grains, the phytate content reduces due to enzymatic breakdown of phytate which improves iron availability.

Tannins

These are condensed polyphenolic compounds present in high amount in seed coat of most legumes and certain vegetables and fruits. Tannins bind with iron irreversibly and interfere with iron absorption.

Oxalates

Green leaf vegetables like spinach, Amaranths and some legumes are rich source of oxalates and are known to interfere with calcium absorption by forming insoluble salts with calcium (Table 6.6). Complexation of calcium by oxalate may result in calcium deficiency. Dietary oxalates can be absorbed and contribute to increased exeretion of oxalates in urine. High oxalate excretion may predispose to oxalate crystals leading to urinary stones. Thus, stone patients are advised to avoid high oxalate containing foods. However, vegetables rich in calcium *e.g.* fenugreek, colocasia, sweet potato, ridge gourd and snake gourd rich in insoluble calcium oxalate are not likely to be harmful.

Table 6.6: Nitrate and Oxalate Content in Vegetables

Vegetables	Nitrate Content mg/kg fresh wt.	Oxalic Acid Content mg/100 g
Cabbage	390	3
Lettuce	1200	–
Spinach	2100	658
Beet root	1500	40
Radish (white)	1100	9
Broccoli	410	–
Brussels sprout	7-12	4
Cauliflower	37	19
Celery leaves	1200	37
Amaranths	–	772
Chillies	–	67
Tomato ripe	–	4
Drum stick	–	101

Maximum oxalate concentration is found in Amaranths and Spinach leaf (44-157 g/kg dry wt.). A substantial portion of it (39-48 percent) may be in water soluble form. The most obvious method of reducing the risk of oxalate poisoning is to cook the leaves which reduce the water soluble oxalates.

Nitrates

Vegetables are the single largest source of nitrates in human diet (Table 6.6). Nitrate accumulation can have serious deleterious effects. Within the gastrointestinal tract nitrate (NO_3) is reduced to nitrite (NO_2) which is absorbed into the blood stream where it binds with haemoglobin, oxidizing ferrous iron to Ferric iron to form methaemoglobin. This form of haemoglobin complex is incapable of O_2 transport. The result is anoxia, specifically referred to as methaemoglobinemia. Organically grown vegetables have less nitrate content. As leaf vegetables are rich sources of nitrate, consumer should go for organically grown leaf vegetables.

Glucosinolates

Glucosinolates are thioglucosides which were widely distributed in plants, particularly among members of cruciferae and leaf vegetables. Over 100 different glucosinolates are identified. In cabbage, the main glucosinolates are Sinigrin, progitrin and glucobrassicin. These three glucosinolates also predominate in a variety of other brassica vegetables including Brussels sprouts, broccoli and cauliflower. Upon hydrolysis by plant microbial thioglucosidase (myrosinase) isothiocyanate and nitrites are released. The isothiocyanate residues possess goitrogenic properties which inhibit iodine uptake by thyroid glands.

Conclusions

Vegetable play a significant and important role in human nutrition providing not only essential nutrients but also other compounds for health promotion and disease prevention. Under nutrition is a serious problem in under developing and developing countries. Forty percent of the world's malnourished children and 35 per cent of the developing world's low birth weight infants live in the India. Micronutrient deficiencies often go unnoticed despite their insidious effects on immune system and growth cognitive development. Micronutrient deficiencies are referred to as "hidden hunger" and include iodine deficiency disorder, iron deficiency anemia, and vitamin A deficiency. Most people subsist on cereals based starchy staple-diets lacking in diversity, which contribute to micronutrient deficiency and result in severe diseases, especially in young, pregnant women and children. The most popular approaches to address malnutrition are supplementation and food-based strategies. Vegetables, the cheapest source of vitamins and minerals, are high value food sources for the poorest families and can be incorporated in home gardens.The content and role of nutrients and bioactive substances in vegetable are discussed specially in the context of reduced risk of cardiovascular disease, cancer, diabetes, skin disease, cataracts and age related functional decline. The health effects of nutrients and consumption of bioactive substances depend both on their intake and bioavailability. To establish evidence for the effects of nutrients and bioactive substances on human health and to better identify which nutrients provide the greatest effectiveness in disease prevention. It is first of all essential to determine the nature and distribution of these nutrients in our diet.

References

1.   Acikgoz, F.E. and Deveci, M. (2011). Comparative analysis of vitamin C, crude protein, elemental nitrogen and mineral content of canola greens (*Brassica napus* L.) and kale (*Brassica oleracea* var. *acephala*). *African J. Biotech*. 10(83): 19385-19391.

2. Adegoke, G. O., Allamu, A. E. and Akingbala, J. O. (1976). Influence of sundrying on the chemical composition, aflatoxin content and fungal count of two pepper varieties- *Capsicum annuum* and *C. frutescens. Plant Foods for Hum.Nutr.* 49: 113-117.

3. Adeniyi, B.A., Oluwole, F.S. and Anyiam, F.M. (2006). Antimicrobial and antiulcer activities of methanol extract of *Allium sativum* on *Helicobacter pylori. J. Boil. Sci.* 6: 521-526.

4. Adesina, S.K., Olatunji, O.A., Berg-enthal, D. and Reisch, J. (1988). New biogenetically significant constituents of *Clausena anisata* and *Murraya koenigii. Pharmazie.* 43(3): 221-222.

5. Aguiyi, J.C., Uguru, M.O., Johnson, P.B. and Obi, C.I. (1997). Effect of *Mucuna pruriens* seed extract on smooth and skeletal muscle preparations. *Fitoterpia.* 68(4): 366 - 370.

6. Ahmed, S. and Beigh, S.H. (2009). Ascorbic acid, carotenoids, total phenolic content and antioxidant activity of various genotypes of *Brassica oleracea. J. Med. and Biol. Sci.* 3 (1): 1-8.

7. Akanitapichat, P., Phraibung, K., Nuchklang, K. and Prompitakkul, S. (2010). Antioxidant and hepatoprotective activities of five eggplant varieties. *Food Chem. Toxicol.* 48: 3017–3021.

8. Alibas, I. (2009). Microwave, vacuum and air drying characteristics of collard leave. *Drying Technology* 27: 1266-1273.

9. Amin, K.M.Y., Kham, M.N., Zillur - Rehmans and Khan, N.A. (1996). Sexual function improving affected *Mucuna pruriens* in sexually normal male rats. *Fitoterpia.* 67 (1): 53 - 58.

10. Amjad, I., Khalil, I.A., Nadia, A., Khan, M.S. (2006). Nutritional quality of important food legumes. *Food Chem.* 97(2): 331-335.

11. Andrews, J. (1995). *Peppers: The Domesticated Capsicums*, University of Texas Press, Austin. 186 pp.

12. Anonymous. (1991). Komunikasi pene-litian dan pengembangan Tanaman Industri. 8: 32 - 35.

13. Anu, A. and Peter, K.V. (2000). The chemistry of paprika. *Indian Spices.* **37**(2): 15–18.

14. Arrgusti, K.T. (1990). In: *Onion and Allied Crops*, Vol.III, (Eds. J.L.brewster and H.D. Rabinowitch). Boca Raton, Florida, CRC Press, pp 93-108.

15. Au, T.K., Collins, R.A., Lam, T.L, Ng, T.B., Fong, W.P. and Wan D.C. (2000). The plant ribosome inactivating proteins luffin and saporin are potent inhibitors of HIV-1 integrase. *FEBS Letters* 471: 169–172.

16. Augusti, K.T. and Sheela, C.G. (1996). Antiperoxide effect of S-allyl cysteine sulfoxide, an insulin secretagogue, in diabetic rats. *Experientia.* 52: 115–120.

17. Aung, L.H., Fouse, D.C. and Kushad, M. (1991). The distribution of carbohydrates in the organs of *Sechium edule* Sw. *J. Hort. Sci.* 66(2): 253-257.

18. Awasthi, C.P. and Tandon, P.K. (1988). Biochemical composition of some unconventional Indian leafy vegetables. *Narendra Deva Journal of Agricultural Research.* 3(2): 161-164.

19. Azevedo, L., Lima, P. A., Gomes, J. C., Stringheta, P. C., Ribeiro, D. A. and Salvadori, D. M. F. (2007). Differential response related to genotoxicity between eggplant (*Solanum melanogena*) skin aqueous extract and its main purified anthocyanin (delphinidin) *in vivo. Food Chem. Toxicol.* 45: 852–858.

20. Bangash, J. A., Arif, M., Khan, F., Khan, F., Rahman A., and Hussain, I. (2011). Proximate composition, minerals and vitamins content of selected vegetables grown in Peshawar. *J. Chem. Soc. Pak.* 33(1): 118-122.

21. Bañuelos, G.S. and Meek, D. W. (1989). Selenium accumulation in selected vegetables. *J. Plant Nutri.* 12: 1255–1272.

22. Baruah, A.B. and Goswami, B.C. (1979). Carotenoids of *Cephalandra indica* (*Coccinia indica*). *Current Sci.* 48(14): 630-632.

23. Benjamin, BH, Ihrig, KH and Roth, DA. (1951). The use of okra as a plasma replacement. *Rev Canad Biol,* 10: 215 -21.

24. Bhatnagar, S.S., Santapau, H., Desai, J.D.H., Yellore, S. and Rao, T.N.S. (1961). Biological activity of Indian medicinal plants. Part 1.Antibacterial, antitubercular and antifungal action. *Indian J.Med. Res.* 49: 799–805.

25. Bhattacharya, L., Mukhopadhyay, M. and Chakraborty, D.P. (1989). 3(1', 1'-Dimethyl allyl) Xanthyletin and Scopo-letin, two plant growth inhibitors from *Murraya koenigii* Spreng. *Plant Physiol. Biochem.* 16(1): 23-26.

26. Bhattacharya, L., Roy, S.K. and Chak-raborty, D.P. (1982). Structure of the carbozole alkaloid isomurrayazoline from *Murraya koenigii* (Stembark). *Phytochemistry.* 21(9): 2432-2433.

27. Bone, R. A., Landrum, J. T., Dixon, Z., Chen, Y. and Llerena, C. M. (2000). Lutein and zeaxanthin in the eyes, serum and diet of human subjects. *Experimental Eye Research.* 71(3): 239–245.

28. Boonkerd, T., Songkhla, B. Na and Thephuttee, W. (1993). *Coccinia grandis* (L.) Voigt. In PROSEA. PlantResources of South- East Asia 8. Vegetables, Pudoc Scientific Publishers, Wageningen. pp. 150-151.

29. Bosland, P.W. (1994). Chilli: History, cultivation, and uses. *In* G. Charalambous, (*ed.*), *Spices, Herbs and Edible Fungi.* Developments in Food Science Vol. 34. Elsevier, Amsterdam, pp. 347–366.

30. Bosland, P.W. and Votava, EJ. (2000). *Peppers: Vegetable and Spice Capsicums. Crop Production Science in Horticulture.* CAB International Publishing, Wallingford, England, UK. 204 pp.

31. Brat, P., George, S., Bellamy, A., Du Chaffaut, L., Scalbert, A., Mennen, L., Arnault, M. and Amiot, M.J. (2006). Daily polyphenol intake in France from fruit and vegetables. *J. Nutri.* 136: 2368–2373.

32. Bridle, P., and Timberlake, C. F. (1997). Anthocyanins as natural food colours— Selected aspects. *Food Chem.* 58(1): 103–109.

33. Brinkworth, R. I., Chen, J., Leuenberger, P. M., Freiburghaus, A. U. and Follath, F. (1992). Flavones are inhibitors of HIV proteinase. Biochem.Biophys. Res. Commun., 188, 631-637.

34. Bueno, R. R., Moura, S.and Fonseca, O. M. (1970). Preliminary studies on the pharmacology of *Sechium edule* leaves extracts. *An. Acad. Bras. Cienc.* 40: 285–289.

35. Buttery, R.C., Selfort, R.M., Gudanini, D.G. and Ling, L.C. (1969). Characterization of some volatile constituents of bell peppers. *J. Agric. Food Chem.* 17: 1322–1327.

36. Cabello-Hurtado, F., Gicquel, M., and Esnault, M. (2012). Evaluation of the antioxidant potential of cauliflower (*Brassica oleracea*) from a glucosinolate content perspective. *Food Chem.* 132: 1003–1009.

37. Caceres, A., Saravia, A., Rizzo, S., Zabala, L., Leon, E.D. and Nave, F. (1992). Pharmacologic properties of *Moringa oleifera*: 2: Screening for antispasmodic, anti-inammatory and diuretic activity. *J. Ethnopharmacol.* 36: 233–237.

38. Cai, R., Hettiarachchy, N.S. and Jalaluddin, M. (2003). High-performance liquid chromatography determination of phenolic constituents in 17 varieties of cowpeas. J. Agric. Food Chem. 51(6): 1623-1627.

39. Campbell, L.D., Slominski, B.A., Nugon-Baudon, L., Rabot, S., Lory, S., Quinsac, A., Krouti, M. and Ribaillier, D. (1995). Studies on intestinal tract glucosinolate content, xenobiotic metabolizing enzymes and thyroid status in germ-free and conventional rats fed rapeseed meal. *Proc. 9th Int. Rapeseed Congr.* 1: 209–211.

40. Carlson, D.G., Daxenbichler, M.E. and Tookey, H.L. (1987). Glucosinolates in turnip tops and roots: cultivars grown for greens and / or roots. *J. Am. Hort. Sci.* 112: 179-183.

41. Castro, W. and Neuwirth, J. (1971). Reducing fluid friction with okra. *Chem Tech*, nov: 697-701.

42. Cerny, K. and Addy, H. A. (1973). The winged bean (*Psophocarpus tetragonolobus* Desv.) in the treatment of Kwashiorkor. *British J. Nutri.* 29: 105 - 107.

43. Chakrabarti, A.K. (2003). Importance of vegetables. In: *Vegetables, Tubercrops and Spices* (Eds. S Thamburaj and N Singh). DIPA, ICAR, New Delhi pp.1-9.

44. Chakraborty, D.P., Bhattacharya, P., Islam, A. and Roy, S. (1974). Structure of Murrayacinine: a new carbozole alkaloid from *Murraya koenigii* Spreng. *Chemistry and Industry*, 165-166.

45. Chakraborty, D.P., Bhattacharya, P., Roy, S., Bhattacharya, S.P. and Biswas, A.K. (1978). Structure and synthesis of mukonine a new carbozole alkaloid from *Murraya koenigii* (Stembark.). *Phytochemistry*, 17(4): 831-835.

46. Chakraborty, D.P., Ganguly, S.N., Maji, P.M., Mitra, A.R., Das, K.C. and Weinstein B. (1973). Murra-yazp; omome: a carbozole alkaloid from *Murraya koenigii* Spreng. *Chemistry and Industry*. 7: 322-323.

47. Chakravarty, H.L. (1982). Fascicles of Flora of India. Botanical Survey of India, Calcutta. Vashista, P.C. (1974). *Taxonomy of Angiosperms*. P.B.M. Press, New Delhi, India.

48. Chang, K.J. and Cheong, S.H. (2008). Volatile organosulfur and nutrient compounds from garlic by cultivating areas and processing methods. *Fed. Am. Soc. Exp. Bio. J.*, 22: 1108-1112.

49. Chaudhury, B.K. and Chakraborty, D.P. (1971). Mukoic acid, the first carbazole carboxylic acid from a plant source. *Phytochemistry*. 10(8): 1961-1970.

50. Choudhary, B. (1967). *Vegetables*. National Book Trust, India, New Delhi.

51. Chun, O. K., Kim, D., Smith, N., Schroeder, D., Han, J.T. and Lee, C.Y. (2005). Daily consumption of phenolics and total antioxidant capacity from fruit and vegetables in American diet. *J. Sci. Food Agric*. 85: 1715-1724.

52. Cichewicz, R.H. and Thorpe, P.A. (1996. The antimicrobial properties of chile peppers (*Capsicum* species) and their uses in Mayan medicine. *J. Ethnopharmacology*. **52**: 61–70.

53. Collera-Zúñiga, O., Jiménez, F.G. and Gordillo, R.M. (2005. Comparative study of carotenoid composition in three Mexican varieties of *Capsicum annuum* L. *Food Chem*. 90: 109–114.

54. Collins, JK., Wu, G., Perkins-Veazie, P., Spears, K., Claypool, PL., Baker, RA. and Clevidence, B.A. (2007). Watermelon consumption increases plasma arginine concentrations in adults. *Nutrition,* 23: 261-266.

55. Cornwell, T., Cohick, W. and Raskin, I. (2004). Dietary phytoestrogens and health. Phytochemistry. 65: 995-1016.

56. Dangi, S.Y., Jolly, C.I. and Narayana, S. (2002). Antihypertensive activity of the total alkaloids from the leaves of *Moringa oleifera*. *Pharm Biol.*, 40: 144–148.

57. Deokule, S.S. (1991). Phytochemical studies on roots of [*Mucuna pruriens* (L.) DC.] *Biovigyanam*.17 (2): 111-114.

58. Deshmukh, S.K., Jain, P.C. and Agrawal, S.C. 1986. A note on mycotoxicity of some essential oils. *Fitoterpia*. 57(4): 295-297.

59. Devi, R., Arcot, J, Sotheeswaran, S. and Ali, S. (2008). Folate contents of some selected Fijian foods using tri-enzyme extraction method. *Food Chem*. 106: 1100–1104.

60. Ding, Y., Kinjo Yang, C.R. and Novara, T. (1991). Triterpenes from *Mucuna birdwoodiana*. *Phytochemistry*. 30 (1): 3703 - 3707.

61. Du, Q., Xu, Y., Li, L., Zhao, Y., Jerz, G. and Winterhalter, P. (2006). Antioxidant constituents in the fruits of *Luffa cylindrica* (L.) Roem. *J. Agric. Food Chem*. 54: 4186-90.

62. Duenas, M., Fernandez, D., Hernandez, T., Estrella, I. and Munoz, R. (2005). Bioactive phenolic compounds of cowpeas (*Vigna sinensis* L). Modifications by fermentation with natural microflora and with *Lactobacillus plantarum* ATCC 14917. *J. Sci. Food. Agr.* 85(2): 297-304.

63. Duke, J.A. (1981). *Hand book of legumes of world economic importance*. Plenum Press, New York. p. (199-265.

64. Duke, James A., Duke, Peggy-Ann K., DuCellie and Judith L. (2007). *Duke's Handbook of Medicinal Plants of the Bible*. CRC Press. p. 232

65. Duthie, S. J., Whalley, L.J., Collins, A. R., Leaper, S., Berger, K., and Deary, I. J. (2002). Homocysteine, B vitamin status, and cognitive function in the elderly. *Am. J. Clin. Nutr.* 75 (5): 908-913.

66. Dyrby, M., Westergaard, N. and Stapelfeldt, H. (2001). Light and heat sensitivity of red cabbage extract in soft drink medel systems. *Food Chem.* 72: 431–437.

67. Eilert, U., Wolters, B. and Nahrstedt, A. (1981). Antibiotic principle of seeds of *Moringa oleifera* and *Moringa stenopetala. Planta Medica,* 42(1): 55-61.

68. Faizi, S., Siddiqui, B. S., Saleem, R., Siddiqui, S., Aftab, K. and Gilani, A. H. (1994). Novel hypotensive agents, niazimin A, niazimin B, niazicin A and niazicin B from *Moringa oleifera.* Isolation of first naturally occurring carbamates. *J. Chem. Soc., Perkin Trans.* 1: 3035-3040.

69. Faizi, S., Siddiqui, B. S., Saleem, R., Siddiqui, S., Aftab, K. and Gilani, A. H. (1995). Fully acetylated carbamate and hypotensive thiocarbamate glycosides from *Moringa oleifera. Phytochemistry.* 38(4): 957-963.

70. Faizi, S., Siddiqui, BS., Saleem, R., Aftab, K., Shaheen, F. and Gilani, AH. (1998). Hypotensive constituents from the pods of *Moringa oleifera. Planta Med.* 64: 225–228.

71. Fan, S., Meng, Q., Saha, T., Sarkar, F.H. and Rosen, E.M. (2009). Low concentrations of diindolylmethane, a metabolite of indole-3-carbinol, protect against oxidative stress in a BRCA1-dependent manner. *Cancer Res.* 69(15): 6083-6091.

72. Fanelli, S.L., Castro, Toranzo, G.D., De, E.G. and Castro, J.A. (1998). Mechanisms of the preventive properties of some garlic components in the carbon tetrachloride-promoted oxidative stress. Diallylsulfide; diallyldisulfide; allylmercaptan and allyl methyl sulfide. *Res. Commun. Mol. Pathol. Pharmacol.* 102: 163-174.

73. Fenwick, G.R. and Hanley, A.B. (1985). The genus *Allium. Crit. Rev. Food Sci. Nutr.* 23: 1-73.

74. Fernandes, F., Valentao, P., Sousa, C., Pereira, C.A., Seabra, R.M. and Andrade, P.B. (2007). Chemical and Antioxidative assessment of dietary turnip (*Brassica rapa* var. *rapa*). *Food Chem.* 105: 1003-1010.

75. Fiebig, M., Pezzuto, J.M., Soejarto, D.D. and Kinghorn, A.D. (1985). Koenoline a further cytotoxic carazole alkaloid from *Murraya koenigii. Phytochemistry.* 24(12): 3041-3043.

76. Finley and Penland (1998). Adequacy or deprivation of dietary selenium in healthy men: Clinical and psychological findings. *J. Trace Elem. Exp. Med.* 11 (1): 11–27.

77. Flick, G.J., Aung, L.H., Ory, R.L. and St. Angelo, A.J. (1977). Nutrient composition and selected enzyme activities in *Sechium edule. J. Food Sci.* 42(1): 11-13.

78. Francis, F. J. (1999). *Anthocyanins and betalains.*In F.J Francis (Ed.), Colorants (pp. 55–66). St Paul, MN: Eagan Press.

79. Franke, A. A., Custer, L. J., Arakaki, C. and Murphy, S. P. (2004). Vitamin C and flavonoid levels of fruits and vegetables consumed in Hawaii. *J. Food Comp. Anal.* 17: 1–35.

80. Fujihara, S., Nakashima, T., Kurogochi, Y. and Yamaguchi, M. (1986). Distribution and metabolism of sym. Homospermidine and Canavaline in the Sword bean (*Canavalia gladiate*) Cr. Shironata. *Plant Physiol.* 82 (3): 795-800.

81. Fujii, Y. (1990). Mucuna of the Leguminosae (1) A reassessment of its value and new possibilities. *Agri. Hort.* 65(7): 835-840.

82. Garcia, Y.P. (1998). Physico-chemical characteristics of candied chayote (*Sechium edule,* Swartz). Lligan city (Philippines).

83. Gardzielewska, J., K. Pudyszak, T. Majewska, M. Jakubowska and Pomianowski, J. (2003). Effect of plant-supplemented feeding on fresh and frozen storage quality of broiler chicken meat. *Electronic J. Polish Agric. Univ.* 6: 12-12.

84. Gautam, M.P. and Purohit, R.M. (1974). Antimicrobial activity of the essential oil of the leaves of *Murraya koenigii* (L.) Spreng. (Indian Curry Leaf). *Journal of Pharmacy.* 36(1): 11-12.

85. Gbeassor, M., Kedjangani, A. Y., Koumaglo, K., Souza, C. De, Agbo, K., Alkikokou, K. and Amegbo, K.A. (1990). *In vitro* antimalaria activity of six medicinal plants. *Phytotherapy Research.* 4(3): 115-117.

86. Gbenga, O.E., Adebisi, O.E., Fajemisin, A.N. and Adetunji, A.V. (2009). Response of broiler chickens in terms of performance and meat quality to garlic *Allium sativum* supplementation. *Afr. J. Agric. Res.* 4: 511-517.

87. George, B., Kaur, C., Khurdiya D. S. and Kapoor, H. C. (2004). Antioxidants in tomato (Lycopersicon esculentum) as a function of genotype. *Food Chemistry.* 84: 45–51.

88. Ghasi, S., Nwobodo, E. and Oli, J.O. (2000). Hypocholesterolemic effects of crude extract of leaf of *Moringa oleifera* Lam in high-fat diet fed Wistar rats. *J. Ethnopharmacol.* 69: 21–25.

89. Gill, N. S., Dhiman, K., Bajwa, J., Sharma, P. and Sood, S. (2010). Evaluation of free radical scavenging, anti-inflammatory and analgesic potential of *Benincasa hispida* seed extract. *International Journal of Pharmacol.* 6: 652-657.

90. Gill, N.S., Arora R. and Kumar, S.R. (2011). Evaluation of antioxidant, anti-inflammatory and analgesic potential of the Luffa acutangula Roxb.Var. amara. *Research Journal of Phytochemistry*. 5: 201-208.

91. Girdhar, S.,Wanjari,M.M., Prajapati, S. K. and Girdhar, A. (2010). Evaluation of anti-compulsive effect of methanolic extract of Benincasa hispida Cogn. fruit in mice. *Acta Poloniae Pharmaceutica-Drug Research*. 67: 417"421.

92. Goet, R. K., Pandey, V. B., Dwivedi, S. P. D. and Rao, Y. V. (1988). Antiinflammatory and antiulcer effects of kaempferol, a flavone, isolated from Rhamnus procumbens. *Indian J. Exp. Biol*. 26: 121-124.

93. González, E., Montenegro, M. A. and Nazareno, M. A. (2001). Carotenoid composition and vitamin A value of an Argentinian squash (Cucurbita moschata). *Archivos Latinoamericanos de Nutrición*. 51(4): 395–399.

94. Gordon, E. A., Guppy, L. J. and Nelson, M. (2000). The antihypertensive effects of the Jamaican Cho-Cho. *West Indian Medical Journal*. 49: 27–31.

95. Gowda, P.H.R., Shivasankar, K.T. and Gowda, J.V.N. (1989). A note on chimera in chow-chow (*Sechium edule*). *Crop Research. Hisar*. 2(2): 233-234.

96. Gratacos-Cubarsi, M., Ribas-Agusti, A., Garcia-Regueiro, J. A. and Castellari, M. (2010). Simultaneous evaluation of intact glucosinolates and phenolic compounds by UPLC-DAD-MS/MS in *Brassica oleracea* L. var. *botrytis*. *Food Chem*. 121: 257-263.

97. Gupta, K., Barat, G.K., Wagle, D.S. and Chawla, H.K.L. (1989). Nutrient content and anti-nutritional factors in conventional and unconventional leafy vegetables. *Food Chemistry*, 31(2): 105-116.

98. Gupta, R.C., Dixit, B.S. and Benerji, R. (1995). Oils of some conventional legume seeds. *Journal of the Oil Technologists' Association of India.* 27 (4): 239 - 241.

99. Ha, T J., Lee, M H., Park, C H., Pae, S. B., Shim, K.B., Ko, J. M., Shin, S. O., Baek, I. Y., and Park, K.Y. (2010). Identification and Characterization of Anthocyanins in Yard-Long Beans (*Vigna unguiculata* ssp. *sesquipedalis* L.) by High-Performance Liquid Chromatography with Diode Array Detection and Electrospray Ionization/Mass Spectrometry (HPLC"DAD"ESI/MS) Analysis. *J. Agric. Food Chem*. 58 (4): 2571–2576.

100. Han, S.W., Tae, J., Kim, J.A., Kim, D.K., Seo, G.S., Yun, K.J., Choi, S.C., Kim, T.H., Nah, Y.H. and Lee, Y.M. (2003). The aqueous extract of *Solanum melongena* inhibits PAR2 agonist-induced inflammation. *Clinical Chemiica. Acta*. 328: 39–44.

101. Hanley, A.B. and Fenurick, G.R. (1985). *Journal of Plant Foods*. 6: 211-256.

102. Heimler, D., Vignolini, P., Dini, M.G., Vincieri, F.F. and Romani, A. (2006). Antiradical activity and polyphenol composition of local edible varieties. *Food Chem*. 99: 464-469.

103. Hempel, J. and Bohm, H. (1996). Quality and quantity of prevailing flavonoid glycosides of yellow and green French beans. *J. Agric. Food Chem*. 44: 2114-2116.

104. Hornero-Mendez, D., Costa-Garcia, J. and Mínguez-Mosquera, M.I. (2002). Characterization of carotenoid high-producing *Capsicum annuum* cultivars selected for paprika production. *J. Agric. Food Chem.* **50**: 5711–5716.

105. Hsu, S. (2005). Green tea and the skin. *J. Am. Acad. Dermato.* 52(6): 1049-1059.

106. Huang, H. Y., Huang, J. J., Tso, T. K., Tsai, Y. C. and Chang, C. K. (2004). Antioxidant and angiotension-converting enzyme inhibition capacities of various parts of *Benincasa hispida* (wax gourd). *Nahrung.* 48: 230"233.

107. Iauk, L., Galati, E.M., Kirjavainen, S., Forestieri, A.N. and Trovato, A. (1993). Analgesic and antipyretic etleas of Mucuna pruriens. *International Journal of Pharmacognosy*, 31 (3): 213-216.

108. Ige, M. M., Ogunsua, A. O. and Oke, L. O. (1984). Functional properties of some Nigerian Oil seeds. Conophor seeds and three varieties of melon seeds. *J. Agric Food Chem.* 32: 822-825.

109. Isabelle, M., Lee, B. L., Lim, M. T., Koh, W.-P., Huang, D. and Ong, C. N. (2010). Antioxidant activity and profiles of common vegetables in Singapore. *Food Chemistry*, 120(4): 993–1003.

110. Iwalokun, B.A., Ogunledun, A., Ogbolu, D.O., Bamiro, S.B. and Jimi-Omojola, J. (2004). *In vitro* antimicrobial properties of aqueous garlic extract against multidrug-resistant bacteria and *Candida* species from Nigeria. *J. Med. Food.,* 7: 327-333.

111. Jain, A., Soni, M., Deb, L., Jain, A., Rout, S.P., Gupta, V.B. and Krishna, K.L. (2008). Antioxidant and hepatoprotective activity of ethanolic and aqueous extracts of *Momordica dioica* Roxb.Leaves. *J. Ethnopharmacol.* 115: 61–66.

112. Jiratanan, T. and Liu, R. H. (2004). Antioxidant activity of processed table beets (Beta vulgaris var, conditiva) and green beans (*Phaseolus vulgaris* L.). *J. Agr. Food Chem.* 52 (9): 2659–2670.

113. Kala, A. and Jamuna, P. (2006). The comparative evaluation of the nutrient composition and sensory attributes of four vegetables cooked by different methods. *Int. J. Food. Sci. Tech.* 41: 163-171.

114. Kanner, K., Harel, S. and Granit, R. (2001). Betalains–a new class of dietary cationized antioxidants. *J. Agr. Food Chem.* 49: 5178–5185.

115. Kaur, C. and Kapoor, H. C. (2002). Anti-oxidant activity and total phenolic content of some Asian vegetables. *Int.J. Food Sci.* 37: 153-161.

116. Khanum, F., Swamy, M. S., Krishna, K.R.S., Santhana K. and Viswanathan, K. R. (2000). Dietary fiber content of commonly fresh and cooked vegetables consumed in India. *Plant Foods Hum Nutr.* 55: 207–218.

117. Kooi, G. (1993). [*Canavalia gladiata* (Jacq.) DC.] In PROSEA. Plant Resources of South East Asia-8 Vegetables. J. S. Siemensma and Kasem Piluek (eds). Pudoc Scientific Publishers, Wagen-ingen.

118. Kopsell, D.E., Kopsell, D.A., Lefsrud, M.G. and Curran, C.J. (2004). Variability in elemental accumulations among leafy *Brassica oleracea* cultivars and selections. *J. Plant Nutr.* 27: 1813-1826.

119. Korus, A. (2010). Level of vitamin C, polyphenols and antioxidant and enzymatic Activity in three varieties of kale (*Brassica Oleracea* L. var. *acephala*) at different stages of maturity. *Int J. Food Properties*. 14: 1069-1080.

120. Krishna, De A. (2003). *Capsicum: The Genus Capsicum*. Medicinal and aromatic plants-industrial profiles Vol. 33. Taylor and Francis, London and New York. 275 pp.

121. Kubola, J. and Siriamornpun, S. (2008). Phenolic contents and antioxidant activities of bitter gourd (*Momordica charantia* L.) leaf, stem and fruit fraction extracts *in vivo*. *Food Chem*. 110: 881–890.

122. Kumar, A. and Vimalavathini, R. (2004). Possible anorectic effect of methanol extract of *Benincasa hispida* (Thunb.) Cong, fruit. *J. Pharmacol*. 36: 348"350.

123. Kurilich, A.C., Tsau, G.J., Brown, A., Howard, L., Klein, B.P., Jeffery, E.H., Kushad, M., Wallig, M.A. and Juvik, J.A. (1999). Carotene, tocopherol, and ascorbate contents in subspecies of *Brassica oleracea*. *J. Agric. Food Chem*. 47: 1576-1581.

124. Kushawa, S.K., Jain, A., Jain, A., Gupta, V.B., Patel, J.R. and Dubey, P.K. (2005). Hepatoprotective activity of fruits of *Mormordica dioica* Roxb. *Plant Archive*. 5: 613–616.

125. Kwon, Y. I., Apostolidis, E. and Shetty K. (2008). *In vitro* studies of eggplant (*Solanum melongena*) phenolics as inhibitors of key enzymes relevant for type 2 diabetes and hypertension. *Bioresource Tech*. 99: 2981–2988.

126. Lam, S.K. and Ng, T.B. (2010). Isolation and characterization of a French bean hemagglutinin with antitumor, antifungal, and anti-HIV-1 reverse transcriptase activities and an exceptionally high yield. *Phytomedicine*, 17(6): 457–462.

127. Larson, R.L. (1988). The antioxidants of higher plants. *Phytochemistry*. 4: 969-978.

128. Laurena, A.C., Revilleza, M.J.R. and Mendoza, E.M.T. (1994). Polyphenols, phytate, cyanogenic glycosides and tripsin inhibitor activity of several Philippine indigenous food legumes. *J. Food Comp Anal*. 7: 194-202.

129. Lee, Y., Howard, L. R. and Villalon, B. (1995). Flavonoid and antioxidant activity of fresh pepper (*Capsicum annum*) cultivars. *J. Food Sci*. 60 (3): 473-476.

130. Li, H., Deng, Z., Liu, R., Loewen, S. and Tsao, R. (2012). Ultra-performance liquid chromatographic separation of geometric isomers of carotenoids and antioxidant activities of 20 tomato cultivars and breeding lines. *Food Chem*. 132: 508–517.

131. Li, Q. H., Fu, C.L., Rui, Y.K., Hu, G.H. and Cai, T.Y. (2005). Effects of protein bound polysaccharide isolated from pumpkin on insulin in diabetic rats. *Plant Foods Hum Nutr.*, 60: 13–16.

132. Li, Q.H., Tian, Z. and Cai, T.Y. (2001). Study on the hypoglycemic action of pumpkin extract in diabetic rat. *Acta Nutr. Sin.* 25(1): 34–36.

133. Lin, K.W., Yang, S.C. and Lin, CN. (2011). Antioxidant constituents from the stems and fruits of *Momordica charantia*. *Food Chem*. 127: 609–614.

134. Lipipun, V., Kurokawa, M. and Suttisri, R. (2003). Efficacy of Thai medicinal plant extracts against herpes simplex virus type 1 infection *in vitro* and *in vivo*. Antiviral Res. **60**: 175–180.

135. Lo Scalzo, R., Genna, A., Branca, F., Chedin, M. and Chassaigne, H. (2008). Anthocyanin composition of cauliflower (*Brassica oleracea* L. var. *botrytis*) and cabbage (*B. oleracea* L. var. *capitata*) and its stability in relation to thermal treatments. *Food Chem.* 107: 136-144.

136. Losoya, X. (1980). Mexican medicinal plants used for treatment of cardiovascular diseases. *Am. J. Chin. Med.* 8: 86–95.

137. Macleod, A.J. and Pieris, N.M. 1982. Analysis of the volatile essential oils from *Murraya koenigii* and *Pandanus latifolius*. *Phytochemistry*. 21 (7): 1653-1657.

138. Maiti, S. and Mishra, T. K. (2000). Anti venom drug of Santals, Savars and Mantos of Midnapore district of West Bengal, India. *Ethno-botany*. 12: 77 - 80.

139. Makhal, S., Mandal, S. and Kanawjia, S.K. (2006). Phytosterols and stanols: the new age in designing novel functional dairy foods. *Indian Food Ind.* 25: 44–53.

140. Makonnen, E., Hunde, A., Damecha, G. (1997). Hypoglycaemic effect of Moringa stenopetala aqueous extract in rabbits. *Phytother Res.* **11**: 147–148.

141. Mallorca, R., Leon, S.Y. De and Lim Sylianco, C.Y. (1992). Free glutamic acid in some Philippine fruits and vegetables. *ASEAN Food Journal*, 7(2): 108-110.

142. Manchalia, S., Murthy, K. N. C. and Patil, B. S. (2011). Crucial facts about health benefits of popular cruciferous vegetables. *J. Func. Food*. 4: 94 –106.

143. Marin, A., Ferreres, F., Tomas Barberan, F.A. and Gil, M. (2004). Characterization and quantitation of antioxidant constituents of sweet pepper (*Capsicum annuum* L.). *J. Agric. Food Chem*. 52: 3861–3869.

144. Martin, F. (1982). Okra, potential multi-purpose crop for the temperate zones and tropics. *Economic Bot.* 36(3): 5-340.

145. Masaadeh, H.A., Hayajneh, W.A. and Momani, N.M. (2006). Microbial ecology of dental plaques of Jordanian patients and inhibitory effects of (*Allium sativum* and *Allium cepa* L.) extracts. *J. Medical Sci.* 6: 650-653.

146. Mathad, V. S. B., Chandanam, S., Setty, S. R. T., Ramaiyan, D., Veeranna, B. and Lakshminarayanasettry, A. B. V. (2005). Antidiarrheal evaluation of *Benincasa hispida* (Thunb.) Cogn. Fruit extract. *J Pharmacol Ther.*, 4: 24-27.

147. Matsubara, K., Kaneyuki, T., Miyake, T. and Mori, M. (2005). Antiangiogenic activity of nasunin, an antioxidant anthocyanin, in eggplant peels. *J. Agric. Food Chem*. 53: 6272–6275.

148. Mazza, G. and Miniati, E. (1993). *Anthocyanins in fruits, vegetables, and grains.* (pp. 283–288). Boca Raton FL: CRC Press.

149. Meena, B.A., Umapathy, K.P., Pankaja, N. and Prakash, J. (1987). Soluble and insoluble oxalates in selected foods. *J. Food Sci. Tech.* 24(1): 43-44.

150. Miean, K.H. and Mohamed, S. (2001). Flavonoid (myricetin, quercetin, kaempferol, luteolin, and apigenin) content of edible tropical plants. *J. Agri. Food Chem*. 49: 3106–3112.

151. Miller-Cebert, R.L., Sistani, N.A. and Cebert, E. (2009a). Comparative mineral composition among canola cultivars and other cruciferous leafy greens. *J. Food Composit. Anal*. (22): 112-116.

152. Miller-Cebert, R.L., Sistani, N.A. and Cebert, E. (2009). Comparative protein and folate content among canola cultivars and other cruciferous leafy vegetables. *J. Food Agric. Environ*. 7(2): 46-49.

153. Minguez-Mosquera, M.I. and Hornero-Mendez, D. (1994). Comparative study of the effect of paprika processing on the carotenoids in peppers (*Capsicum annuum*) of the Bola and Agridulce varieties. *J. Agric Food Chem*. 42: 1555–1560.

154. Minorsky, PV. (2002). Lycopene and prevention of prostate cancer. The love apple lives up to its name. *Plant Physiology*. 130: 1077-1078.

155. Moreno, D.A., Carvajal, M., Lopez-Berenguer, C. and Garcia-Viguera, C. (2006). Chemical and biological characterisation of nutraceutical compounds of broccoli. *J. Pharm. Biomed. Anal*. 41: 1508-1522.

156. Morton, J.F. (1991). The horseradish tree, *Moringa pterigosperma* (Moringaceae). A boon to arid lands. *Econ Bot*. 45: 318–333.

157. Mukherjee, M., Mukherjee, S., Shaw, A.K. and Ganguly, S.N. (1983). Muko-nicine, a carbazole alkaloid from leaves of Murraya koenigii. *Phytochemistry*. 22(10): 2328-2329.

158. Muresan, R. and Popescu, H. (1993). *Abelmoschus esculentus* (L.) Moench. cultivat la Cluj ca sursa de poliholozide *Clujul Medical*. 66 (4): 201–209.

159. N.A.P. (1981). National Academy Press, Washington DC. The winged bean - a high Protein crop for the tropics, 2nd Edition.

160. Nadkarni, K.M. (1927). *Indian Material Media*, Nadkarni and Co., Bombay. Linney, G. (1986). "*Coccinia grandis* (L.) Voight: A new cucurbitaceous weed in Hawai'i". *Hawaii Botanical Society Newsletter*. **25** (1): 3-5.

161. Nadkarni, K.M. (1994). In: Nadkarni, A.K (Ed.), *Indian Materia Medica*, vol. I, p. 755.

162. Nandi, Biplab K. and Bhattacharjee, Lalita (2002). *Proc. Intern. Conf. Veg*., nov-11-14, banglore, pp.539-552.

163. Nautiyal, B.P. and Venkataraman, K.G. (1987). *Moringa* - an ideal tree for social forestry - growing conditions and uses. Part-I. My Forest. 23(1): 53-58.

164. Neuhauser, M. L. (2004). Dietary flavonoids and cancer risk: evidence from human population studies. *Nutr. Cancer*. 50: 1-7.

165. Nicoli, M.C., Anese, M. and Parpil, M. (1999). *Trends in Food Sci. Tech*. 10: 94-100.

166. Nielsen, J.K., Norbaek, R. and Olsen, C.E. (1998). Kaempferol tetraglucosides from cabbage leaves. *Phytochemistry*. 49: 2171-2176.

167. Nielsen, J.K.; Olsen, C.E. and Petersen, M.K. (1993). Acylated flavonol glycosides from cabbage leaves. *Phytochemistry*, 34: 539-544.

168. Nisha, P., Abdul Nazar, P. and Jayamurthy, P. (2009). A comparative study on antioxidant activities of different varieties of *Solanum melongena*. *Food Chem Toxicol*. 47: 2640-2644.

169. Noda, Y., Kaneyuki, T., Igarashi, K., Mori, A. and Packer, L. (2000). Antioxidant activity of nasunin, an anthocyanin in eggplant peels. *Toxicology*. 148: 119–123.

170. Olivera-Castillo, L., Pereira-Pacheco, F., Polanco-Lugo, E., Olvera-Novoa, M., Rivas-Burgos, J. and Grant, G. (2007). Composition and bioactive factor content of cowpea (*Vigna unguiculata* L. Walp) raw meal and protein concentrate. *J. Sci. Food Agr*. 87(1): 112-119.

171. Olmedilla, B., Granado, F., Blanco, I., Vaquero, M. and Cagigal, C. (2001). Lutein in patients with cataracts and age-related macular degeneration: A long-term supplementation study. *J. Sci. Food Agric*. 81, 904–909.

172. Ordonez, A. A. L., Gomez, J. D., Cudmani, N., Vattuone, M. and Isla, M. I. (2003). Antimicrobial activity of nine extracts of *Sechium edule* (Jacq.) Swartz. *Microb. Ecol. Health Dise*. 33–39.

173. Ordonez, A. A. L., Gomez, J. D., Vattuone, M. and Isla, M. I. (2006). Antioxidant activities of *Sechium edule* (Jacq.) Swartz extracts. *Food Chem*. 97: 452–458.

174. Osagie, A. U. and Odutuga, A. A. (1986). Chemical characterisation and edibility of the oils extracted from four Nigerian oil seeds. Nig.

175. Pal, S.K., Mukherjee, P.K. and Saha, B.P. (1995). Studies on the antiulcer activity of *Moringa oleifera* leaf extract on gastric ulcer models in rats. *Phytother Res*. 9: 463–465.

176. Pedren, O., M. A. and Escribano, J. (2000). Studying the oxidation and the antiradical activity of betalain from beetroot. *J. Bio. Edu*. 35: 49–51.

177. Penninx, B.W.J.H., Guralnik, J. M., Ferrucci, L., Fried, L P., Allen, R H. and Stabler, S.P. (2000). Vitamin B12 Deficiency and depression in physically disabled older women: epidemiologic evidence from the women's health and aging study. *J. Psych*. 157: 715-721.

178. Perera, C.O. and Yen, G.M. (2007). Functional properties of carotenoids in human health. *Int.J. Food Prop*. 10 (2): 201–230.

179. Perez-Galvez, A., Martin, H.D., Sies, H. and Stahl, W. (2003). Incorporation of carotenoids from paprika oleoresin into human chylomicrons. *British J. Nutr*. 89: 787–793.

180. Perkins-Veazie, P., Maness, N. and Roduner, R. (2002)Composition of orange, yellow, and red-fleshed watermelons. *Cucurbitacea*. 436–440.

181. Persson, K., Falt, A.S. and Bothmer, V.R. (2001). Genetic diversity of allozymes in turnip (*Brassica rapa* var. *rapa*) from the Nordic area. *Hereditas*. 134: 43-52.

182. Philip, J. and Peter, K.V. and Gopalakrishnana, P.K. (1981). Curry leaf - a mineral packed leafy vegetable. *Indian Hort.* 25(4): 2, 27.

183. Phillips, R.D., McWatters, K.H., Chinnan, M.S., Hung, Y.C., Beuchat, L.R., Sefa-Dedeh, S., Sakyi-Dawson, E., Ngoddy, P., Nnanyelugo, D., Enwere, J., Komey, N.S., Liu, K.S., Mensa-Wilmot, Y., Nnanna, I.A., Okeke, C., Prinyawiwatkul, W. and Saalia, F.K. (2003). Utilization of cowpeas for human food. *Field Crops Research* 82(2-3): 193-213.

184. Piattelli, M. (1981). The betalains: Structure, biosynthesis, and chemical taxonomy. In E. E. Conn, *The biochemistry of plants: A comprehensive treatise* (Vol. 7) (pp. 557–575). New York: Academic Press.

185. Piluek K., (1994). The Importance of Yardlong Bean. In: *Proc. 2nd Symp. Vegetable Legumes.* Kasetsart University Research and development Institute and ARC–AVRDC, Bangkok, Thailand.

186. Pisarikova, B., Peterka, J., Trckova, M., Moudry, J., Zral, Z. and Herzia. I. (2006). Chemical Composition of the Above-ground Biomass of *Amaranth cruentus* and *A. hypochondriacus. Acta Vet. Brno.* 75: 133–138.

187. Pizzocaro, F., Ferrari, V., Acciarri, N., Morelli, R., Russo-Volpe, S. and Prinzivalli, C. (2000). Antioxidant and antiradical activities in green and violet cauliflower ecotypes with different maturity stages. *Workshop of VI Giornate Scientifiche SOI, Sirmione* (pp. 34–35).

188. Price, K. R., Casuscelli, F., Colquhoun, I. J. and Rhodes, M. J. C. (1997). Hydroxycinnamic acid esters from broccoli florets. *Phytochemistry.* 45: 1683–1687.

189. Rachchh, M. A. and Jain, S. M. (2009). Gastroprotective effect of *Benincasa hispida* fruit extract. *Indian J. Pharmacol.* 40: 271-275.

190. Raghu, K.L., Ramesh, C.K., Srinivasa, T.R. and Jamuna, K.S. (2011). Total antioxidant capacity in aqueous extracts of some common vegetables. *Asian J. Exp. Biol. Sci.* 2(1): 58-62.

191. Rahman, A., Laidi, R. and Firdous, S. (1988). NMR studies on mahanine. *Fitoterpia.* 59(6): 594-595.

192. Rahman, K. and Lowe, G. M. (2006). Garlic and cardiovascular disease: A critical review. *J. Nutr.* 136 (3): 736-740.

193. Rai, P. K., Jaiswal, D., Diwakar, S. and Watal, G. (2008b). Antihyperglycemic profile of *Trichosanthes dioica*.seeds in experimental models. *Pharm. Biol.* 46(5) 360-365.

194. Rai, P. K., Jaiswal, D., Singh, R. K., Gupta, R. K. and Watal, G. (2008a). Glycemic Properties of *Trichosanthes dioica* Leaves. *Pharm. Biol.* 46(12): 894-899.

195. Ramassamy, C. (2006). Emerging role of polyphenolic compounds in the treatment of neurodegenerative diseases: a review of their intracellular targets. *Eur. J. Pharmacol.* 545: 51-64.

196. Rao, A. V., Ray, M. R. and Rao, L. G. (2006). Lycopene. In: L. T. Steve (Ed.) *Advances in food and nutrition research* (Vol. 51, pp. 99–164). Academic Press.

197. Rao, A.V.R., Bhide, K.S. and Majumdar, K.B. (1980). Mahanimbinol. *Chemistry and Industry*, 17: 697-698.

198. Rao, A.W. and Agarwal, S. (2000). Role of antioxidant lycopene in cancer and heart disease. *J. Am. Coll. Nutr.* 19 563– 569.

199. Rao, B.K., Kesavulu, M.M., Giri, R. and Apparao, C. (1999). Antidiabetic and hypolipidemic effects of *Momordica cymbalaria* Hook fruit powder in alloxan-diabetic rats. *J. Ethnopharmacol.* 67: 103–109.

200. Rao, V.A., Devi, P.U. and Kamath, R. (2001). In vivo radioprotective effect of *Moringa oleifera* leaves. *Indian J Exp Biol.* 39: 858–863.

201. Reddy, B.P., Goud, R.K., Mohan S.V. and Sarma P. N. (2009). Antiproliferative and antiangiogenic effects of partially purified *Luffa acutangula* fruit extracts on human lung adenocarcinoma epithelial cell line (A-549). *Curr Trends Biotechnol Pharm.* 3(4): 396 – 404.

202. Reilly, C.A., Crouch, D.J. and Yost, G.S. (2001). Quantitative analysis of capsaicinoids in fresh peppers, oleoresin capsicum and pepper spray products. *J. Forensic Sci.* 46: 502–509.

203. Reisch, J., Abedajo, A.C., Aladesanmi, A.J., Adesina, K.S., Bergenthal, D. and Meve, U. (1994). Chemotypes of *Murraya koenigii* growing in Sri Lanka. *Planta Medica.* 60(3): 295-296.

204. Reuter, H.D., Koch, H.P. and Lawson, L.D. (1996). Therapeutic Effects and Applications of Garlic and its Preparations. In: *Garlic: The Science and Therapeutic Application of Allium sativum L. and Related Species,* Koch, H.P. and L.D. Lawson (Eds.). Williams and Wilkins, Baltimore, MD., pp: 135-213.

205. Ribeiro, R., De, A., Barros, F. De, Fiuza, De Melo, M.M.R, Muniz, C., Chieia, S., W.M.D., Gomes and C. Trolin, G. (1988). Acute diuretic effect in conscious rates produced by some medicinal plants used in the state of Sao Paulo, Brasil. *J. Ethnopharmacol.* 24 (1): 19-29.

206. Ricardo Bressani, Luiz G. Elias and Arnoldo, Garcia-Soto (1989). Limiting amino acids in raw and processed Amaranth grain protein from biological tests". *Plant foods for human nutrition* (Kluwer Academic Publishers) 39 (3): 223–234.

207. Rodrigues, B. F. and Torne, S.G. (1992). Effect of VAM, Rhizobium interaction on nodulation and total biomass of [*Canavalia gladiata* (Jacq.) DC.]. *Plant Cell Incompatibility Newsletter,* 24: 47 - 50.

208. Rodriguez-Amaya, D. B., Kimura, M. and Amaya-Farfan (2008). J. Fontes Brasileiras de Carotenóides.Tabela Brasileira de Composição de Carotenoides em Alimentos. *Ministerio de Meio Ambiente,* 58–59.

209. Ross, I.A. (1999). *Medicinal Plants of the World.* Humana Press, New Jersey, USA, pp. 213–219.

210. Roy, C., Ghosh, T. K. and Guha, D. (2007). The antioxidative role of Benincasa hispida on colchicine induced experimental rat model of Alzheimer's disease. *Biogenic Amines.* 21: 1"2.

211. Roy, S., Gosh, S. and Chakraborty, D.P. (1979). Structure of mahanimboline. *Chemistry and Industry*. 19: 669-670.

212. Ruckmani, K., Kavimani, S., Anandan, R. and Jaykar, B. (1998). Effect of *Moringa oleifera* Lam on paracetamol-induced hepatoxicity. *Indian J. Pharm Sci*. 60: 33–35.

213. Sadilova, E., Stintzing, F.C. and Carle, R. (2006). Anthocyanins, colour and antioxidant properties of eggplant (*Solanam melengona* L.) and violet pepper (*Capsicum annuum* L.) peel extracts. *Z. Naturforsch*. 61: 527–535.

214. Saeed, M.K., Anjum, S., Ahmad, I., Alim-un-Nisa, Ali. S., Zia, A. and Ali, S. (2012). Nutritional facts and free radical scavenging activity of turnip (*Brassica rapa*) from Pakistan. *World. Appl. Sci. J.* 19 (3): 370-375.

215. Salama, A. M., Polo, N. A., Enrique, M., Contreras, C. R. and Maldonado, R. L. (1986). Preliminary phytochemical analisis and determination of the antiinflammatory and cardiac activities of the fruit of *Sechium edule*. *Revista Colombiana en Ciencias Quimica Farmaceuticas*. 15: 79–82.

216. Saluja, M.P., Kapil, R.S. and Popli, S.P. (1978). Studies in medicinal plants: Part-VI. Chemical constituents of *Moringa oleifera* Lam. (Hybrid variety) and isolation of 4-hydroxymellein. *Indian J. Chem*. 16(11): 1044-1045.

217. Sarkar, F.H., Rahman, K.M.W. and Li, Y. (2003). Bax Translocation to Mitochondria Is an Important Event in Inducing Apoptotic Cell Death by Indole-3-Carbinol (I3C) Treatment of Breast Cancer Cells. *J. Nutr*. 133: 2434S–2439S.

218. Sarker, R.H., Yesmin, S. and Hoque, M.I. (2006). Multiple shoot formation in eggplant (*Solanum melongena* L.). *Plant Tissue Cult. Biotech*. 16: 53–61.

219. Schneeman O. (1999). Building scientific consensus: the importance of dietary fiber. *Am. J. Clin. Nutr*. 25: 691–699.

220. Scott JM, Rebeille, F. and Fletcher, J. (2000). Folic acid and folates: the feasibility of nutritional enhancement in plant foods. *J. Sci. Food Agric*. 80: 795-824.

221. Sharma, G., and Pant, M.C. (1988). Effects of feeding *Trichosanthes dioica* (parval) on blood glucose, serum triglyceride, phospholipid, cholesterol, and high density lipoprotein-cholesterol levels in the normal albino rabbit. *Current Sci*. 57: 1085–1087.

222. Shekhawat, N., Soam, P.S., Singh, T. and Vijaybergia, R. (2010). Assessment of free radical scavenging activity of crude extracts of some medicinal plants. *Middle East J Sci Res*. 5(4): 298-301.

223. Shingade, M.Y. and Chavan, K.N. (1996). Unconventional leafy vegetables as source of minerals. *Van Vigyan* 34(1/2): 1-6.

224. Shweta, Gautam, Priya, Singh and Yogesh, Shivhare (2011). *Praecitrullus fistulosus*: A miraculous plant. *Asian J. Pharm. Tech*. 1(1): 9-12.

225. Siddhuraju, P. and Becker, K. (2007). The antioxidant and free radical scavenging activities of processed cowpea (*Vigna unguiculata* (L.) Walp.) seed extracts. *Food Chem*. 101(1): 10-19.

226. Siegers, C.P., Steffen, B., Robke, A. and Pentz, R. (1999). The effects of garlic preparations against human tumor cell proliferation. *Phytomedicine*, 6: 7-11.

227. Siemonsma, J. S. and Piluck, K. (1993). *Plant Resources of South East Asia* No 8 Vegetable Pudoc Scientific Publishers, Wageningen.

228. Sikora, E., Cieslik, E., Leszczynska, T., Filipiak-Florkiewicz, A. and Pisulewski, P.M. (2008). The antioxidant activity of selected cruciferous vegetables subjected to aquathermal processing. *Food Chem.* 107: 55-59.

229. Simopoulos, (2002). Omega-3 fatty acids in wild plants, nuts and seeds. *Asia Pacific Journal Clinical Nutrition* 11(6): 163–173.

230. Singh, A. P., Luthria, D. L., Wilson, T., Nicholi, V., Singh, V. and Banuelos, G. S. (2009). Polyphenols content and antioxidant capacity of eggplant pulp. *Food Chem.* 114: 955–961.

231. Singh, J., Upadhyay, A.K., Bahadur, A. and Singh, K.P. (2004). Dietary antioxidant and minerals in Crucifers. *J Veg Crop Product.* 10(2): 33-41.

232. Smith, C.J.S., Watson, C.F., Ray, J., bird, C.R., Morris, P.C., Schuch, W. and Grierson, D. (1998). *Nature.* 334: 724-726.

233. Sri Kantha, S., Hettiarachehy, N.S., Herath, H.M.W. and Wickramanayake, T.W. (1978). Studies on the nutritional characteristics of winged bean (*Psophocarpus tetragonolobus*), tubers and leaves. *Proceedings of the 34th Annual Session of the Sri Lanka Association for the Advancement of Science*, Colombo, Sri Lanka.

234. Srivastava, Y., Venkatakrishna-Bhatt, H., Verma, Y., Venkaiah, K. and Raval B. H. (1993). Antidiabetic and adaptogenic properties of *Momordica charantia* extract: An experimental and clinical evaluation. *Phytother. Res.* 7(4): 285–289.

235. Strack, D., Steglich, W. and Wray, V. (1993). Betalains. In: P.M. Dey, and J. B. Harborne (Eds.), *Methods in Plant Biochemistry: Alkaloids and Sulphur Compounds* (Vol. 8) (pp. 421–450). London: Academic Press.

236. Sudheesh, S., Presannakumar, G., Vijayakumur, S., and Vijayalakshmi, N. R. (1997). Hypolipidemic effect of flavonoids from *Solanum melongena*. *Plant Food Hum Nutr.* 51: 321–330.

237. Svobodova, A., Psotova, J. and Walterova, D. (2003). Natural phenolics in the prevention of UV-induced skin damage: a review. *Biomed Pap Med.* 147: 137-145.

238. Szmitko, P. E. and Verma, S. (2005). Antiatherogenic potential of red wine: clinician update. *Am. J. Physiol. Heart and Circulatory Physiol.* 288: 2023-2030.

239. Tahiliani, P. and Kar, A. (2000). Role of *Moringa oleifera* leaf extract in the regulation of thyroid hormone status in adult male and female rats. *Pharmacol Res.* 41: 319–323.

240. Tan, M. J., Ye, J. M., Tumer, N., Hohnen-Behrens, C., Ke, C.-Q. and Tang, C.T. (2008). Antidiabetic activities of triterpenoids isolated from bitter melon associated with activation of the AMPK pathway. *Chem. Bio.* 15: 263–273.

241. Tang, L., T. Jin, X. Zeng, and J.S. Wang. (2005). Lycopene inhibits the growth of human androgen-independent prostate cancer cells *in vitro* and in BALB/C nude mice. *J. Nutr.* 135: 287-290.

242. Taur, D.J. and Patil, R.Y. (2011). Mast cell stabilizing, antianaphylactic and antihistaminic activity of *Coccinia grandis* fruits in asthma. *Chin. J. Nat. Med.* 9 (5): 359-362.

243. Toshiro, H.; Takahiko, A. and Seiichiro, N. (1986). The composition and vitamin A value of the carotenoids of pumpkins of different colors. *J. Food Biochem.* 11: 59–68.

244. Trewavas, A. J. and Stewart, D. (2003). Paradoxical effects of chemicals in the diet on health. *Curr. Opin. Plant Biol.* 6: 185-190.

245. USDA nutrient database, search for: cooked Amaranth, wheat germ, oats.

246. Vadde, R., Rani, P. J. and Rao, P. R. (2007). Hypocholesterolemic effect of diet supplemented with Indian bean (*Dolichos lablab* L. var lignosus) seeds. *J Nutr. Food Sci.* 37 (6): 452–456.

247. Vaidya MV and Nanoti, MV. (1989). Bhindi seed powder as coagulant in removal of turbidity from water. *Indian J. Environ. Hlth.* 31: 8-43.

248. Vaishnav, M.M. and Gupta, K.R. (1995). A new saponin from *Coccinia indica* roots. *Fitoterpia*, 66 (6): 546-547.

249. Vaishnav, M.M. and Gupta, K.R. (1996). Ombuin 3-0-arabinofuranoside from *Coccinia indica*. *Fitoterpia.* 67 (1): 80.

250. Vallejo, F., Tomas-Barberan, F.A. and Ferreres, F. (2004). Characterisation of flavonols in broccoli (*Brassica oleracea* L. var. *italica*) by liquid chromatography-UV diode-array detection-electrospray ionisation mass spectrometry. *J. Chromatogr.* 1054: 181-193.

251. Vallejo, F., Tomas-Barberan, F.A. and Garcia-Viguera, C. (2003). Phenolic compound contents in edible parts of broccoli inflorescences after domestic cooking. *J. Sci. Food Agric.* 83: 1511-1516.

252. Vanderslice, J.T., Higgs, D.J., Hayes, I.M. and Block, G., (1990). Ascorbic acid and dehydroascorbic acid content of foods-as-eaten. *J. Food Com. Anal.* 3: 105–118.

253. Venkateswaran, S. and Pari. J. (2003). Effect of *Coccinia indica* leaf extract on plasma antioxidant in streptozotocin induced experimental diabetes in rats. *Phytotherapy Res.* 17(6): 605-608.

254. Verma, S.C., Banerji, R., Misra, G. and Nigam, S.K. (1976). Nutritional value of *Moringa. Current Sci.* 42(21): 769-770.

255. Vinson, J. A., Dabbagh, Y. A., Serry, M. M. and Jang, J. (1995). Plant flavonoids, especially tea flavonols, are powerful antioxidants using an *in vitro* oxidation model for heart disease. *J. Agric. Food Chem.* 43: 2800-2804.

256. Vinson, J.A., Hao, Y., Su, X. and Zubik, L. (1998). Phenol antioxidant quantity and quality in foods: vegetables. *J Agri.Food Chem.* 46: 3630–3634.

257. Visanji, J.M., Thompson, D.G., Padfield, P.J., Duthie, S.J. and Pirie, L. (2004). Dietary Isothiocyanates Inhibit Caco-2 Cell Proliferation and Induce G2/M Phase Cell Cycle Arrest, DNA Damage, and G2/M Checkpoint Activation. *J. Nutr.* 134 3121–3126.

258. Wang, H., Cao, G., and Prior, R. L. (1997). Oxygen radical absorbing capacity of anthocyanins. *J. Agri.Food Chem.* 45, 304–309.

259. Watanabe, Y., Sarumaru, H. and Shimada, N. (1983). Distribution of urease in higher plants. Technical Bulletin, Faculty of Horticulture, Chiba University. 32: 37 - 43.

260. Watts, Donald (2007). *Dictionary of Plant Lore.* Academic Press. p. 226.

261. Wen, T.N., Prasad, K.N., Yang, B. and Ismail, A. (2010). Bioactive substance contents and antioxidant capacity of raw and blanched vegetables. *Innov. Food Sci. Emer. Techn.* 11: 464-469.

262. Weniger, B. and Robineau, L. (1988). Elements pour unepharmacopee Caraibe-Se´minaire Tramil 3, p. 145.

263. West, C.E. and Poortvliet, E.J. (1993). The carotenoid content of foods with special reference to developing countries. International Science and Technology Institute, Arlington, Virginia.

264. WOI (1972). *Wealth of India. Raw Material.* Council of Scientific and Industrial Research, New Delhi.

265. Wong, J. H. and Ng, T. B. (2005). Sesquin, a potent defensin-like antimicrobial peptide from ground beans with inhibitory activities toward tumor cells and HIV-1 reverse transcriptase. *Peptides.* 26, 1120–1126.

266. Woolfe, LM, Chaplin, M and Otchere, G. (1977). Studies on the mucilages extracted from okra fruits and baobab leaves. *J. Sci. Food Agr.* 28: 519-29.

267. Wu, G., Collins, J.K., Perkins-Veazie, P., Siddiq, M., Dolan, K.D., Kelly, K.A., Heaps, C.L. and Meininger, C.J. (2007). Dietary supplementation with watermelon pomace juice enhances arginine availability and meliorates the metabolic syndrome in Zucker diabetic fatty rats. *J. Nutr.* 137: 2680–2685.

268. Wu, S. J. and Ng, L. T. (2008). Antioxidant and free radical scavenging activities of wild bitter melon (*Momordica charantia* Linn. var. *abbreviata* Ser.) in Taiwan. *LWT.* 41: 323–330.

269. Wu, T. H., Chow, L. P. and Lin, J. Y. (1998). Sechiumin, a ribosome inactivating protein from the edible gourd *Sechium edule* Swartz purification, characterization, molecular cloning and expression. *European J. Biochem.* 255(2): 400–408.

270. Wu, X., Beecher, G.R., Holden, J.M., Haytowitz, D.B., Gebhardt, S.E. and Prior, R.L. (2004). Lipophilic and hydrophilic antioxidant capacities of common foods in the United States. *J. Agri. Food Chem.* 52(12): 4026-4037.

271. Yang, R.Y. (2006). Application of Antioxidant Activity Analytical Methods for Studies on Antioxidant Activities of Vegetables. Ph.D. Dissertation. Institute of

Tropical Agriculture and International Cooperation of National Ping-tung University of Science and Technology, Taiwan.

272. Yang, R.Y., Tsou, S.C.S., Lee, T.C., Hanson, P.M. and Lai, P.Y. (2005). Antioxidant Capacities and Daily Antioxidant Intake from Vegetables Consumed in Taiwan. Paper presented at the Symposium on Taiwan-America Agricultural Cooperative Projects, Taipei, Taiwan, 15th November (2005.

273. Yen, G. C., Chen, H. Y. and Peng, H. H. (2001). Evaluation of the cytotoxicity, mutagenicity and antimutagenicity of emerging edible plants. *Food Chem Toxicol.* 11: 1045–1053.

2015, Horticulture for Nutrition Security
Editor: **Prof. K.V. Peter**
Published by: **DAYA PUBLISHING HOUSE, NEW DELHI**

Pages **125–139**

Chapter 7

Advances in Plant Sciences for Nutritional Security

Mehanathan Muthamilarasan and Manoj Prasad

Malnutrition and hunger among the world population at alarming rates pose serious threat to global food security. Further, the FAO Hunger Report (2012) depicts that, about 12.5 per cent of the global population (one in eight people) is starving, excluding 100 million children under the age of five. Irrespective of the adults, about 2.5 million children die every year due to starvation and malnutrition which ultimately hinder human potential (FAO Hunger Report 2012). Since, plants are the primary producers in the food chain, they serve as versatile biochemical factories capable of producing almost complete complement of essential dietary micronutrients. However, the dietary micronutrients are unevenly disseminated among different plant parts. For instance, iron content in a rice leaf is as high as 100–200 ppm (parts per million), but very low in the polished rice grain (~3 ppm) (Mayer *et al.,* 2008). Similarly, provitamin A carotenoids are present only in rice leaves but not in its edible part. Unfortunately, economically backward people rely predominantly on starchy staples such as rice, wheat, maize, or cassava, but these crops do not supplement the biochemical diversity needed for a healthy life which leads to micronutrient malnutrition (MNM). Plant science has a central role in addressing these issues of both hunger and malnutrition. Since, MNM affects more than half of the world population, biofortification offers an economical and sustainable approach of delivering micronutrients via micronutrient-dense crops to the human population. Hence, this book chapter summarizes the strategies of generating biofortified-crop plants along with the significant achievements reported in biofortification of major crop plants such as orange sweet potato, maize, cassava, rice, wheat and other crops like lentils, banana, cowpea, sorghum and potato.

Strategies for Developing Nutrient-Enriched Lines

Agronomic Biofortification

Soil is composed of mineral elements in the form of free ions, surface-adsorbed ions, dissolved compounds or precipitates, which are the parts of lattice structures or present within the soil biota (White and Broadley, 2009). If the soil is devoid of mineral micronutrients, then the plants flourishing in the soil will also have low mineral and micronutrient content. To rectify this, agronomic efforts have been focused towards the application of mineral fertilizers and the enhancement of the solubility and movement of nutrients in the soil. Though this strategy is relatively simple and produces immediate results, this method can only be applied for fortifying plants with mineral elements and not with organic nutrients (*e.g.* vitamins). These organic nutrients should be synthesized by the plants *in vivo*. Furthermore, the viability of this biofortification strategy depends on several factors like soil composition, mineral mobility in the soil and in the plant, and its accumulation site (Zhu *et al.,* 2007, Hirschi 2009). Hence, application of fertilizers comprising essential mineral micronutrients could not be deemed as a universal approach for enhancing the micronutrient levels in edible crop tissues. Generally, mineral elements with a good mobility in the soil and in the plant are good candidates for a fruitful agronomic biofortification (White and Broadley 2009), and so the use of inorganic fertilizers for selenium (Se), iodine (I) and zinc (Zn) was particularly successful (Dai *et al.,* 2004, Hartikainen 2005, White and Broadley 2005). This has been evidenced in several countries like Finland, New Zealand and France, where the application of inorganic Se fertilizers resulted in increased crop Se concentration up to 10-fold (Eurola *et al.,* 1989, 1991). In China and Thailand, enrichment of crops with inorganic I and Zn have also been proved successful (Carvalho and Vasconcelos 2013). At the same time, biofortification for iron (Fe) is proved unsuccessful due to the low mobility of Fe in soil. This is because of the rapid binding of ferrous sulfate to soil particles and conversion into Fe (III), thus becoming unavailable to plant roots (Grusak and DellaPenna 1999).

To overcome this constraint, synthetic metal chelators such as EDTA are used to enhance the efficacy of micronutrient fertilizers supplemented to deficient soils (Zhao and McGrath 2009). Fe- and Zn-chelates were reported to be effective in increasing mineral concentration in edible cereal, vegetable and fruit tissues (Rengel *et al.,* 1999). Alternatively, foliar fertilization method of applying soluble inorganic fertilizers to edible tissues is also practised during the situations where mineral elements are not readily translocated. This method is commonly used in horticultural crops to prevent Ca-deficiency disorders (Ho and White 2005) and also for the application of magnesium sulfate in some crops (Metson 1974). Interestingly, exploiting soil micro-organisms like mycorrhizal fungi and nitrogen-fixing bacteria for increasing the phytoavailability of mineral elements is also reported (Rengel *et al.,* 1999). Several studies established that the mycorrhizal associations increased Fe, Se, Zn and Cu concentrations in crop plants (Rengel *et al.,* 1999, Cavagnaro 2008).

The prime bottleneck of the fertilization strategy for plant biofortification is the frequent need for regular applications. This makes the approach costly, difficult in logistic terms and potentially hazardous for the environment (Hirschi, 2009; White and Broadley 2009). Further, the availability of some nutrients would become limited because of the risk of reserve-exhaustion (White and Broadley, 2009).

Considering the above, the agronomic biofortification through fertilizer application will be deemed feasible only if more cost-effective and long-term strategies to improve micronutrient density in edible plant portions are developed. Nevertheless, this strategy should be combined with other biofortification approaches, particularly when the phytoavailability of minerals constrains their concentration in the edible plant parts (Graham *et al.,* 2007, White and Broadley 2009).

Conventional Plant Breeding

Although the traditional plant breeding is mostly used for increasing crop yields and enhancing crop resistance to environmental stresses, several studies supported the implementation of breeding for enhancing the nutrient content. Unfortunately, reports had also shown that the increases in crop yield over the last four decades have been accompanied by decreases in the mineral concentrations including Fe, Zn, Cu and Mg in the edible plant tissues (Garvin *et al.,* 2006, Fan *et al.,* 2008, White and Broadley 2009). Since conventional plant breeding discovers the inherent characteristics of the diverse crop varieties, this strategy for improving the nutrient content received widespread public acceptance and a simple legal framework (Bouis 2000, Hirschi 2009). Moreover, this biofortification approach only represents a one-off cost as it involves a single initial subsidized distribution and the seeds can be harvested and used for future years (Carvalho and Vasconcelos 2013). These merits made this cost-effective and long-term biofortification strategy as the most expedient solution for improving the micronutrient density of edible plant tissues (Hirschi 2009).

On the other hand, this strategy also encompasses certain demerits, such as; (i) the long development time, (ii) the dependence on the phytoavailability of the mineral nutrients in the soil and (iii) the need for sufficient genetic variation of a given trait within species. However, many traits needed in biofortification programmes can be identified by exploring the genetic variation in germplasm collections or by exploiting transgressive segregation or heterosis (Mayer *et al.,* 2008, Carvalho and Vasconcelos 2013).

Numerous studies reported that a large within-species genetic variation does exist in various crops, both in terms of the concentration of nutrients in the edible tissues and their bioavailability to human gastric system (White and Broadley 2009). For example, various rice genotypes exhibit a 4-fold variation in Fe and Zn levels and up to a 6.6-fold variation has been observed in beans and peas (Gregorio *et al.,* 2000). Apparently, this genotypic variation is generally more reduced in tubers (White and Broadley 2009) and in fruits (Hakala *et al.,* 2003; Carvalho and Vasconcelos 2013). The cultivated wheat germplasm had limited genetic variation in the Se content, but this bottleneck has been overcome by crossing the cultivated wheat with wild wheat genotypes (Lyons *et al.,* 2005; Carvalho and Vasconcelos 2013). Promisingly, seed banks and germplasm storages would play a major role in the future biofortification processes. The strengths and opportunities associated with the biofortification through conventional breeding have encouraged the commencement of many international programs to enrich the nutrient content of several crops, both to improve health and to prevent MNM.

Transgene-Based Approaches

Genetic engineering serves as a straight-forward approach and a valid alternative for increasing the concentration and bioavailability of micronutrients in the edible crop tissues when there is a lack of sufficient genotypic variation for the desired trait within the species or if the crop is not amenable to conventional breeding (Mayer *et al.*, 2008). With the advent of next-generation sequencing and high-throughput analysis platforms, the genomes of many staple crops have been sequenced (Muthamilarasan *et al.*, 2013) and this opens new opportunities for biofortification strategy. Redistributing micronutrients between tissues, enhancing the efficiency of biochemical pathways in edible tissues, and reconstruction of selected pathways are the potential targets of transgenes. Instead of increasing the production or accumulation of micronutrients, some strategies are established for the removal of 'antinutrients' or inclusion of 'promoter' substances to improve the bioavailability of micronutrients.

Transgene-based approaches to increase the mineral content of plants have mainly focused on Fe and Zn, the micronutrients often deficient in human diets (Curie and Briat 2003, Palmgren *et al.*, 2008). Overexpression of ferritin, an iron-storage protein had resulted in 3- to 4-fold increase of Fe levels in rice (Goto *et al.*, 2000, Vasconcelos *et al.*, 2003). Though the mineral levels get declined during polishing of rice, the iron content and bioavailability of transgenic polished rice was still significantly higher in ferritin-enhanced lines (Murray-Kolb *et al.*, 2002). Further, there are many examples where transgene-based approach had led to generation of transgenic crops with increased concentration of vitamins or mineral elements (White and Broadley 2009). 'Golden Rice' is the most popular genetically modified (GM) rice variety, where the carotenoid biosynthetic pathway has been reconstituted in non-carotenogenic endosperm tissue to produce β-carotene (pro-vitamin A) to help combat vitamin A deficiency (Paine *et al.*, 2005). Promisingly, these GM rice varieties would supplement the recommended daily requirement of vitamin A (in the form of β-carotene) in 100 to 200 g of rice. This approach is also being successfully demonstrated in other crops such as maize, orange, cauliflower, tomato, yellow potatoes and golden canola (White and Broadley 2009). Similarly, GM carrot expressing high levels of a deregulated transporter which accumulated about 2-fold more Ca in the edible tissues was also developed (Morris *et al.*, 2008, Carvalho and Vasconcelos 2013). In contrast with the 'Golden Rice', the feeding trials using this labeled carrot proved that Ca absorption was considerably increased in both animal models with diets having the GM carrot but not all the increased Ca was bioavailable (Murray-Kolb *et al.*, 2002). This exemplifies the fact that increase in nutrient content may not be directly translated into similar increase in bioavailability (Carvalho and Vasconcelos 2013).

Although the transgene-based biofortification strategy has several common strengths and weaknesses with the conventional plant breeding, this approach faces the threat of a low public acceptance, and consequently a complex legal framework. In spite of these drawbacks, the genetic engineering is now attempting towards 'multigene transfer', where several micronutrients can be added to the same plant (Carvalho and Vasconcelos 2013). 'Multivitamin corn' is a very good example of this 'multigene transfer' strategy, where GM corns were developed with high levels of β-carotene, ascorbate (vitamin C) and folate (vitamin B9) (Naqvi *et al.,* 2009).

Biofortification in Orange Sweet Potato (OSP)

The International Potato Center (CIP) and National Agriculture Research and Extension System (NARES) had screened a vast germplasm for identifying the varieties which produce 2-fold provitamin A (30 ppm target level) (http://r4d.dfid.gov.uk/pdf/outputs/croppostharvest/r7036_file7a_manilaconf.pdf). Further, the chosen varieties were improved to suit local tastes and agronomic conditions. Nutrition research showed that provitamin A retention was more than 80 per cent after boiling or steaming and minimum of 75 per cent after solar or sun drying, typical types of preparation (Bengtsson *et al.,* 2008, Bechoff *et al.,* 2010). Interestingly, provitamin A from OSP is highly bioavailable and its intake can lead to a considerable increase in vitamin A reserves across age groups (Haskell *et al.,* 2004, Van Jaarsveld *et al.,* 2005, Low *et al.,* 2007).

Enriching Maize with Micronutrients

International Maize and Wheat Improvement Center (CIMMYT) along with International Institute of Tropical Agriculture (IITA) and National Agricultural Research and Extension Systems (South Africa) had initiated maize breeding programmes for provitamin A. Germplasm screening discovered genetic variation for the target level (15 ppm) of provitamin A carotenoids in temperate maize, which was then bred into tropical varieties (Carvalho and Vasconcelos, 2013). Recent developments in marker-assisted selection have expedited the promptness and precision of identifying genes regulating the traits of interest in maize.

Noteworthy, food processing and cooking methods result in provitamin A losses below 25 per cent (Li *et al.,* 2007), whereas drying and dark storage at 25° C for ~4 months may lead 25–60 per cent decay of provitamin A (Burt *et al.,* 2010). Bioavailability (the conversion rate of β-carotene to retinol) was originally assumed to be 12 to 1, but nutrition studies have found more efficient bioconversion rates of 3 to 1 and 6.5 to 1 (Li *et al.,* 2010, Muzhingi *et al.,* 2011). Nutritional efficacy studies are underway in worldwide and the results are expected, which would provide clues to proceed further research in this area.

Biofortification of Cassava

The IITA and the International Center for Tropical Agriculture (CIAT) have screened the cassava germplasm for high provitamin A content and breeding programmes are underway with an initial target of 15 ppm provitamin A (Saltzman *et al.,* 2013). As with maize, the bioavailability of provitamin A is much better than assumed; conversion of β-carotene to retinol has been measured at 3.8 to 1 and 4.3 to 1, with and without added oil, respectively (Saltzman *et al.,* 2013). Researches on the stability of provitamin A in cassava during traditional cooking processes highlight high losses (65–80 per cent) associated with the most common form of cassava consumption (Chavez *et al.,* 2007, Thakkar *et al.,* 2009). Given that retention is in part determined by cassava variety and great variability in processing methods exists at the household level, a study of retention during gari (a creamy-white, granular flour of cassava) production and storage with recently released varieties has been commissioned to confirm the tentative revision of the breeding target to 10 ppm (Saltzman *et al.,* 2013).

Rice with Enhanced Nutrient Contents

Rice provides up to 80 per cent of the energy intake of the economically-backward people in the Asian countries. Recently, International Rice Research Institute (IRRI) and the Bangladesh Rice Research Institute (BRRI) had developed high-zinc rice varieties for Bangladesh and India (Saltzman *et al.,* 2013). Noteworthy, high-yielding rice varieties with more than 75 per cent of the target are undergoing field trials in Bangladesh and India and are expected to be released in time being. Micronutrient-retention studies depicted that the zinc content of rice is not significantly reduced by parboiling and less so by milling relative to iron, as zinc is distributed more homogenously throughout the brown rice grain (Liang *et al.,* 2008). Controlled studies performed by BRRI quantified the loss of zinc from rice during milling and washing before cooking, which showed that ~10 per cent of the zinc in the milled grain was lost during washing prior to cooking (Juliano 1985). Another 10–14 per cent of the zinc may be lost during boiling of rice in an excessive volume of water, which is discarded prior to serving (Dipti 2012). Till date, there is no data on bioavailability and efficacy of the zinc-biofortified rice due to poor sensitivity of serum zinc concentration in response to relatively low amounts of additional zinc intake.

The Swiss Federal Institute of Technology was the first to develop 'Golden Rice' and its research was progressed by Syngenta towards commercialization. Rice cultivars with higher levels of provitamin A, up to 37 ppm were produced and donated for use by the 'Golden Rice' Network (Al-Babili and Beyer 2005). Currently, the research on 'Golden Rice' is being furthered by IRRI (Beyer 2010). Interestingly, bioavailability testing has confirmed that 'Golden Rice' is an effective source of vitamin A for humans, with an estimated conversion rate of β-carotene to retinol of 3.8:1 (Tang *et al.,* 2009).

In addition, transgenic rice lines with high iron content were developed by the University of Melbourne and IRRI. It contains 14 ppm iron in the white rice grain and translocate iron to accumulate in the endosperm (Johnson *et al.,* 2011). Since the iron is unlikely to be bound by phytic acid in the endosperm, the iron would be bioavailable. Still, it has to pass numerous check-points and field trials before reaching the commercial market.

Status of Biofortification in Wheat

Wheat is a major source of dietary energy and protein for the global population. Its potential to contribute towards eliminating micronutrient-related malnutrition was recently realized and numerous efforts have been invested towards biofortification of wheat for high Zn, Fe and Se (Velu *et al.,* 2013). Targeted breeding of these biofortified varieties was initiated by exploiting available genetic diversity for Zn and Fe from wild relatives of cultivated wheat and synthetic hexaploid progenitors (Velu *et al.,* 2013). The most promising and convincing results from the performance of competitive biofortified wheat lines demonstrated excellent adaptation in target environments without compromising essential core agronomic traits. The Punjab Agricultural University, India is now assessing the Fe and Zn losses associated with traditional milling and cooking methods (ftp://ftp.fao.org/docrep/fao/005/y8346m/y8346m06.pdf). Further, an independent study on absorption of Zn among Mexican women showed that total absorbed Zn was significantly higher from the biofortified variety of wheat as compared to non-biofortified wheat (Rosado *et al.,* 2009). Additional absorption and efficacy research for Zn, Fe and Se are requisites to identify the bioavailability and validate the cultivars for genotype-specific variations.

Biofortification of other Crops

In addition to the above, biofortification research had also been initiated for several other crops. Firstly, the International Center for Agricultural Research in the Dry Areas (ICARDA) is working towards biofortification of lentils for higher levels of Fe and Zn (http://www.icarda.org/south-asia-and-china-regional-program). In 2009, it commenced the multilocation testing in Bangladesh, Ethiopia, India, Nepal, and Syria. Noteworthy, mineral-dense lentil varieties identified in early screening are already promoted for wide-scale cultivation in Bangladesh (http://www.icarda.org/south-asia-and-china-regional-program). Govind Ballabh Pant University of Agriculture and Technology, India is actively engaged in cowpea biofortification and has released two early-maturing high-iron and zinc cowpea varieties, Pant Lobia-1 (2008) and Pant Lobia-2 (2010) (http://www.gbpuat.ac.in/research/DES per cent 20Web per cent 20Page.htm). These varieties are enrolled in the national seed multiplication system and the seeds are available to farmers.

International Crops Research Institute for the Semi-Arid Tropics (ICRISAT) has developed zinc- and iron-dense sorghum hybrids which are expected to undergo multilocation testing and on-farm adaptation trials in India (http://www.icrisat.org/crop-sorghum.htm). CIP developed high-iron potato lines. Queensland University of Technology and the National Agricultural Research Organization of Uganda are working towards developing transgenic banana enriched with provitamin A and iron (http://www.banana.go.ug/index.php/news/39-gm-bananas-could-cut-blindness-anaemia-in-east-africa). Bananas with up to 20 ppm provitamin A have been developed and trials are underway in Uganda (Namanya 2011). Still, attempts have not been made towards the biofortification of other cereals, millets and orphan crops.

Role of 'OMICS' in Biofortification

In the post-genomic era, deciphering the functional connections between genes, transcripts, proteins, metabolites and nutrients remain the biology's greatest challenges. The recent technological advances in genomics, transcriptomics, proteomics and metabolomics will highly benefit the plant biofortification processes. Recently, the term 'Nutrigenomics' has been coined to study of the effects of foods and food constituents on gene expression. A schematic representation on the role of 'omics' in ensuring nutritional security is shown in Figure 6.1. The integration of this knowledge would be more fruitful in defining the appropriate biofortification strategies (Carvalho and Vasconcelos 2013). Genes controlling the tissue-specific nutrient accumulation and nutrient concentration have been identified using genomic tools although the precise roles of these genes remain elusive. Certain genes may co-ordinate the uptake and transport of more than one mineral, and so the 'omic' studies should be carefully applied in order to maintain mineral homeostasis.

As mentioned earlier, performing identification and characterization of the genes that regulate nutrient uptake, transport, storage and bioavailability was a tedious procedure. The recent advances in high-throughput sequencing/transcriptomics may expedite the study. For example, the high-throughput technologies such as pyrosequencing, microarray, serial analysis of gene expression (SAGE), suppression subtractive hybridization (SSH) and macroarray technology will provide a non-targeted, full spectrum analysis of all the genes expressed by a tissue at a given time point.

Proteomic research has assisted the researchers in understanding the effects of proteins on plant mineral nutrient homeostasis. It seeks to observe the protein fluctuations under variable developmental and environmental influences, as programmed by the genome, and mediated by the transcriptome (Carvalho and Vasconcelos, 2013). Of note, a recent proteomics technology termed iTRAQ (isobaric tags for relative and absolute quantification)-based quantitative proteomics was used to analyze the microsomal proteins from *Arabidopsis* roots (Fukao *et al.*, 2011). Other proteomic technologies used for studying the protein chemistry are SDS-PAGE, mass spectrometry, protein chips etc.

In addition to genomics, transcriptomics and proteomics, metabolomics is also gaining momentum in biofortification studies. It provides a better understanding of the pathways responsible for the biosynthesis of nutritionally relevant metabolites. Plants are reservoirs of chemically diverse metabolites, which are usually present in a large range of concentrations (De Vos *et al.*, 2007). Hence no single analytical method is able to extract and detect all the metabolites. The available techniques such as gas chromatography-mass spectrometry (GC-MS), high-performance liquid chromatography (HPLC), capillary electrophoresis (CE) and nuclear magnetic resonance (NMR) can be used for metabolite profiling.

Further, it is worth mentioning that plants require at least 17 elements for proper growth and development and that there are 92 elements identified on earth (Karley and White 2009). To study all these elements and identify the mechanisms that coordinately regulate these elements in response to genetic and environmental factors was performed using ionomics (Williams and Salt 2009). This is achieved using high-throughput inductively coupled plasma optical emission spectroscopy (ICP-OES) and inductively coupled plasma mass spectrometry (ICP-MS). In addition, the *in silico* protein analysis tools provided through ExPASy (http://www.expasy.org),

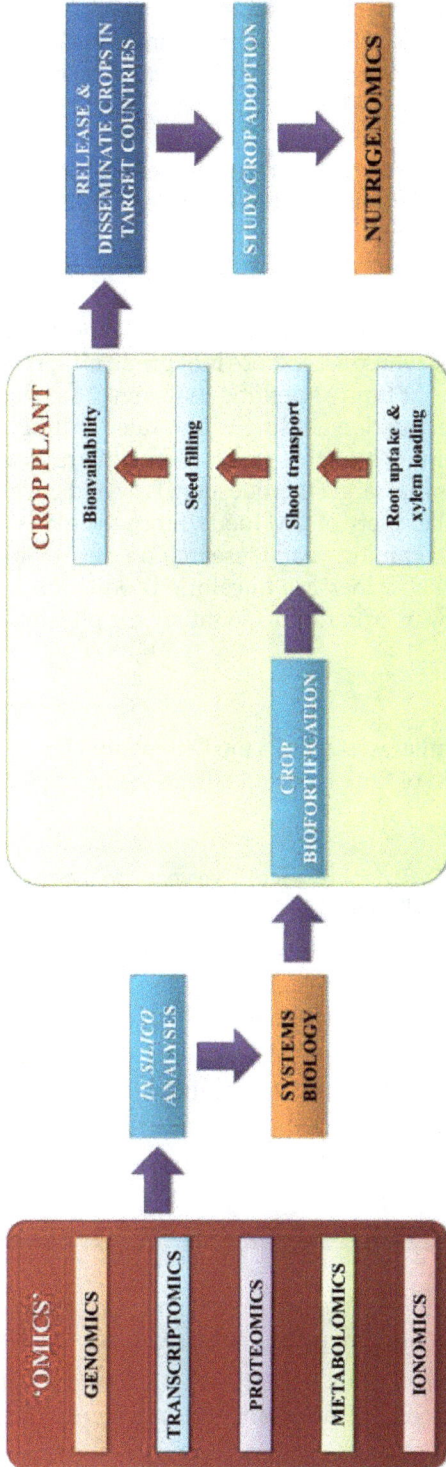

Figure 7.1: 'OMICS' Pipeline for Generating Elite Cultivars with Appropriate Micronutrient Contents.

MIRA (Chevreux *et al.,* 2004), Myrna (Langmead *et al.,* 2010), MG-RAST (Meyer *et al.,* 2008), Blast2Go (B2G) (Conesa *et al.,* 2005), InterPro (Hunter *et al.,* 2008) and Mascot (Perkins *et al.,* 1999) will also expedite the proteomic research on nutritional aspects.

Conclusion

Biofortification is a lengthy and multifaceted process, which involves a great deal of scientific and economical input. This book chapter exemplifies the different approaches of biofortification along with certain examples of success in increasing the micronutrient-density and/or bioavailability in many plant-based foods. Promisingly, in the near future many notable achievements will be produced with the assistance of 'omics' techniques. But the real success of biofortified crops relies on the hands of farmers and public whose acceptance is of prime importance in evading nutritional insecurity. In terms of future scientific challenges, the major dogmas in biofortification is the imperfect understanding of the rate limiting steps that are involved in translocating minerals to the seeds. Hence there is an immediate requirement for further studies that target this question. Law and governance should also be stringent in assuring the safety of GM foods. Since the global population is expected to reach 10 billion in coming years, research on developing biofortified crops that is seeking a sustainable increase in calorie production and furnishing staple crops with the necessary micronutrients to satisfy the physiological needs of the poor is demanded.

Acknowledgment

Mr. Mehanathan Muthamilarasan thanks the University Grants Commission, New Delhi, India for the award of Junior Research Fellowship.

References

1. Al-Babili S and Beyer P (2005). Golden rice - five years on the road - five years to go? *Trends in Plant Science* **10**, 565-573.

2. Bechoff A, Westby A, Owori C, Menya G, Dhuique-Mayer C, Dufour D and Tomlins K (2010). Effect of drying and storage on the degradation of total carotenoids in orange-fleshed sweet potato cultivars. *Journal of the Science of Food and Agriculture* **90**, 622-629.

3. Bengtsson A, Namutebi A, Alminger ML and Svanberg U(2008). Effects of various traditional processing methods on the all-trans-b-carotene content of orange fleshed sweet potato. *Journal of Food Composition and Analysis* **21**, 134-143.

4. Beyer P (2010). Golden rice and 'golden' crops for human nutrition. *New Biotechnology* 27, 478-481.

5. Bouis H E (2000). Enrichment of food staples through plant breeding: A new strategy for fighting micronutrient malnutrition. *Nutrition* **16**, 701-704.

6. Burt A, Grainger C, Young JC, Shelp B and Lee E (2010). Impacts of post-harvest handling on carotenoid concentration and composition in high-carotenoid

maize (*Zea mays* L.) kernels. *Journal of Agricultural and Food Chemistry* **58**, 8286-8292.

7. Carvalho S and Vasconcelos MW (2013). Producing more with less: Strategies and novel technologies for plant-based food biofortification. *Food Research International* **54**, 961-971.

8. Cavagnaro TR (2008). The role of arbuscular mycorrhizas in improving plant zinc nutrition under low soil zinc concentrations: A review. *Plant and Soil* **304**, 315-325.

9. Chavez AL, Sanchez T, Ceballos H, Rodriguez-Amaya DB, Nestel P, Tohme J and Ishitani M (2007). Retention of carotenoids in cassava roots submitted to different processing methods. *Journal of Science of Food and Agriculture* **87**, 388-393.

10. Chevreux B, Pfisterer T, Drescher B, Driesel AJ, Müller WEG, Wetter T and Suhai S (2004). Using the miraEST assembler for reliable and automated mRNA transcript assembly and SNP detection in sequenced ESTs. *Genome Research* **14**, 1147-1159.

11. Chowdhury S, Meenakshi JV, Tomlins K and Owori C, (2011). Are consumers in developing countries willing to pay more for micronutrient-dense biofortified foods? Evidence from a field experiment in Uganda. *American Journal of Agricultural Economics* **93**, 83-97.

12. Conesa A, Götz S, García-Gómez JM, Terol J, Talón M and Robles M (2005). Blast2GO: A universal tool for annotation, visualization and analysis in functional genomics research. *Bioinformatics Applications Note* **21**, 3674-3676.

13. Curie C and Briat JF (2003). Iron transport and signaling in plants. *Annual Review of Plant Biology* **54**, 183-206.

14. Dai J -L, Zhu Y -G, Zhang M and Huang M -Z (2004). Selecting iodine-enriched vegetables and the residual effect of iodate application to soil. *Biological Trace Element Research* **101**, 265-276.

15. Darnton-Hill I, Webb P, Harvey PW, Hunt JM, Dalmiya N, Chopra M, Ball MJ, Bloem MW and de Benoist B (2005). Micronutrient deficiencies and gender: social and economic costs. *American Journal of Clinical Nutrition* **81**, 1198S-1205S.

16. De Vos RC, Moco S, Lommen A, Keurentjes JJ, Bino RJ and Hall RD (2007). Untargeted large-scale plant metabolomics using liquid chromatography coupled to mass spectrometry. *Nature Protocols* **2**, 778-791.

17. Dipti SS(2012). Bioavailability of selected minerals in different processing and cooking methods of rice (*Oryza sativa* L). Human Nutrition. Doctoral Thesis University of the Philippines.

18. Eurola M, Ekholm P, Ylinen M, Koivistoinen P and Varo P T (1989). Effects of selenium fertilization on the selenium content of selected Finnish fruits and vegetables. *Acta Agriculturae Scandinavica* **39**, 345-350.

19. Eurola MH, Ekholm PI, Ylinen ME, Koivistoinen PE and Varo PT (1991). Selenium in Finnish foods after beginning the use of selenite supplemented fertilisers. *Journal of the Science of Food and Agriculture* **56**, 57-70.

20. Fan MS, Zhao FJ, Fairweather-Tait SJ, Poulton PR, Dunham SJ and McGrath SP (2008). Evidence of decreasing mineral density in wheat grain over the last 160 years. *Journal of Trace Elements in Medicine and Biology* **22**, 315-324.

21. Fukao Y, Ferjani A, Tomioka R, Nagasaki N, Kurata R, Nishimori Y, Fujiwara M and Maeshima M (2011). iTRAQ analysis reveals mechanisms of growth defects due to excess zinc in Arabidopsis. *Plant Physiology* **155**, 1893-1907.

22. Garvin DF, Welch RM and Finley JW (2006). Historical shifts in the seed mineral micronutrient concentration of US hard red winter wheat germplasm. *Journal of the Science of Food and Agriculture* **86**, 2213-2220.

23. Goto F, Yoshihara T and Saiki H (2000). Iron accumulation and enhanced growth in transgenic lettuce plants expressing the iron-binding protein ferritin. *Theoretical and Applied Genetics* **100**, 658-664.

24. Graham RD, Welch RM, Saunders DA, Ortiz-Monasterio I, Bouis HE, Bonierbale, M, de Haan S, Burgos G, Thiele G, Liria R, Meisner CA, Beebe SE, Potts MJ, Kadian M, Hobbs PR, Gupta RK and Twomlow S (2007). Nutritious subsistence food systems. *Advances in Agronomy* **92**, 1-74.

25. Gregorio GB, Senadhira D, Htut H and Graham RD (2000). Breeding for trace mineral density in rice. *Food and Nutrition Bulletin* **21**, 382-386.

26. Grusak MA and DellaPenna D (1999). Improving the nutrient composition of plants to enhance human nutrition and health. *Annual Review of Plant Physiology and Plant Molecular Biology* **50**, 133-161.

27. Guiro AT, Galan P, Cherouvrier F, Sall MG and Hercberg S (1991). Iron absorption from African pearl millet and rice meals. *Nutrition Research* **11**, 885-893.

28. Hakala M, Lapveteläinen A, Houpalahti R, Kallio H and Tahvonen R (2003). Effects of varieties and cultivation conditions on the composition of strawberries. *Journal of Food Composition and Analysis* **16**, 67-80.

29. Hartikainen H (2005). Biogeochemistry of selenium and its impact on food chain quality and human health. *Journal of Trace Elements in Medicine and Biology* **18**, 309-318.

30. Haskell MJ, Jamil KM, Hassan F, Peerson JM, Hossain MI, Fuchs GJ and Brown KH (2004). Daily consumption of Indian spinach (*Basella alba*). or sweet potatoes has a positive effect on total-body vitamin a stores in Bangladeshi men. *American Journal of Clinical Nutrition* **80**, 705-714.

31. Hirschi KD (2009). Nutrient biofortification of food crops. *Annual Review of Nutrition* **29**, 401-421.

32. Ho L and White PJ (2005). A cellular hypothesis for the induction of blossom end rot in tomato fruit. *Annals of Botany* **95**, 571-581.

33. Hotz C, Loechl C, de Brauw A, Eozenou P, Gilligan D, Moursi M, Munhaua B, van Jaarsveld P, Carriquiry A and Meenakshi JV (2012b). A large-scale intervention to introduce orange sweet potato in rural Mozambique increases vitamin A intakes among children and women. *British Journal of Nutrition* **108**, 163-176.

34. Hotz, C, Loechl C, Lubowa A, Tumwine JK, Ndeezi G, Nandutu Masawi A, Baingana R, Carriquiry A, de Brauw A, Meenakshi JV and Gilligan DO (2012a). Introduction of B-carotene-rich orange sweet potato in rural Uganda results in increased vitamin A intakes among children and women and improved vitamin A status among children. *Journal of Nutrition* **142**, 1871-1880.

35. Hunter S, Apweiler R, Attwood TK, Bairoch A, Bateman A, Binns D, Bork P, Das U, Daugherty L, Duquenne L, Finn RD, Gough J, Haft D, Hulo N, Kahn D, Kelly E, Laugraud A, Letunic I, Lonsdale D, Lopez R, Madera M, Maslen J, 36.McAnulla C, McDowall J, Mistry J, Mitchell A, Mulder N, Natale D, Orengo C, Quinn AF, Selengut JD, Sigrist CJ, Thimma M, Thomas PD, Valentin F, Wilson D, Wu CH and Yeats C (2008). InterPro: The integrative protein signature database. *Nucleic Acids Research* **37**, 211-215.

37. Johnson AAT, Kyriacou B, Callahan DL, Carruthers L, Stangoulis J, Lombi E and Tester M (2011). Constitutive overexpression of the *OsNAS* gene family reveals single-gene strategies for effective iron- and zinc-biofortification of rice endosperm. *PLoS ONE* **6**, e24476.

38. Juliano BO (1985). Rice properties and processing. *Food Reviews International* **1**, 432-445.

39. Karley AJ and White PJ (2009). Moving cationic minerals to edible tissues: Potassium, magnesium, calcium. *Current Opinion in Plant Biology* **12**, 291-298.

40. Langmead B, Hansen KD and Leek JT (2010). Cloud-scale RNA-sequencing differential expression analysis with Myrna. *Genome Biology* **11**, R83.

41. Li S, Nugroho A, Rocheford T and White WS (2010). Vitamin A equivalence of the B- carotene in B-carotene-biofortified maize porridge consumed by women. *American Journal of Clinical Nutrition* **92**, 1105-1112.

42. Liang J, Li Z, Tsuji K, Nakano K, Nout MJR and Hamer R (2008). Milling characteristics and distribution of phytic acid and zinc in long-, medium-, and short-grain rice. *Journal of Cereal Science* **48**, 83-91.

43. Low JW, Arimond M, Osman N, Cunguara B, Zano F and Tschirley D (2007). A food-based approach introducing orange fleshed sweet potato increased vitamin A intake and serum retinol concentrations in young children in rural Mozambique. *Journal of Nutrition* **137**, 1320-1327.

44. Lyons GH, Judson GJ, Ortiz-Monasterio I, Genc Y, Stangoulis JCR and Graham RD (2005). Selenium in Australia: Selenium status and biofortification of wheat for better health. *Journal of Trace Elements in Medicine and Biology* **19**, 75-82.

45. Mayer JE, Pfeiffer WH and Bouis P (2008). Biofortified crops to alleviate micronutrient malnutrition. *Current Opinion in Plant Biology* **11**, 166-170.

46. Meenakshi JV, Banerji A, Manyong V, Tomlins K, Mittal N and Hamukwala P (2012)Using a discrete choice experiment to elicit the demand for a nutritious food: willingness-to-pay for orange maize in rural Zambia. *Journal of Health Economics* **31**, 62-71.

47. Metson AJ (1974). Magnesium in New Zealand soils I Some factors governing the availability of soil magnesium: A review New Zealand. *Journal of Experimental Agriculture* **2**, 277-319.

48. Morris J, Hawthorne KM, Hotze T, Abrams SA and Hirschi KD (2008). Nutritional impact of elevated calcium transport activity in carrots. *Proceedings of the National Academy of Sciences of the United States of America* **105**, 1431-1435.

49. Murray-Kolb LE, Takaiwa F, Goto F, Yoshihara T, Theil EC and Beard JL (2002). Transgenic rice is a source of iron for iron-depleted rats. *Journal of Nutrition* **132**, 957-960.

50. Muthamilarasan M, Theriappan P and Prasad M (2013). Recent advances in crop genomics for ensuring food security. *Current Science* **105**, 155-158.

51. Muzhingi T, Gadaga TH, Siwela AH, Grusak MA, Russell RM and Tang G (2011). Yellow maize with high B-carotene is an effective source of vitamin A in healthy Zimbabwean men. *American Journal of Clinical Nutrition* **94**, 510-519.

52. Namanya P (2011). Towards the Biofortification of Banana Fruit for Enhanced Micronutrient Content. Doctoral Thesis Queensland University of Technology.

53. Naqvi S, Zhu C, Farre G, Remessar K, Bassie L, Breitenbach J, Perez Conesa D, Ros G, Sandmann G, Capell T and Christou P(2009). Transgenic multivitamin corn through biofortification of endosperm with three vitamins representing three distinct metabolic pathways. *Proceedings of the National Academy of Sciences of the United States of America* **106**, 7762-7767.

54. Paine JA, Shipton CA, Chaggar S, Howells RM, Kennedy MJ, Vernon G, Wright SY, Hinchliffe E, Adams JL, Silverstone AL and Drake R (2005). Improving the nutritional value of Golden Rice through increased pro-vitamin A content. *Nature Biotechnology* **23**, 482-487.

55. Palmgren MG, Clemens S, Williams LE, Krämer U, Borg S, Schjørring JK and Sanders D (2008). Zinc biofortification of cereals: Problems and solutions. *Trends in Plant Science* **13**, 464-473.

56. Perkins DN, Pappin DJC, Creasy DM and Cottrell JS (1999). Probability based protein identification by searching sequence databases using mass spectrometry data. *Electrophoresis* **20**, 3551-3567.

57. Phattarakul N, Rerkasem B, Li LJ, Wu H, Zou CQ, Ram H, Sohu VS, Kang BS, Surek H, Kalayci M, Yazici A, Zhang FS and Cakmak I (2012). Biofortication of rice grain with zinc through zinc fertilization in different countries. *Plant and Soil* **361**, 131-141.

58. Rengel Z, Batten GD and Crowley DE (1999). Agronomic approaches for improving the micronutrient density in edible portions of field crops. *Field Crops Research* **60**, 27-40.

59. Rosado JL, Hambidge KM, Miller LV, Garcia OP, Westcott J, Gonzalez K, Conde J, Hotz C, Pfeiffer W, Ortiz-Monasterio I and Krebs NF (2009). The quantity of zinc absorbed from wheat in adult women is enhanced by biofortification. *Journal of Nutrition* **139**, 1920-1925.

60. Saltzman A, Birol E, Bouis HE, Boy E, De Moura FF, Islam Y and Pfeiffer WH (2013). Biofortification: Progress toward a more nourishing future. *Global Food Security* **2**, 9-17.

61. Tang G, Qin J, Dolnikowski GG, Russel RM and Grusak MA (2009). Golden rice is an effective source of vitamin A. *American Journal of Clinical Nutrition* **89**, 1776-1783.

62. Thakkar S, Huo T, Maizya-Dixon B and Failla M (2009). Impact of type of processing on retention and bioaccessibility of B-carotene in cassava. *Journal of Agriculture and Food Chemistry* **54**, 1344-1348.

63. van Jaarsveld PJ, Marais DW, Harmse E, Nestel P and Rodriguez-Amaya DB (2006). Retention of B-carotene in boiled, mashed orange-fleshed sweet potato. *Journal of Food Composition and Analysis* **19**, 321-329.

64. Vasconcelos M, Datta K, Oliva N, Khalekuzzaman M, Torrizo L, Krishnan, S, Oliveira M, Goto F and Datta SK (2003). Enhanced iron and zinc accumulation in transgenic rice with the ferritin gene. *Plant Science* **164**, 371-378.

65. Velu G, Ortiz-Monasterio I, Cakmak I, Hao Y and Singh RP (2013). Biofortification strategies to increase grain zinc and iron concentrations in wheat. *Journal of Cereal Science* doi: 101016/jjcs201309001.

66. Waters, BM and Sankaran RP (2011). Moving micronutrients from the soil to the seeds: Genes and physiological processes from a biofortification perspective. *Plant Science* **180**, 562-574.

67. White PJ and Broadley MR (2005). Biofortifying crops with essential mineral elements. *Trends in Plant Science* **10**, 586-593.

68. White PJ and Broadley MR (2009). Biofortification of crops with seven mineral elements often lacking in human diets - Iron, zinc, copper, calcium, magnesium, selenium and iodine. *New Phytologist* **182**, 49-84.

69. Williams L and Salt DE (2009). The plant ionome coming into focus. *Current Opinion in Plant Biology* **12**, 247-249.

70. Zhao F-J and McGrath SP (2009). Biofortification and phytoremediation. *Current Opinion in Plant Biology* **12**, 373-380.

71. Zhu C, Naqvi S, Gomez-Galera S, Pelacho AM, Capell T and Christou P (2007). Transgenic strategies for the nutritional enhancement of plants. *Trends in Plant Science* **12**, 548-555.

72. Zou CQ, Zhang Y, Rashid A, Ram H, Savasli E, Arisoy, RZ, Oritz-Monasterio, I, Simunji IA, Wang SA, Sohu ZHA, Hassan VA, Kaya MA, Onder YA, Lungu OA, Mujahid OA, Yaqub M, Joshi A, Zelenskiy AKA, Zhang YA and Cakmak FSA (2012). Biofortication of wheat with zinc through zinc fertilization in seven countries. *Plant Soil* **361**, 119-130.

2015, Horticulture for Nutrition Security
Editor: **Prof. K.V. Peter**
Published by: **DAYA PUBLISHING HOUSE, NEW DELHI**

Pages 141–146

Chapter 8

Food Supplements to Complement Urban Food Security

K.V. Peter and Binoo Bonny

Despite tremendous gains in global food production, there is unusual increase in the number of undernourished in recent years with figures crossing 1 billion mark in 2009-10 with only marginal reductions thereafter. The estimates of the global trends of poverty and under nourishment show that the number of malnourished people in 2007 increased by 75 million, over and above FAO's estimate of 848 million undernourished in 2003-05. The Asia-Pacific region with 578 million undernourished account for 62 percent of the world's food insecure population. More importantly, with increased urban migration for higher income non agricultural vocations in the cities, there is a reversal of urban – rural population ratio from 40:60 to 60:40. This has posed tremendous pressure on the carrying capacity of urban areas leading to higher cost of living and reduced access to quality hygienic food.

Journey of Indian Agriculture from the years of acute food crisis to food self sufficiency and food security entitlements has been noteworthy. Enormity of the problems faced by Indian agriculture is evident from the fact that despite the national gains in food self sufficiency (51million tones in 1951 to 258 million tons in 2013) and security through the heralded Green revolution, India still has a population of 250 million food insecure inhabitants which roughly estimate to one fourth of world's food insecure people The Global Hunger Index (GHI), a more comprehensive tool used to measure and track global hunger, ranks India in the category of alarming poverty along with Bangladesh. More disturbing are the rates of growing malnutrition among urban households. The status of urban health in India as projected in the

National Health Survey (2006) provides a bleak picture of rampant anemia, symptoms of different forms of malnutrition like under nutrition and hidden hunger leading to loss of worker productivity. New life style diseases like diabetes, obesity, cardiovascular diseases and diseases of eye are also taking heavy toll in urban areas. Interventions through food supplements for balanced diet rich in fibre, minerals and vitamin sources are recommended in urban food security programs throughout the country.

Kerala, the southernmost state of India, represents a unique dimension of urban food security arising from the narrow rural urban divide on demographic and socio-economic parameters and presence of enormous migrant laborers from northern and north eastern states of the country. The rurban character of the state with the towns and villages forming a continuum provides the opportunity for Metropolitan Agriculture that links rural agricultural entrepreneurs to metropolitan markets. Some major interventions from the state in rural – urban linkages that enabled establishment of market-linked agro-food networks addressing urban food security are discussed.

Urban and peri-urban farming and nutrition garden/kitchen garden/backyard garden programmes are being promoted in cities of Kerala, to combat the problems of malnutrition. Leaf vegetables like amaranth, chekkurmanis, Ceylon spinach, agathi leaf, curry leaf, drum stick leaf and tender cow pea leaves are encouraged to grow in available spaces. Pesticide free and safe to eat vegetables through community linked production from Kudumbasree women SHGs is promoted through urban fare price markets of Horticorp and civil societies. The malady of protein hunger is addressed through protein rich vegetables like winged bean, dolichos bean, cow pea, cluster bean and rice beans grown in homesteads and leased public lands. Cucurbit flowers (pumpkin, ashgourd) rich in iron and minerals popularized through midday meals of Anganvadi children of poor households are ideal to fight hidden hunger. Noni (*Morinda citrifolia* L.) (Figures 8.1–8.2) that comes up well in highly saline soils is another recommended food supplement. Studies have shown its positive food supplementary values, especially of the fruit juice that contains more than 166 nutraceuticals, antioxidants and other phytochemicals (Tables 8.1–8.2). Noni juice in sachets is highly suited for mid day meal program in urban areas. Climate smart cereals like millets-pearl millet, finger millet, sorghum, maize- and biofortified crops like Golden rice, high lycopene tomatoes and high carotene sweet potatoes can alleviate hidden hunger to a great extent. Experiments in food and bio parks in rice and coconut that add value to plant biomass through agro-processing and preparation of a wide range of market linked products also show potential to solve urban food security problems. Protected cultivation of vegetables, herbs and spices is promoted through Governmental interventions.

Availability of quality drinking water is a pre-requisite to healthy urban diet. Food supplements play vital role to make current diets balanced and nutritious. Nutrition education to housewives is undertaken through "Kudumbasree" approach.

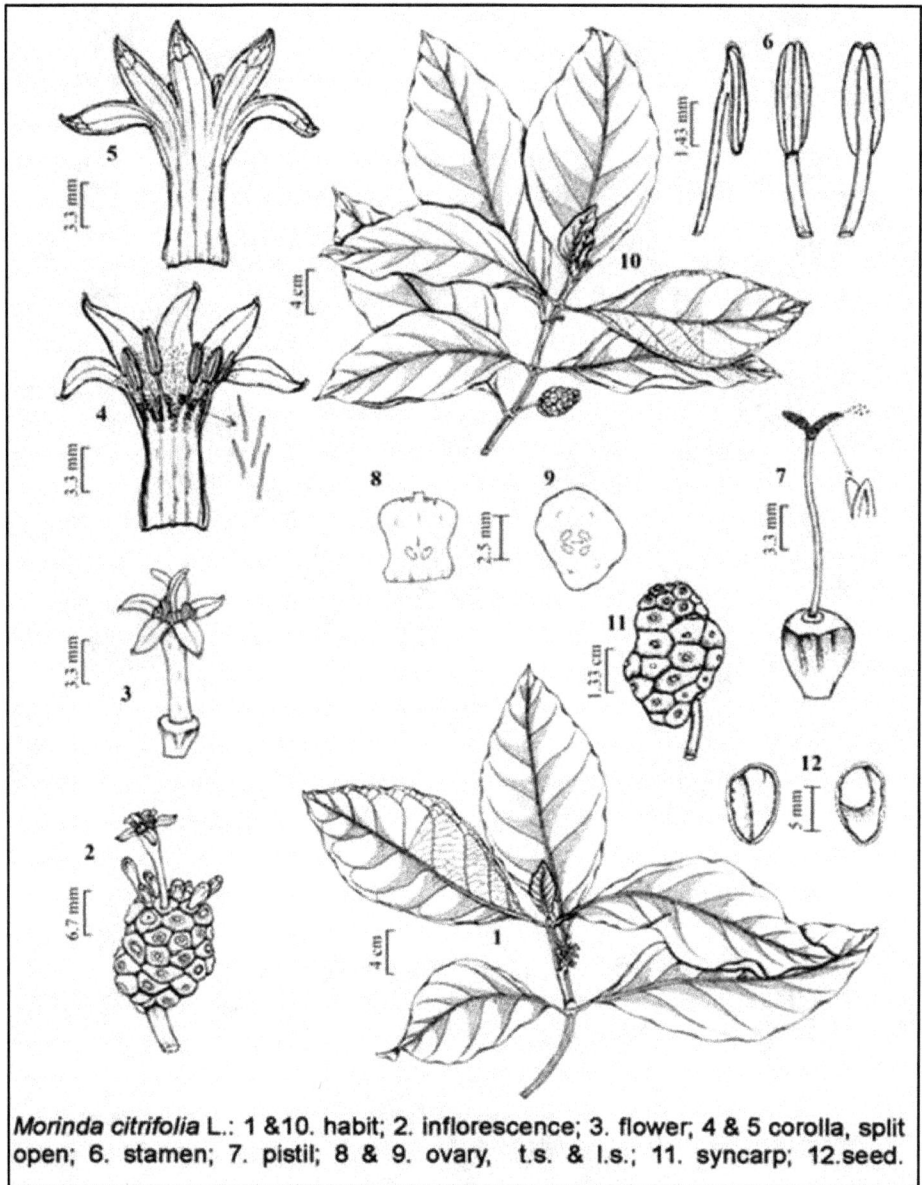

Morinda citrifolia L.: 1 &10. habit; 2. inflorescence; 3. flower; 4 & 5 corolla, split open; 6. stamen; 7. pistil; 8 & 9. ovary, t.s. & l.s.; 11. syncarp; 12.seed.

Figure 8.1: Noni (*Morinda citrifolia* L.).

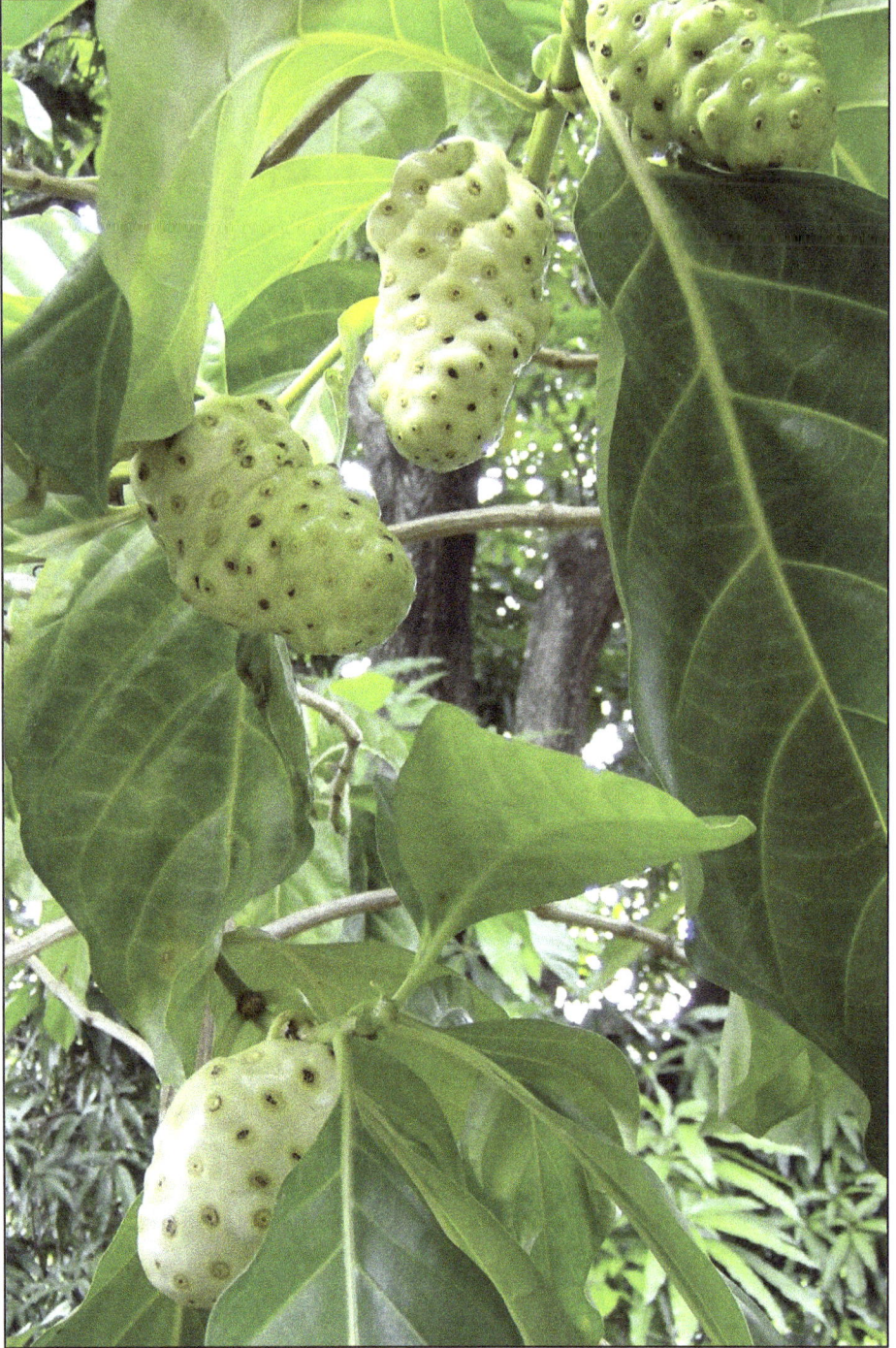

Figure 8.2: Noni (*Morinda citrifolia* L.) – Bearing Tree.

Table 8.1: Important Bioactive Compounds in Noni and *in-vitro* Biological Activity

Bioactive compound	Part	Class	Bioactivity
E-phytol, Cycloartenol Stigmasterol	Fruit	Sterol	Anti-tuberculosis
Damnacanthal	Fruit	Anthraquinone	Anti-HIV
Isoscopoletin Aesculetin Quercetin	Fruit	Phenolics	Antioxidant
Americanin A	Fruit	Lignans	Antioxidant
Narcissoside	Fruit	Flavonoids	Antioxidant
2-Methoxy- 1,3,6-trihydro-xyanthraquinone(5)	Fruit	Anthraquinone	phase II enzyme inducer
Damnacanthol-3-O-β-D-primeveroside Lucidin 3-O- β -D-primeveroside	Root	Anthraquinone	Anti-diabetic
2-O-(β -D-Glucopyranosyl) -1-O-octanoyl-beta-D-glucopyranose	Fruit	Fatty acid ester	Anti-inflammatory
1,4-Dihydroxy-2-methoxy-7-Methylanthraquinone Austrocortinin	Fruit	Anthraquinone	Wound healing activity
(+)-3,4,3',4'-Tetrahydroxy-9,7'alpha-epoxylignano-7 alpha,9'-lactone (+)-3,3'-Bisdemethyltanegnol Quercetin Kaempferol	Fruit	Lignans	Cardiovascular activity
Americanol A 3,3'-Bisdemethylpinoresinol Isoprincepin Morindolin	Fruit	Lignans	Cardiovascular activity
Damnacanthal	Root	Anthraquinone	Anti-cancer
Morindone	Root	Anthraquinone	Anti-cancer
6-O-(β -D-Glucopyranosyl) -1-O-octanoyl-βb-D-glucopyranose Asperulosidic acid	Fruit	Glycosides	Anticancer

Table 8.2: Noni-Rich in Nutrients

☆ Noni contains approximately 200 nutraceuticals and phyto-chemicals including amino-acids, vitamins, minerals, fatty acids, alcohols , phenols, anthraquinone glycosides, carotenoids, esters, flavonoids, iridoids, ketones, lactones, lignans, nucleosides, triterpenoids and sterols

NUTRIENT CONTENT -DIFFERENT PARTS OF *Morinda citrifolia*					
Parts of the plant	K	Ca	Mg	Fe	Cu
(ppm)					
Leaf	1219	5462	570.60	4.47	2.23
Wood	Trace	270.02	44.67	378.54	6.22
Bark	Trace	534.34	47.93	146.67	5.46
Fruit	1226	58.89	196.64	42.44	27.44

References

Peter, P.I. (2007). Monograph on Noni (*Morinda citrifolia* L.). World Noni Research Foundation, Chennai - 96, p. 852.

www.divinenoni.com

2015, Horticulture for Nutrition Security
Editor: **Prof. K.V. Peter**
Published by: **DAYA PUBLISHING HOUSE, NEW DELHI**

*Pages **147–168***

Chapter 9

Fertility Management for Horticultural Crops in Acidic Soils of Warm Humid Tropics

P. Sureshkumar and P. Geetha

Agricultural productivity and agro-biodiversity of an area are largely governed by the overhead climate and qualities of land and soil. In the tropics and subtropics, where sunlight and temperature are not limiting, it is often the precipitation and the capacity of the land and soil to retain water and nutrients that control biological productivity. High rain fall and high temperature in the tropics results in leaching of bases like K, Ca and Mg and accumulation of insoluble sesquioxides of Al, Fe and Mn. There will be loss of organic matter and silica also. The entire process results in acidification and formation of laterites. Thus, the laterite soil which predominate the entire state of Kerala is poor in fertility.

Agro-ecological Zones

Kerala state planning board commissioned the NBSS and LUP, Bangalore to carry out a detailed study of the agro ecology of the state and to delineate the agro ecological zones, units and sub units. Subsequently the NBSS and LUP divided the state into five agro ecological zones based mainly on physiographic features as (1) coastal plains, (2) midland laterites, (3) foothills, (4) high hills and (5) Palakkad plain (Nair *et al.,* 2011). This classification visualizes the entire state as a macro water shed with streams and rivers (44 rivers) originating from the high hill regions, flowing through the foot hills and midland laterites and emptying the water into the coastal plains as lakes, *kayal,* and other water bodies which is connected to marine intrusions. The five AEZs are sub divided into 23 Agro-ecological units (AEUs) (Figure 10.1).

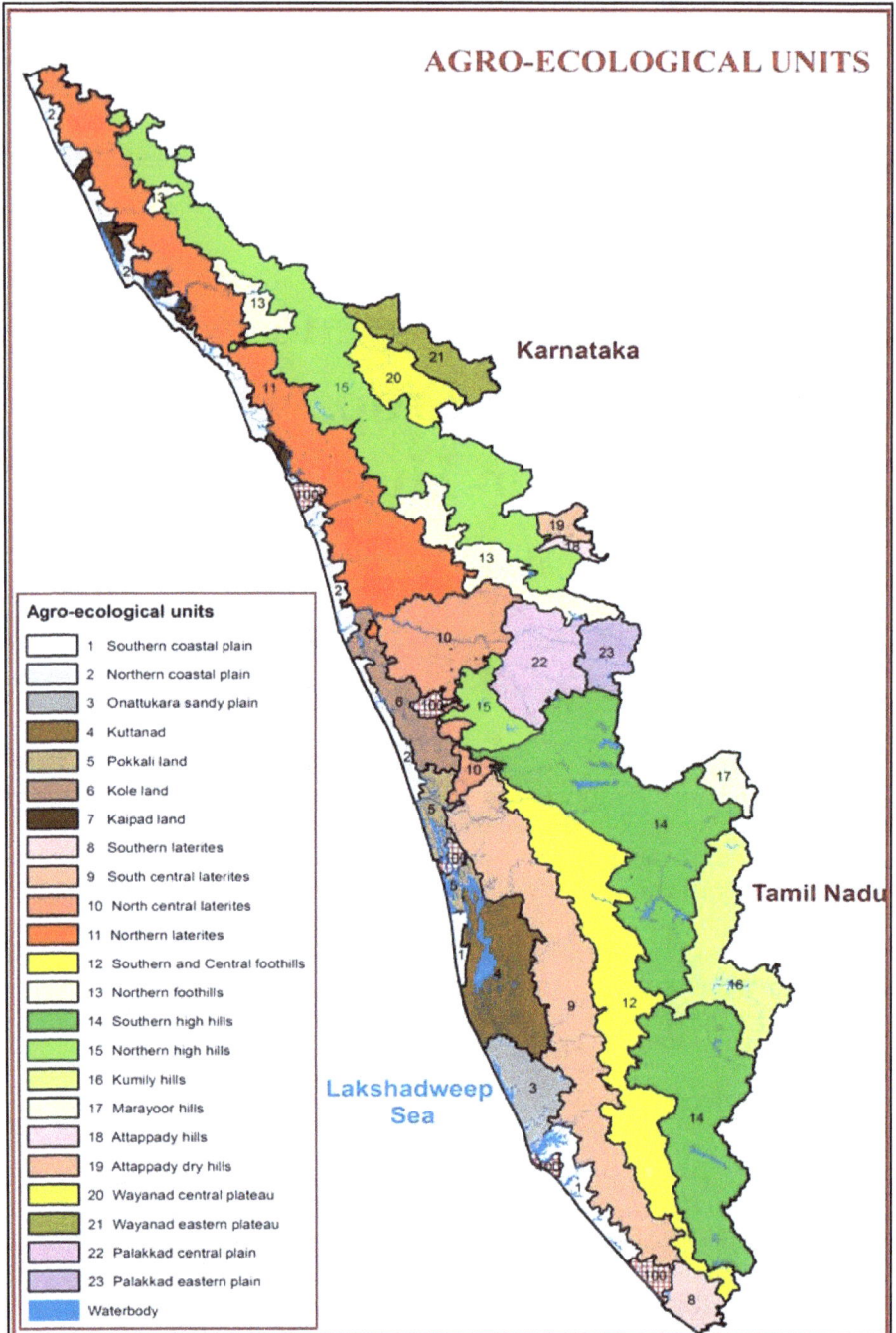

AGRO-ECOLOGICAL UNITS

Agro-ecological units

1 Southern coastal plain
2 Northern coastal plain
3 Onattukara sandy plain
4 Kuttanad
5 Pokkali land
6 Kole land
7 Kaipad land
8 Southern laterites
9 South central laterites
10 North central laterites
11 Northern laterites
12 Southern and Central foothills
13 Northern foothills
14 Southern high hills
15 Northern high hills
16 Kumily hills
17 Marayoor hills
18 Attappady hills
19 Attappady dry hills
20 Wayanad central plateau
21 Wayanad eastern plateau
22 Palakkad central plain
23 Palakkad eastern plain
Waterbody

Karnataka

Tamil Nadu

Lakshadweep Sea

Figure 9.1: Agro-ecological Units of Kerala.

1. Coastal Plain

The coastal plain comprises the nearly level to gently sloping lands along the coast at elevation below 30 metres and lying between Arabian Sea and the mid lands. It includes sandy beaches, sandy plains, coastal laterites and low lying areas such as estuaries, backwaters, submerged lands, swamps, marshes, *kayal* lands and broad valleys. The zone covers 5,09,246 ha (13.10 per cent of the total area) in the state.

2. Mid Land Laterites

This zone covers undulating to rolling lands interspersed with narrow valleys between the coastal plain on the west and foot hills and hills on the east extending from the southern end to the northern end of the state. The elevation ranges from 30 to 300 metres. The zone comprises 10,56,385 ha(27.18 per cent).

3. Foot Hills

The undulating to rolling lands and low hills between the midland laterites on the west and high hills on the east constitute this agro-ecological zone. The terrain has only very narrow valleys. The elevation ranges from 300 to 600 metres. The area covered in this zone is 4,60,074ha (11.84 per cent).

4. High Hills

The hilly region comprising Western Ghats and plateaus extending from south to north constitute this zone. The elevation ranges from 600metres to peaks as high as 1800 metres. The slope of hills can be as high as 80 per cent. The zone covers 15,53,225 ha(39.97 per cent).

5. Palakkad Plain

The Palakkad gap resembling an inland plain with low elevation is a prominent physical feature along the valley of the Bharathapuzha river. The gently sloping lands of Palakkad, east of Kuthiran hills flanked on the south and north by Nelliyampathy hills and Attappady hills respectively and merging to Tamil Nadu uplands through the gap in Western Ghats constitute this zone covering 1,60,006ha (4.12 per cent).

Soils

The soils of Kerala are developed under tropical humid climate. Majority of the area under these five zones are extremely acidic to moderately acidic (pH 3.5 to 6.0) with low organic matter content, low cation exchange capacity and with low nutrient and water holding capacity. The exceptions are neutral to slightly alkaline soils of dry hill areas of Idukki, Attappadi and the black cotton soils of Chitur taluk in Palakkad. Laterization is the predominant soil forming process under Kerala state. The process is characterized by the leaching (eluviation) of bases (Na, K, Ca, and Mg), silica and organic matter due to high rain fall(3000mm) resulted in accumulation of sesquioxides of Al, Fe and to some extent Mn (Buol *et al.,* 1998). The net effect is formation of soils with dominance of low activity kaolinitic clay with lowest possible CEC, low organic matter content and high acidity.

The soils in general are low to medium in organic carbon, and hence in nitrogen, high in P due to continuous application of P fertilizers which are being fixed and accumulated and low in K. Severe deficiency of Ca and Mg are observed. S deficiency is only occasional. Fe and Mn are abundant in plant available form which is to the tune of toxic levels in wet lands submerged with water. Zn deficiency is only marginal. Cu deficiency is usually less and its level is sufficient in soils where Cu based fungicides are used especially under plantations like rubber, cardamom, pepper etc. Boron deficiency is severe and wide spread and symptoms are common in banana, coconut, areca nut, nutmeg, cabbage and cauliflower as well as in rice.

Crops

The state of Kerala is dominated by horticultural crops in the uplands of the above AEZs. These include coconut as plantation as well as in homesteads mainly in the uplands of coastal plain and the midland laterites, followed by rubber. The vegetables and fruit crops especially banana are also cultivated in the midland laterites. The wetlands in the coastal plains and the lateritic brown hydrmorphic soils of midland are covered by rice. The rice lands are being converted to vegetable and banana especially in the brown hydromorphic midland laterites. Pepper is widely cultivated in the midlands as an under crop in midland laterites along with coconut and areca nut. Pepper as plantation is very common in foothills and high hill areas especially in Wayanad Plateau and in high ranges of Idukki. Cardamom is cultivated in the high hill areas of Idukki and Wayanad districts. So also is the case of tea. Coffee is a major crop in foothills and Wayanad plateau.

Summer vegetables like cucurbits, bhindi, brinjal, chillies, and amaranthus are cultivated in the uplands as well as in summer rice fallows in the midland laterite zone. The cool season vegetables like carrot, cabbage, cauliflower, onion etc. are seen in the high altitude dry hill zones of Idukki. These crops especially varieties of cabbage and cauliflower suited for hot humid climate are also grown in plains. The total area under each of the above crops is given in Table 9.1.

Table 9.1: Area Under Different Horticultural Crops in Kerala

Crops	Area Under Cultivation (ha)	Per cent of Net Cultivated Area
Spices and condiments		
Pepper	85335	4.2
Arecanut	104548	5.1
Nutmeg	18161	0.9
Cardamom	41600	2
Turmeric	2970	0.2
Tamarind	14879	0.7
Cashew	6908	0.34
Vanilla	400	0.02
Cloves	939	0.05

Contd...

Table 9.1–*Contd...*

Crops	Area Under Cultivation (ha)	Per cent of Net Cultivated Area
Cinnamon	271	0.01
Garlic	155	0.01
Garcinia	426	0.02
Sugar crops		
Sugarcane	2604	0.13
Palmyrah	3195	0.16
Tubers		
Tapioca	74498	3.65
Other tubers	19900	0.97
Vegetables	41154	2.0
Fresh fruits		
Banana	59069	2.9
Mango	75559	3.7
Plantain	48747	2.4
Jack	90333	4.4
Other fruits	37422	1.8
Oil seeds		
Coconut	820867	40.2
Others	3359	0.16
Fibre, drug and Narcotics		
Cotton	400	0.02
Lemon Grass	371	0.02
Others	370	0.02
Plantation crops		
Rubber	539565	26.5
Coffee	85359	4.2
Tea	37028	1.8
Cocoa	12764	0.63

Source. Department of Economics and Statistics, Government of Kerala, 2013.

Fertility Management

Fertility Management with holistic view of sustainability of fertility as well as soil health is an uphill task under tropical climate since quality deterioration and fertility drains are continuous and naturally occurring due to high rainfall and high temperature. The soils developed under topical humid climate are dominated by low activity kaolinitic clay with low surface area, with low CEC and hence with low nutrient and water holding capacity.

The management of soil fertility in the above said soils should be planned in such a way that it should start first with management of soil acidity, followed by nutrient and water management.

1. Acidity

Construction of a new building should start with a strong basement to support the structure to be built. Similarly any management of fertility in tropical acidic soils should start with correction of soil acidity. Acidity in tropical soils of Kerala is due to accumulation of Al, Fe and Mn and loss of bases due to leaching by rain water. Monitoring and correction of soil pH to near neutral level is the basic criterion to be followed. This would enable the rhizosphere soil solution to be dominated with plant available forms of nutrients. Acidity will be from two sources; the active acidity due to H^+ ions in soil solution and the reserve acidity due to Al^{3+} Fe^{2+} and Mn^{4+} ions on the solid phase (oxides and hydrous oxides of these elements), which generate H^+ ions due to hydrolysis under favorable moisture regimes. This acidity can be neutralized by application of lime followed by washing/leaching with water. Application of lime, if based on the actual lime requirement may come to the tune of 3 to 4 tonnes per ha and may go even to the level of 8 to 9 tonnes due to high levels of reserve/potential acidity. An alternate strategy can be easily followed by farmers which is economically feasible is application of lime in every season just to neutralize the active acidity. This can be done on the basis of pH. The lime requirement based on pH is given in Table 9.2. The correction of pH to neutral level also helps in controlling the fungal diseases. Fungi prefer an acidic soil reaction for favorable growth. Near neutral to alkaline pH restricts fungal growth. Correction of soil acidity thus controls many of the devastating fungal diseases like that caused by *Phytophthora; viz.* Quick wilt or foot rot in pepper, bud rot in coconut, Mahali or fruit rot in arecanut, rot (Azhukal) in cardamom etc. Incidence of bacterial wilt also can be restricted by lime application.

Table 9.2: Soil pH and Lime Requirement

Sl.No.	pH Range	Class	Lime Requirement (kg $CaCO_3$ ha^{-1})
1.	Ultra acid	<3.5	1000
2.	Extremely acid	3.5-4.4	850
3.	Very strongly acid	4.5-5.0	600
4.	Strongly acid	5.1-5.5	350
5.	Moderately acid	5.6-6.0	250
6.	Slightly acid	6.1-6.5	100

The chemical environment of rhisosphere is modified by lime in such a way as the pH is raised to near neutral level to have the following direct effect:

1. The available/soluble forms of Fe, Al and Mn are reduced due to precipitation.
2. This in turn releases precipitated/fixed forms of P thereby increasing P availability (Geetha, 2008).

3. The competitive absorption of Fe and Mn is restricted due to decrease in their concentration thereby enhancing absorption of applied potassium, Mg and Ca (Priya *et al.,* 2007).

4. The biological decomposition of organic matter is hastened resulting in mineralization of N in to available NH_4^+ and later to NO_3^- forms (Haynes, and Swift, 1988).

5. Mo availability is enhanced due to increase in pH. Zn and Cu will be more due to less competition from Fe and Mn.

Thus modifying the pH to near neutrality by lime increases availability and absorption of nutrients by roots, both from native and applied sources, reduces the attack by soil borne pathogens, enhances activity of beneficial microbes like nitrifiers and N-fixers.

2. Organic Matter

The tropical climate warrants the possible depletion of organic matter status in soils of Kerala due to oxidation of organic matter at a faster rate because of favorable moisture and temperature conditions. The low activity clay dominated soil with low organic matter is a major hurdle in water and nutrient retention and supplying capacity of these soils. The organic matter status is an indication of healthy and fertile soil. Hence maintenance and improvement of organic matter needs special attention. Organic matter has multifaceted role in soil as it improves, structure, nutrient and water holding capacity, porosity, infiltration rate, biological activity, soil aeration, N fertility, decreases bulk density and surface crusting (Haynes and Naidu, 1998) and ultimately soil health. The CEC of the tropical soils is mainly due to pH dependent charges as permanent charges of kaolinitic clay dominated soil is very low confined to the broken edges of clay mineral. The main source of pH dependent charges is functional groups of humic compounds. The microbial activity and population of beneficial microbes is governed by the level of organic matter. Most of Kerala soils are texturally sand dominated and hence for improved porosity and water retention, level of organic matter should be improved. Further, N fertility management is closely associated with organic carbon content due to constancy of C : N ratio, improvement in N status can be attained only by improving organic carbon. In brief, any crop production system should be sustainable only by basal application of organic matter in the form of compost or FYM. The practice should be continued in every season. Organic manures are also sources of micro nutrients. Hence regular application of organic manures can meet the micronutrient requirement. Here, care must be taken to enrich the manure with those micronutrients which are inherently deficient in a particular area. This can be done by addition of micronutrient fertilizers at the time of composting or preparation of manure.

3. Nitrogen, Phosphorus and Potassium

Most of the horticultural crops especially vegetables; banana and ginger are grown under high input intensive farming system in the state. Hence high amount of N P K fertilizers are used. This, if not balanced or if not as per the requirements of the crop can cause imbalanced nutrition. The use of high amounts of N fertilizers

Potassium deficiency

Calcium deficiency:
Blossom end rot in tomato

Magnesium deficiency in tomato

Sulphur deficiency: Marbled appearance in leaves(Rape Seed)

Zn deficiency in tomato

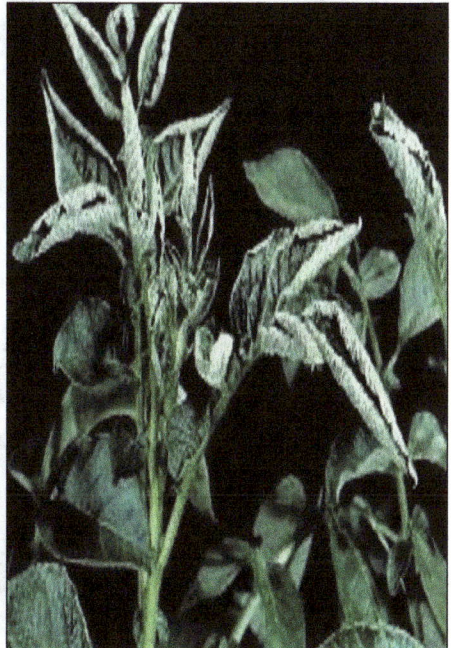

Copper deficiency (marginal curling) in chillies

Boron deficiency in coconut

Boron deficiency in banana

Boron deficiency in cauliflower

Boron deficiency in tomato

especially urea and ammonium phosphates can generate residual acidity since these forms of N fertilizers are acid forming ones. This also focuses to the importance of liming before fertilizer application in every cropping season.

The acidic kaolinitic clay dominated soils with high content of oxides and hydrous oxides of Fe and Al in the state are having high P fixing capacity. As a result, continuous application of P fertilizers like ammonium phosphates, rock phosphates and bone meal for the last 2-3 decades resulted in buildup of P in these soils to a level of saturation. Thus it is advisable that in soils where P was continuously applied for long periods, even skipping of P or at least reducing the levels of P fertilizers must be seriously thought off. With respect to availability of already fixed P, it can get released to available form just by increasing the pH to near neutrality by liming. P solubilizers also will be active at this pH.

The soils of Kerala are poor in K and since it is dominated by 1:1 type Kaolinite, K fixation is practically nil. Further, exchangeable K is also low as the CEC itself is very low. Thus, K retention in these soils is negligible. Hence application of K should be done judiciously as per requirement of crops in correct time. It should be given in as many splits as possible to avoid losses (KAU, 2011).

4. Calcium, Magnesium and Sulphur

The acid soils of Kerala are inherently deficient in these two secondary nutrients. Ca deficiency results in failure of terminal bud to open and get distorted, root tip also fails to grow. Die back starting from the tip is the common symptom. In banana Ca- B deficiency complex is very common resulting in choking of leaves at the crown. Young leaves fail to emerge out. The deficiency of Ca can cause failure of bunch emergence. Blossom end rot is commonly seen in tomato. Application of lime not only reclaims acidity but also corrects Ca deficiency.

Mg is deficient in more than 80 per cent of Kerala soils. Interveinal chlorosis of older leaves is the typical symptom. Application of $MgSO_4$ @ 80 kg ha^{-1} can correct Mg deficiency. If good quality dolomite is available its application can correct both Ca and Mg deficiency as well as acidity. In cowpea, application of Ca and Mg is mandatory since this crop requires near neutral pH for N fixation, and also for flowering and fruit set.

Sulphur deficiency is only to the tune less than 5 per cent in the state. $MgSO_4$ application can correct deficiency of sulphur. Sulphur requirement is more for crops like cabbage and cauliflower as well as for onion.

5. Micro-nutrients (Fe, Cu, Mn and Zn)

In general, the availability of micronutrients which are absorbed by cationic form (Fe^{2+}, Cu^{2+}, Mn^{2+} and Zn^{2+}) will be sufficient to meet the crop requirement under acidic soil environment. However, the soils of the state are very rich in Fe and Mn and their higher concentrations in soils antagonistically affect absorption of Zn and Cu by plants even under sufficiency levels. Thus, it becomes essential to apply Zn and Cu through foliar nutrition to get proper response. The deficiency of Cu in soils of Kerala is to the tune of about 17 per cent. Deficiency of Cu is seldom observed in plantations due to frequent application of Cu based fungicides. Upward cupping and curling of younger leaves in tomato and chillies, stunted growth of vines and crinkling of younger leaves in cucurbitaceous vegetables are typical symptoms of Cu deficiency.

The extent of Zn deficiency is to the tune of 30 per cent in Kerala soils. Rosette appearance and little leaf are the common symptoms of Zn deficiency. High levels of P in soil affect Zn availability as Zn gets precipitated as zinc phosphates. Under such situation foliar spray of $ZnSO_4$ (0.1 to 0.2 per cent) may be adhered with.

6. Boron

Boron deficiency is widespread in the state due to leaching losses of both native and applied boron under high rainfall condition. Boron is essential for translocation of metabolites. It is also needed for pollen fertility. Hence its deficiency at the time of flowering can cause failure in fertilization and fruit set. At vegetative stage, terminal bud fails to open. In palms like coconut and areca nut, boron deficiency starts with transverse translucent streaking on the leaflets and the tip of the leaflets become sharply bent which is commonly called as "hook leaf". These sharp leaflet hooks are quite rigid and cannot be straightened out without tearing the leaflets. In some species, these "hooks" drop off. Drying of inflorescence is seen in coconut under severe boron deficiency. In vegetables like cabbage and cauliflower, curling of leaves, hollow stem and curd formation are the major deficiency symptoms. Application of borax @ 10 kg ha⁻¹ can correct the deficiency of boron (KAU, 2011). The deficiency of boron causes abnormally reduced leaf size for the younger leaves and failure of younger buds to open in severe cases in rubber. Foliar application of 0.5 percent solution of borax is also recommended to correct boron deficiency.

Soil Test Based Recommendations

The Kerala Agricultural University has blanket recommendations for all crops. However, it is always better to adopt need based fertilizer applications depending upon the soil type and the crops grown. Hence soil testing for plant available nutrients becomes mandatory as it can help the farmer to apply nutrients based on the available nutrient status in the soil.

The department of Agriculture, Govt. of Kerala is following a ten class system (class No. 0 to 9) for N, P and K. Accordingly, the soil is classified into 10 classes based on the fertility status with respect to N, P and K and fertilizer recommendations as percentage of POP recommendations are advocated for each class. Even for the highest class, a minimum percentage is recommended so as to sustain the fertility status. The classes and recommendations are given in Table 9.3.

In case of secondary and micronutrients only two classes have been categorized for the time being as either deficient or sufficient. It is recommended to apply secondary and micro nutrients only if the result of analysis shows deficiency. Among the secondary nutrients, calcium is recommended based on soil pH. The ratings for Ca, based on pH are given in Table 9.2. The rating for Mg, S and micro-nutrients is given in Table 9.4.

Table 9.3: NPK Ratings and Recommendations for Field Crops (Fertilizer recommendations on area basis)

Soil Fertility Class	Per cent of Organic Carbon		N as per cent of General Recomm.	Available P (kg ha⁻¹)	Available K) (kg ha⁻¹)	P and K as per cent of General Recommendation
	Sandy	Clayey/Loamy				
0	0.00-0.10	0.00-0.16	128	0.0-3.0	0-35	128
1	0.11-0.20	0.17-0.33	117	3.1-6.5	36-75	117
2	0.21-0.30	0.34-0.50	106	6.6-10.0	76-115	100
3	0.31-0.45	0.51-0.75	97	10.1-13.5	116-155	94
4	0.46-0.60	0.76-1.00	91	13.6-17.0	156-195	83
5	0.61-0.75	1.01-1.25	84	17.1-20.5	196-235	71
6	0.76-0.90	1.26-1.50	78	20.6-24.0	236-275	60
7	0.91-1.10	1.51-1.83	71	24.1-27.5	276-315	48
8	1.11-1.30	1.84-2.16	63	27.6-31.0	316-355	37
9	1.31-1.50	2.17-2.50	54	31.1-34.5	356-395	25

Source: Aiyer and Nair, 1985.

Table 9.4: Ratings and Recommendations for Secondary and Micronutrients

Nutrient	Deficiency	Sufficiency	Recommendation
Magnesium	< 1.0 me. 100g⁻¹ (120 mg kg⁻¹)	>1.0 me. 100g⁻¹ (120mg kg⁻¹)	*80 kg $MgSO_4$ ha⁻¹
Sulphur	< 5mg kg⁻¹	5 -10 mg kg⁻¹	25 kg S ha⁻¹
0.1N HCl-Zn (For Acid soils)	< 1.00 mg kg⁻¹	> 1.00 mg kg⁻¹	20 kg $ZnSO_4$ ha⁻¹
DTPA – Zn (for neutral to alkaline soils)	< 0.60 mg kg⁻¹	> 0.60 mg kg⁻¹	20 kg $ZnSO_4$ ha⁻¹
0.1N HCl-Cu (For Acid soils)	< 1.00 mg kg⁻¹	> 1.00 mg kg⁻¹	1.5 to 2.0 kg $CuSO_4.5H_2O$ ha⁻¹
DTPA – Cu (For neutral to alkaline soils)	< 0.12 mg kg⁻¹	> 0.12 mg kg⁻¹	1.5 to 2.0 kg $CuSO_4.5H_2O$ ha⁻¹
DTPA – Fe/0.1 N HCl Fe (For both acidic and alkaline soils)	< 5.0 mg kg⁻¹	> 5.0 mg kg⁻¹	15 kg $FeSO_4$ ha⁻¹
DTPA – Mn/0.1 N HCl Mn (For both acidic and alkaline soils)	< 1.0 mg kg⁻¹	> 1.0 mg kg⁻¹	Foliar Spray of 0.5 per cent $MnSO_4$
Hot Water Extractable Boron	< 0.5 mg kg⁻¹	>0.5 mg kg⁻¹	10 kg Borax ha⁻¹or 0.5 per cent solution of Borax

*500 g $MgSO_4$ per palm for coconut.

Source: KAU, 2011.

The time of application is important with respect to both compatibility of sources of different nutrients, as well as the requirement of the crop. A general pattern to be followed to get maximum use efficiency is suggested here under. After soil test results are obtained, based on the pH, lime can be applied at the time of land preparation. Lime should be powdered so as to pass through 60 mesh sieve (0.25 mm sieve size) and thoroughly mixed with the soil. After two weeks, the recommended quantity of organic manure can be applied. Basal dose of N, P and K fertilizers can be applied 10 to 15 days after organic matter application. Care should be taken not to apply lime and organic manure together or lime and fertilizers together. At least 10 to 15 days interval should be given between the application of lime organic manure and fertilizers. The other secondary nutrients (Mg and S) and micro-nutrients can be applied along with NPK fertilizers if they are found deficient. In the case of boron, it is recommended to be applied in two split doses first as basal along with N P K fertilizers and next at the time of flowering. It is advisable not to mix Zn and B fertilizers along with phosphatic fertilizers since these two have antagonistic effects with boron. Another option is to apply these two elements as foliar spray.

Fertilizer Management in Precision Farming

One of the features of precision farming is to have maximum possible use efficiency of applied inputs especially water and fertilizers. Hence fertilizers dissolved at appropriate concentration in water and applied through irrigation water by drip irrigation. This practice is known as fertigation. The advantage of this method is that nutrients and water in required quantity at correct times are placed in the root zone so that maximum absorption of applied nutrients and water is assured. Thus, losses are prevented. Hence, the total quantity of fertilizers to be applied can be substantially reduced. However, the quantity should be fixed in such a way to meet the uptake.

The crop wise schedule of fertigation, the number of splits through which the fertilizers are to be applied, the sources and quantity of fertilizers required for vegetable crops under protected cultivation are given in Table 9.6. The above requirements for banana in precision farming under open condition are provided in Table 9.5. The fertigation schedule for ginger is given in Table 9.7. These recommendations can be modified based on soil test. The micro nutrients need to be applied only if deficiency is applied by soil test. Care should be taken to monitor the pH and electrical conductivity in rhisosphere followed by liming at the onset of each crop. This is very much essential in soils of Kerala as the soils are acidic and the acidity will definitely be aggravated due to application of acid forming nitrogen fertilizers. It is also advisable to have soil as a component of the growing media since it can buffer the anticipated problems of acidity and salinity due to fertigation.

Table 9.5: Fertigation schedule for banana (Nendran) rate for 1000 tissue culture plants

Period	No. of Splits	Urea*	MAP*	MOP*
(a) Full dose- 190:115:300 g/plant, 50 per cent P- basal application				
1-60 days	15	4.8	2.5	6.7
61-120 days	15	8.7	3.8	11.7
121-180 days	15	8.3	0	10
181-280 days	25	2.5	0	3.0

* Fertilizer material (kg) per application (every 4th day).

Contd...

Table 9.5–*Contd...*

Period	No. of Splits	Urea*	MAP*	MOP*
(b) Sucker planting				
30-60 days	9	8.1	4.2	11.1
60-120 days	15	8.7	3.8	11.7
120-180 days	15	8.3	0	10.0
180-280 days	25	2.5	0	3.0
(c) Fertigation schedule for banana (Grand naine) for 1000 tissue culture plants				
(Full dose 200:200:400 g/Plant)				
1-60 days	15	4.6	4.4	11.1
61-120 days	15	5.5	6.6	15.5
121-180 days	15	8.7	0	13.3
181-280 days	25	4.4	0	2.7

* Fertilizer material (kg) per application

Soil Nutrient Imbalances – Impact on Animal and Human Health

Most people recognize that soil plays a significant role in food production but fewer are aware of the role of soils in food security from a healthy prospective. Most of the elements that are required for human health come from soil through either plant or animal products consumed by humans. This means soil health and quality influence the quantity and quality of food consumed by animals and human which in turn ultimately influence the human health.

Seventeen elements essential for higher plants include carbon (C), hydrogen (H), oxygen (O), nitrogen (N), phosphorus (P), potassium (K), sulphur (S), magnesium (Mg), calcium (Ca), iron (Fe), manganese (Mn), zinc (Zn), copper (Cu), boron (B), molybdenum (Mo), chlorine (Cl) and nickel (Ni), out of which fourteen except carbon (C), hydrogen (H) and oxygen (O) come directly from soil. All these elements are essential for human health also. Among these elements application of large quantity of N, P and K fertilizers for increasing crop production has led to increase in biomass production but this unbalanced fertilizer application without addition of enough organic matter has enhanced soil nutrient imbalances especially with respect to micronutrients (Fe, Mn, Zn, Cu, B, Mo, Cl and Ni).

Soil nutrient deficiencies may occur in soil for a number of reasons. The first reason among them is lack of that particular nutrient in parent rock from which that soil is formed ie. the inherent incapability of the soil to supply that particular element. The second reason is that, though a particular element is present in soil, its bioavailability may be poor due to fixation (fixation of P in acidic and alkaline environment). Other reasons include continuous cultivation of crops without addition of fertilizers and manures which lead to nutrient mining from soil and finally the soil becomes deficient. The prevailing climate of a particular area also plays a major role in nutrient status of soil present in that area. Leaching of base forming cations from

Table 9.6: Fertigation Schedule for Vegetables

Crops		Total (kg/ha)	Total NPK (kg/Ha)				30 splits											
			Basal	Estb 6	Veg 12	Fruting 22	19/19/19 kg/Ha			13:00:45 kg/Ha			Urea kg/Ha			12:61:00 kg/Ha		
							Estb 6	Veg 12	Fruting 22	Estb 6	Veg 12	Fruting 22	Estb 6	Veg 12	Fruting 22	Estb 6	Veg 12	Fruting 22
			Doses	Doses	Doses	Doses	Doses	Doses	Doses	Doses	Doses	Doses	Doses	Doses	Doses	Doses	Doses	Doses
Cabbage (40 tones ha⁻¹)	N	211		42.20	84.4	84.4	25.8	25.8	25.8	28.0	232.7	232.7	81.1	75.30	141.1	0.00	8.00	8.00
	P	49	24.5	4.90	9.8	9.8												
	K	274		54.8	109.6	109.6												
Cauliflower (30 tones ha⁻¹)	N	150		30.0	60.0	60.0	31.6	31.6	31.6	26.7	173.3	173.3	52.2	39.1	88.0	0.00	9.80	9.80
	P	60	30	6.0	12.0	12.0												
	K	210		42.0	84.0	84.0												
Cucumber (100 tones ha⁻¹)	N	175		35.0	70.0	70.0	65.8	65.8	65.8	55.6	238.9	238.9	48.9	42.3	109.8	2.00	20.50	20.50
	P	125	62.5	12.5	25.0	25.0												
	K	300		60.0	120.0	120.0												
Okra (30 tones ha⁻¹)	N	90		18.0	36.0	36.0	19.7	19.7	19.7	20.0	111.7	111.7	31.0	4.3	35.9	0.00	6.1	6.1
	P	37.5	18.75	3.8	7.5	7.5												
	K	135		27.0	54.0	54.0												

Contd...

Table 9.6—Contd...

Crops		Total (kg/ha)	Basal	Total NPK (kg/Ha)			40 splits 19/19/19 kg/Ha			13:00:45 kg/Ha			Urea kg/Ha			12:61:00 kg/Ha		
				Estb 6 Doses	Veg 12 Doses	Fruting 22 Doses	Estb 6 Doses	Veg 12 Doses	Fruting 22 Doses	Estb 6 Doses	Veg 12 Doses	Fruting 22 Doses	Estb 6 Doses	Veg 12 Doses	Fruting 22 Doses	Estb 6 Doses	Veg 12 Doses	Fruting 22 Doses
Capsicum	N	210		31.5	63.0	115.5	18.9	18.9	34.7	22.0	176.0	322.7	60.7	44.8	208.7	0.00	5.9	10.8
	P	48	24	3.6	7.2	13.2												
	K	276		41.4	82.80	151.8												
Bitter gourd	N	210		31.5	63.00	115.5	29.2	29.2	53.6	5.00	137.7	252.4	50.4	55.7	208.7	0.00	9.1	16.7
	P	74	37	5.55	11.1	20.4												
	K	225		33.75	67.5	123.8												
Snake gourd	N	210		31.5	63.00	115.5	29.2	29.2	53.6	5.00	137.7	252.4	50.4	55.7	208.7	0.00	9.1	16.7
	P	74	37	5.55	11.1	20.4												
	K	225		33.75	67.5	123.8												
Pumpkin	N	120		18.00	36	66.0	23.7	23.7	43.4	10.00	90.00	165.0	29.3	10.48	101.1	0.00	7.4	13.5
	P	60	30	4.5	9.00	16.5												
	K	150		22.5	22.5	82.5												
Ashgourd	N	120		18	36	66	23.8	23.8	43.11	10	90	165	29.3	10.4	101.1	0	7.4	13.5
	P	60	30	4.5	9	16.5												
	K	150		22.5	22.5	82.5												
Brinjal	N	175		26.25	52.5	96.3	15.8	15.8	38.60	41.7	193.3	354	50.5	17.1	166.8	0	4.9	9.
	P	40	20	3	6	11												
	K	300		45	90	165												
Yard long bean	N	170		25.5	51	93.5	41.4	41.4	76	46.7	189.2	346.8	38.3	15	160.9	0	12.9	23.7
	P	105	52.5	7.875	15.75	28.9												
	K	310		46.5	93	170.5												
	K	380		45.6	91.2	167.2												

Contd...

Table 9.6—*Contd...*

50 Splits

Crop	Total NPK	Basal kg/ha	Total NPK (kg/ha) Estb (6 Doses)	Veg (12 Doses)	Fruitig (22 Doses)	Last (10 Doses)	19:19:19 (kg/ha) Estb (6 Doses)	Veg (12 Doses)	Fruiting (22 Doses)	Last (10 Doses)
Tomato (100 tons ha⁻¹) N	280	65.00	33.60	67.20	123.20	56.0	51.30	51.30	94.10	0.00
P	130		9.75	19.50	35.75	0.0				
K	380		45.60	91.20	167.20	76.0				

50 Splits

Crop	Total NPK	13:00:45 (Kg/ha) Estb (6 Doses)	Veg (12 Doses)	Fruiting (22 Doses)	Last (10 Doses)	Urea (kg/ha) Estb (6 Doses)	Veg (12 Doses)	Fruiting (22 Doses)	Last (10 Doses)	12:61:00 (kg/ha) Estb (6 Doses)	Veg (12 Doses)	Fruiting (22 Doses)	Last (10 Doses)
Tomato (100 tons ha-1) N	280	26.70	181.0	331.8	168.9	51.80	52.50	225.40	43.50	0.00	16.00	29.30	0.00
P	130												
K	380												

high rainfall areas lead to deficiency of these cations like Ca, Mg, K, and Na in that area. This is the reason for widespread deficiency of these elements in tropical soils of Kerala state.

Antagonistic interaction among nutrients (interaction between Zn and P, B and P) also reduces the bioavailability and plant uptake of nutrients. Nonspecific ionic competition between elements (Fe and K in low land lateritic soils) also induces nutrient deficiencies. For example, in low land lateritic soils, uptake of K may be low due to competition from excessive Fe^{2+}, even under sufficient available K levels.

Zinc deficiency appears to be the most widespread and frequent micronutrient problem in crop and pasture plants worldwide, resulting in severe losses in yield and nutritional quality. This is particularly the case in areas of cereal production. It is estimated that nearly half the soils on which cereals are grown have levels of available Zn low enough to cause Zn deficiency. Zinc deficiency in humans is a critical nutritional and health problem in the world. It affects, on average, one-third of the world's population, ranging from 4 to 73 per cent in different countries (*Hotz and Brown, 2004,*). The recent analyses made under the Copenhagen Consensus in 2008 (www.copenhagenconsensus.com) identified Zn deficiency, together with vitamin A deficiency, as the top priority global issue.

On a weight basis, an average between 3 and 5 per cent of soil consists of iron, which makes it the fourth most abundant element in the geosphere. Although Fe is abundant in soils and sediments, its phytoavailability usually is very low, mainly because Fe^{3+} tends to form insoluble Fe oxides. Almost 2 billion people in the world suffer from severe Fe deficiency.

Another soil related human sickness is iodine deficiency, which leads to goitre, severe cognitive and neuromotor deficiencies and other neuropsychological disorders. In areas where soil iodine content is low, human intake of iodine, through foods and drinking water, will be low, and without some form of iodine supplementation, iodine deficiency disorders may develop. Human population largely depending on the food produced in these areas suffers from Iodine Deficiency Disorder (IDD). This is the reason for the occurrence of wide spread IDD in the population of mountainous or sub-mountainous as well as flood-plains and sandy leached soil tracts of the world such as India, Bangladesh, Nepal, Myanmar etc (Singh, 2009).

Calcium and Magnesium deficiencies are most critical in tropical acid soils of Kerala. This unique problem is reflected in human health especially in females as Ca-deficient food results in acute Ca deficiency especially after child birth causing osteoporosis.

Selenium is not an essential micronutrient in plant nutrition, but it has been classified under the group of beneficial elements. These are elements which, although not essential, can stimulate plant growth and improve plant health. The deficiency of selenium in soil leads to low accumulation in crops. There are two well-defined disorders that are caused by selenium deficiency. They are *Keshan disease* and *Kaschin–Beck disease*. *Keshan disease*, named after the Chinese province where it was first described, occurs mainly in children and women of childbearing age, and impairs cardiac functioning. *Kaschin–Beck* disease is an *osteo-arthropathy*, causing deformity of joints and cartilage dis-functioning (Lyons *et al.,* 2004).

Modern agricultural science often treats the soil as a physical medium for anchoring plant roots that can then be bathed in nutrient and growth regulator solutions. As soil is an important constituent of the human biosphere, any harmful change to this segment of the environment seriously affects the overall quality of human life. Soil health degradation has increased nutritional imbalances in the society. Further, this has lead to public health problems also. Maintaining soil health is necessary for the health and wealth of a society. Nutritional security for a society in a sustainable way can only be achieved through management of health of the soil of that region.

References

1. Aiyer, R. S. and Nair, H. K. (1985). *Soils of Kerala and their management.* In Soils of India and their management. The fertilizer association of India, New Delhi. 445p.

2. Buol, S. W., Hole, R. D. McCraken, R. J. and Southard, R. J. (1998). *Soil Genesis and Classification.* Panama Publishing Corporation, New Delhi, 527p.

3. Geetha, P. (2008). *Quantity Intensity Relations of Phosphorus with Reference to its Bioavailability in Lateritic Soils of Kerala.* M.Sc(Ag.) thesis. Kerala Agricultural University, Thrissur, p.

4. Government of Kerala (2013). *Agricultural statistics 2011-2012.* Department of Economics and Statistics, Thiruvananthapuram.231p.

5. Haynes, R. J. and Naidu, R. (1998). Influence of lime fertilizer and manure applications on soil organic matter content and soil physical conditions: a review. *Nutrient Cycl. Agroecosyst.* 51: 123-137.

6. Haynes, R. J. and Swift, R. S. (1988). Effect of lime and phosphate addition on changes in enzyme activities, microbial biomass and levels of extractable nitrogen, sulphur and phosphates in an acid soil. *Biol. Fertil. Soils.* 6: 153-158.

7. Hotz C and Brown KH (2004) International zinc nutrition consultativegroup (IZiNCG) technical document #1. Assessment of the risk of zinc deficiency in populations and options forits control. *Food Nutr Bull* 25, S99–S199.

8. KAU [Kerala Agricultural University] (2011). *Package of Practices Recommendations: crops* (14th Ed.), Kerala Agricultural University, Thrissur, 360p.

9. Lyons, G. H., Stangoulis, J. C. R., and Graham, R. D. (2004). Exploiting micronutrient interaction to optimize biofortification programs: the case for inclusion of selenium and iodine in the HarvestPlus program. *Nutri Rev.* 62: 247–252.

10. Nair, K. M. Kumar, K.S., Srinivas, S.,Sujatha, K.,Venkatesh, D. H., Naidu, L. G. K., Dipak Sarkar and Rajasekharan P. (2011). *Agro-ecology of Kerala.* NBSS Publ. No. 1038, National Bureau of Soil Survey and Land Use Planning, Nagur, India. 395p.

11. Priya, P., Sureshkumar, P. and Mariam, K. A.(2007). Net Ionic Equilibrium in soil plant system. *J. Indian Soc. Soil Sci.* **55(3)**: 421-425.

12. Singh, M. V. (2009). Micronutrient nutritional problems in soils of India and improvement for human and animal health. *Indian J. Fert.* 5(4): 11-16, 19-26 and 56.

2015, Horticulture for Nutrition Security
Editor: **Prof. K.V. Peter**
Published by: **DAYA PUBLISHING HOUSE, NEW DELHI**

Pages **169–188**

Chapter 10

Chemistry of Macronutrients Fixation in Acidic Soils

B.B. Basak and Rajiv Rakshit

Soil acidity is an economic and natural resource threat throughout the world. Acidification of soil is a natural process with major ramifications on plant growth. As soils become more acidic, particularly when the pH drops below 4.5, it becomes increasingly difficult to produce food crops. In our country, about 25 million hectares of cultivated lands with pH value less than 5.5 are very poor in physical, chemical and biological characteristics. Naturally, nutrient availability and fixation in soil are governed by the chemical condition of soil. Knowledge and understanding about the fate of nutrient applied in soil as fertilizer are very important for improving use efficiency of applied nutrients in soil. Chemistry associated with macronutrients fixation in soil is specifically discussed here along with a brief account on how best the fixed nutrients can be utilized for plant nutrition.

Acid Soil Environment: Concept and Understanding

In aqueous systems, an acid is a substance that donates hydrogen ions or protons (H^+) to some other substance. Conversely, a base is any substance that accepts H^+. The H^+ ions or active acidity, increase with the strength of the acid. The undissociated H^+ contributes to a soil potential acidity. Buffer systems can maintain the pH of a solution within narrow range when small amounts of an acid or a base are added. Buffering defines the resistance to change in pH. Soils differ in terms of active and potential acidity. Also soils behave like buffered weak acids, with the H^+ in cation exchange complex (CEC) of humus and clay minerals providing the buffer for soil solution pH.

Soil acidity is of three kinds namely a) active b) exchangeable c) reserve acidity. The hydrogen ions in soil solution contribute to the active acidity. It may be defined as the acidity developed due to concentration of H^+ and Al^{3+} ions in the soil solution. The concentration of hydrogen ion in soil solution due to active acidity is very small, implying that only a meager amount of lime would be required to neutralize active acidity. Normal measurement of pH of the soil denotes the active acidity. Inspite of smaller concentration, active acidity is important since the plant root and the microbes around the rhizosphere are influenced by it and because a dynamic equilibrium exists among active, exchange and reserve acidities of soil. The exchangeable Al and H ions contribute to the exchangeable acidity. It may be defined as the acidity developed due to adsorbed H^+ and Al^{3+} ions on soil colloids. However this exchangeable aluminium and hydrogen ions concentration is meagre in moderately acid soils. The reserve acidity includes the hydrogen and aluminium ions present in the non-exchangeable form with clays and organic matter. It is measured by titrating a soil suspension up to a certain pH normally about 8.0, the amount of acidity in the soil being equivalent to the amount of NaOH used.

The important sources of soil acidity are: exchangeable H^+ and Al^{3+}, Fe and Al oxides, soil organic matter and clay minerals. Two adsorbed cations- H^+ and Al^{3+} are mainly responsible for soil acidity. The exchangeable hydrogen ions present in soil neutralize the negative charge arising from the isomorphous substation of cations. The hydrogen ions are thus due to permanent charge on the mineral surfaces. The pH dependent charge may arise from the structural OH^- groups at the corners and edges of soil clay minerals, which dissociate into H^+ ions. The Al^{3+} ions displaced from clay minerals by cations are hydrolyzed to monomeric and polymeric hydroxyaluminium complexes. Hydrolysis of the monomeric forms have been illustrated in the following stepwise equations, which in each case liberates H^+ and lower the pH unless neutralized by OH^- present in the system.

1. $Al^{3+} + H_2O \rightarrow Al(OH)^{2+} + H^+$

2. $Al(OH)^{2+} + H_2O \rightarrow Al(OH)^{2+} + H^+$

3. $Al(OH)^{2+} + H_2O \rightarrow Al(OH)_3 + H^+$

4. $Al(H_2O)_6^{3+} + H_2O \rightarrow Al(OH)(H_2O)_5^+ + H_3O^+$

5. $Al(OH)(H_2O)_5^{2+} + H_2O \rightarrow Al(OH)_2(H_2O)_4^+ + H_3O^+$

6. $Al(OH)_2(H_2O)_4^+ + H_2O \rightarrow Al(OH)_3(H_2O)_3^+ + H_3O^+$

7. $Al(OH)_3(H_2O)_3^0 + H_2O \rightarrow Al(OH)_4(H_2O)_2^- + H_3O^+$

Soil acidity is determined by the amount of hydrogen ion (H^+) activity in soil solution and is influenced by edaphic, climatic and biological factors. Acid soils are mostly found in the areas of high rainfall. Rainfall is most effective in causing soils to become acidic if plenty of water move through the soil rapidly. In acid soil regions (ASR), precipitation exceeds evapo-transpiration and hence leaching is predominant causing loss of bases from soil. The iron and aluminium derivatives are relatively insoluble, seldom leached and contributes to surface acidity particularly in laterite soils. Since the effect of rainfall on acid soil development is very slow, it may take hundreds of years for new parent material to become acidic under high rainfall. Decomposition of organic matter produces H^+ ion which is responsible for acidity. The carbon dioxide (CO_2) produced by decaying organic matter reacts with water in the soil to form a weak acid called carbonic acid. This is the same acid that develops when CO_2 in the atmosphere reacts with rain to form acid rain naturally. Several organic acids are also produced by decaying organic matter, but they are also weak acids. Acidic materials (granites, rhyolites, diorites) are basis for acidic soil formation. In these soils, predominant minerals are quartz, feldspar and oxides. Al and Fe are in soluble forms. Some rocks (parent material) are acidic in nature (*e.g.* igneous rocks). After weathering of these rocks, acidic constituents dominate the composition of soil. Nitrogen and phosphorus fertilizers also contribute significantly to the formation of acid soils. Ammonium nitrogen can be a major factor in the acidification of sandy, low buffer-capacity soils, unless a careful liming programme is maintained. When ammonium is converted to nitrate by soil microbes, hydrogen ions are released. Acid rain is mainly a mixture of sulphuric and nitric acids depending upon the relative quantities of oxides of sulphur and nitrogen emissions. Due to the interaction of these acids with other constituents of the atmosphere, protons are released causing increase in the soil acidity.

Macro-nutrients in Soil

Macronutrients or major nutrients are so-called because these are required in large quantities, more than that iron. Nitrogen, Phosphorus and Potassium are termed as primary or major nutrients because of their larger requirement by the plants and correction of their wide-spread deficiencies is often necessary through application of commercial fertilizers of which these are the major constituents. The macronutrients like N, P and K present in diverse form in soil but plant can take up only particular ionic from of that particular nutrient from soil. So fixation of that particular ionic from is very important in respective to plant availability. Nitrogen in soil exists in two major forms: (i) organic N and (ii) inorganic N. The inorganic forms (NH_4^+-N, NO_3^--N and NO_2^--N) are very important from crop nutrition point of view, because plant roots take up N from the soil mostly as NO_3^--N and to some extent as NH_4^+-N. The NO_2^- forms is unstable and is usually present in soil in lesser extent. Both organic and inorganic from phosphorus present in soil but inorganic form of phosphorus ($H_2PO_4^-$ and HPO_4^{2-}) directly available to plant. The type of phosphate ions present in the soil solution depends on the soil pH. In soils having neutral to slightly alkaline pH, the $H_2PO_4^-$ is the most common form. As the soil pH gets lowered and it becomes slightly to moderately acidic, both $H_2PO_4^-$ and HPO_4^{2-} ions prevail. At higher soil acidity, $H_2PO_4^-$ form tends to dominate. Unlike the N and P, K present in soil mainly in inorganic or mineral form and K^+ ions is directly taken by plant. In soil K ion is present in solution, exchangeable, non-exchangeable and mineral form but K ion in soil solution is readily available to plants.

Nitrogen Fixation

Approximately 80 per cent of the air consists of nitrogen gas (N_2), but plants cannot use atmospheric N directly to make protein. The gaseous N must first be converted, or "fixed," into forms plants can use. Biological nitrogen fixation is the only process through which atmospheric N_2 converted into plant available form. Nitrogen fixation is the natural process, either biological or abiotic, by which nitrogen (N_2) in the atmosphere is converted into ammonia (NH_3). This process is essential for life because fixed nitrogen is required to biosynthesize the basic building blocks of life, *e.g.*, nucleotides for DNA and RNA and amino acids for proteins. Nitrogen fixation also refers to other abiological conversions of nitrogen, such as its conversion to nitrogen dioxide.Biological nitrogen fixation (BNF), discovered by Beijerinck in 1901 and carried out by a specialized group of prokaryotes. These organisms utilize the enzyme nitrogenase to catalyze the conversion of atmospheric nitrogen (N_2) to ammonia (NH_3). Plants can readily assimilate NH_3 to produce the aforementioned nitrogenous biomolecules. These prokaryotes include aquatic organisms, such as cyanobacteria, free-living soil bacteria, such as *Azotobacter*, bacteria that form associative relationships with plants, such as *Azospirillum*, and most importantly, bacteria, such as *Rhizobium* and *Bradyrhizobium* which form symbioses with legumes and other plants (Postgate, 1981).

Mechanisms of Biological Nitrogen Fixation

The nitrogen molecule is composed of two nitrogen atoms joined by a triple covalent bond, thus making the molecule highly inert and nonreactive. Nitrogenase catalyzes the breaking of this bond and the addition of three hydrogen atoms to each nitrogen atom. Although the process involves a number of complex biochemical reactions, it may be summarized in a relatively simple way by the following equation:

$$N_2 + 8H_2 + 16ATP \longrightarrow 2NH_3 + 2H_2 + 16ADP + 16P_i$$

The equation above indicates that one molecule of nitrogen gas (N_2) combines with eight protons) ($8H^+$) to form two molecules of ammonia ($2NH_3$) and two molecules of hydrogen gas ($2H_2$). This reaction is conducted by an enzyme known as nitrogenase. The 16 molecules of ATP represent the energy required for the BNF reaction to take place. In biochemical terms 16 ATP represents a relatively large amount of plant energy (Figure 10.1). Thus, the process of BNF is 'expensive' to the plant in terms of energy usage. As ammonia (NH_3) is formed, it is converted to an amino acid such as glutamine. The Nitrogen in amino acids can be used by the plant to synthesize proteins for its growth and development.

1. Host Specificity

There are roughly 1,300 leguminous plant species in the world. Of these, nearly 10 per cent have been examined for nodulation, 87 per cent of which were nodulated. Thus not all legumes are infected by rhizobia. *Gliricidia sepium* and *Vigna unguiculata* (cowpea) nodulate freely but nodules have never been found on roots of *Cassia siamea*. A *Rhizobium* that nodulates cowpea may not nodulate *Leucaena* and vice versa. Leguminous species mutually susceptible to nodulation by a particular group of bacteria constitute a cross-inoculation group. Six cross-inoculation groups were

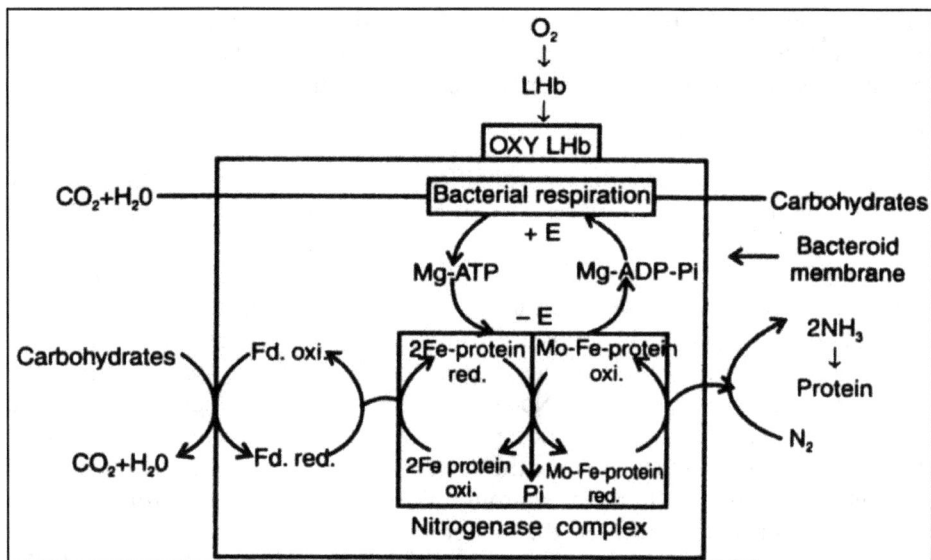

Figure 10.1: A Diagram Illustrating the Interrelationship of Reactions Involved in Nitrogen-Fixing Process in a Legume Nodule (After Evans *et al.,* 1978). In limitations and Potentialities for Biological Nitrogen fixation in the tropics pp.209-222, Eds. J. Dobereiner, H. R. Burries and A. Hollaender, Plenum Press, New York.

defined in the early days of *Rhizobium* research in addition to the cowpea group. This classification scheme is undergoing modifications based on recent research. Table 10.1 gives a short list of rhizobia and their hosts to illustrate the grouping of rhizobia. Not all symbioses fix N_2 with equal effectiveness. This means that a given legume cultivar nodulated by different strains of the same species of *Rhizobium* would fix

Table 10.1: A Short List of Rhizobium Species and their Corresponding Host Plants

Rhizobium Species	Host Plants
Bradyrhizobium japonicum	Glycine max (soybean)
Rhizobium fredii	Glycine max (soybean)
R. phaseoli	Phaseolus vulgaris (common bean)
R. meliloti	Medicago sativa (alfalfa)
	Melilotus sp. (sweet clovers)
R. trifolii	Trifolium sp. (clovers)
R. leguminosarum	Pisum sativum (peas)
	Vicia faba (broad bean)
"Cowpea rhizobia" group or Rhizobium sp.	Vigna unguiculata (cowpea),
	Arachis hypogaea (peanut),
	Vigna subterranea (Bambara groundnut)
	Leucaena sp., Albizia sp.,
Azarhizobium caulinodans	Sesbania sp. Sesbania rostrata (stem nodulating)

different amounts of nitrogen. Selection of elite strains of *Rhizobium is* based on this observation. Similarly, a given strain of *Rhizobium* will nodulate and fix different amount of N_2 in symbiosis with a range of cultivars of the same plant species.

2. Root Nodule Formation

Sets of genes in the bacteria control different aspects of the nodulation process. One Rhizobium strain can infect certain species of legumes but not others *e.g.* the pea is the host plant to *Rhizobium leguminosarum* biovar *viciae*, whereas clover acts as host to *R. leguminosarum* biovar *trifolii*. Specificity genes determine which strain infects which legume. Even if a strain is able to infect a legume, the nodules formed may not be able to fix nitrogen. Such strains are termed ineffective. Effective strains induce nitrogen-fixing nodules. Effectiveness is governed by a different set of genes in the bacteria from the specificity genes. Nod genes direct the various stages of nodulation. The initial interaction between the host plant and free-living rhizobia is the release of chemoreceptors by the root cells into the soil. Some of these encourage the growth of the bacterial population in the area around the roots (the rhizosphere). Reactions between certain compounds in the bacterial cell wall and the root surface are responsible for the rhizobia recognizing their correct host plant and attaching to the root hairs. Flavonoids secreted by the root cells activate the nod genes in the bacteria which then induce nodule formation. The whole nodulation process is regulated by highly complex chemical communications between the plant and the bacteria. Once bound to the root hair, the bacteria excrete nod factors. These stimulate the hair to curl. Rhizobia then invade the root through the hair tip where they induce the formation of an infection thread. This thread is constructed by the root cells and not the bacteria and is formed only in response to infection. The infection thread grows through the root hair cells and penetrates other root cells nearby often with branching of the thread. The bacteria multiply within the expanding network of tubes, continuing to produce nod factors which stimulate the root cells to proliferate, eventually forming a root nodule. Within a week of infection small nodules are visible to the naked eye. Each root nodule is packed with thousands of living Rhizobium bacteria, most of which are in the misshapen form known as bacteroids. Portions of plant cell membrane surround the bacteroids. These structures, known as symbiosomes, which may contain several bacteroids or just one, are where the nitrogen fixation takes place (Figure 10.2).

3. Nitrogen Fixation in Acid Soil

Soil acidity is a significant problem facing agricultural production in many areas of the world and limits legume productivity (Brockwell *et al.,* 1991; Bordeleau and Prevost, 1994; Correa and Barneix., 1997). Most leguminous plants require a neutral or slightly acidic soil for growth, especially when they depend on symbiotic N_2 fixation. Soil acidity constrains symbiotic N_2 fixation in both tropical and temperate soils, limiting *Rhizobium* survival and persistence in soils and reducing nodulation (Graham, 1982; Brockwell *et al.,* 1991: Ibekwe *et al.,* 1997). Some of the Rhizobial strains are tolerant to acidity. It has been found that *R. loti* multiplied at pH 4.5 but *Bradyrhizobium* strains failed to multiply; the acid-tolerant strains of *R. loti* demonstrate a comparative advantage over acid-sensitive strains in the ability to nodulate their

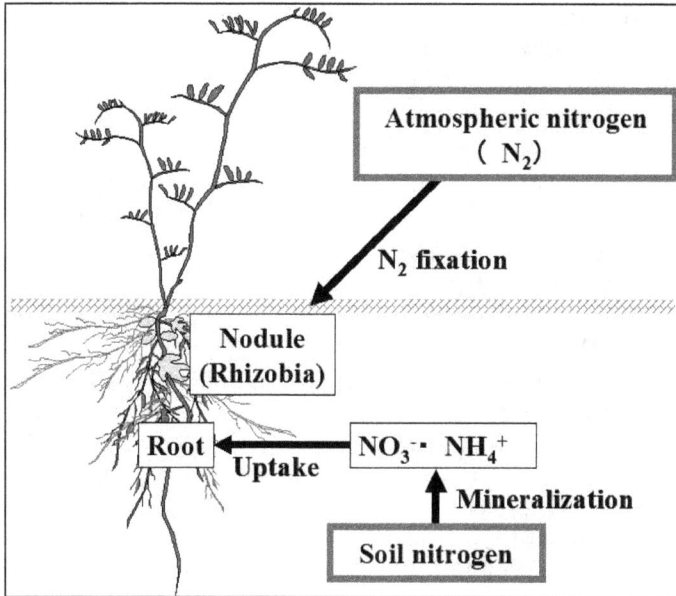

Figure 10.2: Schematic Diagram of Atmospheric N Fixation in the Nodule of Legume Plant.

host legume at pH 4.5 (Cooper *et al.,* 1985). *R. tropici* and *R. loti* are moderately acid tolerant, while *R. meliloti* is very sensitive to acid stress. The fast-growing strains of rhizobia have generally been considered less tolerant to acid pH than have slowly growing strains of *Bradyrhizobium,* although some strains of the fast-growing rhizobia, *e.g.,* *R. loti* and *R. tropici,* are highly acid tolerant (Cooper *et al.,* 1985). The basis for differences in pH tolerance among strains of *Rhizobium* and *Bradyrhizobium* is still not clear. Graham (1982) reported high cytoplasmic potassium and glutamate levels in acid-stressed cells of *R. leguminosarum* bv. *phaseoli,* a response which is similar to that found in osmotically stressed cells.

Nodulation failure under acid-soil conditions is common, especially in soils of pH less than 5.0. The inability of some rhizobia to persist under such conditions is one cause of nodulation failure, but sometimes viable *Rhizobium* population exhibit poor nodulation (Graham, 1982; Carter *et al.,* 1994). The number of nodules, the nitrogenase activity, the nodule ultrastructure, and the fresh and dry weights of nodules were affected to a greater extent at a low medium pH (4.5). Early stage of infection process is more sensitive to soil acidity which had more severe effects on rhizobial multiplication than did Al stress and low P conditions.

The host legume appears to be the limiting factor for establishing *Rhizobium*-legume symbiosis under acidic conditions. Legume species differ greatly in their response to low pH with regard to growth and nodulation because the amount of N_2 fixed by forage legumes on low fertility acidic soil is dependent on legume growth and persistence (Thomas *et al.,* 1997). Legumes like *Trifolium subterranean, T. balansae, Medicago murex,* and *M. truncatula,* showed tolerance to soil acidity. *R. leguminosarum* bv. *viciae* is able to form nodule after inoculation with *Vicia faba* in acid soils (Aarons and Graham, 1991; Carter *et al.,* 1994). Legume species vary markedly in their tolerance to Al³⁺ and Mn²⁺, with some plants being significantly more strongly affected by these ions than are the rhizobia. Therefore, for acid soils with high Al content, improvement is achieved by manipulating the plant rather than the rhizobia (Taylor *et al.,* 1991).

Mineral toxicity (specific ion toxicity) is the most significant characteristic of soil acidity usually accompanied by nutrient deficiency and nutrient disorder. Acidic stress markedly affects ion absorption by and growth of roots; the membrane structure and function of the roots suffer fatal changes under these stress conditions. The requirement of some essential elements, *e.g.,* Ca²⁺ and P, is increased under severe stress conditions. The requirement of Ca²⁺ for growth of *R. meliloti* was increased under osmotic stress. The Ca-depleted cells of *R. leguminosarum* are swollen, lack rigidity, and express an additional somatic antigen normally blocked by side chains of the LPS O antigen. High levels of salinity (up to 10 per cent NaCl) along with decreased Ca²⁺ content of *Rhizobium* cells, greatly distorted the outer membrane structure of the *Rhizobium* cells. In the same way, calcium appears significantly more important in cells exposed to low pH. Calcium plays a vital role in cytoplasmic pH maintenance, phosphorus mobilization and ion transport which are caused mostly by changes in membrane properties, apart from that Ca²⁺ plays an essential role in cell division, elongation, and membrane structure and function. O'Hara *et al.* (1989) found that in acid-sensitive strains of *R. meliloti*, 1.2 mM Ca²⁺ was needed for cytoplasmic pH maintenance. It was found that phosphorus-limited cells or cells grown at low pH needed Ca²⁺ for phosphorus mobilization in the cell. Calcium addition in low pH soil improves both growth and ion uptake by roots and also offset the harmful effects of ions such as K⁺ and H⁺ and control K transport through the control of K⁺ permeability and activation of K⁺ uptake through the acidification of the cytoplasm.

Table 10.2: Estimate of the Amount of Nitrogen Fixed by Various Legumes (FAO, 1984)

Plants	Scientific Name	Nitrogen Fixed (kg N/ha/year)
Horse bean	*Vicia faba*	45–552
Pigeon pea	*Cajanus cajan*	168–280
Cowpea	*Vigna unguiculata*	73–354
Mung bean	*Vigna mungo*	63–342
Soybean	*Glycine max*	60–168
Chickpea	*Cicer arietinum*	103
Lentil	*Lens esculenta*	88–114
Peanut	*Arachis hypogaea*	72–124
Pea	*Pisum sativum*	55–77
Bean	*Phaseolus vulgaris*	40–70
Leucaena	*Leucaena leucocephala*	74–584
Alfalfa	*Medicago sativa*	229–290
Clover	*Trifolium* spp.	128–207

Phosphorus (P) is one of several elements which affects N_2 fixation and is a principal yield-limiting nutrient in many regions of the world. Strains of *Rhizobia* differ markedly in tolerance to phosphorus deficiency. Native soil and rhizospheric P deficiency induces Rhizobial P deficiency especially under acidic conditions, where dissolved phosphorus salts may be precipitated in the presence of aluminum. Slow-growing strains of rhizobia appear more tolerant to low P levels than do fast-growing *R. meliloti*, in particular (Cooper *et al.*, 1985). Both fast and slow growing Rhizobia are greatly influenced by the phosphorus availability because inducible alkaline phosphatase activity was detected in P-limited cells of fast-growing *R. trifolii* strains. Recently, it has been reported that free-living *R. tropici* and bacteroids respond to P stress by increasing their P transport capacity and inducing both acid and alkali phosphatases. Phosphorus appears essential for both nodulation and N_2 fixation. Nodules are strong sinks for P and range in P content from 0.72 to 1.2 per cent; as a consequence, N_2 fixation-dependent plants will require more of this element than those supplied with combined nitrogen. Nodulation, N_2 fixation and specific nodule activity are directly related to the P supply. External application of phosphorus (25 mg of P per kg of soil) to acidic soils significantly increased the percent nodule occupancy of *Trifolium subterranean* by *R. leguminosarum* bv. *trifolii*. The nodulation and N_2 fixation (nitrogenase activity) of *T. vesiculosum* increased significantly after the addition of P (100 mg per kg of soil) and K (300 mg per kg of soil); nitrogenase activity was doubled when the P concentration increased to 400 mg per kg of soil. Nitrogen fixation by the *Frankia*-actinorhizal symbiosis may be limited by low available P in soils. Singleton *et al.* (1991) observed increased N_2 fixation by *Rhizobium japonicum* by adding phosphate to P-deficient soil. So, low P status is a frequent limitation to nodulation of actinorhizal plants. It has been reported that symbiotic N_2 fixation of the *Frankia-Casuarina* association requires higher P levels than those required for plant growth, when the P concentration in soil is low. Genetic variations among species of *Allocasuarina* in relation to P requirement were identified; species showed different nodulation abilities in soils with low P. Mycorrhizal infection of roots of legumes has been reported to stimulate both nodulation and N_2 fixation, especially in soils low in available P. The role of mycorrhizal fungi in the protection of the *Medicago sativa-R. meliloti* symbiosis against salt stress by increased plant available P and generally declined as the salinity in the solution culture increased.

The interaction of P and Zn and their effects on nodulation of legumes under salt stress were studied. Saxena and Rewari (1991) found that application of phosphate (20 and 40 ppm) improved the growth and nodulation of chickpea (*C. arietinum*) in the presence of Zn^{2+} (5 ppm) at different levels of salinity. They suggested that augmentation with Zn^{2+} provided protection to the plant under saline conditions by reducing the Na^+/K^+ ratio in the shoot; the shoot N content after augmentation with Zn^{2+} and in the presence of phosphate was equal to that of nonsaline control. Differences between cultivars of some legume species with regard to phosphorus requirements have been reported. Variability of N_2 fixation under low P availability existed between lines of *P. vulgaris*; high N_2-fixing and high-yielding progeny lines were detected.

Two strategies have been adopted to solve the problem of soil acidity: (i) selecting tolerant plants, and (ii) liming the acidic soil to ameliorate the effects of acidic conditions. It has been suggested that Al-tolerant (acid-tolerant) plant species contain and exude more organic acid and other ligands that form stable chelates with Al and thereby reduce its chemical activity and toxicity. Application of lime (at rate of 2,500 kg ha^{-1}) and superphosphate (at rates up to 20 kg ha^{-1}) to acidic soils increased the soil pH from 4.5 to 4.9, decreased the concentration of extractable Al and Mn, and improved growth and N$_2$ fixation of *T. subterranean* (Almendras and Bottomley, 1987).

4. Phosphorus Fixation

The fixation of P by soils has long been recognized. Thomas Way in 1850 demonstrated that the whole of phosphate was retained when solution of sodium phosphate in water and guano in dilute H$_2$SO$_4$ were poured over a layer of calcareous soil. He suggested that an insoluble calcium phosphate was formed resulting decrease in solubility of applied phosphorus. Phosphate fixation or reversion can be viewed as the conversion of soil solution P to insoluble compounds by the soil components, causing reduction in the amount that plant roots can absorbed. Mechanisms involving sorption and precipitation have been suggested to explain in the P fixation in soils. In fact, P fixation is not an ideal adsorption on soil components, a combination of adsorption, chemisorptions and precipitation. Soil solution pH has a significant effect on phosphate fixation. In neutral to alkaline soil (pH 7 and above), phosphates get adsorbed on calcium carbonate and are slowly converted to insoluble apatites. In acid soils (pH below 7) iron and aluminum react with phosphate to form highly insoluble compounds. In this way, some phosphate of the labile pool is continuously being transferred to the non-labile pool and thus become immobile.

5. Chemistry of Phosphorus Fixation in Acid Soils

The inorganic P-forms in soil are the compound associated mainly with aluminium, iron and calcium. Their relative abundance and solubility are controlled by a number of factors including soil pH (Figure 10.3). From the different study, it was found that phosphorus is the most readily available between pH 6 and 7. The most dominant ionic forms of P in the soil solution are H$_2$PO$_4$ and HPO$_4^{2-}$ at neutral range whereas PO$_4^{3-}$ ion dominates in alkaline condition. The availability of P primarily of H$_2$PO$_4$ and HPO$_4^{2-}$ ions are highly pH dependent. Its availability in many soils is the highest when the pH is neutral or slightly acidic and it declines as soil becomes strongly acidic or strongly alkaline. Therefore in acid soils P is highly susceptible towards fixation and thus rendering it unavailable for plant uptake.

The acid soils are highly dominated by amorphous Fe and Al oxides and their hydroxides which are the potential sites of P adsorption and fixation. Three types of reactions may be considered important in relation to phosphate fixation in soils: i) adsorption ii) isomorphous replacement and iii) double decomposition. Freundlich and Langmuir adsorption isotherms fit well at low concentrations of phosphate in solution whereas the isomorphous replacement of hydroxyl ions with phosphate ions is a possibility. The reaction of Fe and Al hydroxides with the phosphate ions are probably the most significant for phosphate fixation in soils. When H$_2$PO$_4^-$ concentration in soil solution is high, it reacts with these minerals forming precipitates

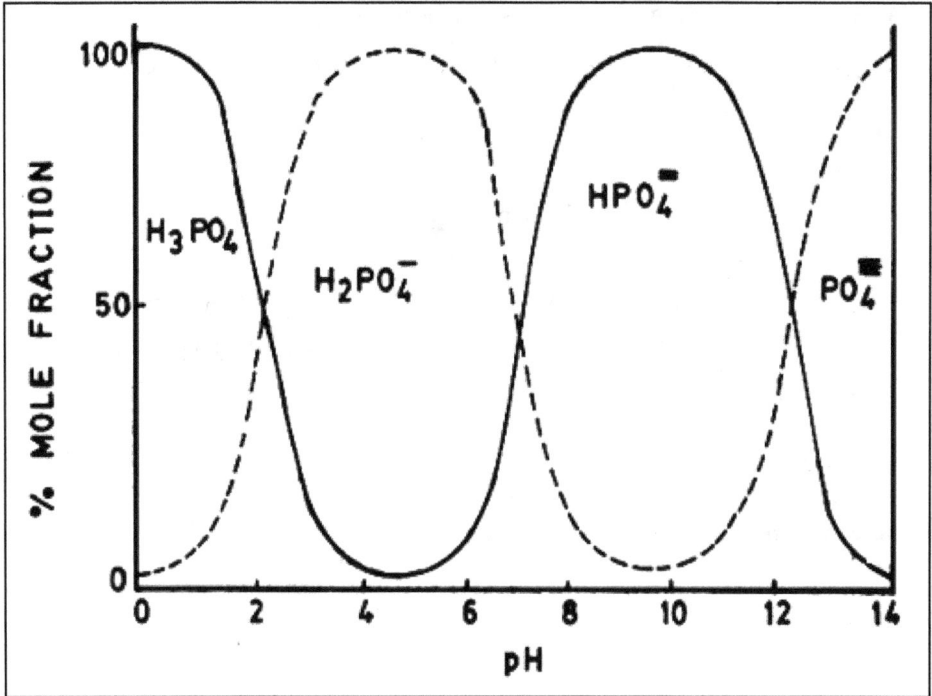

Figure 10.3: Relationship between Solution of pH and the Relative Concentration of Three Soluble Forms of Phosphate.

of Fe and Al hydroxyl phosphates. Most of the P-fixation occurs in acidic soils, where $H_2PO_4^-$ reacts with the surfaces of insoluble oxides of iron, almonium and manganese, involves series of chemical fixation reactions and thus interlocks the P. Some of these reactions are given as under:

a) Precipitation Reaction

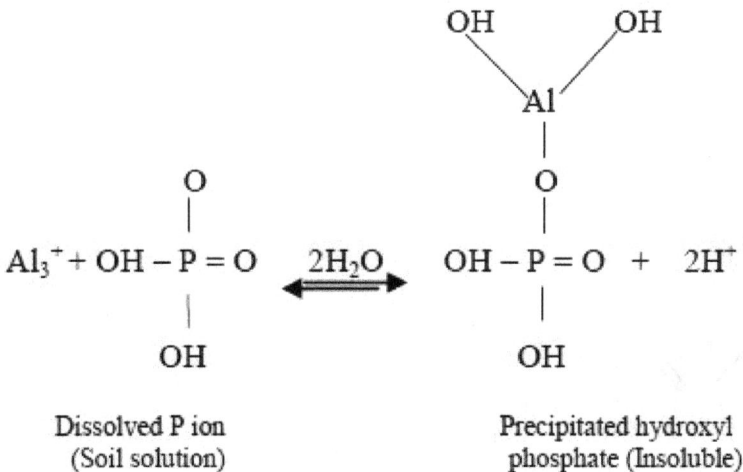

$$Al_3^+ + OH - \underset{\underset{OH}{|}}{\overset{\overset{O}{|}}{P}} = O \quad \underset{2H_2O}{\rightleftarrows} \quad \underset{\underset{OH}{|}}{\overset{\overset{O}{|}}{\underset{\underset{|}{Al}}{OH}}} OH - P = O \quad + \quad 2H^+$$

Dissolved P ion   Precipitated hydroxyl
(Soil solution)   phosphate (Insoluble)

b) Anion Exchange Reactions (Outer Sphere Complex)

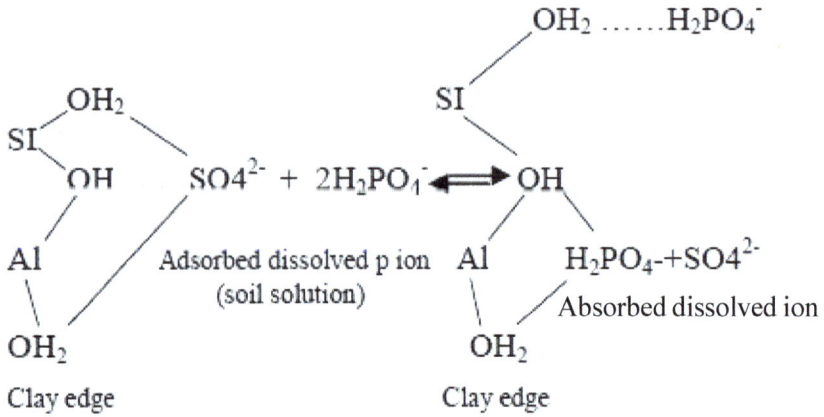

$$
\begin{array}{l}
\text{SI}\diagup\overset{\displaystyle OH_2}{\diagdown OH} \quad SO4^{2-} + 2H_2PO_4^- \;\rightleftharpoons\; \text{SI}\diagup\overset{\displaystyle OH_2 \ldots\ldots H_2PO_4^-}{\diagdown OH} \\[1em]
\text{Al} \qquad \text{Adsorbed dissolved p ion} \quad \text{Al} \qquad H_2PO_4^- + SO4^{2-} \\
\quad \diagdown \qquad\qquad \text{(soil solution)} \qquad \diagdown \qquad\qquad \text{Absorbed dissolved ion} \\
\quad OH_2 \qquad\qquad\qquad\qquad\qquad\qquad OH_2 \\
\quad \text{Clay edge} \qquad\qquad\qquad\qquad\quad \text{Clay edge}
\end{array}
$$

Reaction with Al and Fe Oxide Surface (Inner Sphere Complex)

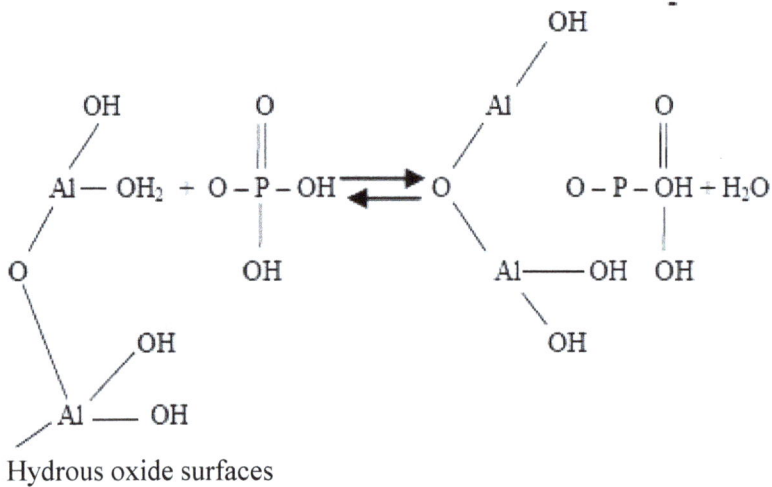

Hydrous oxide surfaces

Formation of Stable Binuclear Bridge (Inner Sphere Complex)

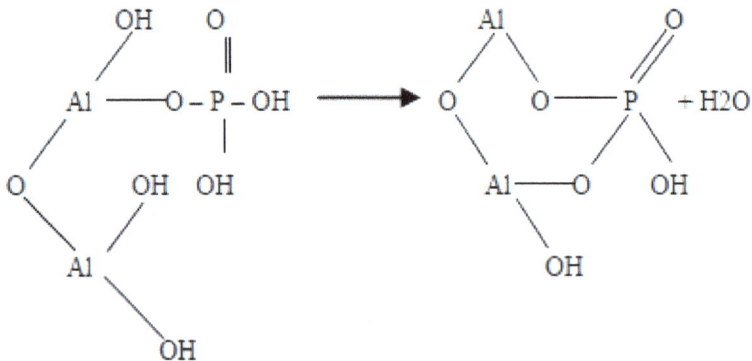

In reaction (A), freshly formed hydroxyl phosphate is slightly soluble, because of having a greater surface area exposed to the soil solution. Therefore, in P it is available initially to some extent to the plants. With advanced time, the precipitated hydroxyl phosphate ages and becomes less soluble and becomes totally unavailable for plants. In reaction (B), phosphate is reversibly adsorbed by anion exchange with broken clay edges of kaolinite clays. In reaction (C), the phosphate ion replaces –OH group in the surface structure of Al oxide minerals and in reaction (D), the phosphate further penetrates the mineral surface by forming a stable binuclear bridge.

Phosphate Management in Acid Soils

The fixation and immobility of P in acid soils of the tropics can either be major problem or blessings in disguise, depending on how soils and P fertilizers are managed. Phosphorus fixation is often high in Oxisols and Ultisols because they are most likely to have P-fixing clay minerals (amorphous and crystalline hydrous oxides of Fe and Al), high Fe and Al, and low pH, all of which are conducive to P fixation (Uexkull, 1989). Fox *et al.* (1989) pointed out that when soil P concentration exceeds a certain level, P uptake by crops will actually be inhibited. Thus, highly concentrated bands of P fertilizer should be avoided. Because of strong interest in rock phosphates for direct application to tree crops in Southeast Asia, one must know that rock phosphates differ significantly in their reactivity. Because of their low cost, high Ca content and residual effects, rock phosphates are especially well suited for amendment of acid soils poor in P and Ca. Several researchers have successfully attempted to reduce the cost of P fertilization in acid soils by direct use of rock phosphates (Misra and Pattanayak, 1997) to soil (pH < 5.5) or use of rock phosphates and single super phosphate mixture in 1:1 ratio to soil (pH 5.6-6.5) or to apply rock phosphates to green manure crops prior to rice crops taken in sequence or use of compacted products of Jhamarkotra rock phosphates (JPR).

Further it is recommended to apply the entire P requirement of the cropping sequence, particularly for groundnut-rice cropping system in form of rock phosphates to the groundnut crop grown during *rabi* season and the residual effect be realized through rice crop grown during *kharif* season. This is because the rock phosphate applied to dry season groundnut gets solubilised to greater extent and the portion that gets fixed during dry season becomes available to rice crop due to soil reduction (Misra and Pattanayak, 1997).

Potassium Fixation

Since the middle of the 17th century, J R Glauker in Netherlands first proposed that saltpeter (KNO_3) was the principle of vegetation, K has been recognized as being beneficial to plant growth (Russell, 1961). The essentiality of k to plant growth has been known since the work of von Liebig published in 1840. Of the major nutrient elements, K is usually the most abundant in soil (Reitemeier, 1951). A mineral soil generally ranges between 0.04-3 per cent K. Soil K exists in four forms in soils: solution, exchangeable, fixed or non-exchangeable and structural and minerals. Soil solution and exchangeable K level comprise a small portion of the total K. Though non-exchangeable K comprises a significant portion of soil potassium, but most of the soil

K present as mineral form (Sparks and Huang, 1985). There are equilibrium and kinetics reaction between four forms of soil K that affect the level of soil solution K at any particular time, and thus, the amount of readily available K for plants.

At a given time, potassium in the solution and exchangeable forms constitutes the fraction readily available to plants. The exchangeable K tends to attain equilibrium rapidly with solution K but, slowly with non-exchangeable K (Figure 10.4). The solution K concentration largely controls the k movements (diffusion) towards the plant roots and thereby the uptake by plants. On depletion of exchangeable K, non-exchangeable K replenishes it and supply of K to plants is maintained. When K is added to the soil through mineral fertilizers or organic manures in excess of its crop removal, it initially increases solution K and subsequently, increases the exchangeable and non-exchangeable K though the shifting of the equilibrium (Figure 10.4).

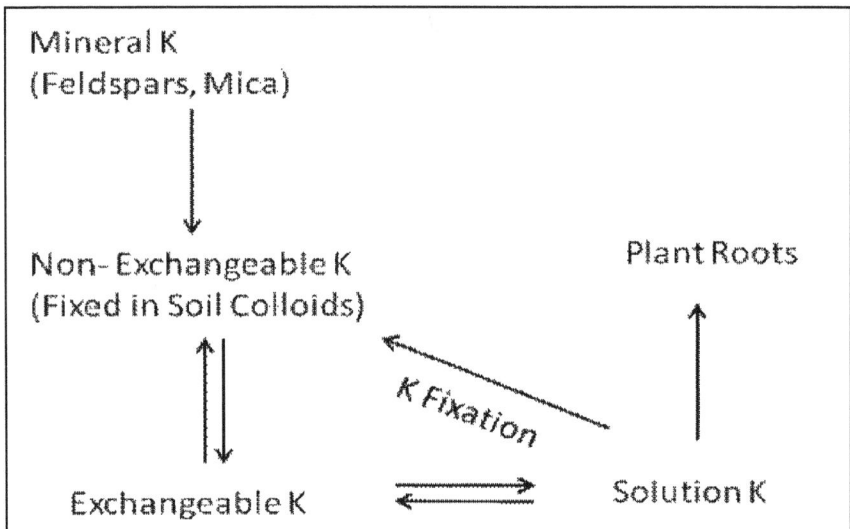

Figure 10.4: Conceptual Model of Potassium Dynamics in Soils and Plant Roots.

Potassium Fixation Mechanisms in Acid Soil

In the processes of potassium fixation in soils, the added soluble K is converted into a form that cannot be extracted with a neutral salt solution. Potassium fixation occurs when K^+ ions form a surface complex with oxygen atoms in the interlayers of certain silicate clay minerals. Potassium fixation processes phenomenon are fixed limited to interlayer ions such as K^+ has been explained in terms of good fit of K^+ ions (crystalline radius and coordination number are ideal) in an area created by holes and adjacent oxygen layer (Barshad *et al.*, 1951). The important forces involved in interlayer reactions in clays are electrostatic attraction between the negatively charged layers and the positive interlayer ions. The degree of K fixation in clays and soils depends on the type of clay mineral and its charge density, degree of interlayring, the moisture content, the concentration of K^+ ions as well as the concentration of competing cations and pH of the ambient solution bathing of clays or soils (Rich, 1968; Spark and Huang, 1985). The major clay minerals responsible for K fixation are montmorillonite, vermiculite and weathered micas. In acid soils, principle clay minerals responsible for K fixation are dioctahedral vermiculite. The degree of K fixation is strongly influenced by the charge density on the layer silicate. Those with high charge density fix more potassium than those with low charge density (Walker, 1957). Martin *et al.* (1946) showed at pH values up to 2.5 there is no fixation; between pH 2.5-5.5, amount of K fixation is very rapidly. Above pH 5.5 fixations increased more slowly. The increase in K fixation between pH can be ascribed to the decreased number of $Al(OH)_x$ species which decrease K fixation (Rich, 1964; Rich and Black, 1964).

Potassium ions are absorbed by clay minerals on the binding sites which differ in their selectivity (Figure 10.5). For 2:1 clay minerals such as illites, vermiculites and weathered micas, three different absorption sites can be distinguished. These sites are at the planner surface (p-position), at the edges of the layer (e-position) and in the interlayer space (i-position). The specificity of these three binding sites for K differs considerably. The binding selectivity for K by organic matter and clays of the kaolinite type are similar to the p-position sites. Here, the K-bond is relatively weak so that K absorbed may easily be replaced by other cations, particularly by Ca^{2+} and Mg^{2+} ions. The i-postion has maximum specificity for K^+. These binding sites largely account for K^+ fixation in soil.

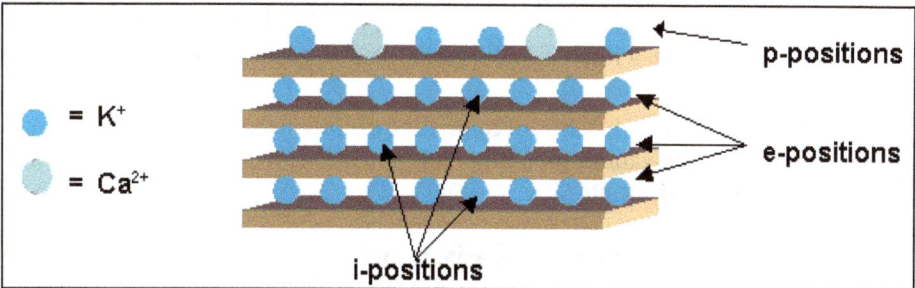

Figure 10.5: Model of Expandable Layer Silicate Structure with Interlayers (p, e and i-positions).

The potassium added through manures and fertilizers initially increase the solution and exchangeable K contents. The saturation of the exchangeable complex with respect to K leads to the entry of K into wedge (partially open) and interlayer spaces (Figure 10.5). This results in fixation of K in non-exchangeable form. According to 'Lattice Hole Theory' (Page and Baver, 1940), the exposed surface and surface between sheets of three layer (2:1) type minerals consist of oxygen ions, arranged hexagonally. The opening within the hexagon is equal to the diameter of an oxygen ion (approximately 2.8 A°). Ion having a diameter of this magnitude (*e.g.* K^+ 2.66 A°) will fit snugly into the lattice holes and such ions will be very tightly as they come in contact with the negative electrical charges within the crystal. As a result of this, the layers are bound together, thus preventing dehydration and re-expansion. Ions having diameter close to 2.8 A° can also be fixed to a considerable extent. Relatively small hydration energy of K^+ result in easy dehydration of the ions and strong retention. In vermiculite or illite, isomorphous substitution in tetrahedral position of their lattice creates negative charges close to the unit layer surface, which explains the strength of the bond between K[+] and the lattice sheet. Hence, in illite and vermiculite even under wet conditions. Potassium fixation is, therefore, a serious problem in soils containing illite and vermiculite clay minerals dominated in acid soil.

Factors Influencing Potassium Fixation

The degree of K fixation depends on a number of factors, such as

1. Charge density of the minerals which means negative charge per unit silicate layer of the mineral. Potassium fixation is high when density is high. Vermiculite and illite tend fix K best under relatively wet conditions, while fixation by montmorillonite and stratified minerals occurs under dried condition. The fixing power of 2:1 type clay minerals usually follows the order: Vermiculite > Illite > Smectite (momtmorillonite).

2. Depletion of the interlayer wedge zone which means what extent the interlayer wedge zone that is depleted of K. If the wedge zone is confined to the edge of the particle, then only small amounts of K can be fixed. On the other hand if the wedge zone penetrates deeply into the mineral, considerable amount of K can be fixed.

3. Moisture content of the medium plays an important role in K fixation. Wetting and drying cycles lead to fixation of K in soils rich in available K.

4. Concentration of K ion in soil solution also determines the K fixation by the clay minerals.

5. Nature and concentration of competing cations in the surrounding medium influence K fixation. Ions like NH_4^+ and H[+] can compete with K[+] for K fixing or binding sites.

Measurement of Potential K Fixation

Clear understanding of soil potassium dynamics is very much needed for defining the potential K fixation in soil. In soil, K maintains dynamic equilibrium between different potassium pools means the reversible transformation from one form to other that is assumed to occur in soil (Figure 10.4). In some soil, low concentrations of K in solution due to crop removal leads to release of fixed K. In other soil, the K present in interlayer fixation sites may be very slowly released and can be a significant source of plant nutrition.

The ammonium acetate extract (1 N NH_4OAc, pH 7) is a widely used soil extractant to estimate both soluble and exchangeable K. However this procedure is inadequate for soils that have micaceous or vermiculitic mineralogy, which can release some non-exchangeable (fixed) K when the solution and exchangeable K pools are depleted. An alternative method for measuring non-exchangeable plant-available K in soils (*i.e.* the plant-available portion of fixed K) is using the sodium tetraphenylboron extraction. A practical version of this procedure involves 5-minute incubation (Cox *et al.*, 1999). They report that in some soils, this procedure extracted 1.5 to 6 times more K than did NH_4OAc and closely correlated with plant uptake of K. However, this method did not adequately measure K fixation capacity universally (Murashkina *et al.*, 2007b). The assessment of non-exchangeable K status of soil used to do by using H- resins and 6 $N H_2SO_4$ and found good correlation with plant uptake (Srinivasa Rao *et al.*, 2001). The capacity of soils to supply K from exchangeable and non-exchangeable form is not easily determined because K maintains dynamic equilibrium in soil. Due to the complexity that exists in the soil system, none of the methods is universally applicable for all the soils. Haylocks (1965) introduced the terms 'Step-K' and 'Constant Rate K' as plant utilizable non-exchangeable potassium reserves in soil. Step-K is the release of potassium with repeated extractions with 1 $N HNO_3$ and the 'Constant Rate-K' which is the rate achieved when release of K with every extraction is equal to the previous one. The constant rate K which takes into account soil type and mineralogy may serve as guide to the long-term K supplying power of soils.

K fixation can be a significant barrier to meeting the nutritional requirements of crops. In most other soils, K-fixation should not be a significant factor to consider. New laboratory techniques for estimating both the K-fixation potential and the release rate will help with K management decisions.

Management of Fixed K

The phenomena of both the fixation of available potassium and/or release of fixed or non-exchangeable K play an important role in the dynamics of soil potassium. Lower concentration of K in the solution due to leaching or crop removal favours release of K. when there is no external addition of K, plants are capable of taking up substantial amount of potassium without much change in exchangeable-K pool. This indicates that in due course of time exchangeable-K gets replenished from other sources. Non exchangeable K contributes substantially to potassium availability and uptake in soil rich in micaceous minerals, especially in vermiculite and illite dominating alluvial soils. Under exhaustive cropping, non-exchangeable K contribution to total K uptake can be as high as 90 per cent in alluvial soils. Highly-weathered red and lateritic soils are poor in K supply because of low non-exchangeable K reserves. Some perennial grasses like ryegrass (*Lolium perenne*) are very efficient user of non-exchangeable K where as legumes like red clover (*Trifolium pretense*) is not efficient user. Thus consideration of soil characteristic and plant type is very important for efficient utilization of fixed (non-exchangeable) K in soil. As result, K fertilization rate to be applied gets reduced in proportion to the amount of non-exchangeable potassium in soil.

References

1. Aarons, S. R. and Graham, P. H. (1991). Response of *Rhizobium leguminosarum* bv. *phaseoli* to acidity. *Plant Soil* 134: 145–151.

2. Almendras, A. S. and Bottomley, P. J. (1987). Influence of lime and phosphate on nodulation of soil-grown *Trifolium subterraneum* L. by indigenous *Rhizobium trifolii*. *Appl. Environ. Microbiol.* 53: 2090–2097.

3. Barshad, I. (1951). Cation exchange in soils.I. Ammonia fixation and its relation to potassium and to determination of ammonia exchange capacity. *Soil Sci.* 77: 463-472.

4. Bordeleau, L. M. and Prevost. D. (1994). Nodulation and nitrogen fixation in extreme environments. *Plant Soil* 161: 115–124.

5. Brockwell, J., Pilka, A. and Holliday. R. A. (1991). Soil pH is a major determinant of the numbers of naturally-occurring *Rhizobium meliloti* in non-cultivated soils of New South Wales. *Aust. J. Exp. Agric.* 31: 211–219.

6. Carter, J. M., Gardner, W. K. and Gibson, A. H. (1994). Improved growth and yield of faba beans (*Vicia faba* cv. *Fiord*) by inoculation with strains of *Rhizobium leguminosarum* biovar *viciae* in acid soils in south-west Victoria. *Aust. J. Agric. Res.* 45: 613–623.

7. Cooper, J. E., Wood, M. and Bjourson, A. J. (1985). Nodulation of *Lotus pedunculatus* in acid rooting solution by fast-and slow-growing rhizobia. *Soil Biol. Biochem.* 17: 487–492.

8. Correa, O. S. and Barneix, A. J. (1997). Cellular mechanisms of pH tolerance in *Rhizobium loti. World J. Microbiol. Biotechnol.* 13: 153–157.

9. Cox, A.E., B.C. Joern, S.M. Brouder, and D. Gao. (1999). Plant-available potassium assessment with a modified sodium tetraphenylboron method. *Soil Sci. Soc. Am. J.* 63: 902–911.

10. Fox, R.L., Bosshart, R.P., Sompongse, D. and Lin, Mu-Lien. (1989). Phosphorus requirements and management of sugarcane, pineapple, and bananas. *Present in the symposium on Phosphorus Requirements for Sustainable agriculture in Asia and the Pacific Region*, 6-10 March 1989, IRRI, Los Banos, Philippines.

11. Graham, P. H., Viteri, S. E., Mackie, F., Vargas, A. T. and Palacios, A. (1982.) Variation in acid soil tolerance among strains of *Rhizobium phaseoli. Field Crops Res.* 5: 121–128.

12. Haylocks, O. F. (1956) A method of estimating the availability of non-exchangeable potassium. Transaction of 6th International Congress Soil Science Paris 2: 403-408.

13. Ibekwe, A. M., Angle, J. S., Chaney, R. L. and Vonberkum, P. (1997). Enumeration and nitrogen fixation potential of *Rhizobium leguminosarum* biovar *trifolii* grown in soil with varying pH values and heavy metal concentrations. *Agric. Ecosyst. Environ.* 61: 103–111.

14. Martin, J. C. Overstreet, R. and Hoagland. (1946). Potassium fixation in soils in replaceable and non replaceable form in relation to chemical reaction in soil. *Soil Sci. Soc. of Am. Proceedings.* 10: 94-101.

15. Misra, U. K. and Pattanayak, S. K(1997). Characterization of rock phosphates for direct use in different cropping sequences. Technical Report of the U.S. India Fund. Project number *In: AES-708. Grant No. F61- IN-744, 1991-1995.*

16. Murashkina, M. A., Southard, R. J. and Pettygrove, G. S. (2007). Potassium Fixation in San Joaquin Valley Soils Derived from Granitic and Nongranitic Alluvium. *Soil Sci. Soc. Am. J.* 71: 125-132.

17. O'Hara, G. W., Goss, T. J., Dilworth, M. J. and Glenn. A. R. (1989.) Maintenance of intracellular pH and acid tolerance in *Rhizobium meliloti*. *Appl. Environ. Microbiol.* 55: 1870–1876.

18. Page, J. B. and Baver, L. D. (1940). Ionic size in relation to fixation of cations by colloidal clay. *Soil Sci. Soc. of Am. Proceedings* 4: 150-1555.

19. Postgate, J. (1981). Microbiology of the free-living nitrogenfixing bacteria, excluding cyanobacteria. In: Gibson AH, Newton WE (eds) Current perspectives in nitrogen fixation. Elsevier/North-Holland Biomedical, Amsterdam, pp 217–228.

20. Reitemeier, R. F. (1951). The chemistry of soil potassium. Advances in Agronomy. 3: 113-164.

21. Rich, C. I. (1964). Effect of cation size and pH potassium exchange in Nason soil. *Soil Sci.* 98: 100-106.

22. Rich, C. I. (1968.) Minerology of soil potassium. p 79-91. (In): V.J. Kilmer *et al.* (Ed.) The role of potassium in agriculture. American Society of Agronomy, Madison, WI.

23. Rich, C. I. and Black, R. W. (1964). Potassium exchange as affected by cation size, pH and mineral structure. *Soil Sci.* 97: 384-390.

24. Saxena, A. K., and Rewari, R. B. (1991). The influence of phosphate and zinc on growth, nodulation and mineral composition of chickpea (*Cicer arietinum* L.) under salt stress. *World J. Microbiol. Biotechnol.* 7: 202–205.

25. Singleton, P. W., Abel Magid, H. M. and Tavares. J. W. (1985). Effect of phosphorus on the effectiveness of strains of *Rhizobium japonicum*. *Soil Sci. Soc. Am. J.* 49: 613–616.

26. Sparks, D. L. and Huang, P. M. (1985). Physical chemistry of soil potassium. p. 201-276. In R.D. Munson (Ed.) Potassium in agriculture. American Society of Agronomy, Madison, WI.

27. Srinivasa Rao, Ch., Rupa, T. R. and Subba Rao, A. (2001). Sub-soil potassium availability in 22 benchmark soils of India. *Commun. Soil Sci. Plant Anal.* 32: 863-876.

28. Taylor, R. W., Williams, M. L. and Sistani, K. R. (1991). Nitrogen fixation by soybean-Bradyrhizobium combinations under acidity, low P and high Al stresses. *Plant Soil* 131: 293–300.

29. Thomas, R. J., Askawa, N. M., Rondon, M. A. and Alarcon, H. F. (1997). Nitrogen fixation by three tropical forage legumes in an acid soil savanna of Colombia. *Soil Biol. Biochem.* 29: 801–808.

30. Torimitsu, K., Hayashi, M., Ohta, E and Sakata, M. (1985). Effect of K^+ and H^+ stress and role of Ca^{2+} in the regulation of intracellular K^+ concentration in mung bean roots. *Physiol. Plant.* 63: 247–252.

31. Uexkull, H. R. and Bosshart, R.P. (1989) Management of Acid Upland Soils in Asia. In: Craswell, E.T., and Pushparajah, E. (Eds.) *Management of Acid Soils in the Humid Tropics of Asia.* ACIAR Monograph No.13 (m SRAM Monograph No.1), Australian Centre for International Agricultural Research, Canberra, pp 2-19.

32. Walker, G. F. (1957). On the differentiation of vermiculite and smectites in clay. *Clay Miner. Bull.* 3: 154-163.

2015, Horticulture for Nutrition Security
Editor: Prof. K.V. Peter
Published by: DAYA PUBLISHING HOUSE, NEW DELHI

Pages *189–216*

Chapter 11

Rootstocks for Abiotic Stress Management in Fruits

S.K. Mitra and T.K.S. Irenaeus

Grafting

Grafting is the union of two or more pieces of living plant tissue, which are forced to develop vascular connection and grown as a single plant.

Scion

Scion is a short piece of detached shoot containing one or several dormant buds which when united with stock forms the upper portion of graft and from which will grow the stem or branches or both of the grafted plant. It would be from a desired cultivar and free from diseases.

Abiotic Stress

Abiotic stress is defined as the negative impact of non-living factors on the living organisms in a specific environment. Plants are especially dependent on environmental factors, so it is particularly constraining and abiotic stress is a major limiting factor for plant growth and food production. Research has also shown that abiotic stressors are the most harmful when they occur together, in combinations of abiotic stress factors (Mittler, 2006). Abiotic stress occurs in different forms. The most common of the stressors are the easiest for people to identify, but there are many other less recognizable abiotic stress factors which affect environments constantly. The most basic stressors include drought, extreme temperatures, high winds, floods, salinity and other natural disasters like tornados and wildfires. The lesser-known stressors generally occur on a smaller scale and so are less noticeable, but they include poor edaphic conditions like rock content and pH of soils, high radiation, compaction, contamination, and other, highly specific conditions like rapid rehydration during seed germination (Palta *et al.,* 2006). Drought, salinity, extreme temperatures and nutrient imbalances are the major environmental stresses to crop productivity.

Abiotic stress conditions cause the accumulation of reactive oxygen species (ROS), such as superoxide radicals (O_2^-), hydroxyl radicals (OH^-) and hydrogen peroxide (H_2O_2) (Mittler, 2002). Reactive oxygen species are strong oxidizing species that cause oxidative damage to membrane lipids and proteins and this accumulation is reported in apple rootstock (Molassiotis *et al.,* 2006).

As sessile organisms, plants must regulate their growth and development in order to respond to numerous external stimuli and in an ever-changing environment. A plant's first line of defence against abiotic stress is in its roots. If the soil holding the plant is healthy and biologically diverse, the plant will have a higher chance of surviving in stressful conditions (Lijbert *et al.,* 2007). Rootstocks are essential for the ability of plants to adapt to abiotic stresses by mediating a wide range of adaptive responses.

Rootstock

Rootstock is the lower portion of the graft which develops into the root system of grafted plant. It may be a seedling, rooted cutting or a layered plant.

Importance of Rootstocks

Rootstocks provide growers with useful tool to manipulate the vigour and performance of orchard tree. Effects on tree size, precocity, fruit production and maturity are achieved through complex relationship between the roots and the canopy of the plant. They are also able to significantly alter the pattern of canopy development and functions such as photosynthesis. Besides giving anchorage to the tree, rootstock is also responsible for the absorption of water and nutrients, storage of photosynthates and synthesis of hormones making the scion part more tolerable. Several characters including leaf nutrient status, vigour and size, depth of rooting, low temperature tolerance, adoption to adverse soil conditions, disease resistance and fruit quality are affected by rootstocks.

Interstock

Interstock is the piece of stem inserted by means of two graft union between scion and rootstock. An inter stem is used for several reasons such as to overcome incompatibility between scion and rootstock, to make use of winter hardy stem and take advantage of winter controlling property.

Influence of Rootstock on Plant

The rootstock and the scion which constitute a grafted or budded plant mutually exert specific effects on the overall growth and performance of the grafted plant. Some of the salient effects are:

1. Size and growth habit – it is the most significant effect manifested in the grafted or budded plant by the influence of rootstock.
2. Induction of precocity or early maturing.
3. Advancement of fruit bud formation.
4. Enhancement of fruit set.
5. Increasing the productivity.
6. Improvement of fruit quality.

There is a growing interest in dwarf orchard trees, but the mechanisms by which the rootstock influences the vegetative growth of the scion are still unclear. Early anatomical studies have led to the hypothesis of rootstocks influencing tree water relations (Beakbane, 1956). Since then, several other studies relating plant anatomy to growth have been conducted reporting the presence of smaller and/or a fewer vessels in the roots and/or graft tissue suggesting lower hydraulic conductivities of these plant regions (Soumelidou *et al.*, 1994). Other mechanisms that have been invoked to explain the dwarfing potentials of rootstocks include low nutrient uptake capacity, reduced hormonal levels, lower net CO_2 assimilation rates and stomatal conductances. Conclusions have been drawn in many crops that plants grafted on dwarfing rootstocks had significantly lower hydraulic conductance of the plant (K_{plant}) like lower leaf-specific K_{plant} (Cohen and Naor, 2002), lower K in fine roots, stems and the graft union region in apple (Atkinson *et al.*, 2003) and lower leaf-specific root hydraulic conductance (K_{root}) in peach (Basile *et al.*, 2003), which is known to influence the plant growth capacity through changes in leaf water status and gas exchange. This was in contrast to that reported in kiwifruit by Clearwater *et al.* (2004) which documented higher leaf specific K_{plant} in low-vigour rootstocks.

Use of tetraploid (4×) rootstocks in citrus has been emphasised during the last decade, in particular with development of somatic hybrids that combine the traits of two parents (Grossner and Gmitter, 2010; Dambier *et al.*, 2011). For example, by combining a genome of *Poncirus trifoliata* with a citrus genome such as Cleopatra mandarin (*Citrus reshni* Hort. Ex Tan.), it was possible to generate plants resistant to biotic stressors such as Citrus tristeza virus and Phytophthora, and abiotic stressors resistant to alkalinity, salinity and drought. Interestingly, allopolyploidization does not seem to be the only way to improve tolerance traits in rootstocks as autotetrapoid citrus seedlings have also shown better tolerance to salt and water stress than the parental diploid (2×) stock (Saleh *et al.*, 2008; Allario, 2009), probably because of higher abscissic acid biosynthesis (Allario, 2009).

Drought

Among environmental stresses, drought has the greatest impact on plant. Increasing crop resistance to this stress would be the most economical approach for improving productivity and reducing agricultural reliance on fresh-water resources. Water stress affects the plant growth and productivity. Results in mango showed that increasing water stress reduced the physiological parameters particularly at the later days of plant growth, due to stomatal and non-stomatal factors. Leaf chlorophyll content however, increased since the chlorophyll pigments may have been resistant to dehydration. Luvaha *et al.* (2010) concluded that increase in water stress reduces the gas exchange parameters of mango rootstock seedlings but slightly increased chlorophyll content.

Rootstock has been used as a source of stress resistance and/or tolerance in apple. *Malus prunifolia*, is one of the most drought-resistant rootstocks among apple rootstocks. *M. hupehensis* which thrives in wet habitats though is highly resistant to water-logging, shade, cold, and disease is vulnerable to drought (Duan *et al.*, 2009).

Different rootstocks of citrus showed variation in their ability to tolerate drought. Kinnow mandarins have become the most favoured choice cultivar amongst citrus growers in India because it has adapted very well under semi-arid climatic conditions, where other citrus varieties failed. The trees of 'Folha Murcha' budded onto 'Cravo FCAV', 'Cravo Limeira', 'Rangpur' limes and 'Sunki' mandarin in Sao Paulo, Brazil (Stuchi *et al.*, 2000; Cantuarias-Aviles *et al.*, 2011) and 'Tahiti' plants grafted onto the 'Cravo FCAV' and 'Cravo Limeira', 'Rangpur' limes showed tolerance to drought (Cantuarias-Aviles *et al.*, 2012). 'Valencia' orange grafted on FA-5 rootstock (*Citrus rashni* × *Poncirus trifoliata*) also showed higher drought tolerance than its parents (Rodriguez-Gamir *et al.*, 2010).

Citrus tree water relations and tolerance to most abiotic stresses vary greatly according to the rootstocks. Sweet orange trees grafted on Cleopatra mandarin had better plant water status under deficit irrigation conditions, stimulating a greater vegetative growth than for those on Carrizo citrange (Perez-Perez *et al.*, 2008). Citrus rootstocks influence drought-stress response differently because of their differences in root distribution pattern, water uptake efficiency, and root hydraulic conductivity. The higher drought tolerance could also be related to the greater osmotic adjustment (OA), which was reflected by smaller reductions in leaf relative water content (RWC) and in higher turgor potentials and leaf gas exchange than the other rootstocks (Rodriguez-Gamir *et al.*, 2010).

The drought resistance classification of rootstocks in grape might vary from country to country. In areas where water is a limiting factor to grapevine productivity, using drought resistant rootstocks '110R', '140Ru' and '1103P' (Ezzahouani and Williams, 1995; Chen, 2000) should be beneficial. Rootstock hybrids developed from *V. berlandieri* × *V. rupestris* were considered to be drought tolerant (Ezzahouani and Williams, 1995). Hybrids of *V. berlandieri* × *V. riparia* were reported to be more tolerant to drought (Kocsis *et al.*, 1998).

Almond cv. Nonpareil plants grafted on wild peach rootstock attained more height, spread, scion girth, annual shoot growth, and leaf area, whereas those on bitter almond rootstock had higher stock girth in plants subjected to different moisture stress conditions while growth and vigour of trees on bitter almond were less influenced by water stress conditions (Sharma and Joolka, 2002).

Flood Tolerance

Soil drainage is an important aspect for site selection of grapevines. The rootstock *V. cinerea* was considered to be flood tolerant (Pongraz, 1983). Striegler *et al.* (1993) found that tolerance was increased slightly by grafting 'Seyval' a flood susceptible cultivar on rootstock '3309C'. The rate of shoot elongation was considered the most sensitive index of flood tolerance.

Commercial annona varieties are generally grown on seedling rootstocks. These trees are very susceptible to soil flooding and root diseases prevalent in saturated soils. Flooding of commercial annona species, even for short periods, reduces net CO_2 assimilation and vegetative growth, may cause defoliation and severely reduces flowering and fruit set. Pond apple (*A. glabra*) native to subtropical and tropical America including southern Florida is not grown commercially but is extremely flood tolerant. However, many commercial annona species, such as atemoya are not graft compatible with *A. glabra*. The major atemoya cultivar of Southern Florida 'Gefner' is not directly compatible with *A. glabra* which may be utilized by using an interstock that are compatible with both the scion and rootstock (Nunez-Elisea *et al.*, 1998). 'Gefner' atemoya can be successfully grafted onto *A. glabra* using "49-11" as an interstock. The resulting scion-interstock-rootstock combination is extremely flood tolerant or flood adapted. In Columbia, *A. reticulata* and *A. montana* are used as rootstock for tolerance to drought and soil moisture.

Temperature

Tolerance to temperature stress is becoming a desirable trait. We are facing an unequivocal global warming which may represent another challenge to plant productivity and geographic distribution. Temperature is one of the major environmental factors affecting plant productivity.

In mango, rootstock Carabao was found tolerant to different cold temperature conditions (Ribeiro *et al.*, 2002). Differences of cane and bud hardiness of grape rootstocks 'K5BB', '3309C' and 'SO4' were also found by Miller *et al.* (1988a) where '3309C' had the most cold hardy canes and buds; its acclimation in fall was faster and declamation in spring was slower than 'K5BB' and 'SO4'. They also found that different rootstock cultivars influenced cold hardiness of scions grafted to them differently (Miller *et al.*, 1988b). For example, grafted 'White Riesling' grape was found significantly hardier than the own-rooted vines. The observed differences in the LT50 values (the low temperature lethal to 50 per cent of primary buds death) ranged from 0.5°C to 3.0°C in cane hardiness; 'White Riesling' grafted to '3309C' had significantly a fewer shoot lets nodes and grafted 'White Riesling' on 'K5BB' had hardier buds in most cases. It was proposed that '3309C' would be the cold hardiest. 'Seyval' grafted on '3309C' appeared to be the most cold hardy compared to K5BB', and own rooted 'Seyval'(Striegler and Howell, 1991) while, 'Cabernet Sauvignon' and 'Chardonnay', on 'K5BB' and '1103P' showed less frost damage than on 'SO4' and '420A' (Palliotti *et al.*, 1991).

Tolerance to cold and yield improvement due to physiological modifications by use of rootstocks has been reported in olive. Increased yield in cv. "Moraiolo", grafted on Lecino, Leccino Compact, Nostrale di Rigali, San Arcangelo and Vocio has been recorded by increasing cold resistance while for "San Felice", increase in yield due to increased frost resistance was recorded when grafted on Leccino and Vocio (Panneli *et al.*, 2002). Perez-Lopez *et al.* (2010) grouped the different cultivars of olive as resistant (Cornicabra), tolerant (Picual, Ascolana Tenera and Arbequina), or sensitive (Frantoio and Changlot Real). They further reported that this response is probably due to delayed stomatal closure. Only Cornicabra and Picual showed a significant reduction in leaf conductance (below 10°C and 6 °C respectively). This absence of stomatal control led to a significantly greater dehydration in Ascolana Tenera. These variations in response to the soil chilling temperature suggest that different mechanisms may be at work. It is therefore necessary to study the influence of rootstock in the frost resistance of olive plants.

Rootstock, cultivar and their interaction significantly affected the extent of frost damage to reproductive and vegetative buds in peach. Frost damage to shoots was found significantly affected only by cultivar. In cv. Gracia, 99.92 per cent reproductive buds and 78.25 per cent vegetative buds werre killed by frost damage, while in Redhaven the corresponding figures were 99.18 per cent and 87.61 per cent, respectively. The lowest damage levels in Gracia were on rootstock of Myrobalan plum and peach species *Persica* [*Prunus*] *davidiana*. The lowest percentage of reproductive buds killed in Redhaven was on the Myrobalan stock MY-KL-A, and the lowest percentage of vegetative buds killed was on almond MN33 and the Myrobalans MY-KL-A, MY-BO1, Dzanka 4 and MY-GA-R (Nitransky, 1994).

Salinity

Salinity is one of the major abiotic stresses that affects plant production and growth in many arid and semi-arid areas throughout the world. Inappropriate crop production and irrigation systems which employ poor quality water further aggravate the situation through generation of additional secondary salinization processes. Soil salinity is often reported to have adverse effects on crop growth and productivity particularly through the alteration of the metabolic processes of the plants by inducing changes in ion toxicity and osmotic pressure of the tissues. The responses to these alterations are often accompanied by a variety of symptoms, such as the reduction of leaf area, abscission of leaves, increase of leaf thickness and succulence, necrosis of shoot and root, and reduction of internode lengths.

Sukkary rootstock of mango proved to be more appropriate rootstock for use in regions irrigated with saline water reaching 4000 ppm (Hafez *et al.,* 2011). In areas with low water quality Gomera-1 rootstock of mango is feasible for use as it proved to be the most adaptable rootstock to saline conditions. Gomera-3 was found more sensitive, taking up higher amounts of Cl^- and Na^+ than Gomera-1 (Duran-Zuazo *et al.,* 2003). The highly salt tolerant mango rootstock 13/1 showed greater vegetative growth with NaCl than with Na_2SO_4. At low salt concentrations, leaf area was reduced but the number of leaves increased. Plant analysis revealed higher Na^+ concentrations in young than old plant parts and saline conditions significantly reduced transpiration rate, especially with NaCl (Schmutz and Ludders, 1993).

Mango cultivars, 13/1, Sabre and Olour can be used as rootstocks for mango cultivation in soils upto 35 ESP (exchangeable sodium percentage), whereas Bappakai, Nakkare and Kurukkan can be used as rootstock in soils upto 25 ESP though, all polyembryonic mango genotypes can be used as rootstock in a slightly sodic soil of 15 ESP (Yadav and Singh, 2006). Seedlings from stone of Kesar variety of mango were better with significantly higher survival percentage, germination percentage and growth parameters in saline conditions while more mortality of seedling was observed in Totapuri variety. In case of electrical conductivity (EC) level of water, significantly higher survival percentage was registered only at 1.20 dSm^{-1} EC level. Poor seedling survival (14.12 per cent) was recorded at 4.00 dSm^{-1} EC level of water. Germination percentage, number of leaves, plant height and root length were increased with decreasing EC level. The accumulation of sodium was higher in leaves, whereas, potassium and Na:K ratio were noted lower with higher EC level (Varu and Barad, 2010).

Soil salinity is an important problem in citrus production and citrus is the most salt sensitive perennial horticultural crop causing tissue burning, loss of yield, leaf abscission and finally death (Romero-Aranda *et al.*, 1998). In sweet oranges, reductions of up to 6 per cent in yield may be expected for every additional 100 $\mu\Omega$/cm EC increase in irrigation water salinity above 400 EC (Cole, 1983). Physiological disturbances and eventually visible damage occur due to high leaf chloride concentrations (Cl$^-$) arising from root zone salinity (Cooper and Shull, 1953), and high leaf sodium concentrations (Na$^+$) caused detrimental effects on photosynthesis and transpiration (Behboudian *et al.*, 1986). These physiological disturbances may lead to growth and yield reductions long before visible symptoms of salt damage are evident. Ability of citrus trees to tolerate rootzone salinity mostly depends on rootstocks and one way of improving the salt tolerance of citrus is to graft scions on to salt tolerant rootstocks. Citrus rootstocks vary widely in their ability to restrict accumulation of Cl$^-$, Na$^+$ or both from either their own shoot or that of a scion (Syvertsen and Yelenosky, 1988; Zekri, 1991). Strains of trifoliate orange were found to accumulate low levels of Cl$^-$ and Na$^+$ ions in the shoot (Sykes, 2011). Rootstocks differ in their salinity tolerance as judged by the ability to inhibit the accumulation of Cl$^-$ and/or Na$^+$ in leaves of the scion. Ion accumulation in shoots is rootstock-dependent but Na$^+$ and Cl$^-$ accumulation also can be scion dependent. The exclusion of Cl$^-$ from shoots is related to the ability of cell membranes to restrict the movement of Cl$^-$ across the root to the vascular tissue and by the degree of Cl$^-$ accumulation in the roots. High concentration of salts in the root zone reduces soil water potential and the availability of water, however increased uptake of Na$^+$ and/or Cl$^-$ can create a more negative leaf osmotic potential leading to increased turgor in leaves. The lower Cl$^-$ and Na$^+$ concentration in leaves of 'Sunburst' mandarin grafted on Cleopatra than on Carrizo, suggest that the salinity tolerance of Cleopatra is associated with ion sequestration in roots with less transport to leaves (Garcia-Sanchez *et al.*, 2002). Other studies showed differing salt tolerances between rootstocks mainly due to differences in their ability to exclude Cl$^-$, Na$^+$ or both ions from the scion (Banus and Primo-Millo, 1992; Garcia-Sanchez *et al.*, 2006).

The Cl$^-$ and Na$^+$ exclusion traits in citrus rootstocks are heritable and can be transmitted to breeding progeny (Ream and Furr, 1976). The Cl$^-$ excluding ability of Rangpur lime may be expressed in its hybrids (Ream and Furr, 1976; Sykes, 1992). Similarly, the Na$^+$ excluding trait found in *Poncirus trifoliata* (L.) Raf. at low salinities (Walker, 1986) may be expressed in its hybrids (Sykes, 1992). Rootstock such as FA-5 (*Citrus rashni × Poncirus trifoliata*) seems to be tolerant to salinity (Forner-Giner *et al.*, 2009) and to calcareous soils (Gonzalez-Mas *et al.*, 2009).

Traditional grape cultivation in India entailed growing commercial varieties of grape on its own roots. A decline in the productivity of table grapes in the major grape-growing states of Maharashtra, Karnataka and Andhra Pradesh led the way to the utilization of rootstocks in grape cultivation. Most of the table grape-growing districts in these states experience severe drought conditions during the critical growth stages, such as fruit bud differentiation, shoot maturity and full bloom. Another problem in some areas of these states is an increase in soil salinity which severely reduced the vine growth when grown on its own roots (Satisha *et al.,* 2010). Grapes appear to be more sensitive to salinity as the plants grow older. With succeeding years, the chloride and sodium accumulate more rapidly in the leaves, causing leaf burn to develop (Varalakshmi and Ganeshamurthy, 2013). Sodium accumulation in grapes leaf did not significantly increase until the leaves were already severely damaged by chloride accumulation (Raghupathi and Ganeshmurthy, 2011). Chloride toxicity has been the principle limiting factor for grapevines grown on their own root. However, a significant reduction in chlorine accumulation occurs in chloride-sensitive scions when grown on Dogridge or 1613-3 rootstocks (Bernstein *et al.,* 1969). The salt tolerance of these two rootstocks would probably be limited by soil osmotic effects long before chloride reached toxic levels.

The use of salt tolerant rootstocks is an efficient strategy to alleviate the adverse effects of salinity in grapevines (Walker *et al.,* 1997). *V. champini* and *V. vinifera* are tolerant to salinity (Leon *et al.,* 1969). Grown at high salinity conditions, 'Shiraz' grape had higher wine K+, pH and colour hue on 'Ramsey', '1103 Paulsen' and '140 Ru' than on their own roots (Walker *et al.,* 2000). Under a relatively high saline soil condition in South Africa, different rootstocks showed different symptoms of stress; rootstocks '101-14 Mgt' and '143-B Mgt' were recommended for salinity tolerance (Southey and Jooste, 1992). Rootstock effects on salt tolerance of 'Sultana' were reported by Walker *et al.* (2002); the best performing rootstocks were 'Ramsey', '1103P' and 'R2', which could impart most vigour to the scions. Accumulation of sodium and chloride in leaves and petioles of Dogridge, Salt Creek and 1623 rootstocks are usually lower. The different salt tolerant *Vitis* species *V. berliendieri, V. riparia* and *V. champini* may be exploited to overcome the salinity problem (Mitra, 2013).

Salinity tolerance may be influenced by the root distribution pattern of rootstocks and may involve the contribution of chloride-exclusion on promising rootstocks. Grapevines salinity tolerance has been strongly associated with their ability to restrict Cl- entry in the shoot and it has been used as a screening method for classifying genotypes according to their salt tolerance (Walker *et al.,* 1997; Mullins *et al.,* 1996). When 'Cardinal' and 'Thompson Seedless' were grafted to rootstock 'Dog Ridge', '1163-3', and 'Salt Creek', the accumulated chloride contents in leaves were only 1/3, 1/10, and 1/16 of those on their own-rooted vines, respectively (Leon *et al.,* 1969). However, contradictions can be found in the literature in terms of the salt tolerance of grapevine rootstocks (Antcliff *et al.,* 1983; Arbabzadeh and Dutt, 1987) which implies that additional factors, as well as to salt accumulation in the shoot, are involved and eventually determine grapevine response to salt stress (Paranychianakis and Chartzoulakis, 2005).

Grapevine rootstocks possess additional mechanisms operating under salt stress which may compensate for the differences in salt accumulation in the leaves. Thus leaf-Na^+ and -Cl^- content are not the only parameters that should be considered when screening genotypes to evaluate their salt tolerance or to predict the performance of grapevines grown in saline environments. In addition to genetic factors regulating salt uptake and accumulation in the shoot, differences in morphological factors appear to also have an apparent influence on salt accumulation in the shoot. Rootstocks that produce higher root biomass usually have lower Cl^- uptake and therefore its higher capacity for salt sequestration. Preferential sequestration of salts into roots has been suggested as a mechanism of salt tolerance (Munns, 2002).

Paranychianakis and Angelakis (2008) suggested that leaf salt content alone should not be the only criteria to classify genotypes according to their tolerance to salinity and that salinity-induced damage is linked with prevailing environmental conditions. Furthermore, it can be inferred that grapevines have additional mechanisms to cope with salt stress which may counteract differences in salt uptake and accumulation in the shoot. A distinct superiority in terms of salinity tolerance among rootstocks was only observed at the 0.50ET irrigation level, where Soultanina vines grafted on 41B developed earlier and more acute leaf burns than on the other rootstocks like 103P and110R. Irrigation of grapevine at sub-optimum levels (0.50 and 0.75 ET) exacerbated the impact of salinity on vine performance, hence deficit irrigation should not be practiced when irrigating with waters of elevated salt concentration.

In apple, *Malus prunifolia* Borkh cv. Dongbeihuanghaitang, *M. sieboldii* Rehd cv. Daguohongsanyehaitang, *M. prunifolia* Borkh cv. Qiuzi, and *M. xiaojinensis* Cheng et Jia were reported as salt-tolerant rootstocks, whereas *M. prunifolia* Borkh cv. Yingyehaitang, *M. micromalus* Hemsl, and *M. sieboldii* Rehd cv. Lushihongguo were salt-sensitive rootstocks. These differences in sensitivity were associated with variations in the activities of anti-oxidation enzymes and in the amount of organic osmotica (Yin *et al.,* 2010).

Variations in the sodium and chloride uptake in plant organs of *Pistacia* spp. were noted and seemed to be controlled more efficiently in *P. atlantica* than in *P. vera*. In both species, the K^+ content was noted to undergo a significant decrease when salinity increased. While when the K^+/Na^+ ratio was maintained above 2 at low NaCl treatments, it was sharply decreased at high NaCl conditions, suggesting a failure of K–Na selectivity mechanism. The Ca^{2+}/Na^+ ratio decreased significantly at 60 and 80 mM NaCl in *P. vera* and at 60 mM NaCl for *P. atlantica*. In both *Pistacia* species, high NaCl treatments (131–240 mM NaCl) induced a significant increase in proline content (Chelli-Chaabouni *et al.,* 2010).

The *in vitro* responses of *P. vera* and *P. atlantica* to NaCl treatment reflected similar behaviours to those achieved under *in vivo* conditions and indicated some reactions which are activated from early developmental stages and provide further information on the possible mechanisms employed by these species to avoid salt toxicity. The higher salt tolerance of *P. atlantica* observed in *in vitro* seems to be correlated with higher survival rate, a lower growth reduction, and a lesser reduction in terms of the

K⁺/Na⁺ and Ca²⁺/Na⁺ selectivity ratios in growing tissues (Chelli-Chaabouni *et al.*, 2010).

In olive, Sanna (2004) reported that Kalamata grafted on Picual rootstock exhibited the greatest salt tolerance followed by those grafted on Fratoio, whereas Kalamata on Koroneiki showed the lowest salt tolerance. There are differences in tolerance to iron chlorosis grown in calcareous soils among olive cultivars, and that tolerance is mainly determined by the genotype of the rootstock. Use of tolerant cultivars may be adopted for those conditions where iron chlorosis could become a problem (Alcantara *et al.*, 2003).

Nutrient Imbalances

Iron chlorosis one of the major nutritional imbalances in fruit tree orchards in many areas such as Mediterranean area occurs due to the limited iron bioavailability in aerobic and alkaline pH environments (Romheld and Nikolic, 2007) which was attributed to the presence of high levels of bicarbonate ions in the calcareous soils. It is estimated that from 20 to 50 per cent of fruit trees grown in the Mediterranean basin suffer from Fe deficiency (Jaegger *et al.*, 2000). These soils often have more than 20 per cent of calcium and magnesium carbonates and are strongly buffered, with a pH between 7.5 and 8.5. The high level of bicarbonate ions in the soil affects metabolic processes in roots and leaves, decreasing soil and plant Fe availability (Mengel, 1995), leading to the condition known as lime-induced iron chlorosis. Rootstocks were also reported to have greatly influenced the scion's tolerance to B toxicity in *Prunus* (El-moutaium *et al.*, 1994) and deficiency in apple (Wojcik *et al.*, 2003). Use of different rootstocks can be effective to improve the boron uptake and movement across the graft union (Nartvaranant *et al.*, 2003).

One unfavourable soil factor to plant growth is aluminium (Al) toxicity, a limiting factor in many acid soils, which extend over 40 to 70 per cent of the world's arable soils (Rengel, 1992). Use of high Al tolerant plants can be one option since plants have different degrees of adaptation to Al variability as observed among plants of same genus or among cultivars of the same species, *e.g.*, citrus (Lin and Myhre, 1991).

In citrus orchards, B-deficiency is widespread and is responsible for considerable loss of productivity and poor fruit quality (Han *et al.*, 2008). The efficiency of B-acquisition is dependent mainly on the root system of citrus rootstocks. Root morphology of the rootstocks is the key factor in citrus productivity and can provide useful information regarding the B-absorbing capacity of the roots and their tolerance to low B-availability. From a breeding perspective, improvement in B-acquisition by changing the root morphology or identification of both B-efficient and B-inefficient citrus rootstock genotypes according to the root-morphological traits is very important in resolving the problems of B-deficiency (Rerkasem and Jamjod, 2004).

Genotypes variations in response to B deficiency were reported by Mei *et al.* (2011). Carrizo citrange and Red tangerine were found as B-efficient, whereas Fragrant citrus and Sour orange were B-inefficient genotypic rootstocks. In another study, the growth of root, stem of scion and leaves were found less affected by low B treatments when 'Newhall' navel orange scion was grafted on Carrizo citrange than on Trifoliate orange. Thus, the growth of scions under low B conditions was mainly depended on the rootstock used, *i.e.*, Carrizo citrange-grafted plants were more tolerant to low B compared to the plants grafted on Trifoliate orange (Sheng *et al.,* 2009).

Lin and Myhre (1991) concluded that rootstocks used in citrus cultivation could be classified for Al resistance based on their fresh weight of whole plant material: *Citrus reshni* > *Citrus jhambiri* > *Citrus aurantium* > 'Swingle' (*Citrus paradisi* × *Poncirus trifoliata*) > 'Carrizo' (*Citrus sinensis* × *Poncirus trifoliata*).

Grape rootstocks have a "strategy" to overcome lime induced chlorosis with higher root iron uptake and greater reducing capacity (Bavarresco *et al.,* 1991). Bavaresco *et al.* (1993) reported the effect of rootstocks on the occurrence of lime-induced chlorosis of potted *V. vinifera* 'Pinot Blanc', a susceptible cultivar to lime-induced chlorosis. They found that the iron-efficient rootstock '140 Ruggeri' (*V. berlandieri* × *V. rupestris*) did not induce chlorosis when grown on calcareous soil, while the iron-inefficient rootstock '101-14' (*V. riparia x V. rupestris*) did. It was later confirmed that the chlorosis in 'Pinot Blanc' was reduced when grafted to rootstocks 'SO4' and '3309C', which was thought to be related to different hydraulic conductivities between the rootstock and the own-rooted vines (Bavaresco and Lovisolo, 2000). For example, the own-rooted 'SO4' showed the highest specific conductivity, associated with the highest rate of shoot growth and leaf chlorophyll content. Grape species reported to be tolerant to growing on lime soil are *V. berlandieri* and *V. cinerea*; some representative rootstocks include '41 B', '333 EM' and 'Fercal' (Zimmermann and Zimmermann, 1973; Kocsis *et al.,* 1998; Mullins *et al.,* 1992). 'Fercal' – an INRA hybrid rootstock (*V. berlandieri* × *V. vinifera* cv. Colombard No. 1 × 333 EM) is resistant to chlorosis induced by calcareous soil.

Mechanism of Plant Response to Stress

Plants have developed mechanisms to withstand drought, such as higher root-shoot ratios, fewer and smaller leaves, concentrated solutes osmotic adjustment or increased activity of oxidative stress enzymes in leaf cells (Lei *et al.,* 2006). Adjustments in cell wall elasticity can help to maintain water uptake: a decrease in cell wall elasticity will enable a decrease in the pressure potential, thus decreasing the plant water potential (Bowman and Roberts, 1985; Marshall and Dumbroff, 1999). On the other hand, an increase in cell wall elasticity can help to maintain turgor pressure at lower relative water volumes, as the cell walls are able to shrink around the cell contents (Ruiz-Sanchez *et al.,* 1997). Ion toxicity and osmotic stress have been found to alter cell wall elasticity, contributing to increased stress tolerance in a number of woody species. Increased cell wall elasticity during stress benefitted lemon trees (Ruýz-Sanchez *et al.,* 1997). In contrast, decreased cell wall elasticity in combination with osmotic adjustment allowed a positive water balance to be maintained during NaCl stress (Nabil and Coudret, 1995). Net photosynthesis, stomatal conductance and carboxylation efficiency were affected by the rootstock genotype combination, so that under water stress only some rootstock genotypes transferred drought tolerance to the scion as observed in grapes (Iacona *et al.,* 1998).

Stomatal closure is one of the earliest responses to drought, playing an important role in water loss control in plants. Although being modulated by water stress in leaves, among several factors, stomata may close in response to drought before any change in leaf water potential or leaf water content is detectable (Gollan *et al.,* 1985). It is recognised that leaf water status interacts with stomatal aperture and, under stress conditions, a positive relationship is often found between leaf water potential and stomatal conductance (Medrano *et al.,* 2002). However, this relationship is highly dependent on the plant species, the drought history of the individuals, the size of pots in which they are rooted and the environmental conditions during drought (Flexas *et al.,* 1999).

Plants regulate their osmotic potential and compartmentalise toxic ions to cope with the primary effects of salinity. The regulation of osmotic potential to maintain turgor pressure (despite a lower water potential) involves several processes such as the uptake of K^+, the compartmentalization of Na^+ and Cl^- into the vacuole, and the synthesis of compatible solutes such as proline, glycine betaine, polyol, sugars etc. (Ashraf, 1994).

Colmenero Flores *et al.* (2007) identified a chloride transporter presumably involved in long-distance transport and plant development. A secondary effect of toxic ions during salt stress (Gomez-Cadenas *et al.,* 1996), along with other abiotic stresses such as drought (Chaves and Oliveira, 2004) and temperature (Sala, 1998), is the triggering of oxidative stress (Tanou *et al.,* 2009), which causes damage to the leaf photosynthetic machinery (Allen and Ort, 2001).

In mango, increase in plant resistance to salt-stress was associated with the anti-oxidant activity of ascorbic acid and a partial inhibition of salt-induced (Schmutz and Ludders, 1993). Difference in salinity tolerance was probably based on the ability to protect leaves from excessive Na^+ and to accept higher Cl^- contents in the leaves without severe growth damage as in rootstocks of 13-1. Significantly higher Ca^{2+} and Mg^{2+} contents were found in leaves and roots of 13-1 compared to Turpentine, which might explain the higher tolerance of Cl^- in leaf tissues of 13-1 as well as its higher Na^+ retention potential of roots and stems. Low (30 mM NaCl) salinity levels and increasing root zone temperature promoted tolerance mechanisms such as higher growth, active exclusion under optimum root temperatures and higher Ca^{2+} uptake. It was observed that Na or Cl was individually excluded from leaf tissue depending on cultivar. However, no cultivar was outstanding in the combined exclusion of Na and Cl (Hoult *et al.,* 1997). Survival under flooded conditions appeared to be related to the tree's ability to form hypertrophied stem lenticels; without these lenticels, trees died shortly after flooding (Larson *et al.,* 1993). To improve flooding tolerance, mango trees could be selected for: (1) development of hypertrophied lenticels, which allow transport of oxygen from the atmosphere into the root; and (2) an alternative respiration pathway as well as residual respiration. It is suggested that non-cytochrome pathways might play an important role in the energy charge of roots under restricted soil oxygen (Zude *et al.,* 1998).

The adverse effect of salinity on commonly used genotypes includes symptoms of leaf injury, growth suppression and yield decline. The primary effects of salinity in citrus are decreased stomatal conductance leading to reduced CO_2 diffusion and ultimately decreased net photosynthesis (Garcia-Sánchez and Syvertsen, 2006) and increased ion accumulation (Brumos *et al.,* 2010).

Results demonstrated that the most salt sensitive genotypes accumulated high concentrations of Na⁺ and Cl⁻ and maintained a fair growth and photosynthetic rate. By contrast, salt-tolerant genotypes accumulated less Na^+ and Cl^- and decreased their growth and gas exchange. *Poncirus commun* citron and *Marumi kumquat* were found as sensitive species, while mandarins, pummelo and Australian sour orange were the most tolerant (Hussain *et al.,* 2012a). Among the genotypes studied by them, Engedi pummelo presented a specific trait for salt tolerance that has not been previously reported. Their results suggest that low leaf chloride content can be used as an indicator of salt stress tolerance in citrus genotypes.

The antioxidant defense system in the plant cell includes both enzymatic antioxidants such as superoxide dismutase (SOD; EC1.15.1.1), catalase (CAT; EC 1.11.1.6) and ascorbate peroxidase (APX, EC 1.11.1.11), and non-enzymatic antioxidants such as ascorbate, glutathione and a-tocopherol. As a major scavenger, SOD catalyzes the dismutation of superoxide to hydrogen peroxide and oxygen. However, H_2O_2 is also toxic to the cells and has to be further scavenged by CAT or peroxidase, or both, to water and oxygen (Zhu *et al.,* 2004). In the ascorbate-glutathione cycle, APX reduces H_2O_2 using ascorbate as an electron donor. Altered activities of both antioxidant enzymes and non-enzymatic antioxidants have been commonly reported, and are used frequently as indicators of oxidative stress in plants (Mittler, 2002).

Under stress conditions, plants not only produce antioxidants, but also accumulate compatible solutes such as proline that originally were thought to function as osmotic buffers. However, apart from osmotic adjustment they seem to play a role in maintaining the functional state of macromolecules, probably by scavenging ROS (Xiong and Zhu, 2002). Similarly, several workers reported that the accumulation of solutes such as glycine and proline are linked to water stress, salinity and other abiotic plant stresses (Ashraf and Harris, 2004; Munns and Tester, 2008; Lu *et al.,* 2009). Several studies have associated high proline levels and tolerance to abiotic stresses, such as drought, salinity and high temperatures (Hong *et al.,* 2000; De Ronde *et al.,* 2004; Molinari *et al.,* 2007). Proline accumulates under salt stress in both leaf and root tissues (Aziz *et al.,* 1999) and putatively protects against the osmotic potential generated by salt (Watanabe *et al.,* 2000; Chen *et al.,* 2007; Hoque *et al.,* 2008). There is a strong positive relationship between stress tolerance and proline accumulation in higher plants (Ashraf and Fooland, 2007). Transgenic 'Carrizo' citrange (Molinari *et al.,* 2004) and 'Swingle' citrumelo (Marilia *et al.,* 2011) rootstocks were able to better cope with water deprivation due to osmotic adjustment provided by high proline content since the high endogenous proline level acted not only by mediating osmotic adjustment, but also by contributing to gas exchange parameters and ameliorating deleterious effects of drought-induced oxidative stress (Marilia *et al.,* 2011).

In drought stress, accumulation of proline in many plants represents a general response (Garcia-Sanchez *et al.,* 2007). However, contrary to other reports, salt-stressed citrus leaves do not accumulate proline (Syvertsen and Yelenosky, 1988) but drought-stressed citrus leaves can accumulate proline (Syvertsen and Smith, 1983) and other betaines but probably not glycine betaine (Nolte *et al.,* 1997). Fu *et al.* (2011) noted that overexpression of the *betaine aldehyde dehydrogenase* (AhBADH) gene in transgenic trifoliate orange enhanced salt stress tolerance and may be correlated with low levels of lipid peroxidation, protection of the photosynthetic machinery, and increase in K^+ uptake.

Alterations in physiological process in response to stress occurred in lemon (Perez-Perez *et al.*, 2009). Osmotic adjustment was the main tolerance mechanism for maintenance of turgor under salt stress, and was achieved by the uptake of Cl^- ions as observed in lemon. Gas-exchange parameters were reduced by drought stress but not by salinity, stomatal closure being the main adaptive mechanism for avoidance of water loss and maintenance of leaf turgor. Immature leaves exhibited major stress symptoms in drought conditions, whereas salinity mainly affected mature leaves.

Mature leaves had high leaf K^+, Ca^{2+} and Mg^{2+} concentrations, which could have alleviated the negative effect of drought (Perez-Perez *et al.*, 2009). Immature leaves had low leaf mineral concentrations, which may have been related to the increased synthesis of proline. A high leaf proline concentration, despite the low leaf mineral concentration, could have protected the photosynthetic apparatus (Lawlor, 2001) which was supported by the fact that the A_{CO2} reduction was similar in both mature and immature leaves (Perez-Perez *et al.*, 2009). Although A_{CO2} was not affected by salinity in either type of leaf, the total chlorophyll concentration was reduced in mature leaves, apparently due to high leaf Cl^- concentration (Perez-Perez *et al.*, 2007).

Plants of *Malus prunifolia* which are drought tolerant maintained their structural cell integrity longer than drought sensitive species *Malus hupehensis*. *M. hupehensis* was more vulnerable to drought than was *M. prunifolia*, resulting in larger increases in the levels of H_2O_2, O_2, and malondialdehyde (MDA) from the former. Experimental results of Wang *et al.* (2012) demonstrated that in order to minimize oxidative damage, both the activities of antioxidant enzymes and antioxidant concentrations were increased in the leaves of *M. prunifolia* and *M.hupehensis* in response to water stress. Moreover, plants of *M. prunifolia* exhibit higher antioxidant capacity and a stronger protective mechanism, such that their cell structural integrity is better maintained during exposure to drought. Except for catalase (CAT) and monodehydroascorbate reductase (MDHAR), the activities of superoxide dismutase (SOD), peroxidase (POD), ascorbate peroxidase (APX), glutathione reductase (GR), and dehydroascorbate reductase (DHAR) were enhanced to a greater extent in *M. prunifolia* than in *M. hupehensis* in response to drought. This was also true for levels of ascorbic acid (AsA) and glutathione (GSH). Under well-watered conditions, changes in lipid peroxidation and relevant antioxidant parameters were not significantly different between the two species.

At the ultrastructural level, chloroplasts usually had normal oblong shapes and a typical membrane system of stroma and grana thylakoids. However, when water stress was imposed, those thylakoids became considerably swollen and vesiculated, and eventually were destroyed (Wang *et al.*, 2012). This was especially true for leaves from *M. hupehensis*, in which the cells showed very little grana-stacking after 12 days of restricted irrigation. *M. prunifolia* has a lower rate of O_2 production and reduced concentrations of H_2O_2 and MDA. This leads to less detrimental changes to its chloroplasts and mitochondrial structures. Therefore, compared with *M. hupehensis*,

M. prunifolia appears to possess a better mechanism for protecting its organelles when plants undergo sustained periods of water deficit. This greater performance by our drought-tolerant apple rootstock can be attributed to a smaller increase in O_2 production, lower H_2O_2 and MDA concentrations, and higher values for leaf relative water content (RWC), as well as greater antioxidative capacity.

Ozden *et al.* (2009) suggested that proline and H_2O_2 could play an important role in oxidative stress injury of grapevine leaves grown in *in vitro* culture. Also, proline might have a direct positive effect on antioxidant enzyme system, membrane phase change, malondialdehyde (MDA) and electrolyte leakage (EL).

In vitro studies in Prunus rootstocks ('Masto de Montadana', GF677 and 'Adesoto 101') have shown that proline concentration in root tissues and root exudates from all rootstocks increased as salt concentration in the medium increased, following a trend similar to that of whole plant tissues (Marin *et al.*, 2010) and culturing excised roots has proven to be a very good experimental model for the early detection of tolerance to abiotic stresses such as salinity (Marin and Marin, 1998).

The accumulation of the amino acid proline in plant tissues in response to different abiotic stresses may play an important role against oxidative damages caused by reactive oxygen species (ROS) (Marilia *et al.*, 2011). Due to its action as singlet oxygen quencher (Alia *et al.*, 2001) and scavenger of hydroxyl radicals (Smirnoff and Cumbes, 1989), proline is able to stabilize DNA, proteins and membranes (Alia *et al.*, 2001). Besides being a ROS scavenger, proline plays several other important roles during stress adaptation, by acting as an osmotic adjustment mediator (Zhang *et al.*, 1995) and storage of carbon, nitrogen and energy (Hare and Cress, 1997). Another role of proline might be that it protects against osmotic stress due to its capacity in increasing antioxidant systems (Zhang *et al.*, 1995). In addition, its synthesis and degradation can provide a way to buffer cytosolic pH and balancing cell redox (Venekamp, 1989).

Some shoot characteristics of cashew plantlets that are favourable for dealing with high salinity (transpiration, transport and accumulation of salt ions and organic solute accumulation involved with cell protection and osmotic adjustment, such as proline and free amino acids in leaves) are dependent on rootstocks. It was concluded that physiological disturbances induced by salinity in cashew plantlets were more influenced by rootstock than by scion and these changes were also dependent on compatibility between scion and rootstock (Ferreira-Silva *et al.*, 2010).

Similarly, plant hormones play an important role in response to unfavourable environmental conditions. They are involved in the signalling response to drought and salinity by the activation of acclimation processes such as stomatal closure, regulation of hydraulic conductivity and regulation of developmental processes which affect stress tolerance, such as senescence abscission (Sakamoto *et al.*, 2008).

Changes in concentration of root metabolites occur which could indicate tolerance levels to iron deficiency. Iron deficiency induces root metabolic changes in addition to FC-R activity induction and rhizosphere acidification to sustain the increased iron uptake capacity of Fe-deficient plants. Carbohydrates, amino acids and especially organic acid concentrations often increase with iron deficiency in herbaceous plants

(Abadia *et al.,* 2002; Zocchi, 2006; M'Sehli *et al.,* 2008). Carbohydrate concentrations and rates of carbohydrate catabolism are reported to increase under iron deficiency to sustain energetic requirements of the stressed plant (Zocchi, 2006). Amino acid concentrations are reported to increase in order to sustain the major protein synthesis occurring under iron deficiency (Zocchi, 2006). Organic acids that make the limited soluble soil iron available to plants when they are excreted can facilitate iron translocation and may be associated with protein extrusion and Fe^{3+} reduction activity (Abadia *et al.,* 2002). Fixation of CO_2 leading to organic acid biosynthesis is catalysed by the enzyme phosphoenolpyruvate carboxylase (PEPC; EC4.1.1.31). PEPC activity stimulation has been observed in iron-deficient roots of several species (Rombola *et al.,* 2002; Ollat *et al.,* 2003; Jimenez *et al.,* 2007; Anduluz *et al.,* 2009). In grapevine, organic accumulation in roots is greater in Fe-efficient than in Fe-inefficient genotypes (Brancadoro *et al.,* 1995; Ollat *et al.,* 2003; Jimenez *et al.,* 2007).

In *Prunus,* the tolerant rootstock Adesoto showed higher total organic and amino acid concentrations. In contrast, the susceptible rootstock Barrier showed lower total amino acid concentration and phosphoenolpyruvate carboxylase activity values. These results suggest that the induction of this enzyme activity under iron deficiency indicates the tolerance level of rootstocks to iron chlorosis. The analysis of other metabolic parameters, such as organic and amino acid concentrations, provides complementary information for selection of rootstock genotypes tolerant to iron chlorosis (Jimenez *et al.,* 2011).

Higher root: shoot (R/S) ratios and longer lateral-root ratios are the main factors that help understand the contribution of roots to B-deficiency tolerance in Carrizo citrange and Red tangerine genotypes.

Tetraploidy is known to affect various phenotypic traits like stomatal density, cell size, division rate, organelle composition, root and leaf morphology, growth and fruit quality. All of these traits can affect tree physiology dramatically. When compared to 2× citrus rootstocks, 4× genotypes show lower stomatal conductance and decreased photosynthesis rates, which result in lower rates of whole-plant transpiration, reduced growth and poorer fruit yields (Perez-Perez *et al.,* 2007).

Tetraploids possess stronger antioxidant defence systems in leaves, which lead to increased tolerance for heat (Zhang *et al.,* 2010) and water deficit (Chandra and Dubey, 2010). Higher hesperidin contents has been measured in fruits grown on 4× rootstocks, and these 4× rootstocks may have triggered antioxidant defence systems in the leaves and fruit (Hussain *et al.,* 2012b). The 4× rootstock enhanced the electron flow rate in leaves under saturated light, which is a way to overcome stress. Hence, the presence of hesperidin in fruit may help to reduce the stress effects of environment when 4× rootstocks are used (Hussain *et al.,* 2012b).

Some rootstocks of grape were responsive to silicon under stress condition (Soylemezoglu *et al.,* 2009). Studies on different grape rootstocks showed the role of Si in regulating the salinity and B toxicity stress responses of grapevine rootstocks, and indicated that Si could protect plants against oxidative damage. These effects of Si on physiological aspects of the plant growth can therefore be seen in conjunction with the plants' own endogenous stress responses. Silicon decreases H_2O_2 concentration and lipid peroxidation in grapevine rootstocks, presumably through the observed increase in CAT and SOD (Soylemezoglu *et al.,* 2009).

References

1. Abadia, J., Lopez-Milan, A.F., Rombola, A. and Abadia, A. (2002). Organic acids and Fe deficiency: a review. *Pl. Soil*, **241**: 75-86.

2. Alcantara, E., Cordeiro, A.M. and Barranco, D.(2003). Selection of olive varieties for tolerance to iron chlorosis. *J. Plant Physiol.*, **160**(12): 1467-1472.

3. Alia, J.M., Mohanty, P. and Matysik, J. (2001). Effect of proline on the production of singlet oxygen. *Amino Acids*, **21**: 195-200.

4. Allario, T. (2009). Identification de déterminants physiologiques et moléculaires de la tolérance à la contrainte saline et au déficit hydrique de porte-greffes autotétraploïdes d'agrumes, UFR Sciences et techniques. Thèse de l'Université de Corse, Pascal Paoli, Corte, p. 201.

5. Allen, D. and Ort, D. (2001). Impacts of chilling temperatures on photosynthesis in warm-climate plants. *Trends Plant Sci.*, **6**: 36-42.

6. Andaluz, S., Rodriguez-Celma, J., Abadia, A., Abadia, J. and Lopez-Milan, A.F. (2009). Time course induction of several key enzymes in Meicago truncatula roots in response to Fe deficiency. *Pl. Physiol. Biochem.*, **47**: 1082-1088.

7. Antcliff, A.J., Newman, H.P. and Barret, H.C.(1983). Variation in chloride accumulation in some American species of grapevine. *Vitis,* **22**: 357-362.

8. Arbabzadeh, F. and Dutt, G. (1987). Salt tolerance of grape rootstocks, under greenhouse conditions. *Am. J. Enol. Viticult.*, **38**: 95-101.

9. Ashraf, M. and Harris, P. (2004). Potential biochemical indicators of salinity tolerance in plants. *Plant Sci.*, **166**: 3-16.

10. Ashraf, M. (1994). Breeding for salinity tolerance in plants. *Critical Rev. Plant Sci.*, **13**: 17-42.

11. Atkinson, C.J., Else, M.A., Taylor, L. and Dover, C.J. (2003). Root and stem hydraulic conductivity as determinants of growth potential in grafted trees of apple (*Malus pumila* Mill). *J. Exp. Bot.*, **54**: 1221-1229.

12. Aziz, A., Martin-Tanguy, J. and Larher, F. (1999). Salt stress-induced proline accumulation and changes in tyramine and polyamine levels are linked to ionic adjustment in tomato leaf discs. *Plant Sci.*, **145:** 83-91.

13. Banus, J. and Primo-Millo, E. (1992). Effects of chloride and sodium on gas exchange parameters and water relations of citrus plants. *Physiol. Plant*, **86**: 115-123.

14. Basile, B., Marsal, J., Solari, L.I., Tyree, M.T., Bryla, D.R. and DeJong, T.M. (2003). Hydraulic conductance of peach trees grafted on rootstocks with differing size-controlling potential. *J. Hort. Sci. Biotechnol.*, **78**: 768-774.

15. Bavaresco, L. and Lovisolo, C. (2000). Effect of grafting on grapevine chlorosis and hydraulic conductivity. *Vitis*, **39**: 89-92.

16. Bavaresco, L., Fraschini, P. and Perino, A. (1993). Effect of the rootstock on the occurrence of lime-induced chlorosis of potted *Vitis vinifera* L. cv. 'Pinot Blanc'. *Pl. Soil*, **157**: 305-311.

17. Bavaresco, L., Fregoni, M. and Fraschini, P.(1991). Investigations on iron uptake and reduction by excised roots of different grapevine rootstocks and *V. vinifera* cultivar. *Pl. Soil*, **130**: 109-113.

18. Beakbane, A.B.(1956). Possible mechanisms of rootstocks effect. *Ann. Appl. Biol.*, **44**: 517-521.

19. Behboudian, M.H., Torokfalvy, E. and Walker, R.R. (1986). Effects of salinity on ionic content, water relations and gas exchange parameters in some citrus scion-rootstock combinations. *Sci. Hort.*, **28**: 105-116.

20. Bernstein, L., Ehlig, C.F. and Clark, R.A. (1969). Effect of grape rootstocks on chloride accumulation in leaves. *J. Amer. Soc. Hort. Sci.*, **94**: 584-590.

21. Bowman, W.D. and Roberts, S.W. (1985). Seasonal changes in elasticity in chaparral shrubs. *Physiol. Plant*, **65**: 233-236.

22. Brancadoro, L., Raboti, G., Scienza, A. and Zocchi, G. (1995). Mechanisms of Fe deficeincy in roots of *Vitis* spp. in response to iron-deficiency stress. *Pl. Soil*, **171**: 229-234.

23. Brumos, J., Talon, M., Bouhlal, R. and Colmenero, Flores, J. (2010). Cl-homeostasis in includer and excluder citrus rootstocks: transport mechanisms and identification of candidate genes. *Plant Cell Environ.*, **33**: 2012-2027.

24. Cantuarias-Aviles, T., Mourao-Filho, F.A.A., Stuchi, E.S., Rodrigues da Silva, S. and Espinoza-Nuzez, E. (2011). Horticultural performance of 'Folha Murcha' sweet orange onto twelve rootstocks. *Sci. Hort.*, **129**: 259-265.

25. Cantuarias-Avilés, T., Mourão-Filho, F.A.A., Stuchi, E.S., Rodrigues-da-Silva, S., Espinoza-Núñez, E. and Neto, H.B. (2012). Rootstocks for high fruit yield and quality of 'Tahiti' lime under rain-fed conditions. *Sci. Hort.*, **142**: 105-111.

26. Chandra, A. and Dubey, A. (2010). Effect of ploidy levels on the activities of [Delta]1- pyrroline-5-carboxylate synthetase, superoxide dismutase and peroxidase in Cenchrus species grown under water stress. *Plant Physiol. Biochem.*, **48**: 27-34.

27. Chaves, M.M. and Oliveira, M.M. (2004). Mechanisms underlying plant resilience to water deficits: prospects for water-saving agriculture. *J. Exp. Bot.*, **55**: 2365-2384.

28. Chelli-Chaabouni, A., Mosbah, A.B., Maalej, M., Gargouri, K., Gargouri-Bouzid, R. and Drira, N. (2010). *In vitro* salinity tolerance of two pistachio rootstocks: *Pistacia vera* L. and *P. atlantica* Desf. *Environ. Exp. Bot.*, **69**: 302-312.

29. Chen, J.F.(2000). The status of research on grape rootstock varieties and its prospect. *J. Fruit Sci.*, **17**: 138-146 (in Chinese).

30. Chen, Z., Cuin, T.A., Zhou, M., Twomey, A., Naidu, B.P. and Shabala, S. (2007) Compatible solute accumulation and stress mitigating effects in barley genotypes contrasting in their salt tolerance. *J. Exp. Bot.*, **58**: 4245-4255.

31. Clearwater, M.J., Lowe, R.G., Hofse, B.J., Barclay, C., Mandemaker, A.J. and Blattman, P. (2004). Hydraulic conductance and rootstock effects in grafted vines of kiwifruit. *J. Exp. Bot.*, **55**: 1371-1382.

32. Cohen, S. and Naor, A. (2002). The effect of three rootstocks on water use, canopy conductance and hydraulic parameters of apple trees and predicting canopy from hydraulic conductance. *Pl. Cell Environ.*, **25**: 17-28.

33. Cole, P.J. (1983). Some of the effects of salinity of citrus production. *Aust. Citrus News* 59, November, 10-11.

34. Colmenero Flores, J., Martinez, G., Gamba, G., Vazquez, N., Iglesias, D., Brumos, J. and Talon, M. (2007). Identification and functional characterization of cation-chloride cotransporters in plants. *The Plant J.*, **50**: 278-292.

35. Cooper, W.C. and Shull, A.V. (1953). Salt tolerance of and accumulation of sodium and chloride ions in grapefruit on various rootstocks grown in naturally saline soil. *Proc. Rio Grande Valley Hort. Inst.*, **7**: 107-117.

36. Dambier, D., Benyahia, H., Pensabene-Bellavia, G., Aka Kacar, Y., Froelicher, Y., Belfalah, Z., Lhou, B., Handaji, N., Printz, B. and Morillon, R. (2011). Somatic hybridization for citrus rootstock breeding: an effective tool to solve some important tissues of the Mediterranean citrus industry. *Plant Cell Rep.*, 1-18.

37. De Ronde, J.A., Cress, W.A., Krüger, G.H., Strasser, R.J. and Van Staden, J. (2004). Photosynthetic response of transgenic soybean plants, containing an Arabidopsis P5CR gene, during heat and drought stress. *J. Plant Physiol.*, **161**: 1211-1224.

38. Duan, K., Yang, H., Ran, K., You, S., Zhao, H. and Jiang, Q. (2009). Characterization of a novel stress-response member of the MAPK family in *Malus hupehensis* Rehd. *Plant Mol. Biol. Rep.*, **27**: 69-78.

39. Duran-Zuazo, V.H., Martinez-Raya, A. and Aguilar-Ruiz, J. (2003). Salt tolerance of mango rootstocks (*Mangifera indica* L. cv. Osteen). *Spanish J. Agric. Res.*, **1**(1): 67-78.

40. El-moutaium, R., Hu, H. and Brown, P.H. (1994). The relation tolerance of six prunus rootstocks to boron and salinity. *J. Amer. Soc. Hort. Sci.*, **119**: 1169-1175.

41. Ezzahouani, A. and Williams, L.E. (1995). Influence of rootstock on leaf water potential, yield, and berry composition of Ruby Seedless grapevines. *Amer. J. Enol. Viticult.*, **46**: 559-563.

42. Ferreira-Silva, S.L., Silva, E.N., Carvalho, F.E.L., Lima, de, C.S., Alves, F.A.L. and Silveira, J.A.G. (2010). Physiological alterations modulated by rootstock and scion combination in cashew under salinity. *Sci. Hort.*, **127**: 39-45.

43. Flexas, J., Escalona, J.M. and Medrano, H. (1999). Water stress induces different levels of photosynthesis and electron transport rate regulation in grapevines. *Plant Cell Environ.*, **22**: 39-48.

44. Forner-Giner, M.A., Primo-Millo, E. and Forner, J.B.(2009). Performance of Forner-Alcaide 5 and Forner-Alcaide 13, hybrids of Cleopatra mandarin × Poncirus trifoliata, as salinity-tolerant citrus rootstocks. *J. Am. Pomolog. Soc.*, **63**: 72-80.

45. Fu, X.Z., Khan, E.U., Hu, S.S., Fan, Q.J. and Liu, J.H. (2011). Overexpression of the *betaine aldehyde dehydrogenase* gene from *Atriplex hortensis* enhances salt tolerance in the transgenic trifoliate orange (*Poncirus trifoliata* L. Raf.) *Environ. Exp. Bot.*, **74**: 106-113.

46. Garcia-Sanchez, F. and Syvvertsen, J.P. (2006). Salinity tolerance of Cleopatra mandarin and Carrizo citrange citrus rootstock seedlings is affected by CO_2 enrichment during growth. *J. Amer. Soc. Hort. Sci.*, **131**: 24-31.

47. Garcia-Sanchez, F., Jifon, J.L., Carvajal, M. and Syvertsen, J.P. (2002). Gas exchange, chlorophyll and nutrient contents in relation to Na^+ and Cl^- accumulation in 'Sunburst' mandarin grafted on different rootstocks. *Pl. Sci.*, **162**: 705-712.

48. Garcia-Sanchez, F., Perez-Perez, J.G., Botia, P. and Martinez, V. (2006). The response of young mandarin trees grown under saline conditions depends on the rootstock. *Eur. J. Agron.*, **24**: 129-139.

49. Garcia-Sanchez, F., Syvertsen, J.P., Gimeno, V., Botia, P. and Perez-Perez, J.G. (2007). Responses to flooding and drought stress by two citrus rootstock seedlings with different water-use efficiency. *Physiol. Plant*, **130**: 532-542.

50. Gollan, T., Turner, N.C. and Schulze, E.D. (1985). The responses of stomata and leaf gas exchange to vapour pressure deficits and soil water content. *Oecologia*, **65**: 356-362.

51. Gomez-Cadenas, A., Tadeo, F.R., Talon, M. and Primo-Millo, E. (1996). Leaf abscission induced by ethylene in water stressed intact seedlings of (*Citrus reshmi* Hort. Ex Tan.) requires previous abscissic acid accumulation in roots. *Plant Physiol.*, **112**: 401-408.

52. Gonzalez-Mas, M.C., Llosa, M.J., Quijano, A. and Forner-Giner, M.A. (2009). Rootstock effects on leaf photosynthesis in "Navelina" trees grown in calcareous soil. *HortSci.*, **44**: 280-283.

53. Grossner, J. and Gmitter, F. (2010). Protoplast fusion for production of tetraploids and triploids: applications for scion and rootstock breeding in citrus. *Plant Cell Tiss. Organ Cult.*, **104**: 343-357.

54. Hafez, O.M., Malaka, A.S., Ellil, A.A.A. and Kassab, O.M. (2011). Impact of ascorbic acid in salt tolerant of some mango rootstock seedlings. *J. Applied Sci. Res.*, **Nov.** 1492-1500.

55. Han, S., Chen, L.S., Jiang, H.X., Smith, B.R., Yang, L.T. and Xie, C.Y. (2008). Boron deficiency decreases growth and photosynthesis, and increases starch and hexoses in leaves of citrus seedlings. *J. Plant Physiol.*, **165**: 1331-1341.

56. Hare, P.D. and Cress, W.A. (1997) Metabolic implications of stress-induced proline accumulation in plants. *Plant Growth Regul.*, **21**: 79-102.

57. Hong, Z., Lakkineni, K., Zhang, Z. and Verma, D.P.S. (2000). Removal of feedback inhibition of delta-1-pyrroline-5-carboxylate synthetase results in increased proline accumulation and protection of plants from osmotic stress. *Plant Physiol.*, **122**: 1129-1136.

58. Hoque, M.A., Banu, M.N., Nakamura, Y., Shimoishi, Y. and Murata, Y. (2008). Proline and glycinebetaine enhance antioxidant defense and methylglyoxal detoxification systems and reduce NaCl-induced damage in cultured tobacco cells. *J. Plant Physiol.*, **165**: 813-824.

59. Hoult, M.D., Donnelly, M.M. and Smith, M.W. (1997). Salt exclusion varies amongst polyembryonic mango cultivar seedlings. *Acta Hort.*, **455**: 455-458.

60. Hussain, S., Luro, F., Costantino, G., Ollitrault, P. and Morillon, R. (2012a). Physiological analysis of salt stress behaviour of citrus species and genera: Low chloride accumulation as an indicator of salt tolerance. *South African J. Bot.*, **81**: 103-112.

61. Hussain, S., Curk, F., Dhuique-Mayer, C., Urban, L., Ollitrault, P., Luro, F. and Morillon, R. (2012b). Autotetraploid trifoliate orange (*Poncirus trifoliata*) rootstocks do not impact clementine quality but reduce fruit yields and highly modify rootstock/scion physiology. *Sci. Hort.*, **134**: 100-107.

62. Iaconoa, F., Buccella, A. and Peterlunger, E. (1998). Water stress and rootstock influence on leaf gas exchange of grafted and ungrafted grapevines. *Sci. Hort.*, **75**: 27-39.

63. Jaegger, B., Goldbach, H. and Sommer, K. (2000). Release from lime induced iron chlorosis by CULTAN in fruit trees and its characterisation by analysis. *Acta Hort.*, **531**: 107-113.

64. Jimenez, S., Gogorcena, Y., Hévin, C., Rombolà, A.D. and Ollat, N. (2007). Nitrogen nutrition influences some biochemical responses to iron deficiency in tolerant and sensitive genotypes of *Vitis. Pl. Soil,* **290**: 343-355.

65. Jimenez, S., Ollat, N, Deborde, C., Maucourt, M., Rellan-Alvarez, R., Moreno, M.A. and Gogorcena, Y. (2011). Metabolic response in roots of *Prunus* rootstocks submitted to iron chlorosis. *J. Plant Physiol.*, **168**: 415-423.

66. Kocsis, L., Lehoczky, E., Bakonyi, L., Szabo, L., Szoke, L. and Hajdu, E. (1998). New lime and drought tolerant grape rootstock variety. *Acta Hort.*, **473**: 75-82.

67. Larson, K.D; Schaffer, B. and Davies, F.S. (1993). Physiological, morphological and growth responses of mango trees to flooding. *Acta Hort.*, **341**: 152-159.

68. Lawlor, D.W. (2001). *Photosynthesis,* (Eds.), Oxford: Bios Scientific Publishers.

69. Lei, Y.B., Yin, C.Y. and Li, C.Y. (2006). Differences in some morphological, physiological and biochemical responses to drought stress in two contrasting populations of *Populus przewalskii. Physiol. Plant,* **127**: 182-191.

70. Leon, B., Ehlig, C.F. and Clark, R.A. (1969). Effects of grape rootstocks on chloride accumulation in leaves. *J. Amer. Soc. Hort. Sci.*, **94**: 584-590.

71. Lijbert, B., Ruiter, de, P.C. and Brown, G.G. (2007). Soil biodiversity for agricultural sustainability. *Agric. Ecosyst. Environ.*, **121**: 233-244.

72. Lin, Z. and Myhre, D.L. (1991). Differential response of citrus rootstocks to aluminium levels in nutrient solutions. I. Plant growth. *J. Pl. Nutr.*, **14**(11): 1223-1238.

73. Lu, S.Y., Chen, C.H., Wang, Z.C., Guo, Z.F. and Li, H.H. (2009). Physiological responses of somaclonal variants of triploid bermudagrass (*Cynodon transvaalensis* × *Cynodon dactylon*) to drought stress. *Plant Cell Rep.*, **28**: 517-526.

74. Luvaha, E., Netondo, G.W. and Ouma, G. (2010). Effect of water deficit on the physiological and morphological characteristics of Mango (*Mangifera indica*) rootstock seedlings. *Amer. J. Plant Physiol.*, **5**(1): 7-21.

75. M'Sehli, W., Youssfi, S., Donnini, S., Dell'Orto, M., De Nisi, P., Zocchi, G., *et al.* (2008). Root exudation and rhizosphere acidification by two lines of Medicago ciliaris in response to lime-induced iron deficiency. *Pl. Soil*, **312**: 151-162.

76. Marilia, K.F. de C., Kenia de Carvalho, Souza, de, F.S. Marur, C.J., Pereira, L.F.P, Bespalhok Filho, J.C. and Vieira, L.F.E. (2011). Drought tolerance and antioxidant enzymatic activity in transgenic 'Swingle' citrumelo plants over-accumulating proline. *Environ. Exp. Bot.*, **72**: 242-250.

77. Marin, J.A., Andreu, P., Carrasco, A. and Arbeloa, A. (2010). Determination of proline concentration, an abiotic stress marker, in root exudates of excised root cultures of fruit tree rootstocks under salt stress. Revue des Régions Arides – Numéro spécial – 24 (2/2010). Actes du 3ème Meeting International "Aridoculture et Cultures Oasisennes: Gestion et Valorisation des Ressources et Applications Biotechnologiques dans les Agrosystèmes Arides et Sahariens" Jerba (Tunisie) 15-16-17/12/2009.

78. Marin, M.L. and Marin, J.A. (1998). Excised rootstock roots cultured in vitro. *Plant Cell Rep.*, **18**: 350-355.

79. Marshall, J.G. and Dumbroff, E.B. (1999). Turgor regulation via cell wall adjustment in white spruce. *Plant Physiol.*, **119**: 313-319.

80. Medrano, H., Escalona, J.M., Bota, J., Gulias, J. and Flexas, J. (2002). Regulation of photosynthesis of C_3 plants in response to progressive drought: stomatal conductance as a reference parameter. *Ann. Bot.*, **89**: 895-905.

81. Mei, L., Sheng, O., Peng, S., Zhou, G., Wei, Q. and Li, Q. (2011). Growth, root morphology and boron uptake by citrus rootstock seedlings differing in boron-deficiency responses. *Sci. Hort.*, **129**: 426-432.

82. Mengel, K. (1995). Iron availability in plant tissues – iron chlorosis on calcareous soils. In: J. Abadia, (Eds.), *Iron Nutrition in Soils and Plants*. Kluwer Academic Publishers, Dordrecht, The Netherlands, pp. 389-397.

83. Miller, D.P., Howell, G.S. and Striegler, R.K. (1988a). Cane and bud hardiness of selected grapevine rootstocks. *Amer. J. Enol. Viticult.*, **39**: 55-59.

84. Miller, D.P., Howell, G.S. and Striegler, R.K. (1988b). Cane and bud hardiness of own-rooted white Riesling and scion of White Riesling and Chardonnay grafted to selected rootstocks. *Amer. J. Enol. Viticult.*, **39**: 60-86.

85. Mitra, S.K.(2013). Rootstocks for fruit crops: current scenario and their thrust areas. In: K.L. Chadha, A.K. Singh, S.K. Singh and W.S. Dhillon, (Eds.), *Horticulture for Food and Environmental Security*. Westville Publishing House, New Delhi. pp. 232-242.

86. Mittler, R.(2002). Oxidative stress, antioxidants and stress tolerance. *Trends Plant Sci.*, **7**: 405-410.

87. Mittler, R.(2006). Abiotic stress, the field environment and stress combination. *Trends Plant Sci.*, **11**(1): 15-19.

88. Molassiotis, A., Sotiropoulos, T., Tanou, G., Diamantidis, G. and Therios, I. (2006). Boron-induced oxidative damage and antioxidant and nucleolytic responses in shoot tips culture of the apple rootstock EM9 (*Malus domestica* Borkh). *Environ. Exp. Bot.*, **56**: 54-62.

89. Molinari, H.B.C., Marur, C.J., Bespalhok Filho, J.C., Kobayashi, A.K., Pileggi, M., Leite Junior, R.P., Pereira, L.F.P. and Vieira, L.G.E. (2004). Osmotic adjustment in transgenic citrus rootstock Carrizo citrange (*Citrus sinensis* Obs. x *Poncirus trifoliata* L. Raf.) overproducing proline. *Plant Sci.*, **167**: 1375-1381.

90. Molinari, H.B.C., Marur, C.J., Daros, E., Campos, M.K.F., Carvalho, J.F.P.R., Filho, J.C., Pereira, L.F.P. and Vieira, L.G.E. (2007). Evaluation of the stress-inucible production of proline in transgenic sugarcane (*Saccharum* spp.): osmotic adjustment, chlorophyll fluorescence and oxidative stress. *Physiol plant*, **130**: 218-229.

91. Mullins, G.M., Bouquet, A. and Williams, L.E. (1992). *Biology of The Grapevines*. Cambridge University Press, NY.

92. Mullins, M.G., Bouquet, A. and Williams, L.E. (1996). *Biology of Grapevine*. Press Syndicate of the University of Cambridge, Cambridge, pp. 239.

93. Munns, R. (2002). Comparative physiology of salt and water stress. *Plant Cell Environ.*, **25**: 239-250.

94. Munns, R. and Tester, M. (2008). Mechanisms of salt tolerance. *Annu. Rev. Plant Biol.*, **59**: 651-681.

95. Nabil, M. and Coudret, A. (1995). Effects of sodium choride on growth, tissue elasticity and solute adjustment in two *Acacia nilotica* subspecies. *Physiol. Plant*, **93**: 217-268.

96. Nartvaranant, P. Whiley, A.W. and Subhadrabandhu, S. (2003). Effects of selected rootstock/scion combinations on B uptake and plant growth of mango (*Mangifera indica* L.). *Thai J. Agric. Sci.*, **36**(2): 193-205.

97. Nitransky, S. (1994). Influence of rootstocks on the frost resistance of peach cultivars Gracia and Redhaven during the winter 1992-93. *Rocenka Geneticke-zdroje-rastlin,* 47-51.

98. Nolte, K.D., Hanson, A.D. and Gage, D.A. (1997). Proline accumulation and methylation to proline betaine in citrus: implications for genetic engineering of stress resistence. *J. Amer. Soc. Hort. Sci.*, **122**: 8-13.

99. Nunez-Elisea, R., Schaffer, B., Crane, J.H. and Colls, A.M. (1998). Leaf gas exchange and growth responses of young container grown *Annona* trees to flooding. *HortSci.*, **33**: 541 (Abstr).

100. Ollat, N., Laborde, B., Neveux, M., Diakou-Verdin, P., Renaud, C. and Moing, A. (2003). Organic metabolism in roots of various grapevine (*Vitis*) rootstocks submitted to deficiency and bicarbonate nutrition. *J. Pl. Nutr.*, **26**: 2165-2176.

101. Ozden, M., Demirel, U. and Kahraman, A. (2009). Effects of proline on antioxidant system in leaves of grapevine (*Vitis vinifera* L.) exposed to oxidative stress by H_2O_2. *Sci. Hort.*, **119**: 163-168.

102. Palta, Jiwan, P. and Farag, Karim. (2006). Methods for enhancing plant health, protecting plants from biotic and abiotic stress related injuries and enhancing the recovery of plants injured as a result of such stresses. United States Patent 7101828, September 2006.

103. Palliotti, A., Cartechini, A. and Proietti, P. (1991). Influence of rootstock and height of training system on spring frost sensibility of Chardonnay and Cabernet Sauvignon grape cultivars in the Umbria region. *Annali della Facolta di Agraria.*, **45**: 283-291.

104. Pannelli, G., Rosati, S. and Rugini, E. (2002). The effect of clonal rootstocks on frost tolerance and on some aspects of plant behaviour in Moraiolo and S. Felice olive cultivars. *Acta Hort.*, **586**: 247-250.

105. Paranychianakis, N.V. and Angelakis, A.N. (2008). The effect of water stress and rootstock on the development of leaf injuries in grapevines irrigated with saline effluent. *Agric. Water Management,* **95**: 375-382.

106. Paranychianakis, N.V. and Chartzoulakis, K.S. (2005). Irrigation with saline water in Mediterranean region: from physiology to management practices. *Agric. Ecosyst. Environ.*, **106**: 171-187.

107. Perez-Lopez, D., Gijon, M.C., Marino, J. and Moriana, A. (2010). Water relation response to soil chilling of six olive (*Olea europaea* L.) cultivars with different frost resistance. *Spanish J. Agric. Res.*, **8**(3): 780-789.

108. Perez-Perez, J.G., Robles, J.M., Tovar, J.C. and Botia, P. (2009). Response to drought and salt stress of lemon 'Fino 49' under field conditions: Water relations, osmotic adjustment and gas exchange. *Sci. Hort.*, **122**: 83-90.

109. Perez-Perez, J.G., Romero, P., Navaro, J.M. and Bota, P. (2008). Response of sweet orange cv. 'Lane late' to deficit irrigation in two rootstocks. II. Flowering, fruit growth, yield and fruit quality. *Irrig. Sci.*, **26**(6): 519-529.

110. Perez-Perez, J.G., Syvertsen, J.P., Bota, P. and Garcia-Sanchez, F. (2007). Leaf water relations and net gas exchange responses of salinized Carrizo citrange seedlings during drought stress and recovery. *Ann. Bot.*, **100**: 335-345.

111. Pongraz, D.P. (1983). *Rootstocks for Grapevines*. David Phillip Publisher, Cape Town, USA.

112. Raghupathi, H.B. and Ganeshamurthy, A.M. (2011). Farmers Samples Analysis Database, Dept. of Soil Science and Agricultural Chemistry, IIHR Bangaluru.

113. Ream, C.L. and Furr, J.R. (1976). Salt tolerance of some citrus species, relatives and hybrids tested as rootstocks. *J. Am. Soc. Hort. Sci.*, **101**: 265-267.

114. Rengel, Z. (1992). Role of calcium in aluminium toxicity. *New Phytol.*, **121**: 499-513.

115. Rerkasem, B. and Jamjod, S. (2004). Boron deficiency in wheat: a review. *Field Crop. Res.*, **89**: 173-186.

116. Ribeiro, I.J.A., Soares, N.B., Pettinelli, Jr., A. and Dudienas, E.C. (2002). Behaviour of rootstocks of mango (*Mangifera indica* L.) subjected to low temperature conditions. *Revista Brasileira de Fruticultura*, **24**(1): 249-250.

117. Rodríguez-Gamir, J., Primo-Millo, E., Forner, J.B. and Forner-Giner, M.A. (2010). Citrus rootstock responses to water stress. *Sci. Hort.*, **126**: 95-102.

118. Rombola, A.D., Bruggemann, W., Lopez-Milan, A.F., Tagliavini, M., Abadia, J., Marangoni, B. *et al.* (2002). Biochemical responses to iron deficiency in kiwifruit (*Actinidia deliciosa*). *Tree Physiol.*, **22**: 869-875.

119. Romero-Aranda, R., Moya, J.L., Taeda, F.R., Legaz, F., Primo-Millo, E. and Talon, M. (1998). Physiological and anatomical disturbances induced by chloride salts in sensitive and tolerant citrus; beneficial and detrimental effects of cations. *Plant Cell Environ.*, **21**: 1243-1253.

120. Romheld, V. and Nikolic, M. (2007). Iron. In: Barker, A.V., Pilbeam, D.J. (Eds.). *Handbook of Plant Nutrition*, Boca Raton, FL, USA, CRC Press, Taylor and Francis Group. pp. 329-350.

121. Ruiz-Sanchez, M.C., Domingo, R., Save, R., Biel, C. and Torrecillas, A. (1997). Effects of water stress and rewatering on leaf water relations of lemon plants. *Biol. Plant*, **39**(4): 623-631.

122. Sakamoto, M., Munemura, I., Tomita, R. and Kobayashi, K. (2008). Involvement of hydrogen peroxide in leaf abscission signalling, revealed by analysis with an *in vitro* abscission system in Capsicum plants. *Plant J.*, **56**: 13-27.

123. Sala, J.M. (1998). Involvement of oxidative stress in chilling injury in cold-stored mandarin fruits. *Post Biol. Tech.*, **13**: 255-261.

124. Saleh, B., Allario, T., Dambier, D., Ollitrault, P. and Morillon, R. (2008). Tetraploid citrus rootstocks are more tolerant to salt stress than diploid. *C. R. Biol.*, **331**: 703-710.

125. Sanaa, I.L. (2004). Effect of irrigation with salinized water on growth and chemical constituents of "Kalamata" olive cultivar grafted onto different olive rootstocks. *Arab Univ. J. Agric. Sci.*, **13**(2): 399-417.

126. Satisha, J., Somkumar, R.G., Sharma, J., Upadhyay, A.K. and Adsule, P.G. (2010). Influence of rootstock on growth, yield and fruit composition of Thompson Seedless grapes grown in the Pune region of India. *S. Afr. J. Enol. Vitic.* **31**: 1-8.

127. Schmutz, U. and Ludders, P. (1993). Physiology of saline stress in one mango (*Mangifera indica* L.) rootstock. *Acta Hort.*, **341**: 160-167.

128. Sharma, M.K. and Joolka, N.K. (2002). Effect of rootstocks and soil moisture stress on growth and vigour of almond cv. Nonpareil. *Hort. J.*, **15**(1): 1-7.

129. Sheng, O., Song, S., Peng, S. and Deng, X. (2009). The effects of low boron on growth, gas exchange, boron concentration and distribution of 'Newhall' navel orange (*Citrus sinensis* Osb.) plants grafted on two rootstocks. *Sci. Hort.,* **121**: 278-283.

130. Smirnoff, N. and Cumbes, Q.J. (1989). Hydroxyl radical scavenging activity of compatible solutes. *Phytochem.*, **28**: 1057-1060.

131. Soumelidou, K., Morris, D.A., Battey, N.H., John, P. and Barnett, J.R. (1994). The anatomy of the developing bud union and its relationship to dwarfing in apple. *Ann. Bot.*, **74**: 605-611.

132. Southey, J.M. and Jooste, J.H. (1992). Physiological response of *Vitis vinifera* L. (cv. Chenin blanc) grafted onto different rootstocks on a relatively saline soil. *South African J. Enol. Viticul.*, **13**: 10-22.

133. Soylemezoglu, G., Demir, K., Inal, A. and Gunes, A. (2009). Effect of silicon on antioxidant and stomatal response of two grapevine (*Vitis vinifera* L.) rootstocks grown in boron toxic, saline and boron toxic-saline soil. *Sci. Hort.,* **123**: 240-246.

134. Striegler, R.K. and Howell, G.S. (1991). The influence of rootstock on the cold hardiness of Seyval grapevines I. Primary and secondary effects on growth, canopy development, yield, fruit quality and cold hardiness. *Vitis,* **30**: 1-10.

135. Striegler, R.K., Howell, G.S. and Flore, J.A. (1993). Influence of rootstock on the response of Seyval grapevines to flooding stress. *Amer. J. Enol. Viticult.*, **44**: 313-319.

136. Stuchi, E.S., Donadio, L.C. and Sempionato, O.R. (2000). Tolerancia a seca da laranjeira 'Folha Murcha' em 10 porta-enxertos. *Revista Brasileira de Fruticultura*, **22**: 454-457.

137. Sykes, S.R. (1992). The inheritance of salt exclusion in woody perennial fruit species. *Pl. Soil*, **146**: 123-129.

138. Sykes, S.R. (2011). Chloride and sodium excluding capacities of citrus rootstock germplasm introduced to Australia from the People's Republic of China. *Sci. Hort.*, **128**: 443-449.

139. Syvertsen, J.P. and Smith, M.L. (1983). Environmental stress and seasonal changes in proline concentration of citrus tree tissues and juice. *J. Amer. Soc. Hort. Sci.*, **108**: 861-866.

140. Syvertsen, J.P. and Yelenosky, G. (1988). Salinity can freeze tolerance of citrus rootstock seelings by modifying growth, water relations, and mineral-nutrition. *J. Amer. Soc. Hort. Sci.*, **113**: 889-893.

141. Tanou, G., Job, C., Rajjou, L., Arc, E., Belghazi, M, Diamantidis, Gr., Molassiotis, A. and Job, D. (2009). Proteomics reveals the overlapping roles of hydrogen peroxide and nitric oxide in the acclimation of citrus plants to salinity stress. *The Plant J.*, **60**: 795-804.

142. Varalakshmi, L.R. and Ganeshmurthy, A.M. (2013). Managing horticultural crops in saline and sodic soils. In: K.L. Chadha, A.K. Singh, S.K. Singh and W.S. Dhillon, (Eds.), *Horticulture for Food and Environmental Security*. Westville Publishing House, New Delhi. pp. 317-340.

143. Varu, D.K. and Barad, A.V. (2010). Standardization of mango rootstock for mitigating salt stress. *Indian J. Hort.* **67**(Special Issue): 79-83.

144. Venekemp, J.H. (1989). Regulation of cytosol acidity in plants under conditions of drought. *Physiol. Plant*, **76**: 112-117.

145. Walker, R.B., Blackmore, D.H., Clingeleffer, R.P and Ray, C.L. (2002). Rootstock effects on salt tolerance of irrigated field-grown grapevines (*Vitis vinifera* L. cv. Sultana). I. Yield and vigor inter-relationships. *Aust. J. Grape Wine Res.*, **8**: 3-14.

146. Walker, R.R.(1986). Sodium exclusion and potassium-sodium selectivity in salt-treated trifoliate orange (*Poncirus trifoliata*) and Cleopatra mandarin (*Citrus reticulata*) plants. *Aust. J. Plant Phys.*, **13**: 293-303.

147. Walker, R.R., Blackmore, D.H., Clingeleffer, P.R. and Iakono, F. (1997). Effect of salinity and Ramsey rootstock on ion concentrations and carbon dioxide assimilation in leaves of drip-irrigated, field-grown grapevines (*Vitis vinifera* L. cv. Sultana). *Aust. J. Grape Wine Res.*, **3**: 66-74.

148. Walker, R.R., Read, P.E. and Blackmore, D.H. (2000). Rootstock and salinity effects on rates of berry maturation, ion accumulation and colour development in Shiraz grapes. *Aust. J. Grape Wine Res.*, **6**: 227-239.

149. Wang, S., Liang, D., Li, C., Hao, Y., Ma, F. and Shu, H. (2012). Influence of drought stress on the cellular ultrastructure and antioxidant system in leaves of drought-tolerant and drought-sensitive apple rootstocks. *Pl. Physiol. Biochem.*, **51**: 81-89.

150. Watanabe, A., Kojima, K., Ide, Y. and Sasaki, S. (2000). Effects of saline and osmotic stress on proline and sugar accumulation in *Populus euphratica in vitro*. *Plant Cell Tissue Organ Culture*, **63**: 199-206.

151. Wojcik, P., Wojcik, M. and Treder, W. (2003). Boron absorption and translocation in apple rootstocks under conditions of low medium boron. *J. Plant Nutr.*, **26**: 961-968.

152. Xiong, L. and Zhu, J.K. (2002). Molecular and genetic aspects of plant response to osmotic stress. *Plant Cell Environ.*, **25**: 131-139.

153. Yadav, V.K. and Singh, H.K. (2006). Effect of exchangeable sodium on mineral composition of mango (*Mangifera indica* L.) rootstock. *Scientific Hort.*, **10**: 103-124.

154. Yin, R., Bai, T., Ma, F., Wang, X., Li, Y. and Yue, Z. (2010). Physiological responses and relative tolerance by Chinese apple rootstocks to NaCl stress. *Sci. Hort.*, **126**: 247-252.

155. Zekri, M. (1991). Effects of NaCl on growth and physiology of sour orange and Cleopatra mandarin seedlings. *Sci. Hort.*, **47**: 305-315.

156. Zhang, C.S., Lu, Q. and Verma, D.P.S. (1995). Removal of feedback inhibition of delta-1-pyroline-5-carboxylate synthetase, a bifunctional enzyme catalyzing the first 2 steps of proline biosynthesis in plants. *J. Biol. Chem.*, **270**: 20491-20496.

157. Zhang, X., Hu, C. and Yao, J. (2010). Tetraploidization of diploid *Dioscora* results in activation of the antioxidant defense system and increased heat tolerance. *J. Plant Physiol.*, **167**: 88-94.

158. Zhu, Z., Wei, G., Li, J., Qian, Q. and Yu, J. (2004). Silicon alleviates salt stress and increase antioxidant enzymes activity in leaves of salt-stressed cucumber (*Cucumis sativus* L.) *Plant Sci.*, **167**: 527-533.

159. Zimmermann, J. and Zimmermann, H. (1973). The effect of newly bred rootstocks with the *Vitis cinerea* type Arnold as their pollen parent on the growth and yield of *Vitis vinifera* cultivar Gutedel clone Fr 36-5. *Mitteilungen Rebe und Wein.*, **23**: 1-20.

160. Zocchi, G. (2006). Metabolic changes in iron-stressed dicotyledonous plants. In: L.L. Barton and J. Abadia, (Eds.), *Iron Nutrition in Plants and Rizhospheric Microorganisms*. Dordrecht, Netherlands: Springer; 2006. pp. 359-370.

161. Zude, M., Ebert, G. and Ludders, P. (1998). Influence of flooding on growth and gas exchange of mango rootstocks (*Mangifera indica* L.) and proposed selection criteria for flood tolerance. *Angewandte Botanik*, **72**(3/4): 148-151.

2015, Horticulture for Nutrition Security
Editor: **Prof. K.V. Peter**
Published by: **DAYA PUBLISHING HOUSE, NEW DELHI**

Pages 217–260

Chapter 12

Abiotic Stress Tolerance in Horticultural Crops

Anant Bahadur and Amit Kumar Singh

Horticultural crops are widely cultivated worldwide for their high value products (essential food, minerals and vitamins) that are essential for human nutrition. **Stress** is usually defined as an external factor that exerts a disadvantageous influence on the plant. Under both natural and field conditions, plants are frequently exposed to environmental stresses usually called as 'abiotic stresses'. Abiotic stresses such as drought, heat, soil salinity, tropospheric ozone and excess UV radiation are already causing significant agricultural yield losses, and will become more prevalent in the coming decades due to the effects of global climate change. The changing environments pose serious challenges to global agriculture and place unprecedented pressures on the sustainability of horticulture industry. Adapting strategies that can cope with heat, cold, drought, salinity and other climate extremes is essential to meet the need of growing population and increasing demand for fruits, vegetables and other horticultural products.

Environmental factors, like air temperature can become stressful in just a few minutes; others like soil water content may take days to weeks, and factors such as soil mineral deficiencies can take months to become stressful. Abiotic factors are the major limitation to crop production worldwide. The abiotic stresses like temperature (heat, cold chilling/frost), water (drought, flooding/hypoxia), radiation (UV, ionizing), chemicals (mineral deficiency/excess, pollutants/heavy metals/pesticides, gaseous toxins), mechanical (wind, soil movement, submergence), etc are responsible for over 50 per cent reduction in agricultural production (Wang *et al.,* 2007). These comprise mostly of high temperature (40 per cent), salinity (20 per cent), drought (17 per cent), low temperature (15 per cent) and other forms of stresses (Ashraf *et al.,* 2008). Only 9 per cent of the world area is conducive for crop production, while 91 per cent is afflicted by various stressors. As per the ICAR estimates (2010), 120.8 million ha constituting 36.5 per cent of geographical area in India are degraded due to soil erosion, salinity/alkalinity, soil acidity, waterlogging, and other edaphic problems. The various biotic, abiotic and edaphic factors that affect agriculture production have been shown in Figure 12.1.

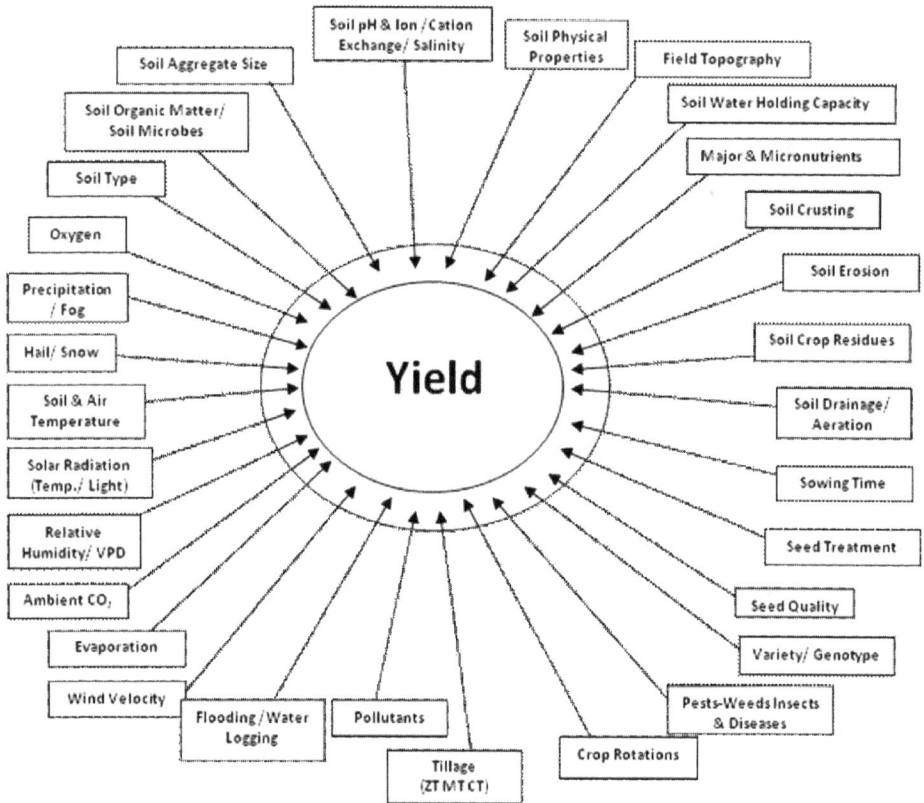

Figure 12.1: Various Factors Affecting Crop Production.

Variables of Abiotic Stresses

1. High temperature (heat stress)
2. Low temperature (chilling and frost)
3. Drought (water deficit)
4. Water logging/flooding
5. Salinity (salt stress)
6. Light (high/low light intensity)
7. UV-radiations

8. Nutrient deficiency in soil

9. Chemicals and pollutants (heavy metals, pesticides and aerosols)

10. Oxidative stress (ROS, ozone)

11. Wind (high wind velocity, sand and dust particles)

12. Extreme weather (avalanche, hail storm, thunder storm, dust storm, cyclones, tidal waves, etc.)

In this chapter, environmental or abiotic factors which produce stress in plants will be discussed, although biotic factors such as pathogens (viruses, bacteria and fungi), insects, herbivores and rodents also cause stress in the plant.

Temperature Stress

Temperature is an important factor influencing seed germination, growth, flowering, fruit set and fruit ripening in horticultural crops. Both high and low temperatures (temporary or constant), can induce morpho-anatomical, physiological and biochemical changes in plants, leading to reduced horticulture production. Furthermore, the 'global warming' resulted in rise of earth annual temperature by 0.7°C between 1906 and 2005. IPCC (2007) projected that the global annual temperature is likely to increase in the range of 1.4 to 4.5°C by the end of this century. Projections by IPCC and other agencies, global warming will lead to 10-40 per cent loss in crop production in India by 2080-2100, if suitable combat strategies are not being adapted.

High Temperature or Heat Stress

High temperature (HT) or heat stress is a major concern in many regions of the tropics and subtropics, since high temperature can affect morpho-anatomical, physiological and biochemical changes, which ultimately affect the plant growth and development and significant reduction in economic yields. Plants generally grow the best at the higher end of their optimal temperature range. In the temperate zone, the minimum temperature for growth is about 4.4°C. Photosynthesis and respiration increase as temperatures rise until the energy used in respiration equals photosynthetic capacity, when growth ceases. For most plants, this temperature is around 35°C. For many cool-season crops, growth may cease at temperatures considerably lower than 35°C. Damage such as sunburns on leaves, branches and stems, anticipated leaf senescence and abscission, shoot and root growth inhibition and fruit discoloration and damage are also consequences of HT stress. Reproductive stage is highly affected by heat stress in most plants. At very high temperatures, severe cellular injury or cell death may occur within minutes, which could be attributed to a catastrophic collapse of cellular organization (Schoffl *et al.*, 1999). At moderately high temperatures, injuries or death may occur only after long-term exposure. Direct injuries due to high temperatures include protein denaturation and aggregation, and increased fluidity of membrane lipids. Indirect or slower heat injuries include inactivation of enzymes in chloroplast and mitochondria, inhibition of protein synthesis, protein degradation and loss of membrane integrity (Howarth, 2005). Heat stress also affects the organization of microtubules by splitting and/or elongation of spindles, formation of microtubule asters in mitotic cells, and elongation of phragmoplast microtubules. These injuries eventually lead to starvation, inhibition of growth, reduced ion flux, production of toxic compounds and reactive oxygen species (ROS) (Schoffl *et al.*, 1999; Howarth, 2005).

As in anatomical changes are concerned, the symptoms observed under heat stress conditions are generally similar to those of water stress. These are reduced cell size, closure of stomata, curtailed water loss, increased stomatal and trichome densities and greater xylem vessels in both root and shoot. HT can cause damage to components of leaf photosynthesis, thereby reducing CO_2 assimilation rates. Sensitivity of photosynthesis to heat is mainly due to damage of components of photosystem II located in the thylakoid membranes of the chloroplast. HT induces the acclimation of photosynthesis by changing the photosynthetic capacity, the temperature response of photosynthesis or both. Changes in several photosynthetic characteristics under high temperatures are good indicators of plant tolerance to heat stress (Wahid *et al.*, 2007).

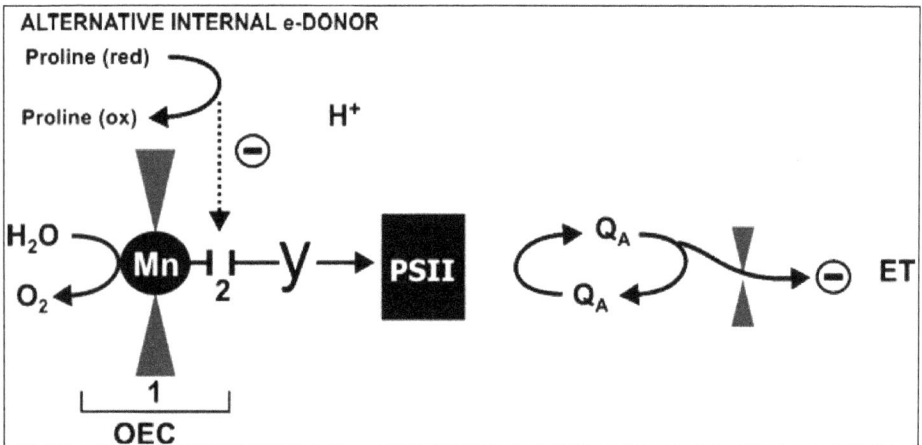

Figure 12.2: Heat Induced Inhibition of Oxygen Evolution and PSII Activity. Heat stress leads to either (1) dissociation or (2) inhibition of the oxygen evolving complexes (OEC). This enables an alternative internal e-donor such as proline instead of H_2O to donate electrons to PSII (adopted from De Ronde *et al.*, 2004).

High temperature affects vegetable crops in several ways. The optimum temperatures for growth and development of tomato are 22–26°C, and exposure to higher temperatures can markedly alter several metabolic processes. The photosynthetic rate in tomato leaf is reduced with short (20 min) increase in leaf temperature to 35°C. In addition, when tomato plants were exposed to a heat-shock temperature of 45°C for 3 h, heat injury in tomato leaf may involve chlorophyll photooxidation mediated by activated oxygen species and more severe alterations in the photosynthetic apparatus (Camejo *et al.*, 2007, 2010). It causes reduction in pollen formation or viability in tomato at temperature above 37°C. Fruit set in tomato occurs only at night temperatures ranging between 12.8-24°C. The typical red colour of most tomato cultivars does not develop when temperatures go above 30°C, but yellow pigment continues to develop. Fruit cracking in tomato occurs, if high temperature (above 32°C) is accompanied with high humidity. Also extreme high or low temperatures which interfere with pollination; low light, excessive nitrogen and heavy rainfall contribute to puffiness in tomato. Sun burning in tomato occurs if fruits exposed above 45°C for 4 hours. In sweet pepper, the optimum temperature for growth ranges between 20 and 25°C. When temperature falls below 15°C or exceeds 32°C, growth is usually retarded and yield decreases. High temperature (29/23°C D/N) reduces percent fruit set and size significantly in sweet pepper as compared to optimum (24/18°C D/N) temperature condition (Saha *et al.,* 2010). In pepper, there is blossom drop if day temperature is 33°C or above or night temperature remains above 26.5°C. Hot chilli does not set fruit well when night temperatures are greater than 24°C. Moderate HT stress (32/26°C D/N) in bell pepper, one week before anthesis remarkably reduced the pollen germination and seeds in fruits. HT increases capsaicin biosynthesis in capsicum but floral abortion occurs when temperatures exceed 30°C. During ripening, red colour development in capsicum is inhibited above 27°C. In common bean, HT stresses (35/20°C D/N) during anthesis reduce pollen germination, pollen tube growth, fertilization, pod and seed set. In broccoli and cauliflower, temperature above 29°C can trigger to flower too soon and causes heat injury, yellowing/browning and loosening of head. If temperature is more than 25°C, there is bolting in European cauliflowers, Asiatic carrots and broccoli. Soil temperature above 21°C hampers the colour development in carrots. Also prolonged ambient HT (more than 30°C) brings about pungent flavour, tough and coarse roots with thick shoulders. At HT, poor pollination occurs in many crops grown for seed (lettuce, sweet corn, cucurbits and carrot). HT induces floral differentiation and flower stalk development in lettuce plants, both of which decrease yield and quality of the crop. In onion and garlic, very high temperature (>42°C) during bulb maturity (April-May) causes reduction in bulb size and poor post harvest life in storage. In potato, HT brings about marked morphological changes like etiolated growth with smaller size leaflets in addition to reduction in tuber number and size. Tuber productions in potato totally stop at temperature of about 29°C. Sudden rise in temperature during January and February advanced the maturity of seed spices which led to improper seed setting and seed filling.

At high temperatures, fruit set in tomato plants failed due to the disruption of proline and sugar metabolism transport during the narrow window of male reproductive development (Sato *et al.,* 2006). The most noticeable effect of high temperatures on reproductive processes in tomato is the production of an exerted style (*i.e.,* stigma is elongated beyond the anther cone), which may prevent self-pollination. Poor fruit set at high temperature has also been associated with low levels of carbohydrates and growth regulators released in plant sink tissues. Reproductive phases most sensitive to high temperature are; gametogenesis (8-9 days before anthesis) and fertilization (1-3 days after anthesis) in various vegetables (Foolad, 2005). Both male and female gametophytes are sensitive to high temperature and response varies with genotype; however, ovules are generally less heat sensitive than pollen. Erickson and Markhart (2002) demonstrated that in pepper, high temperature inhibits the development of pollen grains during final tetrad formation to tetrad dissolution, when flower buds are 16-18 days before anthesis, resulting in pollen sterility. The reduction of pollen viability reduces fruit size and fruit set. In contrast, exposure to high temperatures just before anthesis does not cause injury to either the female or male organs. Instead, inhibition of fertilization or early fruit development occurs after pollination and is responsible for the reduction of fruit set during the period between anthesis and fruit development.

In fruit crops like apple, colour development occurs through the production of anthocyanin. Anthocyanin production is reduced by high temperatures. In banana, the optimum temperature for plant growth is 30°C and leaf emergence stops below 10°C. The leaf production increases by one leaf per month for every 3.3-3.7°C rise in minimum or mean temperature from 10-20°C or 13.5-25°C, respectively. The optimum temperature for bunch growth is 21-22°C and temperature above 30°C causes choking or rosetting in plant. Mango grows the best in temperature range of 24-30°C; however, it can tolerate temperature as high as 48°C during fruit development and maturity. Cashew requires relatively dry atmosphere and mild winter (15-20°C minimum temperature) coupled with moderate dew nights for profuse flowering. High temperature (34.4°C or more) and low RH (< 20 per cent) during afternoon causes drying of flowers resulting in reduced yield. In strawberry, high temperature stress negatively affect the number of inflorescences, flowers and fruits. In cherimoya, warm temperatures determined the production of low-viability pollen; and therefore of asymmetrical and small fruits containing a few seeds.

Low Temperature or Cold Stress

Low temperature (chilling/freezing) injury can occur in all plants, but the mechanisms and types of damage vary considerably. Many fruit, vegetable and ornamental crops of tropical origin experience physiological damage when subjected to temperatures below 12°C (well above freezing temperatures). Chilling injury is damage to plant parts caused by temperatures above the freezing point (0°C). Tomato plant when exposed to temperatures below 10°C suffers from chilling injury, and extended exposure below 6°C can destroy the plants (Park *et al.,* 2004). Frost and freeze injury are closely related. Frost damage occurs during a *radiation freeze*, while freeze damage occurs during an *advection freeze*. In both cases, ice crystals form in plant tissues, dehydrating cells and disrupting membranes. It is believed that intracellular ice formation causes a 'mechanical disruption of the protoplasmic structure' (Levitt, 1980). Crop plants that develop in tropical climates, often experience serious frost damage when exposed to temperature slightly below zero, whereas most crops that develop in colder climates often survive with little damage if the freeze event is not much severe. Some exceptions are lettuce, which originated in a temperate climate, but can be damaged at temperatures near 0°C and some subtropical fruit trees that can withstand temperatures to -5 to -8°C. Frost damage occurs when ice forms inside the plant tissue and injures the plant cells. It may have a drastic effect upon the entire plant or affect only a small part of the plant tissue, which reduces yield or product quality.

Plants fall into four freeze-sensitivity categories: (i) tender; (ii) slightly hardy; (iii) moderately hardy, and (iv) very hardy (Levitt, 1980). Tender plants are those that have not developed avoidance of intracellular freezing (*e.g.* mostly tropical plants). Slightly hardy plants include most of the subtropical fruit trees, deciduous trees during certain periods and horticultural crops that are sensitive to freezing down to about -5°C. Moderately hardy plants include those that can accumulate sufficient solutes to resist freeze injury to temperatures as low as -10°C mainly by avoiding dehydration damage, but they are less able to tolerate lower temperatures. Very hardy plants are able to avoid intracellular freezing as well as damage due to cell desiccation.

Fruit plants can be affected either by winter injury, which occurs when the trees are dormant; and spring frost injury, which occurs when the trees are no longer dormant but in various stages of flower, fruit, and/or leaf development. Both types of injury occur when temperatures drop below certain threshold levels. The injury threshold temperature is lower for dormant than non-dormant tissues, and varies for different species, varieties, and stages of development. Trees that go into dormancy in a weakened condition, due to over-cropping, drought, poor nutrition, pests, etc are more susceptible to winter injury. Table 12.1 lists temperatures for each stage of development at which 10 per cent and 90 per cent bud killing occurs after 30 minutes exposure. The percentage bud kill which causes crop reduction will vary with each crop. For example, to have a full crop of cherries requires well over 50 per cent bud survival, while apples, pears and peaches may need only 10-15 per cent bud survival.

Table 12.1: Critical Temperature (*Tc*, °C) Values for Deciduous Temperate Fruit Trees

Fruit Crops	Stage	10 per cent Kill	90 per cent Kill
Apple	Silver tip	−11.9	−17.6
	Green tip	−7.5	−15.7
	½" green	−5.6	−11.7
	Tight cluster	−3.9	−7.9
	Full pink	−2.7	−4.6
	Full bloom	−2.9	−4.7
	Post bloom	−1.9	−3.0
Apricot	Tip separates	−4.3	−14.1
	Red calyx	−6.2	−13.8
	First white	−4.9	−10.3
	Full bloom	−2.9	−6.4
	In shuck	−2.6	−4.7
	Green fruit	−2.3	−3.3
Peaches (Elberta)	First swell	−7.4	−17.9
	Caylx green	−6.1	−15.7
	Caylx red	−4.8	−14.2
	First pink	−4.1	−9.2
	First bloom	−3.3	−5.9
	Post bloom	−2.5	−3.9

Contd...

Table 12.1–*Contd...*

Fruit Crops	Stage	10 per cent Kill	90 per cent Kill
Pears (Bartlett)	Scales separate	−8.6	−17.7
	Blossom buds exposed	−7.3	−15.4
	Tight cluster	−5.1	−12.6
	Full white	−3.1	−6.4
	First bloom	−3.2	−6.9
	Post bloom	−2.7	−4.0
Cherries (Bing)	First swell	−11.1	−17.2
	Green tip	−3.7	−10.3
	Tight cluster	−3.1	−7.9
	Open cluster	−2.7	−6.2
	First white	−2.7	−4.9
	First bloom	−2.8	−4.1
	Full bloom	−2.4	−3.9
	Post bloom	−2.2	−3.6

The 10 per cent kill and 90 per cent kill imply that 30 minutes at the indicated temperature is expected to cause 10 per cent and 90 per cent kill of the plant part affected during the indicated phenological stage.

Source: Proebsting and Mills (1978).

Low temperatures can affect plants in several ways. Temperatures near or below threshold minimum reduce the metabolic activities and plant growth. If the temperature remain low for an extended period, plant quality also suffers and death may occur. Certain vegetables, such as cole crops and root vegetables produce 'bolt' or 'seed stalks' in response to exposure at low temperatures for extended periods. Vegetable crops differ in their hardiness to cold temperatures, depending upon their genetics and origin. Warm season crops, such as tomato, French bean and cucurbits originated in tropical areas can be severely injured by even a light frost. On the other hand, cool season crops, such as broccoli, cabbage, pea and onion, originated in northern areas can tolerate frost and light freezes of short durations with little damage. Potato is highly sensitive to frost and a complete loss of foliage is reported below 2°C temperature for 2-3 consecutive nights. In Rajasthan, there was heavy loss in production of coriander, cumin, fennel and kalajeera due to frost in 2007-08. About 50 per cent loss is reported of temperature below 1°C for exposure to one night. Low temperature (<10°C) at bulb development leads to bolting of onion and clove and sprouting of garlic. Low temperature in banana causes chilling symptoms on leaves. The newly emerged parts of lamina turn brown with water-soaked area underneath. In mango, cool temperature and frosting (0 to -0.5°C) during inflorescence development reduce the number of perfect flowers. There is 50-60 per cent damage of young mango plant, and 20-50 per cent of old trees at temperature -0.5°C.

Table 11.2: Susceptibility of Fresh Fruits, Vegetables and Flowering Annuals to Freezing Injury

Most Susceptible	Moderately Susceptible	Least Susceptible/Hardy
Apricot, Asparagus, Avocado, Balsam, Banana, Begonia, Cockscomb (*Celosia*), Cosmos, Cucumber, Eggplant, French bean, Lemons, Lettuce, Limes, Marigold, Melons, Okra, Peach, Phlox, Plums, Portulaca, Potato, Pumpkin, Salvia, Squashes, Sweet pepper, Sweet potato, Tomato	Apple, Aster, Broccoli, Carrot, Cauliflower, Grape, Grapefruit, Onion, Celery, Orange, Parsley, Pear, Petunia, Pea, Radish, Spinach, Sweet alyssum, Verbena	Beet, Brussels sprout, Bells of Ireland (*Molucella*), Cabbage, Calendula, Coreopsis, Cornflower, Dates, Dianthus, Endive, Kale, Kohlrabi, Pansy, Parsnip, Primrose, Rutabaga, Salsify, Snapdragon, Stock (*Matthiola incana*), Sweet pea, Turnip, Violet

Source: Wang and Wallace (2003), Caplan (1988).

Common Injury Symptoms in Edible Part of Vegetables Exposed to Freezing

☆ **Artichoke**: Epidermis becomes detached and forms whitish to light tan blisters. When blisters are broken, underlying tissue turns brown.

☆ **Asparagus**: Tip becomes limp and dark; the rest of the spear is water-soaked. Thawed spears become mushy.

☆ **Beet**: External and internal water-soaking; sometimes blackening of conducting tissue.

☆ **Broccoli**: The youngest florets in the center of the curd are the most sensitive to freezing injury. They turn brown and give off strong odors upon thawing.

☆ **Cabbage**: Leaves become water-soaked, translucent and limp upon thawing; epidermis separates.

☆ **Carrot**: Blistered appearance, jagged length-wise cracks. Interior becomes water-soaked and darkened upon thawing.

☆ **Cauliflower**: Curds turn brown and have a strong off-odor when cooked.

☆ **Celery**: Leaves and petioles appear wilted and water-soaked upon thawing. Petioles freeze more readily than leaves.

☆ **Garlic**: Thawed cloves appear greyish-yellow and water-soaked.

☆ **Lettuce**: Blistering; dead cells of the separated epidermis on outer leaves become tan; increased susceptibility to physical damage and decay.

☆ **Onion**: Thawed bulbs are soft, greyish-yellow and water-soaked in cross-section; often limited to individual scales.

☆ **Potato**: Freezing injury may not be externally evident, but shows as gray or bluish-gray patches beneath the sink. Thawed tubers become soft and watery.

☆ **Radish**: Thawed tissues appear translucent; roots soften and shrivel.

☆ **Sweet pepper**: Dead, water-soaked tissue in part or all of pericarp surface; pitting, shrivelling and decay follow thawing.

☆ **Sweet potato**: A yellowish-brown discoloration of the vascular ring and a yellowish-green water-soaked appearance of other tissues. Roots soften and become very susceptible to decay.

☆ **Tomato**: Water soaked and soft upon thawing. In partially frozen fruits, the margin between healthy and dead tissue is distinct, especially in green fruits.

☆ **Turnip**: Small water-soaked spots or pitting on the surface. Injured tissues appear tan or gray and give off an objectionable odour.

Water Stress

1. Drought Stress/Water Deficits

Water is undoubtedly the single most important constituent of the plant, comprising more than 90 per cent of the fresh weight of most herbaceous plants. Water availability is one of the major limitations to plant productivity, and is one of the major factors regulating the distribution of plant species. Over 35 per cent of the world's land surface are considered to be arid or semiarid, experiencing precipitation that is inadequate for most agricultural uses. Regions that experience adequate precipitation may also suffer to water limiting environments. Unfavourable climatic factors such as erratic rainfall, high evaporative demand, and several droughts, among others, contribute to the increasing water scarcity enhancing stress resilience are the most demanding areas in agricultural research. India has faced 26 drought years in the last 130 years; with 1987 and 2002 and 2012 being the major drought years in recent times. About 58 per cent (80 M ha) of the net sown area in India continue to be rainfed that contributes 40 per cent of the total food grain production.

Water has an essential role in plant metabolism, at both the cellular and whole-plant levels, and any decrease in water availability has an immediate effect on plant growth and processes ranging from photosynthesis to solute transport. Plants are generally subjected to shortages in water availability varying in length from hours to days. Water deficit has been defined as the induction of turgor pressure below the maximal potential pressure. The magnitude of such stress is determined by the extent, stage and duration of the deprivation. Therefore, plant responses depend on the nature of the water shortage and may be classified as, (a) physiological responses to short-term changes, (b) acclimation to a certain level of water availability, and (c) adaptations to drought. Short-term responses to water stress, acting within seconds after the onset of stress, and are primarily linked to stomatal regulation, thereby reducing water loss by transpiration and maximizing CO_2 intake. An optimum efficiency in this process would lead to a constant ratio of transpiration to photosynthesis. Mid-term responses (acclimation) include the adjustment of the osmotic potential by solute accumulation, changes in cell wall elasticity and morphological changes. Long-term adaptation to drought includes genetically fixed patterns of biomass allocation, specific anatomical modifications, and sophisticated physiological mechanisms with an overall growth reduction to balance resource acquisition (Chapin, 1991). In vegetables, water deficit at critical growth stages reduces the production severely, besides hampering product quality (Table 12.3).

Table 12.3: Critical Stages of Water Deficit and its Impact on Vegetable Crops*

Vegetable Crops	Threshold Soil Moisture		Critical Stage of Water	Impact of Water Deficit Requirement
	SWT¹ (Bars)	ASM²		
Tomato	−0.45	50 per cent	Flowering and period of rapid fruit enlargement	Flower shedding, lack of fertilization, reduced fruit size, fruit splitting and development of calcium deficient disorder blossom end rot (BER).
Eggplant	−0.45	50 per cent	Flowering and fruit development	Reduces yield with poor colour development in fruits.
Chilli and Capsicum	−0.45	50 per cent	Flowering and fruit set	Shedding of flowers and young fruits, reduction in dry matter production and nutrient uptake.
Cabbage and cauliflower	−0.34	60 per cent	Head/curd formation and enlargement	Tip burning and splitting of head in cabbage; browning and buttoning in cauliflower.
Carrot, radish and turnip	−0.45	50 per cent	Root enlargement	Distorted, rough and poor growth of roots, strong and pungent odour in carrot, accumulation of harmful nitrates in roots.
Cucumber	−0.45	50 per cent	Flowering as well as throughout fruit development	Deformed and non-viable pollen grains, bitterness and deformity in fruits.
Onion	−0.25	70 per cent	Bulb formation and enlargement	Splitting and doubling of bulb, poor storage life.
Okra	−0.70	40 per cent	Flowering and pod development	Considerable yield loss, development of fibres in pods, high mite infestation.
Melons	−2.00	40 per cent	Flowering and evenly throughout fruit development	Poor fruit quality in muskmelon due to decrease in TSS, reducing sugar and ascorbic acid; increase nitrate content in watermelon fruit.
Lettuce	−0.34	60 per cent	Consistently throughout development	Toughness of leaves, poor plant growth, tip burning.
Pea	−070	40 per cent	Flowering and pod filling	Reductions in root nodulation and plant growth; poor grain fill.
Leafy vegetables	−0.25	70 per cent	Throughout growth and development of plant	Toughness of leaves, poor foliage growth, accumulation of nitrates.

1 SWT: Soil water tension, 2 ASM: Available soil moisture at rooting depth (25-40 cm).

* Bahadur (2014).

Drought Stress and Gas Exchanges

On the basis of stomatal conductance as an integrative parameter for the degree of drought, three phases of photosynthetic response can be differentiated along a water deficit gradient in grape (Cifre *et al.*, 2005).

Table 12.4: Physiological and Biochemical Changes in Plants Induced by Drought Stress

Physiological Responses	Biochemical Responses	Molecular Responses
Recognition of root signals	Transient decrease in photochemical efficiency	Stress responsive gene expression
Loss of turgor and osmotic adjustment	Decrease efficiency of Rubisco	Increased expression in ABA biosynthetic genes
Reduced leaf water potential ($q^\bullet$)	Accumulation of stress metabolites like MDHA, Glutathione, Glycinebetaine, proline, polyamines, and α-tocopherol	Synthesis of specific proteins like LEA, DSP, RAB, dehydrins
Decrease in stomatal conductance to CO_2	Increase in antioxidative enzyme like, SOD, CAT, APX, POD, GR and MDHAR	Drought stress tolerance
Decline in net photosynthesis	Reduced ROS accumulation	
Reduced growth rates		

SOD: Superoxide dismutase; CAT: Catalase; APX: Ascorbate peroxidise; POD: Peroxidase; GR: Glutathione reductase; MDHAR: Monodehydroascorbate reductase; LEA: Late embryogenesis abundant.

Source. Shao *et al.* (2008).

1. A phase of mild water stress is defined for a decreasing range of stomatal conductance from 0.5-0.7 to 0.15 mol H_2O $m^{-2}s^{-1}$. This is characterized by a relatively small decline of net CO_2 assimilation, which results in a decline of sub-stomatal CO_2 availability in the mesophyll, the rate of photorespiration increases, which enables the maintenance of the thylakoid electron transport rate (ETR). At this stage, stomatal closure is probably the only limitation to photosynthesis.

2. A moderate drought stress is characterised by intermediate stomatal conductance values (0.05-0.15 mol H_2O $m^{-2}s^{-1}$). During this phase, a further reduction in net CO_2 assimilation occurs and water use efficiency (WUE) usually increases. Ci (leaf CO_2) still decreases, but ETR and the carboxylation efficiency characteristically decline during this phase. Non-photochemical quenching (NPQ), a chlorophyll fluorescence parameter indicative of thermal dissipation in the antenna of PSII, increases under these conditions, and steady state chlorophyll fluorescence drops under high light.

3. A phase of severe drought stress takes place when stomatal conductance is very low (<0.05 mol H_2O $m^{-2}s^{-1}$). During this phase, steeper reduction of net CO_2 assimilation, WUE, ETR and carboxylation efficiency occur. NPQ further increases and the excitation capture of PSII (Fv/Fm) are eventually reduced, especially during very hot days. Further, WUE decreases and Ci steeply increases, indicating that non-stomatal limitation to photosynthesis become dominant resulting in non-recovery of photosynthesis even after irrigation.

Drought Stress and Production of ROS

One of the major factors responsible for impaired plant growth and productivity under drought stress is the production of reactive oxygen species (ROS) in organelles including chloroplasts, mitochondria and peroxisomes. The ROS target the peroxidation of cellular membrane lipids and degradation of enzyme proteins and nucleic acids. Drought stress disturbs the balance between the production of ROS and antioxidant defence mechanisms, causing more accumulation of ROS, which induces oxidative stress. Upon reduction in the amount of available water, plants close their stomata (plausibly *via* ABA signalling), which decreases the CO_2 influx. Reduction in CO_2 not only reduces the carboxylation directly but also directs more electrons to form ROS. Severe drought conditions limit photosynthesis due to a decrease in the activities of Rubisco, phosphoenolpyruvate carboxylase (PEPCase), NADP-malic enzyme (NADP-ME), fructose-1,6-bisphosphatase (FBPase) and pyruvate orthophosphate dikinase (PPDK). Reduced tissue water contents also increase the activity of Rubisco binding inhibitors. Moreover, non-cyclic electron transport is down regulated to match the reduced requirements of NADPH production and thus reduces the ATP synthesis (Figure 12.3).

Strategies for Growing Horticultural Crops in Drought Prone Areas

Plants adapted to arid environments possess inherent drought escape or drought avoidance mechanisms and can be grown in drought hit areas. Drought escape is a phenological phenomenon of plants achieved by early maturity and completion of

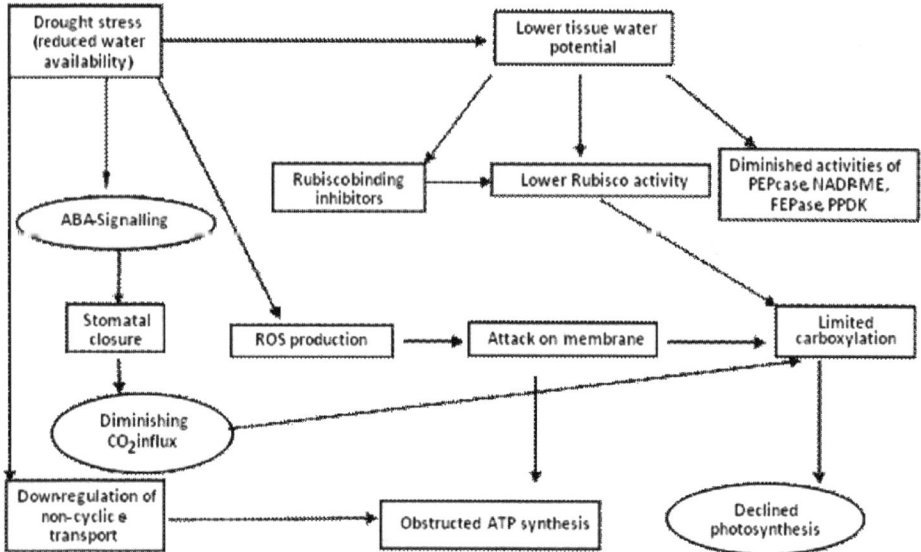

Figure 12.3: Possible Mechanisms of Reduced Photosynthesis under Drought Stress.

life cycle, while drought avoidance mechanisms enable the plants to maintain high water potential so as to avoid the damage due to water deficit. Plants using drought avoidance mechanism have deeper and dense root system, greater root penetration ability, higher stomatal conductance and cuticular resistance to prevent water loss, higher pre-dawn leaf water potential, and avoid leaf rolling for longer intervals. In naturally dry habitats, some plant species rapidly mature and produce seeds before the onset of dry season or start reproducing soon after rainfall. California poppy (*Escholtzia californica* Cham.) completes its life cycle in a few weeks before drought stress starts, while Coffee (*Coffea arabica* L.) and cacao (*Theobroma cacao* L.) flower and fruit when rains follow a drought period. However, some horticultural plant species such as agave (*Agave deserti*) and cactus species store water in their buds, stems or leaves during water deficit period, utilize this stored water under conditions of severe drought. Other plant species avoid water stress by developing deep root system and/ or mechanism involved in low transpirational water loss. Among crops, arid legumes such as cluster bean (*Cyamopsis tetragonoloba*), dew bean (*Vigna aconitifolia*), cowpea (*Vigna unguiculata*) and chickpea (*Cicer arietinum*) are characterized by their deep taproot system with slow growth. Similarly, drought tolerance in *Brassica carinata*, *B. napus* and *B. campestris* is related to their better-developed root system. Thus, the drought tolerance/avoidance in crops usually depends on one or more of the following components include, (i) the capacity of plant roots to extract water from soil (ii) osmotic adjustment capacity, and (iii) efficient water use efficiency. Therefore, crop plants or wild plants having these traits are capable to tolerate water stress and thus they can be grown on drought hit areas.

Table 12.5: Fruit and Vegetables Cultivars having Tolerance against Drought Stress

Crop	Cultivars
Fruits	
Ber (*Ziziphus jujuba*)	Gola, Mundia, Kaithali, Banarasi Karaka, Umran
Aonla (*Phyllanthus emblica*)	Kanchan, Krishna, Balwant, NA -6, NA - 7
Pomegranate (*Punica granatum*)	P-23, P-26, IIHR Selection, Mridula, GKVK 1, Ganesh, Ruby
Custard apple (*Annona reticulata*)	Balanagar, Mammoth Red, Arka Sahan
Guava (*Psidium guajava*)	Allahabad Safeda, L-49, Kohir Safed, Safed Jam
Bael (*Aegle marmelos*)	NB-5, NB-9
Sapota (*Manilkara achras*)	Kalipatti, Cricket Ball
Fig (*Ficus carica*)	Poona, Black Ischia, Deanna, Excel
Mango (*Mangifera indica*)	Banglora, Neelam, Keshar, Bombay Green
Sweet orange (*Citrus sinensis*)	Blood Red, Malta, Mosambi, Pineapple, Valencia
Sour lime (*Citrus aurantiifolia*)	Kadayam, Promalini, Vikram, Sai Sarbati
Tamarind (*Tamarindus indica*)	PKM 1, No. 263, Pratisthan,Yogeshwari
Vegetables	
Tomato (*Solanum lycopersicum*)	Pusa Ruby, Pusa Early Dwarf, S-12, Sel. 7
Chilli (*Capsicum annuum*)	Pusa Jwala, Mathania, Sindhur, Pant C-1, Arka Mohini, Arka Gaurav, Arka Basant, Bharat, Indira, NP-46A, Titan (USA)
Cowpea (*Vigna unguiculata*)	Pusa Dofasali, Pusa Phalguni, Pusa Barsati, Pusa Rituraj, Kashi Kanchan, Kashi Shyamal, Kashi Gauri
Cluster bean (*Cyamopsis tetragonoloba*)	Pusa Sadabahar, Pusa Mausami, Pusa Navbahar, Durga Bahar
Brinjal (*Solanum melongena*)	Pusa Purple Long, Pusa Kranti, Pusa Anmol, Punjab Sadabahar, Arka Sheel, Arka Kusumakar, Arka Navneet, Arka Shirish
Okra (*Abelmoschus esculentus*)	Kashi Pragati, Kashi Vibhuti, Varsha Uphar, Hisar Unnat
Pumpkin (*Cucurbita moschata*)	Arka Chandan, Kashi Harit, Narendra Amrit, CO-1, CO-2
Amaranth (*Amaranthus* spp.)	Chhoti Chaulai, Badi Chaulai, CO-1, CO-2, CO-3
Muskmelon (*Cucumis melo*)	Pusa Sharbati, Pusa Madhuras, Hara Madhu, Punjab Sunehari, Durgapur Madhu, Kashi Madhu, Arka Rajhans, Arka Jeet
Watermelon (*Citrullus lanatus*)	Sugar Baby, Arka Manik, Arka Jyoti, Durgapur Meetha, Durgapur Kesar, Mateera
Ash gourd (*Benincasa hispida*)	Pusa Ujjwal, Kashi Dhawal
Pointed gourd (*Trichosanthes dioica*)	Narendra Parwal-260, Narendra Parwal-307, Rajendra Parwal-1, Rajendra Parwal-2
Snap melon (*Cucumis melo* var. *momordica*)	Pusa Shandar
Long melon (*Cucumis melo* var. *utilissimus*)	Arka Sheetal, Punjab Long Melon-1, Pant Kakri-1
Round melon (*Praecitrullus fistulosus*)	Arka Tinda, Hisar Tinda, Punjab Tinda
Drumstick (*Moringa oleifera*)	PKM-1, PKM-2, Kokan Ruchira

2. Flooding or Waterlogging

Waterlogging is one of the most hazardous natural abiotic stresses. Depending upon the moisture or water level in soil, waterlogging conditions can also be described as flood, submergence, soil saturation, anoxia and hypoxia. Normally plant roots are in contact with oxygen at a partial pressure equivalent to the gaseous atmosphere. The reduction of oxygen below optimal levels, termed *hypoxia*, is the most common form of stress in wet soils and occurs during short-term flooding when the roots are submerged under water but the shoot remains in the atmosphere. The complete lack of oxygen, termed *anoxia*, occurs in soils that experience long-term flooding, in plants completely submerged by water, in deep roots below flooded water. Waterlogging and flooding are common in rainfed ecosystems, especially in soils with poor drainage. In India, about 12 m ha area is waterlogged and floods prone, where productivity of arable crops gets severely affected. Both flooding and waterlogging can seriously reduce yield. Flooding results in yield reduction from 10 per cent upto 40 per cent in severe cases (Hodgson and Chan, 1982). As a consequence of disturbed physiological functioning, vegetative and reproductive growth of plants is negatively affected by flooding. Roots get oxygen for growth and mineral uptake from air pockets in the soil, but when roots are partially submerged (waterlogged) or completely submerged (flooded), the anoxia conditions prevent root growth and send signals to the rest of the plant to reduce shoot growth and plant productivity (Bennett, 2003). Amongst the obvious symptoms of flooding injury is yellowing and death of the leaves, from the lower ones to the stem. This chlorosis seemingly to some extent looks like nitrogen deficiency, and often appears between four and six days after flooding. Reduced stomatal activity and net photosynthesis have also been noted in flooded plants. Reduction in net photosynthesis may be due to limited stomatal aperture during anoxic conditions. Stomatal closure is a response of plants to depleted soil oxygen in the root zone; flooding cripples leaf gas exchange and impedes in transport of soil water and solutes. Therefore, mineral nutrient uptake and water uptake are reduced or stopped in flooded conditions, and plants often wilt. There is also a build-up of ethylene in flooded soils, the plant hormone that in excess amounts can cause leaf drop and premature senescence. Under flooded conditions transpiration is reduced due to closure of stomata, resulting decrease in water absorption by roots.

For most of the horticultural crops, soils should normally have 10 to 30 per cent of their volume composed of larger pores that are filled with air, and 10 per cent is considered the minimum air content for healthy root growth depending on plant species. Waterlogging occurs when both the small and large air pores in the soil become filled with water, usually as a consequence of the water failing to drain away quickly enough from the large pores- resulting in a soil which has little or no oxygen, an environment referred to as anaerobic. When the plant roots do not receive an adequate supply of O_2 their growth is slowed or stopped. Root tips can start to die after 24-48 hours without oxygen. In flooded soils, the oxygen concentration drops to near zero within 24 hours because water replaces most of air in the soil pore space. It was reported that apple roots died at O_2 concentrations less than 3 per cent and the growth of existing roots required soil O_2 levels of at least 5-10 per cent. New root growth initiated only at O_2 concentrations greater than or equal to 12 per cent (Boynton, 1940). Waterlogging and reduced O_2 levels in the soil affect the roots ability to absorb water and nutrients and when this happens, the root sends a signal to the tree which triggers the leaf stomata to close to reduce water loss. Because the roots cannot take up water the leaves begin to wilt. Most research concludes that low O_2 and not excess CO_2 is likely to be the major source of damage associated with short term soil flooding. The effects of waterlogging can be less severe if the water is flowing, because moving water carries dissolved oxygen and also carries away any toxins. Waterlogging in the soil causes death of root hairs, reduces absorption of nutrients and water, increases formation of compounds toxic to plant growth, and finally retards growth of the plant. Waterlogging plants are also susceptible to soil-borne diseases.

Waterlogging also reduces the nutrient content in different plant parts. Oxygen deficiency in the rootzone causes a marked decline in the sensitivity of K^+/Na^+ uptake, and impedes the transport of K^+ to the shoots (Armstrong and Drew, 2002). Hypoxic conditions cause decrease in the permeability of root membrane to Na^+ (Burrett-Lennart *et al.,* 1999). Waterlogging cause deficiency of essential nutrients such as nitrogen, phosphrous, potassium, magnesium and calcium (Smethurst *et al.,* 2005). Prolonged flooding usually results in a cessation of root and shoot growth, wilting, decreased nutrient uptake and often tree death. Some subtropical fruit trees, such as mango and carambola can adapt to short-term flooding, although growth is often reduced. Avocado is a very flood-sensitive fruit crop, and short periods of flooding often result in leaf abscission and inhibition of leaf expansion resulting in smaller leaves. In general, if flooding or waterlogging last for less than 48 hours, most vegetable crops can recover. Longer periods will lead to high amounts of root death and lower chances of recovery. There has not been much research on flooding effects on vegetables; however some physiological effects that have been documented are given below:

☆ Oxygen starvation in tuber crops like potato leads to cell death in tubers and storage roots. This appears as dark or discolored areas in the tubers or roots. In carrots and other root vegetables, the tap roots often die leading to the formation of unmarketable fibrous roots.

☆ Lack of root function and movement of water and calcium in the plant lead to calcium related disorders; most notably incidence of blossom end rot (BER) in tomato, pepper, watermelon and several other susceptible crops.

☆ Leaching and denitrification losses of nitrogen and limited nitrogen uptake in flooded soils also lead to nitrogen deficiency in most vegetable crops.

☆ In bean crops, flooding or waterlogging has shown to decrease flower production and increase flower and young fruit abscission or abortion.

☆ Ethylene buildup in saturated soil conditions can cause leaf drop, flower drop, fruit drop, or early plant decline in many vegetable crops.

☆ In tomato and other vegetables, flooding stress causes deleterious symptoms such as epinasty, leaf chlorosis, necrosis and reduced fruit yield.

Plant Tolerance to Waterlogged Condition

Plant tolerance to waterlogging soil condition may be influenced by a number of factors; (i) soil type; porosity and chemistry; (ii) degree and duration of anaerobiosis (iii) soil microbe and pathogen status (iv) VPD and root zone, and air temperature (v) plant age, stage of development or season of the year, and (vi) plant preconditioning. The importance of each of these variables may differ with plant species. The tolerance of fruit crop to waterlogged condition is mainly determined by the rootstock, although specific foliar symptom of flooding injury may vary with the scion. In oxygen-deprived condition plants shift its metabolism to anaerobic from aerobic mode. Plants which can withstand waterlogging condition have mechanisms such as increased availability of soluble sugar, aerenchyma formation, formation of adventitious roots at shoot base, stem hypertrophy, greater activity of glycolytic pathway and fermentation enzymes, and involvement of antioxidant defence mechanism to cope with the oxidative stress. Ethylene plays an important role in change of mechanisms of plants in deficiency of oxygen. Ethylene initiates and regulates many adaptive molecular, chemical and morphological responses which allow the plant to avoid anaerobiosis by increasing oxygen availability to the roots in a flooded or waterlogged soil, such as development of aerenchyma. (Sairam *et al.*1998). Aerenchyma tissue forms when air spaces develop in the roots and shoots allowing for diffusion of oxygen from above ground portions of plant to the roots. Roots possessing aerenychma tissue are able to penetrate lower anaerobic soil layers. Diffused oxygen in the root zone aids in the detoxification of soil nutrients by oxidizing reduced chemical compounds. The aerenchyma tissue itself acts as a pathway for the removal of unstable compounds such as methane and ethylene from the root zone. Flood tolerant species avoid uptake of toxic compounds with well developed aerenchyma root tissue which aerates the rhizosphere and oxidizes toxic forms of Mn and Fe rendering those insoluble. Adventitious roots may develop at the shoot base in aerated soil layers as a result of flooding and signal severe damage to a plant's original root system. Surface rooting is common in flood tolerant plants and allows roots to function in a hypoxic state where they can undergo partial aerobic respiration.

In fruit crop, Rowe and Beardsell (1973) ranked water logging fruit trees as follows: extremely tolerant- quince (*Cydonia oblonga*); very tolerant- pear (*Pyrus* spp.); moderately tolerant- apple, citrus and plums (*P. domestica* and *P. cerasifera*); sensitive- plum (*P. salicina*), and very sensitive- cherry, apricot and peach.

Light Stress

Sunlight is not only the energy source for photosynthesis, but also the most important factor affecting productivity in horticultural crops. Carbon exchange rate (CER) is strongly dependent on irradiance, absorption, and utilization of photon energy. Low irradiance or insufficient light penetration into the canopy, influences CER directly by reducing photon energy utilization, thus decreasing productivity. Canopy management as a routine activity in horticultural crops is aimed at increasing light interception and productivity, stabilizing yield and improving fruit quality.

Light Intensity

Light intensity refers to the total amount of light, *i.e.* number of quanta or photon falling on any surface. The solar radiation on a clean cloudless day at noon varies

Table 12.6: Waterlogging Tolerant Rootstocks in Fruit Crops

Crop/cultivars	Expression
Pear	
Pyrus betulaefolia	Extremely tolerant
P. calleryana	Very tolerant
Cydonia oblonga (Quince)	Very tolerant
P. communis (Bartlett)	Very tolerant
P. pyrifolia	Moderately tolerant
Apple	
M 1, M 3, M 7, M 13, M 15, M 16	Moderately tolerant
MM 106, M 111, M9, M 4	Moderately sensitive
MM 104, MM 109, M 2, MM 106	Sensitive
M26, Delicious Seedlings	Very sensitive
Northern Spy Seedlings	Extremely sensitive
Cherry	
Prunus cerasus, Stockton Morello	Sensitive
P. avium, Mazzard	Very sensitive
P. mahaleb	Extremely sensitive
Plum and Prune	
Prunus japonica	Moderately tolerant
P. cerasifera, GF 8-1	Moderately sensitive
P. domestica, Damas GF 1869	Moderately tolerant
P. domestica, Cirule 43, Brompton	Moderately sensitive
P. salicina, S37, S300, S2540	Sensitive
P. salicina, S573, 2508, 2514	Very sensitive
P. persica, Lovell, Halford Nemaguard	Extremely sensitive
Grape	Moderately tolerant
Fig	Intolerant
Trifoliate orange	Moderately tolerant
Rough lemon (*C. Jambhiri*)	Moderately tolerant
Cleopatra mandarin	Moderately sensitive
Sweet orange	Moderately sensitive
Sour orange	Sensitive
Rangpur lime	Sensitive
Mango	Moderately tolerant
Guava	Moderately tolerant
Avocado	Sensitive
Loquat	Sensitive
Papaya	Sensitive

Source: Rowe and Beardsell (1973).

from about 1.50-1.75 g cal/cm^2/min/(108000-130000 lux). On overcast cloudy condition about 1000 ft-c (one ft-c = 10.76 lux) reach to the plant, thus on sunny days a little over 10 per cent of the available sunlight are used in photosynthesis, while on cloudy day all the light overcast on plant are utilized. In general, 10000 lux is regarded as low light intensity, whereas, 50000 lux or more are known as high intensity. Most crop plants use about 1,200 foot-candles of light, but they will grow better in light up to 4,000 foot-candles because of the shading from surrounding leaves. Plants and leaves adapted to low light intensity will sunburn, wither, or die if they are suddenly exposed to higher light intensity. Light intensity can be decreased through shading or increased with supplemental lights, reflective material, or white backgrounds.

The rate of photosynthesis in plant is proportional to the intensity of light up to about 1000 lux. Photosynthesis is negligible at about 5 lux, and the light compensation points for many crop species is about 1000 lux or 1200-foot candles. The compensation point is the light intensity at which the rate of photosynthesis equals the rate of respiration. At this light intensity, the rate of net photosynthesis is zero. Optimum light intensity is the range in which rate of gross photosynthesis is high and rate of respiration is normal resulting in high net photosynthesis. Light intensity requirement varies with the crops. Most of the vegetables including root crops require high light intensity of 3000-8000 foot candles. Leafy vegetables can be grown in partial shading but vegetables producing fruits require full sunlight. Over-crowding population in the field should be avoided as mutual shading of the leaves reduces the interception of solar radiation. An individual leaf may be light saturated at an intensity of 30,000 lux, but the entire plant may not be light saturated even at higher intensities, because lower leaves may be below the light compensation point.

Table 12.7: Required Indoor Light Intensity for Exposure of different Horticultural Crops

Type of Plant	Required light (in lux)
Cut flower	
Rose	4840-6450
Chrysanthemum	3760
Lilies	3760
Ornamental plant	
Begonia	3760
Cyclamen	3760
Orchids	4840
Nursery stocks	4840
Vegetable	
Lettuce	4840
Tomato	11000
Cucumber	10000-18000

Insufficient light will cause plants to stretch and become "gangly" or unusually long. It increases internodal shoot length, leaves become broad and thin, and the plants have a loose, open structure. Reduced light intensity can also induce succulence. The other consequences of growing vegetable crops under the light intensity below the optimum range are:

i) Reduction in chlorophyll content of the leaves.

ii) Reduction in the number of palisade cells in the leaves.

iii) Reduction in the rate of photosynthesis.

iv) Reduction in growth and yield.

Similarly, when the vegetable crops are grown in excess light intensity above the optimum range the probable consequences are:

i) Reduction in chlorophyll content of the leaves.

ii) Reduction in the rate of photosynthesis.

iii) In most of the cases, high light intensity is of little use photosynthetically because light harvesting system captures light energy much faster than its utilization for transformation to chemical energy.

iv) Increase in rate of respiration resulting in less net photosynthesis.

v) Increase in leaf temperature induces high rate of transpiration resulting in high water demand.

vi) In some cases, strong irradiance inactivates some enzyme system.

vii) Reduction of yield in most of the cases.

viii) Increase in number of male flowers in monoecious cucurbits.

Light Quality

Light is utilized by the plants through its absorption and simultaneous selective translation of its energy into organized biochemical systems like photosynthesis, seed germination, photoperiodism, pigmentation responses etc. In fact, when crop plants are adequately supplied with water and nutrients, both water use and dry matter production depend upon the receipt and utilization of solar radiation. Most crops grow that best in visible spectrum of light in wavelength range from 400 to 700 nm. This range of wavelengths is called photosynthetically active radiation (PAR). Wavelength of the visible spectrum (400-700 nm) are selectively absorbed by plants through chlorophyll and various pigments, mainly phytochrome, carotenoids and flavones. Radiation in the wavelength range of 0.15 to 3 µm is designated as short wave radiation, whereas that in the range of 3 to 100 µm is designated as long wave, infrared, or thermal radiation. In the earth atmosphere about 95 per cent of the solar radiation is received in the 0.4-2.8 µm-wavelength band. Ultraviolet (15-390 nm) and infrared (760 nm or more) rays are not at all beneficial to crop growth, rather these are damaging.

Table 12.8: Physiological Response of Plant to certain Light Wavelengths

Wavelength (nm)	Impact
720-1000	Stem elongation, germination inhibition of certain seeds, stimulation of onion bulbing
650-690	Suppression of onion bulbing, lycopene synthesis in tomato, flower initiation in long day plants, flower inhibition of short day plant, promotion of germination, promotion of anthocyanins
440 655	Photosynthesis occurs
445-660	Chlorophyll formation takes place
350-500	Phototropism response

Table 12.9: Light Compensation and Saturation Points of some Fruit Crops

Crops	Light Compensation Point[1] ($\mu mol\ m^{-2}\ s^{-1}$)	Light Saturation Point[2] ($\mu mol\ m^{-2}\ s^{-1}$)	A max[3] ($\mu mol\ m^{-2}\ s^{-1}$)
Almond (*Prunus dulcis*)	60	1130-1330	15-20
Apple (*Malus domestica*)	57	1800-1900	16
Fig (*Ficus carica*)	49	1100	15
Grape (*Vitis vinifera*)	67	1800-1900	25
Orange (*Citrus sinensis*)	17	750-1000	15-22
Papaya (*Carica papaya*)	29	1900	25-30
Peach (*Prunus persica*)	40	1300	16-17

Adopted from Restrepo-Díaz *et al.* (2010)

1. The compensation point is the light intensity at which the rate of photosynthesis equals the rate of respiration. At this light intensity, the rate of net photosynthesis is zero.

2. At the light saturation point, increasing the light no longer causes an increase in photosynthesis.

3. *A* max is the maximum photosynthesis rate that occurs at an optimum light intensity.

Effects of Shading on Horticultural Crops

☆ **Muskmelon:** Reduced photosynthetic rate, fresh weight and flesh firmness. Low accumulation of sucrose. Accelerated the formation of the 'water-soaked' symptom in the flesh.

☆ **Sweet pepper:** Enhanced flower abortion and thus reduced fruit yield.

☆ **Cauliflower:** Growth and development after curd initiation decreased with increasing shade levels.

☆ **Carrot:** Reduced photosynthetic rate, stomatal conductance, transpiration and water use efficiency.

☆ **Lettuce:** Decreased leaf thickness and leaf dry matter percentage.

☆ **Pear:** Decreased area per spur leaf, specific leaf mass and fruit diameter.

☆ **Olive:** Reduced percentage of inflorescence buds, number of fruits per tree, and fruit mass.

☆ **Grape:** Affected dry-matter partitioning and photosynthesis.

☆ **Hazelnut:** Reduced yield primarily by decreasing number of nuts and secondarily by decreasing nut size.

☆ **Pecan:** Diminished photosynthesis, stomatal and trichome density.

Salinity Stress

In the past, stress caused by high concentrations of salts had little importance because these situations arose only in coastal areas. However, the development of agricultural techniques in recent years has made salt stress one of the chief problems in agriculture today. Low rainfall, high evaporation, poor water management and indiscriminate use of huge quantities of chemical fertilizers and the overexploitation of aquifers have dramatically multiplied the surface area affected by salinity. More than 800 million hectares (m ha) of land throughout the world are salt-affected (including both saline and sodic soils), corresponding to more than 6 per cent of the world's total land area (FAO, 2009). In addition, out of 230 m ha of irrigated land, 45 m ha ($\approx$20 per cent) are salt affected. However, the intensity of salinity stress varies from place to place. In India, the crop productivity of 6.73 m ha land is limited by the existence of salinity/alkalinity. Similarly, about 12 m ha of acidic soils (pH <5.5) suffer from deficiencies as well as toxicities of certain nutrients and have very low productivity. Generally, dryland salinity has categorized into three different types: low salinity (EC_e 2-4 dS/m), moderate salinity (EC_e 4-8 dS/m) and high salinity (EC_e > 8 dS/m) (Rogers *et al.,* 2005). Depending upon the type of source from which soil became salinized; soil salinity can be categorized as primary and secondary salinization. *Primary* or natural salinization results from weathering of minerals and soil derived from saline parent rocks, and *secondary* slalinization that is caused by human interference such as irrigation, deforestation, overgrazing, or intensive cropping. According to (FAO, 2008) estimate, 32 m ha (2 per cent) out of 1500 m ha are affected by secondary salinity to varying degrees depending upon the type of factors causing salinity. Most of the horticultural crops are glycophytes (grow in soils with low content of sodium salts) and, therefore, highly susceptible to soil salinity even at low electrical conductivity in the saturated soil extract. Classification of vegetable crops as per tolerance/susceptibility to salinity and its effect on yields have been shown in Figure 12.4, and threshold salinity levels of vegetables, fruits, ornamental shrubs and trees have given under Tables 12.10 and 12.11.

A heavy environmental concentration of salts unleashes various types of physical and chemical stress in plants, provoking complex responses that involve changes in plant morphology, physiology and metabolism. It is commonly accepted that growth inhibition by salt stress is associated with alterations in the water relationships within the plant, caused by osmotic effects with specific ionic consequences (excesses or deficits) or energy availability related to carbohydrate concentrations. Under natural conditions, terrestrial higher plants encounter high concentrations of salts close to the seashore and in estuaries where seawater and freshwater mix or replace each other with the tides. Far inland, natural salt seepage from geologic marine deposits can wash into adjoining areas, rendering them unusable for agriculture. However, a

Table 12.10: Salt Tolerances of Vegetable and Fruit Crops[1]

Crops		Salt Tolerance Parameters		
Common Name	Botanical Name	Threshold EC_e $(dS\,m^{-1})^a$	Slope (per cent per $dS\,m^{-1}$)	Rating
VEGETABLES				
Sugar beet	*Beta vulgaris* L.	7.0	5.9	T
Broad bean	*Vicia faba* L.	1.6	9.6	MS
Labiab bean	*Lablab purpureus* (L.) Sweet	–	–	MS
Asparagus	*Asparagus officinalis* L.	4.1	2.0	T
Common bean	*Phaseolous vulgaris* L.	1.0	19	S
Lima bean	*P. lunatus* L.	–	–	MT*
Broccoli	*Brassica oleracea* L. (botrytis group)	2.8	9.2	MS
Brussels sprouts	*B. oleracea* L. (gemmifera group)	–	–	MS*
Cabbage	*B. oleracea* L. (capitata group)	1.8	9.7	MS
Carrot	*Daucus carota* L.	1.0	14	S
Cassava	*Manihot esculenta* Crantz	–	–	MS
Cauliflower	*Brassica oleracea* L. (botrytis group)	–	–	MS*
Celery	*Apium graveolens* L. var. dulce	1.8	6.2	MS
Cowpea	*Vigna unguiculata* (L.) Walp.	4.9	12	MT
Cucumber	*Cucumis sativus* L	2.5	13	MS
Eggplant	*Solanum melongena* L.	1.1	6.9	MS
Garlic	*Allium sativum* L.	1.7	10	MS
Lettuce	*Lactuca sativa* L.	1.3	13	MS
Muskmelon	*Cucumis melo* L. (reticulatus group)	1.0	8.4	MS
Okra	*Abelmoschus esculentus* (L.) Moench	–	–	MS
Onion (bulb)	*Allium cepa* L.	1.2	16	S
Pea	*Pisum sativum* L.	3.4	10.6	MS
Pepper	*Capsicum annuum* L.	1.5	14	MS
Potato	*Solanum tuberosum* L.	1.7	12	MS
Pumpkin	*Cucurbita pepo* L.var. Pepo	–	–	MS*
Radish	*Raphanus sativus* L.	1.2	13	MS
Spinach	*Spinacia oleracea* L.	2.0	7.6	MS
Sweet potato	*Ipomoea batatas* (L.) Lam.	1.5	11	MS
Tomato	*Lycopersicon lycopersicum* (L.) Karst. ex Farw. (syn. *Solanum lycopersicum*)	2.5	9.9	MS
Turnip	*Brassica rapa* L. (rapifera group)	0.9	9.0	MS

Contd...

Table 12.10–*Contd...*

Crops		Salt Tolerance Parameters		
Common Name	Botanical Name	Threshold EC_e (dS m^{-1})[a]	Slope (per cent per dS m^{-1})	Rating
FRUIT CROPS				
Almond	*Prunus duclis* (Mill.) D. A.	1.5	19	S
Apple	*Malus sylvestris* Mill.	–	–	S
Apricot	*Prunus armeniaca* L.	1.6	24	S
Banana	*Musa acuminata* Colla	–	–	S
Blackberry	*Rubus macropetalus* Doug.	1.5	22	S
Coconut	*Cocos nucifera* L.	–	–	MT
Date palm	*Phoenix dactylifera* L.	4.0	3.6	T
Grape	*Vitis vinifera* L.	1.5	9.6	MS
Guava	*Psidium guajava* L.	4.7	9.8	MT
Indian Jujube	*Ziziphus mauritiana* Lam.	–	–	MT
Lemon	*Citrus limon* (L.) Burm.	1.5	12.8	S
Lime	*Citrus aurantiifolia* (Swingle)	–	–	S
Mango	*Mangifera indica* L.	–	–	S
Orange	*Citrus sinensis* (L.) Osbeck	1.3	13.1	S
Papaya	*Carica papaya* L.	–	–	MS
Peach	*Prunus persica* (L.) Batsch	1.7	21	S
Pineapple	*Ananas comosus* (L.) Merrill	–	–	MT
Plum	*Prunus domestica* L.	2.6	31	MS
Pomegranate	*Punica granatum* L.	–	–	MS
Strawberry	*Fragaria* x *ananassa* Duch.	1.0	33	S
Walnut	*Juglans* spp.	–	–	S

S: Sensitive; MS: Moderately sensitive; MT: Moderately tolerant; T: Tolerant.

a: In gypsiferous soils, plants tolerate EC_e about 2 dS m^{-1} higher than indicated.

1: Pessarakli (1999).

much more extensive problem in agriculture is the accumulation of salts from irrigation water.

The deleterious effects of salinity on plant growth are associated with (i) low water potential of the root medium which causes a water deficit within the plant; (ii) toxic effects of ions mainly Na$^+$, Cl$^-$, and SO$_4^{2-}$; (iii) nutritional imbalance caused by reduced nutrient (*e.g.*, K$^+$, Ca^{2+}, Mg^{2+}) uptake and/or transport to the shoot. During the onset and development of salt stress within a plant, all the major processes such as photosynthesis, protein synthesis and energy and lipid metabolisms are affected. The earliest response is a reduction in the rate of leaf surface expansion followed by

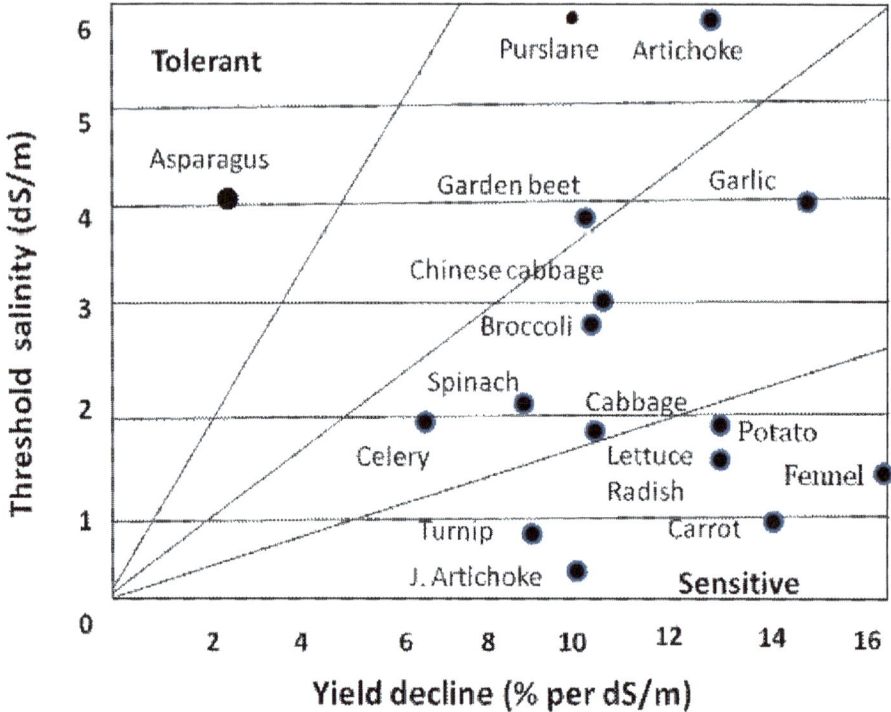

Figure 12.4: Effect of Salinity on Vegetable Crops.
Adopted from Shannon and Grieve (1999).

Table 12.11: Salt Tolerance of Ornamental Shrubs and Trees[1]

Category	Example
Very sensitive (Maximum permissible soil salinity; EC_e (1-2 dS m^{-1})	Star jasmine (*Trachelospermum jasminoides*), Pyrenees cotoneaster (*Cotoneaster congestus*), Oregon grape (*Mahonia aquifolium*)
Sensitive (Maximum permissible soil salinity; EC_e (3-4 dS m^{-1})	*Rosa* spp., *Podocarpus macrophyllus*, *Hedera canariensis*, *Nandina domestica*, *Hibiscus rosa-sinensis*, *Lagerstroemia indica*
Moderately sensitive (Maximum permissible soil salinity; EC_e (4-6 dS m^{-1})	*Lantana camara*, *Bauhinia purpurea*, *Magnolia grandiflora*, *Juniperus chinensis*, *Platycladus orientalis*
Moderately tolerant (Maximum permissible soil salinity; EC_e (6-8 dS m^{-1})	*Callistemon viminalis*, *Nerium oleander*, *Cordyline indivisa*, *Chamaerops humilis*, *Rosmarinus officinalis*
Tolerant (Maximum permissible soil salinity; EC_e (8-10 dS m^{-1})	*Syzygium paniculatum*, *Leucophyllum frutescens*, *Carissa grandiflora*, *Bougainvillea spectabilis*, *Pinus pinea*
Very tolerant (Maximum permissible soil salinity; EC_e (>10 dS m^{-1})	White iceplant (*Delosperma alba*) Rosea iceplant (*Drosanthemum hispidum*) Purple iceplant (*Lampranthus productus*) Croceum iceplant (*Mesembryanthemum croceus*)

1: Pessarakli (1999).

cessation of expansion as the stress intensifies but growth resumes when the stress is relieved (Parida and Das, 2005). Salinity reduces the growth rate resulting in smaller leaves, shorter stature and a fewer leaves. The initial effect of salinity, especially at low to moderate concentrations, is due to its osmotic effects. Roots are also reduced in length and mass but may become thinner or thicker. The severity of salinity response is also mediated by environmental interactions such as relative humidity, temperature, radiation and air pollution. Depending upon the composition of the saline solution, ion toxicity or nutritional deficiencies may arise because of predominance of a specific ion or competition effects among cations or anions. The osmotic effects of salinity contribute to reduced growth rate, changes in leaf colour and developmental characteristics such as root/shoot ratio and maturity rate. High concentrations of Na or Cl may accumulate in leaves or portions thereof and result in 'scorching' or 'firing' of leaves. All salinity effects may not be negative; salinity may have some favourable effects on yield, quality and disease resistance. In spinach, yields may initially increase at low to moderate salinity, cabbage heads are more solid at low salinity levels, but are less compact as salinity increases. Sugar contents increase in carrot and starch content decreases in potato as salinity increases.

Salinity Tolerance in Plant

The mechanisms of salinity tolerance fall into three categories (Munns and Tester, 2008):

1. *Tolerance to osmotic stress:* The osmotic stress immediately reduces cell expansion in root tips and young leaves, and causes stomatal closure. A reduced response to the osmotic stress would result in greater leaf growth and stomatal conductance, but the resulting increased leaf area would benefit only plants that have sufficient soil water. Greater leaf area expansion would be productive when a supply of water is ensured such as in irrigated food production systems, but could be undesirable in water-limited systems, and cause the soil water to be used up before the grain is fully matured.

2. *Na^+ exclusion from leaf blades:* Na^+ exclusion by roots ensures that Na^+ does not accumulate to toxic concentrations within leaves. A failure in Na^+ exclusion manifests its toxic effect after days or weeks, depending on the species, and causes premature death of older leaves.

3. *Tolerance of tissues to accumulated Na^+ or Cl^-:* Tolerance requires compartmentalization of Na^+ and Cl^- at the cellular and intracellular level to avoid toxic concentrations within the cytoplasm, especially in mesophyll cells in the leaf. Toxicity occurs with time, after leaf Na^+ increases to high concentrations in the older leaves. The chloride tolerance limits of rootstocks of some fruit crops have given in Table 12.12.

Measures to Mitigate the Effect of Abiotic Stresses

Measures to adapt under abiotic stresses are critical for sustainable horticulture production. Until now, the scientific information on the effect of environmental stresses on horticulture is only on few vegetable crops like tomato. There is a need for more research on how other vegetable and fruit crops are affected by increased abiotic

Table 12.12: Chloride Tolerance Limits of some Fruit Crop Rootstocks [1]

Fruit Crops	Rootstocks	Maximum Permissible Cl⁻ in Soil Water without Leaf Injury (mol m^{-3})
Citrus (*Citrus* spp.)	Mandarin (Sunki, Cleopatra), grapefruit, Rangpur lime	50
	Rough lemon, tangelo (Sampson, Minneola), sour orange, Ponkan mandarin	30
	Citrumelo 4475, Calamondin, sweet orange, trifoliate orange, Cuban shaddock, Citrange (Savage, Rusk, Troyer)	20
Grape (*Vitis* spp.)	Salt Creek, 1613-3	80
	Dog Ridge	60
	Thompson Seedless, Perlette	40
	Cardinal, Black Rose	20
Stone fruit	Marianna	50
(*Prunus* spp.)	Lovell, Shalil	20
	Yunnan	15
Avocado	West Indian	15
(*Persea americana*)	Guatemalan	12
	Mexican	10

1: Maas (1990).

stresses as direct potential threat from climate change. Various measures can be adopted to alleviate or minimize the harmful effect of abiotic stresses. There are two basic strategies for cope up with all abiotic stresses in horticultural crops *i.e.*, i) evolving abiotic stresses tolerant cultivars and ii) adopting sustainable production technologies.

Superior genotypes adapted to a wider range of climatic conditions could result from the discovery of novel genetic variation for tolerance to different abiotic stresses. Genotypes with improved attributes conditioned by superior combinations of alleles at multiple loci could be identified and advanced. Improved selection techniques are needed to identify these superior genotypes and associated traits, especially from wild, related species that grow in environments which do not support the growth of their domesticated relatives. Germplasms of the major vegetable crops which are tolerant to high temperatures, flooding and drought have been identified and advanced breeding lines are being developed. Efforts are also underway to identify nitrogen-use efficient germplasms in vegetable crops. The development of drought-resistant cultivars/lines of crops through selection and breeding is of considerable economic value for increasing crop production in areas with low precipitation or without any proper irrigation system. In order to develop drought tolerant cultivars, it is imperative to develop efficient screening method and suitable selection criteria. Various agronomic, physiological and biochemical selection criteria for drought tolerance are being employed to select drought tolerant plants, such as seed yield, harvest index, shoot fresh and dry weight, leaf water potential, osmotic adjustment, accumulation of compatible solutes, water use efficiency, stomatal conductance, chlorophyll fluorescence, etc. Stress physiology research identifies mechanism of stress tolerance and provides approach, method and traits for screening stress-resistant genotypes. Molecular biology and genomic studies lead to a better understanding of the structural organization and functional properties of genetic variation for stress-related traits, allow gene-based selection through identification of molecular markers and high-throughput genotyping techniques, and increase the gene pool available including new sources of stress-tolerant traits or transgenes. Plant breeders translate these findings into stress-tolerant crop varieties by using all tools available that include germplasm screening, marker-assisted selection, plant transformation, and conventional breeding methods.

The most effective way to address the climate change is to adopt the sustainable development pathway, besides using renewable energy, water conservation, eco-friendly greenhouse technology, reforestation, etc. Judicious water utilization in the form of drip or sprinkler irrigation, raised bed planting, regulated deficit irrigation practice will be the important factors to deal with the drought condition. Various crop management practices such as mulching, use of shelters/shade net and raised beds help to conserve soil moisture, prevent soil degradation, and protect vegetables from heavy rains, high temperatures and flooding. Experiment conducted in tomato revealed that furrow irrigated raised bed planting can save 36 per cent water with marginal increase in fruit yield (Bahadur *et al.,* 2013*a*). In northern Indian plains, organic mulches are very suitable for conserving soil moisture, modifying soil temperatures, improving gas exchanges and yield in vegetable crops (Bahadur *et al.,* 2009, 2013*b*).

1. Strategies against High Temperature Stress

1. *Grafting Technique*: Supra-optimal temperature causes many deleterious effects like growth reduction, decrease in the photosynthetic rate and increase in respiration, assimilate partitioning towards the fruits, osmotic and oxidative damage, reduced water and ion uptake/movement and cellular dehydration. Since 1990, the grafting technology was initiated by the AVRDC for production of tomato and pepper in the Asian lowland tropics. Since eggplants are better adapted to hot arid climate and have a better tolerance against supraoptimal soil temperature, the use of eggplants as rootstocks for tomato at higher temperature is more promising. Although it was noted that eggplant rootstocks enhanced vegetative growth at 28°C, they had no advantage, rather it decreases total fruit dry weight. The use of a heat-tolerant tomato (cv. Summerset) as a rootstock also failed to improve the yield. However, testing eggplants (*S. melongena* cv. Yuanqie) grafted onto a heat-tolerant rootstock (cv. Nianmaoquie) seemed to be promising and resulted in a prolonged growth stage and yield increase up to 10 per cent (Wang *et al.,* 2007). Comparing different lines of chilli pepper rootstocks (*C. chacoense, C. baccatum, C. frutescens, C. annuum*) confirmed higher yields

under high-temperature conditions for rootstocks recommended by the AVRDC (*C. annuum* cv. Toom-1 and 9852-54). It is known that supraoptimal temperatures may develop multiple mineral deficiencies (P and Fe) in roots and shoots, which both can increase ethylene production. Grafting tomato onto a heat tolerant rootstock (*L. esculentum* cv. RX-335) result in a decreased hydrogen peroxide concentration indicating the lower oxidative stress. It has shown that grafting of tomato onto eggplants reduced electrolyte leakage under supra-optimal temperature stress, indicating less membrane damage and a higher ability to retain solutes and water. The tomato grafted onto eggplants exhibits lower proline level but higher ascorbate concentrations compared with self-grafted tomato.

2. *Use of Plant Growth Regulators:* High temperature (HT) accelerates the most of the plant metabolic processes. The harmful effect of HT is usually aggravated by lack of available moisture. Applications of certain growth regulators can alleviate the harmful effect of HT in vegetable crops. In tomato fruit set under high temperature conditions can be achieved by spraying of sodium salt of PCPA (50 ppm) or NAA (10 ppm) at flowering. Similarly foliar spray of triacontanol (1-2 ppm) in chilli improves fruit sets and reduces the flower and fruit drops at high temperature condition. Seed germination in lettuce at higher temperature can be enhanced by soaking seed in ABA at 0.1 to 1.0 ppm solution. Similarly seed germination in celery at higher temperature can be enhanced by imbibing the seeds in GA_{4+7} and subsequently in BA solution. Application of ABA, IAA and ethephon in tomato, capsicum, bean and cucumber plants reduce the stomatal opening and water consumption.

3. *Shading Net:* The harmful effect of heat and intense light can be minimised by using shading devices. A screen made of fabric (polyester and poly propylene) is more convenient for shading a small part of greenhouse. They are available in different density of weave with degree of shading ranging 20-90 per cent, however 50 per cent is more common. Shading nets in vegetables are known to reduce temperature, increase humidity and improve the appearance of some vegetables. The use of shade should be minimal in crops like tomato, which have a dense canopy. Besides, filter films such as fluorescent films, photochromic films and UV-blocking films are also used to reduce light intensity in commercial greenhouses.

Strategies against Cold/Low Temperature Stress

There are several options to protect horticultural crops from cold/low temperatures.

1. *Outdoor heaters* are also used as frost control method in fruit tree orchards, especially in European countries. It is often used in combination with wind machines. High fuel costs have reduced the use of this technique. About 86 heaters with oil burning capacity of 2-4 lit/h are used to warm up for one hectare area. Heaters should be lit when orchard temperature reaches the critical point, particularly during flower bud development stage.

2. *Wind machine* is also used for warming orchards. Wind machines bring warm air down into an orchard from higher elevations during a temperature inversion and also stir the air to prevent cold air from settling into frost pockets are more practical. Wind machine is installed at N-W side with fan height of about 10 m, however it works well when wind speed is less than 1.3 m s^{-1}.

3. *Plastic mulches* increase soil temperature and hasten early plant development. During the day, sunlight warms the soil. At night, the plastic traps the heat, keeping the warmth in the soil. Clear plastic allows greater soil warming than dark coloured (black, brown, gray) plastic (10°F to 20°F warming for clear, 5°F to 10°F for dark). This can increase the earliness of many vegetable crops by speeding up germination and early growth.

4. *Row covers*, which are often used in conjunction with plastic mulch to promote early crop growth while reducing heat loss at night. In newly established vegetable crops, row covers protect plants from temperatures 2° to 5°F below freezing for short durations. Row covers reduce the damage, but do not offer complete protection, when the temperature drops below 0°C. Slitted row covers protect crops from -1.0 to -0.5°C frosts only. The floating row covers protect against frosts of -2.0 to -1.5°C.

5. *Sprinkler irrigation* is sometimes used to protect flowers and vegetables from cold injury. Saturating the soil early in the day may help protect plants, since the water will warm up during the day and release the heat slowly during the night. Sprinkling the plants during frosty nights can also help prevent injury. When water is sprinkled on plants as they cool, the heat of freezing will keep the plant surface at or near 0°C. This technique is often used in orchards during bloom time when frost or cold temperatures are predicted. The water application rate is usually varied between 0.20-0.60 inch/h depending upon ambient temperature and wind speed. The sprinklers should be setup to provide about 30 percent overlap to cover for a gentle breeze. It is observed that when sprinkler irrigation used in combination with row covers can extend frost protection to around -6.0°C. Vegetable and flower transplants should be hardened off before they are planted in the field. This slows the growth of the plants, decreasing the chance of cold injury. The plants should be gradually exposed to the lower temperatures. For cold protection, the plants can also be placed on wagons, which are brought outdoors during the day and returned to the barn at night.

6. *Chemical frost protectants,* including surfactants and combinations of fungicides and bactericides are the possible options of protecting crops from injury. These products help prevent the formation of ice crystals, by destroying the bacteria that cause ice crystals to form (called 'ice-nucleating bacteria'). These products provide some protection, at least for a few degrees below freezing. However, killing the bacteria will not prevent ice crystal

formation caused by dust and other materials. These products should be used in conjunction with other protection measures, and should not be the only preventative measures used.

7. *Grafting on tolerant rootstock:* Since the late sixties of the 20th century rootstocks have been used to enhance fruit yield in open-field and in unheated polythene greenhouses for cucurbits (cucumber, squash and melon) production during the cold seasons in Japan and Korea, when low soil temperatures seriously affect seedling performance. For cucumber, fig-leaf gourd (*Cucurbita ficifolia* Bouché) and bur cucumber (*Sicos angulatus* L.) are used as rootstocks. Fig-leaf gourd is unique among cucurbit species with an optimal root temperature at approximately 15°C, thus 6°C lower than that of cucumber roots. Studies demonstrated that these two rootstocks improved vegetative growth and early yield at suboptimal temperatures, and also when only the roots are subjected to chilling temperatures < 8°C (Ahn *et al.,* 1999). For tomato, rootstocks of the high-altitude accession LA 1777 of *S. habrochaites* (synonym *L. hirsutum* Dunal), 'KNVF' (the inter-specific hybrid of *S. lycopersicum* × *S. habrochaites*), and chill-tolerant lines from backcrossed progeny of *S. habrochaites* LA 1778 × *S. lycopersicum* cv. T5 are able to alleviate low root-temperature stress for different scions. For watermelon, grafting onto Shin-tosa-type (an inter-specific squash hybrid, *Cucurbita maxima* × *C. moschata*) rootstocks is used to advance the planting date during cool periods. The same rootstocks can also be used to improve the vegetative growth rate of eggplants at suboptimal temperatures (Gao *et al.,* 2008).

8. *Growth regulators* are known to increase cold hardiness in vegetable crops. Abscisic acid application has been found to increase cold hardiness in several vegetable crops. The beneficial effect of such treatment is assumed to mediate through its effect on reducing water loss. Spraying of potato haulms with 0.74 kg/ha of CCC causes a short term retardation of growth, stimulates the root growth and increases resistance of plants to frost. Foliar spray of GA_3 also causes protection of plants against frost. The frost resistant plants contain relatively higher concentration of maltose and TSS. Cycocel (500 ppm) and GA_3 (25 ppm) have been found effective in tomato for overcoming damage caused by frost. Tomato plants sprayed with CCC at 0.4-0.5 per cent appreciably increased the cold hardiness. Such physiological response is not attributed to change in plant sugar content or cell protoplasm viscosity, but due to change in leaf ascorbic acid content. Spray of ethephon on tomato nursery 12 days prior to transplanting increase the frost tolerance in plant. In cucumber, ABA is reported to be involved in development of frost tolerance in plant. Gibberellic acid has been used to identify heat tolerance in Chinese cabbage at the seedling stage. Growth substances have also been used for induction of frost tolerance in lima bean, pole bean, snap bean, spinach, lettuce and tomato.

3. Strategies against Drought or Water Deficit

1. *Selection or development of drought tolerant genotypes or cultivars for growing*:
In order to improve crop productivity under water deficit conditions,
selection of a cultivar with short life span (drought escape), incorporation
of traits responsible for well-developed root system, high stomatal
resistance, high water use efficiency (drought avoidance), and traits
responsible for increasing and stabilizing yield during drought period
(drought tolerance) should be given high priorities. The development of
drought-resistant cultivars/lines of crops through selection and breeding
is of considerable economic value for increasing crop production in areas
with low precipitation or without any proper irrigation system. There are
several genotypes/cultivars that can tolerate up to some degree of drought
stress may be used for growing in arid or drought prone areas. The fruits
and vegetables cultivars suitable for growing in these areas have been given
under Tables 12.5 and 12.13.

Table 12.13 Drought tolerant genotypes/cultivars in vegetables

Sl.No.	Vegetables	Drought Tolerant Genotypes/Species
1.	Tomato	*S. habrochaites* (EC- 520061), *S. pennelli* (IIHR 14-1, IIHR 146-2, IIHR 383, IIHR 553, IIHR 555, K-14, EC-130042, EC-104395, Sel-28), *S. pimpinellifoloium* (PI- 205009, EC- 65992, Pan American, LA1579), *S. esculentum* var. cerasiforme, *S. hirsutum*, *S. cheesmanii*, *S. chilense*, *S. habrochaites*, Arka Vikas, RF- 4A, *L. pennellii* (LA0716), *L. chilense* (LA1958, LA1959, LA1972), *S. sitiens* (LA1974, LA2876, LA2877, LA2878, LA2885)
2.	Brinjal	*Solanum microcarpon, S. gilo, S. macrosperma, S. integrifolium,* Bundelkhand Deshi, *S. sodomaeum* (syn. *S. linneanum*), SM- 1, SM- 19, SM- 30, Violette Round, Supreme
3.	Chilli	*Capsicum chinense, C. baccatum* var. pendulum, *C. eximium,* Arka Lohit, IIHR -Sel. 132
4.	Potato	*Solanum acaule, S. demissum, Solanum chacoense, S. stenotonum, S. ajanhuiri, S. curtilobum, S. xjuzepczukii,* Alpha, Bintje, Kufri Sheetman and Kufri Sindhuri
5.	Okra	*Abelmoschus caillei, A. rugosus, A. tuberosus*
6.	Onion	*Allium fistulosum, A. munzii,* Arka Kalyan, MST 42, MST 46
7.	French bean	*Phaseolus acutifolius*
8.	Water melon	*Citrullus colocynthis* (L.) Schrad
9.	Cucumber	INGR-98018 (AHC-13)
10.	Winter squash	*Cucurbita maxima*
11.	*Cucumis* spp.	*Cucumis melo* var. momordica, *Cucumis pubescens, C. melo* var. callosus, *C. melo* var. chat Arya, *C. melo* (SC-15), VRSM-58, INGR-98015 (AHS-10), INGR-98016 (AHS-82), CU 159, CU 196, INGR-98013 (AHK-119), AHK-200, SKY/DR/RS-101
12.	Cassava	CE-54, CE-534, CI-260, CI-308, CI-848, 129, 7, 16, TP White, Narukku-3, Ci-4, Ci-60, Ci-17, Ci-80
13.	Sweet potato	VLS6, IGSP 10, IGSP 14, Sree Bhadra

Adopted from Kumar *et al.* (2012)

2. *Grafting on tolerant rootstock:* Under drought conditions, one way to reduce losses in production and to improve water use efficiency is grafting them on to rootstocks capable of reducing the effect of water stress. Grafting experiments on drought with fruits, such as kiwi and grapes, prooves that drought tolerant rootstocks are available and may be utilized for drought stress condition on commercial scales. Now-a-days grafting technique is also being used for vegetable production in drought condition. Eggplant is more effective to water uptake than tomato root systems. It may be used under water-stress conditions. Grafted mini-watermelons onto a commercial rootstock (PS 1313: *Cucurbita maxima* Duch. × *Cucurbita moschata* Duch.) revealed a more than 60 per cent higher marketable yield when grown under conditions of deficit irrigation compared with un-grafted melons (Rouphael *et al.,* 2008). The higher marketable yield recorded with grafting was mainly due to an improvement in water and nutrient uptake, indicated by a higher N, K, and Mg concentration in leaves, and higher CO_2 assimilation.

3. *Transgenic approach:* The ongoing research on genetic engineering water stress tolerant plants is mainly based on transfer of one or several genes that are either involved in signaling and regulatory pathways, or that encodes enzymes present in pathways leading to the synthesis of functional and structural protectants, such as osmolytes and antioxidants, or that encodes stress tolerance conferring proteins. All these genes are categorized in three major groups: (i) genes involved in signaling pathways and in transcriptional control, (ii) genes involved in protection of membranes and proteins, such as heat shock proteins (Hsps) and chaperones, late embryogenesis abundant (LEA) proteins, osmoprotectants and free-radical scavengers; (iii) genes involved in water and ion uptake and transport such as aquaporins and ion transporters (Wang *et al.,* 2003). Most successful examples of transgenic crops for drought tolerance are transgenics of DREBs/CBFs transcription factors in different crops as in tomato, rice and wheat.

4. Strategies against Waterlogging or Flood Stress

1. *Foliar sprays of nutrients:* Soil waterlogging may cause both nutrient accumulation and limited nutrient availability. Under flooded conditions, soil nitrate is less abundant, and soil N exists in the form of ammonium (NH_4^+). Phosphate generally becomes more available during soil flooding. Under waterlogged conditions, elements such as Mn and Fe are reduced and may become toxic to flood sensitive plants. Under flooding or submerged condition root is unable to absorb plant nutrients. It is adviced to spray low salt liquid fertilizer to supply about 2.0 kg N, 0.5 kg P_2O_5 and 0.5 kg K_2O per acre for instant relief to plant. Water soluble fertilizer mixture (such as 19:19:19 NPK) may be sprayed to plant @ 6 g/lit of water for instant supply of nutrients. As with all foliar applications, keep total salt concentrations to less than 3 per cent solutions to avoid foliage burn. Foliar

spray of DAP 2 per cent + Muriate of potash (1 per cent) is also beneficial for alleviating the short term water stress. Organic manures can also improve soil physical factors and reduce soil surface crusting, enhance plant rooting and increased availability of Fe and Mn. Therefore the use of manures is beneficial in waterlogging-prone environments.

2. *Grafting on tolerant rootstocks:* Grafting can provide tolerance to soil-related environmental stresses like drought, salinity, low soil temperature and flooding if appropriate tolerant rootstocks are used. Grafting is commercially utilized in some developed countries for eggplants, cucumbers and tomatoes. The cultivated area of grafted Solanaceous and Cucurbitaceous vegetables has increased tremendously in recent years. Problems caused by flooding may be solved by growing flood-tolerant crops or grafting in tolerant plants onto tolerant ones. For instance, grafting improved flooding tolerance of bitter melon (*Momordica charantia* L. cv. New Known You 3) when grafted on luffa (*Luffa cylindrica* Roem cv. Cylinder 2). The reduction of the chlorophyll content in cucumber leaves induced by waterlogging may be enhanced by grafting onto squash rootstocks. When grafting watermelon cv. 'Crimson Tide' to bottle gourd SKP (Landrace) the decrease in chlorophyll content is less compared with non-grafted water melons. AVRDC, Taiwan recommends growing tomato on eggplant 'EG195' or 'EG203' and pepper on chilli accessions 'PP0237-7502', 'PP0242-62' and 'Lee B' lowland tropics, where flooding occurs also during the heat period. Suitable waterlogging tolerant rootstocks for fruit crops are given under Table 12.6.

5. Strategies against Salinity Stress

Various strategies can be adopted to cope with salinity stress. However, two major strategies are adopted in salt affected lands, *i.e.*, technological approach and biotic approach. In the technological approach, one can alter the salty soil through reclamative measures and management practices which enable the plants to grow and produce a reasonable yield. However, these methods are expensive and are not always a practical solution to the problem of soil salinity. Biotic approach has considerable promise in mitigating the problem of soil salinity world over. All biological strategies for crop improvement against salt stress are same as for water stress tolerance like screening and selection, breeding and use of transgenics. The biochemical and physiological traits for salt tolerance are different from plant water stress tolerance. It is largely believed that the adverse effects of salt stress on plant growth are mainly due to its toxic and osmotic effects; therefore major focus is on selective ion accumulation or exclusion, control of sodium uptake and its distribution within the plant, compartmentation of ions at cellular or at whole plant level.

1. *Grafting onto tolerant rootstock:* Grafting is an environment-friendly technique for avoiding or reducing losses in vegetable production caused by salinity, particularly in Solanaceae and Cucurbitaceae families. Grafting onto tolerant rootstock is capable of ameliorating salt-induced damage to the shoot. The explanations for grafting-induced salt tolerance are: (i) higher

accumulation of proline and sugar in the leaves; (ii) higher antioxidant capacity in the leaves; (iii) lower accumulation of Na^+ and/or Cl^- in the leaves. Where grafting 'Moneymaker' onto either 'Radja' or 'Pera' improved tomato fruit yield compared to self grafted 'Moneymaker' when plants were grown at 2009 mM Nacl, whereas there was no effect of either rootstocks or of grafting *per se* on fruit yield in the absence of or at 2005 mM Nacl (Martinez-Rodriguez *et al.,* 2008). In eggplant (*Solanum melongena* L.), grafting cultivar 'Suqiqia' on to it's a wild relative, 'Torvum Vigor' (*Solanum torvum* Swartz) improved the growth performance under saline stress conditions (Wei *et al.,* 2007). Goreta *et al.* (2008) reported that when watermelon 'Fantasy' was grafted onto 'Strongtosa' rootstock (*Cucurbita maxima* Duch. × *Cucurbita moschata* Duch.), the reductions in shoot weight and leaf area due to exposure to salinity were lower than in un-grafted plants. Other experiments demonstrated that grafted 'Crimson Tide' watermelon [*Citrullus lanatus* (Thunb.) Matsum et Nakai] onto *C. maxima* and two *Lagenaria siceraria* rootstocks resulted in higher growth performance than un-grafted plants under saline conditions (8.0 dSm^{-1}, Yetisir and Uygur, 2010). Reduction in shoot dry weight was 41 per cent in un-grafted plants while it varied from 22 per cent to 0.8 per cent in grafted plants under the same saline conditions. The threshold salinity levels and its impact on vegetables, fruits, ornamental shrubs and trees are given under Tables 12.10 and 12.11.

2. *Use of growth regulators:* Certain growth substances are known to alleviate the harmful effect of soluble salts on plant. The physiological changes in plants under salt stress are caused by hormonal imbalance. Plant growth, photosynthetic activity and translocation of assimilates are inhibited under salt stress condition. In bean and tomato, application of Cycocel at 5-12 mg a.i./plant as soil application or as foliar spray at 0.3-0.1 per cent improves considerable tolerance to salinity however; no reduction in uptake of salt could be noticed. Cycocel at 500 ppm improves the salt tolerance and fruit yield in okra upto EC of 6 mmhos/cm; however seed treatment with 100 ppm cycocel for 8 h is found more effective than the foliar spray with 500 ppm. Dipping of tomato and onion seedling in 1 per cent cycocel for 8 h induces salt tolerance in these crops.

Important Biochemical and Physiological Parameters Related to Abiotic stresses

Biochemical Parameters

Abscisic acid (ABA): ABA is a stress hormone and its concentration in leaves can increase upto 50 times under drought condition. ABA biosynthesis in roots and transportation to shoot under stress cause stomatal closing to avoid excess water loss. ABA accumulations take place when LWP reaches -1.0 MP.

Ascorbate (ASC): Ascorbic acid is an important chemical antioxidant, which is responsible for the non-enzymatic scavenging of superoxide radical and H_2O_2, regeneration of α-tocopherol in chloroplast and in enzymatic scavenging of H_2O_2 in association with ascorbate peroxidase.

Ascorbate peroxidase (APX): APX is assumed to play the essential role in scavenging ROS and protecting cells. APX is involved in scavenging of H_2O_2 in water stress and ASH-GSH cycles and utilizes ASH as the electron donor. APX has a higher affinity for H_2O_2 than CAT and PODs, and it plays crucial role in the management of ROS during stress. Enhanced expression of APX in plants has observed during different abiotic stresses.

Carotenoids: Carotenoids are tetraterpenes, located in the plastids of both photosynthetic and non-photosynthetic plant tissues. In chloroplasts, the carotenoids functions as accessory pigments in light harvesting. They have also ability to detoxify various forms of activated oxygen and triplet chlorophyll that are produced as a result of excitation of the photosynthetic complexes by light. There are two types of carotenoids, ß-carotene and zeaxanthinin.

Catalase (CAT): Catalase catalyzes the reduction of H_2O_2 to water and molecular oxygen. CATs are tetrameric heme containing enzymes with the potential to directly dismutate H_2O_2 into H_2O and O_2 and are indispensable for ROS detoxification during stress condition. Compared to APX it is considered a less efficient system of H_2O_2 scavenging due to its higher Km value for H_2O_2 than APX. It is also localized in mitochondria and peroxisomes, and absent in chloroplast, one of the important sites of H_2O_2 generation. CAT has one of the highest turnover rates for all enzymes: one molecule of CAT can convert 6 million molecules of H_2O_2 to H_2O and O_2 per minute.

Glutathione (GSH): Glutathione can function as an antioxidant in many ways. It can react chemically with singlet oxygen, superoxide and hydroxyl radicals, and therefore function directly as a free radical scavenger. GSH may stabilise membrane structure by removing acyl peroxides formed by lipid peroxidation reactions. GSH is the reducing agent that recycles ascorbic acid from its oxidised to its reduced form by the enzyme dehydroascorbate reductase.

Glutathione reductase (GR): GR a flavo-protein oxido-reductase, is a potential enzyme of the ASH-GSH cycle. It plays an essential role in defence system against ROS by sustaining the reduced status of GSH. It is localized predominantly in chloroplasts, but small amount also found in mitochondria and cytosol. GR catalyzes the reduction of glutathione (GSH), a molecule involved in many metabolic regulatory and antioxidative processes in plants, *e.g.*, GR catalyses the NADPH dependent reaction of disulphide bond of oxidized glutathione (GSSG) and is thus, important for maintaining the GSH pool.

Glycinebetaine (GB): Glycinebetaine, an amphoteric quaternary amine, plays an important role as a compatible solute in plants under various abiotic stresses. The accumulation of GB might serve as an intercellular osmoticum, and are correlated with elevation of osmotic pressure.

Heat shock protein (HSPs): In response to sudden rise in 5-10°C in temperature, plant produces a unique set of proteins referred to as heat shock proteins. Most HSPs function to help cells withstand heat stress by acting as molecular chaperones. HSPs serve to attain a proper folding of misfolded aggregated proteins. This facilitates proper cell functioning at elevated and stressful temperature.

Hydrogen peroxide (H_2O_2): It acts as a signal molecule involved in acclamatory signal triggering tolerance to various biotic and abiotic stresses. At high concentrations, it leads to program cell death. H_2O_2 has also been shown to act as a key regulator in a broad range of physiological processes, such as senescence, photorespiration, photosynthesis, stomatal movement and growth and development.

LEA proteins: A large group of genes that are regulated by osmotic stress and known to play role in cellular membrane stress and known 'Late embryogenesis protein' or LEA protein. They accumulate in vegetative tissue during osmotic stress. The protective roles of protein encoded by these genes are associated with ability to retain water and prevent crystallization of important cellular proteins and other molecules during desiccation.

Malondialdehyde (MDA): The peroxidation of lipids is considered as the most damaging process occuring every living organism. Sometimes membrane damage is taken as a single parameter to determine the level of lipid destruction under various abiotic stresses. Now, it is known that during lipid peroxidation products such as ketones, MDA, etc are formed from polyunsaturated precursor hydrocarbon fragments. MDA is synthesized due to degradation of polyunsaturated lipids by ROS. The production of this aldehyde is used as a biomarker to measure the level of oxidative stress. Increased MDA accumulation is correlated with reduction of RWC and photosynthetic pigment content under prolonged drought.

Peroxidases (PODs): PODs are a large family of enzymes that detoxify H_2O_2, organic hydroperoxides or lipid peroxides to generate alcohols. Peroxidases are involved in the defence against drought stresses by means of their role in the detoxification of ROS in the apoplast of lignifying tissues.

Proline: Proline occurs widely in higher plants and normally accumulates in large quantities in response to environmental stresses. It is considered as a compatible osmolyte in stress adaptation, recovery and signalling.

Reactive oxygen species (ROS): ROS are partially reduced forms of atmospheric oxygen. They typically result from the excitation of O_2 to form singlet oxygen (O_2^1) or from the transfer of 1, 2 or 3 electrons to O_2 form superoxide radical (O_2^-), hydrogen peroxide H_2O_2 or a hydroxyl radical (OH^-), respectively. The cells are normally protected against ROS by the operation of the antioxidant defence system comprising enzymatic (SOD, CAT, GR, APX, POD) and non-enzymic (ascorbate, α-tocopherol, carotenoids, glutathione) components. The activities of enzymes of the antioxidant system in plants under stress are usually regarded as an indicator of the tolerance of genotypes against stress conditions. ROS causes the peroxidation of membrane lipids, denaturation of proteins and damage to nucleic acids.

Superoxide dismutase (SOD): The SODs catalyze the dismutation of superoxide into oxygen and H_2O_2. In higher plants, SOD isozymes are present in different cell compartments. Mn-SOD is present in mitochondria and peroxisomes, Fe-SOD mainly in chloroplasts, and Cu/Zn-SOD in cytosol, chloroplasts, peroxisomes and the apoplast. An increased SOD activity by drought stress is considered to antagonize harmful actions of superoxide radicals and this indicates that higher activities of SOD are important for drought tolerance.

Physiological Parameters

Canopy temperature depression (CTD): The canopy temperature depression (CTD) relative to the air temperature can be measured using a hand held infra-red thermometer (Infra-red gun). In C_3 and C_4 plants, leaf temperature can readily raise 4 to 5°C above ambient temperature in intense sunlight near midday, when soil water deficit causes partial stomatal closure. Under such conditions the plant canopy temperature increases considerably, which is an indicator of water deficit or high temperature stress. High stomatal conductance also leads to high canopy cooling through transpiration (CTD).

Cell Membrane Stability (CMS): CMS is measured from electrolyte leakage with conductivity electrode. The stability of various cellular membranes is important especially during high temperature stress. Excessive fluidity of membrane lipids at high temperatures is correlated with loss of physiological functions. High temperatures modify membrane composition and structure and can cause leakage of ions.

Chlorophyll content index (CCI): The function of chlorophyll is to absorb light and transfer that light to a specific chlorophyll pair in the reaction centre of the photosystem. Chlorophyll Content Meter (CCM) is used to measure CCI content in leaves.

Chlorophyll Fluorescence: Chlorophyll fluorescence is a rapid, non-destructive and inexpensive technique that has been used successfully in the evaluation of plant photosynthetic activity. The principle underlying chlorophyll fluorescence analysis is light energy absorbed by chlorophyll molecules in a leaf can undergo one of three fates: it can be used to drive photosynthesis (photochemistry), excess energy can be dissipated as heat or it can be re-emitted as light-chlorophyll fluorescence. Chlorophyll fluorescence highlights the photochemical efficiency of photosystem II and the effectiveness of utilization of chlorophyll *a* excitation energy in the photosynthesis process. Chlorophyll fluorescence is measured by Chlorophyll Fluorescence Meters. Chlorophyll fluorescence measured in terms of photochemical efficiency of PS II (Fv/Fm) is a good indicator of abiotic stresses, particularly of high temperatures/light intensity stresses.

Leaf morphology: Under intense solar radiation and high temperature plants avoid excessive heating of their leaves by decreasing absorption of solar radiation. Reflective leaf hairs and leaf waxes, leaf rolling, reduce leaf area with highly dissected leaves and vertical leaf orientation area few of morphological adaptations of leaves against heat and drought stresses.

Leaf water potential (*LWP*): Cell growth, photosynthesis and crop productivity are all strongly influenced by plant water potentials. It is a good overall indicator of plant health especially under drought. The LWP of well-watered plant ranges from -0.2 to -1.0 MPa, plant under mild stress from -1.0 to -2.0 MPa and plant in severe stress or of arid/desert climates ranges LWP from -2.0 to -5.0 MPa.

Light intensity: High or low light intensity is also a limiting factor in plant. For photosynthetic use the light intensity is expressed in terms of photo active radiation (PAR). PAR of 400-1000 $\mu mol/m^2/s$ is considered as an optimum range for photosynthesis in majority of the plants. Light intensity is measured by Line quantum meter/Photometer/Radiometer.

Photosynthesis (P_n): Photosynthesis is measured by 'infra-red gas analyzer' (IRGA). In all kind of abiotic stresses, P_n decreases due to stomatal closure.

Relative leaf water content (RWC): It estimates the current water content of the sampled leaf tissue relative to the maximal water content it can hold at full turgidity. Relatively higher RWC is an indication of good plant health condition. Under drought stress condition, RWC in leaves decreases remarkably.

Stomatal conductance (g_s): It measures the rate of passage of CO_2 entering, or water vapor exiting through the stomata of a leaf. By opening and closing the stomata, plants can regulate the amount of water lost, by sacrificing CO_2 uptake, when the environmental conditions are unfavourable. Stomatal conductance is the good indicator for most kind of stresses, particularly drought stress. It is measured by steady state porometer.

References

1. Ahn S J, Im Y J, Chung, G C Cho, B H and Suh S R. (1999). Physiological responses of grafted-cucumber leaves and rootstock roots affected by low root temperature. *Scientia Horticulture.* **81:** 397–408.

2. Armstrong W and Drew M C. (2002). Root growth and metabolism under oxygen deficiency. *In* Plant Roots: the Hidden Half, Eds Y Waisel, A Eshel and U Kafkafi. pp. 729–761. Marcel Dekker, New York.

3. Ashraf M, Athar H R, Harris P J C and Kwon T R. (2008)Some prospective strategies for improving crop salt tolerance. *Advances in Agronomy* **97:** 45–110.

4. Bahadur Anant.(2014). Water management in Vegetables. In; Bahadur Anant and Singh K.P. (eds.) Book on Olericulture Vol-I: Fundamentals of Vegetable Production. Pp. 305-321. Kalyani Publisher, New Delhi.

5. Bahadur Anant, Chatterjee A, Kumar R, Singh M and Naik P S (2011). Physiological and biochemical basis of drought tolerance in vegetables. *Vegetable Science* **38**: 1-8.

6. Bahadur Anant, Singh A K, and Singh K P. (2013a.) Effect of planting systems and mulching on soil hydrothermal regime, plant physiology, yield and water use efficiency in tomato. *Indian Journal of Horticulture Sciences* **70:** 48-53.

7.   Bahadur Anant, Singh A K and Chaurasia S N S, (2013b). Physiological and yield response of okra (*A. esculentus* Moench) to drought stress and organic mulching. *Journal of Applied Horticulture* **15:** 187-90.

8.   Bahadur Anant, Singh K P, Rai A, Verma A and Rai M. (2009). Physiological and yield response of okra (*Abelmoschus esculentus* Moench) to irrigation scheduling and organic mulching. *Indian Journal of Agriculture Science* **79:** 813-15.

9.   Barrett-Lennard E G, van Ratingen P and Mathie M.(1999). The developing pattern of damage in wheat (*Triticum aestivum* L.) due to the combined stresses of salinity and hypoxia: experiments under controlled conditions suggest a methodology for plant selection, *Australian Journal of Agricultural Research* **50**: 129-136.

10. Bennett J. (2003). Increased water productivity through plant breeding. *CAB International*: 103-126.

11. Boynton D. (1940). Soil atmosphere and production of new rootlets by apple tree root system. *Proc. American Society of Horticulture Science* **37:** 19-26.

12. Camejo, D., Martí M. d.C. Nicolás, E., Alarcón, J.J., Jiménez, A. and Sevilla, F. (2007). Response of superoxide dismutase isoenzymes in tomato plants (*Lycopersicon esculentum*) during thermo-acclimation of the photosynthetic apparatus. *Plant Physiology* **131**: 367–377.

13. Camejo D, Nicolás E, Torres W and Alarcón, J J. (2010). Differential heat-induced changes in the CO_2 assimilation rate and electron transport in tomato (*Lycopersicon esculentum* Mill). *Journal of Horticulture Science and Biotechnology* **85:** 137–143.

14. Caplan, L A. (1988). Effects of cold weather on horticultural plants in Indiana. Purdue University Cooperative Extension Publication, No. HO-203.

15. Chapin F S. III. (1991). Effects of multiple environmental stresses on nutrient availability and use. In: H. A. Mooney, W. E. Winner, and E. J. Peli, eds. Response of Plants to Multiple Stresses. San Diego: Academic Press: 67.

16. Christiane F, Smethurst, Trevor Garnett and Sergey Shabala. (2005). Nutritional and chlorophyll fluorescence responses of lucerne (*Medicago sativa*) to waterlogging and subsequent recovery. *Plant and Soil* **270**: 31-45.

17. Cifre J, Bota J, Escalona J M, Medrano H and Flexas J. (2005). Physiological tools for irrigation scheduling in grapevine (*Vitis vinifera* L.). An open gate to improve water-use efciency. *Agriculture, Ecosystems and Environment* **106**: 159-170.

18. De Ronde J A D, Cress W A, Kruger G H J, Strasser R J and Staden J V.(2004). Photosynthetic response of transgenic soybean plants containing an Arabidopsis P5CR gene, during heat and drought stress. *Journal of Plant Physiology* **61**: 1211–1244.

19. Erickson A N and Markhart A H. (2002). Flower developmental stage and organ sensitivity of bell pepper (*Capsicum annuum* L.) to elevated temperature. *Plant, Cell and Environment* **25:** 123-130.

20. FAO, (2009). FAO land and plant nutrition management service, http: // www.fao.org/ag/agl/agll/spush/.

21. Foolad M R. (2005). Breeding for abiotic stress tolerances in tomato, p. 613-684. In: Ashraf M. and P. J. C. Harris Eds. Abiotic Stresses: Plant Resistance Through Breeding and Molecular Approaches. The Haworth Press Inc., New York, USA,.

22. Gao Q H, Xu K, Wang X F and Wu Y. (2008). Effect of grafting on cold tolerance in eggplant seedlings. *Acta Horticulture* **771**: 167–174.

23. Goreta S, Bucevic-Popovic V, Selak G V, Pavela-Vrancic M and Perica S. (2008). Vegetative growth, superoxide dismutase activity and ion concentration of salt stressed watermelon as influenced by rootstock. *Journal of Agriculture Science* **146**: 695–704.

24. Hodgson A S and Chan K Y. (1982). The effect of short-term waterlogging during furrow irrigation of cotton in a cracking grey clay. *Australian Journal of Agriculture Research* **33**: 109-116.

25. Howarth C J. (2005). Genetic improvements of tolerance to high temperature. In: Ashraf, M., Harris, P.J.C. (Eds.), Abiotic Stresses: Plant Resistance Through Breeding and Molecular Approaches. Howarth Press Inc., New York.

26. ICAR. (2010). Degraded and Wastelands of India Status and Spatial Distribution. Indian Council of Agricultural Research, KAB-I, Pusa, New Delhi.

27. Kumar Rajesh, Solankey S S and Singh M. (2012). Breeding for drought tolerance in vegetables. *Vegetable Science* **39**: 1-15.

28. Levitt J. (1980). Responses of Plants to Environmental Stresses, Vol. 1 (2nd ed). New York NY: Academic Press. 497p.

29. Maas E V. (1990). Crop salt tolerance. In: K.K. Tanji, ed. Agricultural Salinity Assessment and Management. American Society of Civil Engineers, ASCE Manuals and Reports on Engineering Practice No. 71, ASCE, New York, 262-304.

30. Martinez-Rodriguez, M M, Estan M T, Moyano E, Garcia-Abellan J O, Flores F B, Campos J F, Al-Azzawi, M J Flowers, T J and Bolarín, M C. (2008). The effectiveness of grafting to improve salt tolerance in tomato when an 'excluder' genotype is used as scion. *Environmental and Experimental Botany* **63**: 392-401.

31. Munns R and Tester M. (2008). Mechanisms of salinity tolerance. *Annual Review of Plant Biology* **59:** 651-681.

32. Parida A K and Das A B. (2005). Salt tolerance and salinity effects on plants: A review. *Ecotoxicology and Environmental Safety* **60**: 324-349.

33. Park E J, Jekni Z, Sakamoto A, DeNoma J, Yuwansiri R, Murata N and Chen T H H. (2004). Genetic engineering of glycinebetaine synthesis in tomato protects seeds plants, and flowers from chilling damage. *Plant J*. **40**: 474–487.

34. Pessarakli M. (1999). Handbook of plant and crop stress (ed.). Marcel Dekker, Inc. NY. PP: 177-183.

35. Proebsting E L Jr and Mills H H. (1978). Low temperature resistance of developing flower buds of six deciduous fruit species. *Journal of American Society and Horticultural Science* **103:** 192-198.

36. Restrepo-Díaz H, Melgar J C, and Lombardini L. (2010). Ecophysiology of horticultural crops: an overview. *Agronomía Colombiana* **28:** 71-79.

37. Rogers M E, Craig A D, Munns R, Colmer T D, Nichols P G H, Malcolm C V, Barrett-Lennard E G, Brown A J, Semple W S, Evans P M, Cowley K, Hughes S J, Snowball R, Bennett S J, Sweeney G C, Dear B S and Ewing M A. (2005). The potential for developing fodder plants for the salt-affected areas of southern and eastern Australia: an overview. *Australian Journal of Experimental Research* **45**: 301–329.

38. Rouphael Y, Cardarelli M, Colla G and Rea E. (2008). Yield, mineral composition, water relations, and water use efficiency of grafted mini-watermelon plants under deficit irrigation. *Hortscience* **43:** 730–736.

39. Rowe R N and Beardsell D V (1973). Waterlogging of fruit trees. *Horticultural Abstracts* 43: 534-544.

40. Saha S R, Hossain M M, Rahman M M, Kuo C G and Abdullah S. (2010). Effect of high temperature stress on the performance of twelve sweet pepper genotypes. *Bangladesh Journal of Agriculture Research* **35:** 525-534.

41. Sairam R K, Deshmukh P S and Saxena D C. (1998). Role of antioxidant systems in wheat genotypes tolerance to water stress. *Plant Biology* **41**: 387-394.

42. Sato S, Kamiyama M, Iwata T, Makita N, Furukawa H, Ikeda H. (2006). Moderate increase of mean daily temperature adversely affects fruit set of *Lycopersicon esculentum* by disrupting specific physiological processes in male reproductive development. *Annals of Botany* **97**: 731–738.

43. Schoffl F, Prandl R and Reindl A. (1999). Molecular responses to heat stress. In: Shinozaki, K., Yamaguchi-Shinozaki, K. (Eds.), Molecular Responses to Cold, Drought, Heat and Salt Stress in Higher Plants. R.G. Landes Co., *Austin Texas,* pp. 81–98.

44. Shannon M C and Grieve C M. (1999). Tolerance of vegetable crops to salinity. *Scientia Hortic.* **78**: 5-38.

45. Shao H B, Chu L Y, Abdul Jaleel and Zha C X. (2008). Water-deficit stress-induced anatomical changes in higher plants. *Comptes Rendus Biologies* **331:** 215-225.

46. Wahid S Gelani, Ashraf M and Foolad M R. (2007). Heat tolerance in plants: An overview. *Environmental and Experimental Botany* **61:** 199–223.

47. Wang C Y and Wallace H A.(2003). Chilling and freezing injury. *In:* K.C Gross, C.Y. Wang and M. Saltveit (Eds). *The Commercial Storage of Fruits, Vegetables, and Florist and Nursery Stocks.* USDA Handbook Number, No. 66. See: http: // www.ba.ars.usda.gov /hb66/index.html.

48. Wang W, Vinocur B and Altman A. (2007). Plant responses to drought, salinity and extreme temperatures towards genetic engineering for stress tolerance. *Planta* **218:** 1-14.

49. Wei G P, Zhu Y L, Liu Z L, Yang L F and Zhang G W. (2007). Growth and ionic distribution of grafted eggplant seedlings with NaCl stress. *Acta Botanica Boreal-Occident. Sin.* **27:** 1172–1178.

50. Yetisir H and Uygur V. (2010.) Responses of grafted watermelon onto different gourd species to salinity stress. *Journal of Plant Nutrition.* **33:** 315–327.

2015, Horticulture for Nutrition Security   *Pages 261–299*
Editor: **Prof. K.V. Peter**
Published by: **DAYA PUBLISHING HOUSE, NEW DELHI**

Chapter 13

Shelflife of Fruits and Vegetables- Interventions to Prolong

R.K. Pal and A. Maity

Fruits and vegetables have assumed commercial importance as nutritionally essential food commodity. Because they are not only the major sources of vitamins, sugars, organic acids and minerals but also many phyto-chemicals including dietary fiber and antioxidants which are of immense beneficial to human health. Additionally, fruits and vegetables also contain a variety in color, shape, taste, aroma and texture to refine sensory pleasure in human's diet. The demand for fresh produces at the consumer level are increasing at rapid pace because of growing awareness among the people about the superiority of fresh, natural foods than processed products as demonstrated by health agencies and public media as well as several medical researches. But all fresh fruits and vegetables are vulnerable to dehydration, environmental stresses, mechanical injury and pathological breakdown owing to high moisture content, continuing active metabolism and richness in nutrients that make these commodities highly perishable (Kader, 2002, 2005; Kays and Paull, 2004 and Wills *et al.*, 2007). Often they may become unacceptable if not handleed properly after harvesting. Moreover, fresh horticultural products are important items of international commerce after globalization of trade and free-trade agreements. Longer shipment and distribution period may eventually increase heavy losses. Post-harvest losses can occur at any point in the production and marketing chain. It is estimated that the magnitude of these losses due to inadequate post-harvest handling, transportation and storage of fresh fruits and vegetables is relatively higher, 20-50 per cent, in developing countries when compared to 5-25 per cent in developed countries (Kader, 2005). There lies the importance of proper care and techniques for handling fresh produce after harvest. Post-harvest, the connecting link between the growers and the consumers is concerned with understanding of physiology of harvested plant materials and development of effective and feasible handling technologies that will delay the rate of senescence.

Causes of Post-harvest Losses

The horticultural products are diverse in morphological structure, composition, developmental stages and general physiology. So the actual causes of post-harvest losses in fresh vegetables and fruits are numerous and commodity specific. However, the main events responsible for quality deterioration of produce are continuous metabolism and growth, water loss, physiological disorders, mechanical damages and pathological breakdown (Wills *et al.*, 2007; Kader, 2002).

Continuous Metabolism and Growth

At one time it was believed that the chemical transformations which occur in post-harvest fruits and vegetables were the consequence of disorganization and decompartmentation of the cellular milieu. Further it was assumed by some that chemical transformations were strictly catabolic - a falling apart of the tissue constituents. We now recognize that both anabolic and catabolic reactions play a role in post-harvest biochemical transformations (Table 13.1) and that these changes are not the consequences of disorganization but are the consequence of a "biological clock," *i.e.*, are under some form of genetic control. It is well realized that the quality of harvested commodities can not be improved further but it can be retained till their consumption if the rate of metabolic activities are reduced by adopting appropriate post-harvest handling operations. It has been recognized for many years that there is an inverse relationship between respiration rate and storability of post-harvest crops (Figure 13.1). Respiration, involves enzymatic oxidation of organic substrates with energy production resulting in consumption of O_2 and production of CO_2 and water. Hence, crops which exhibit a relatively high respiration rate tend to deteriorate rapidly; whereas, crops which respire slowly can be stored for extended periods of time.

Table 13.1: Metabolic Changes in Post-harvest Fruits and Vegetables

Degradative	Synthetic
Destruction of chloroplast	Synthesis of carotenoids and anthocyanins
Breakdown of chlorophyll	Synthesis of flavor volatiles
Starch hydrolysis	Synthesis of starch
Organic acid catabolism	Synthesis of lignin
Oxidation of substrate	Preservation of selective membranes
Inactivation of phenolic compounds	Interconversion of sugar
Hydrolysis of pectin	Protein synthesis
Breakdown of biological membranes	Gene transcription
Cell wall softening	Formation of ethylene biosynthesis pathways

Source: Biale and Young (1981).

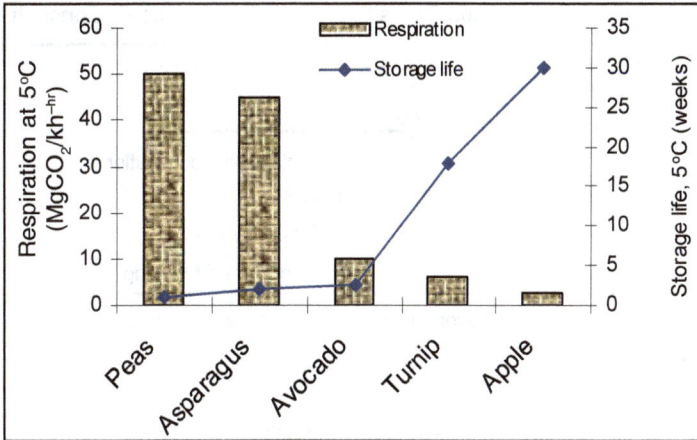

Figure 13.1: Relationship between Respiratory Rate and Storability for Selected Fruits and Vegetables.
Source: Haard, 1983.

Respiration rate of a produce is dependent on wide range of variables, including commodity and environmental factors (Kader, 2002; Kader and Saltveit, 2003; Kays and Paull, 2004; Wills *et al.*, 2007).

Among the environmental factors affecting respiration rate of fresh fruits and vegetables, temperature plays pivotal role in modulating this physiological parameter (Kader, 2002; Kader and Saltveit, 2003; Kays and Paull, 2004; Wills *et al.*, 2007). One means of reducing respiration is lowering of the environmental temperature. The benefits of lowering storage temperature vary from commodity to commodity for at least two reasons. First, the temperature coefficient of respiration is normally about 2.0 but can vary considerably and may range from 1 to 7 for different crops; hence, reducing the temperature in a storage facility from 30°C to 20°C may have a more profound influence on both respiration and storability for certain crops than for others. Second, some crops are sensitive to chilling temperatures and the resulting anomalous metabolism can lead to a wide array of physiological disorders (Haard, 1983). Chilling injury is a result of an imbalance in intermediates arising from respiratory metabolism and other metabolic events.

Similarly, respiration rate and storability respond to modification of the storage atmosphere such as reducing oxygen and increasing carbon dioxide partial pressure. But crops exhibit individuality in their ability to cope with the stress of these changes and for some commodities physiological injury may happen rather than getting extended storage life resulting from such treatments (Table 13.2).

Further, the respiration rate and storability of crops such as apple fruit is related to the amount of endogenous calcium in the tissue. Studies with various crops have shown that manipulation of tissue calcium by either agronomic practice or post-harvest sprays or dips can retard an assortment of physiological injuries and extend storability. The mechanism of calcium action is not clear; however, the concomitant impact on respiration is intriguing.

Table 13.2: Susceptibility of Various Crops to Extreme in CO_2 and O_2 Concentration

Commodity	Minimum O_2 Tolerance (Per cent)	Maximum CO_2 Tolerance (Per cent)	Injury Symptoms
Lettuce	0.5	2-4	Brownish-red discoloration of midrib
Broccoli	0.25	20	Tolerant
Cauliflower	–	5	Excessive softening, discolors on cooking
Potatoes	5	10	Prevents wound healing

a: Values will vary with variety, cultivation practices and storage temperatures.

Source: Haard (1983).

It has long been recognized that ethylene, a gaseous phytohormone can stimulate the respiration of post-harvest fruits and vegetables. In higher plants ethylene is produced from methionine via three enzymatic reactions: (1) methionine is converted to S-adenosyl-L-methionine (S-AdoMet) by S-AdoMet synthetase; (2) the conversion of S-AdoMet to 1- aminocyclopropane-1-carboxylic acid (ACC), the immediate precursor of ethylene, is the rate limiting step catalyzed by ACC synthase (ACS); (3) ACC oxidase (ACO) degrades ACC to release ethylene (Lin *et al.,* 2009; Saltveit, 1999; Wills *et al.,* 2007). The dependency of respiration on ethylene concentration and the type of effect vary with the commodity. "Climacteric" fruits, which exhibit an autonomous burst in respiration coincidence with ripening respond to exogenous ethylene concentrations above certain threshold levels by showing an earlier respiratory climacteric and ripening; that is, ethylene has an all-or-none effect on such crops. Non-climacteric fruits exhibit an ethylene-concentration-dependent respiratory rise, that is, the magnitude of the respiratory burst is higher as ethylene concentration is increased. A recent classification of climacteric and non-climacteric fruits is shown in Table 13.3.

Table 13.3: Classification of Fruits as Climacteric or Non-climacteric

Climacteric		Non-climacteric	
Apple	Muskmelon	Blue berry	Lychee
Apricot	Pawpaw	Cacao	Mountain apple
Avocado	Papaya	Cashew, apple	Olive
Banana	Passion fruit	Cherry, sweet	Orange
Biriba	Peach	Cherry, sour	Pine apple
Breadfruit	Pear	Cucumber	Rose apple
Cherimoya	Persimon	Grape	Strawberry
Chinese goose berry	Plum	Grapefruit	Surinam cherry
Feijoa	Sapota	Javaplum	Tamarillo
Fig	Soursop	Lemon	nor-Tomato
Guava	Tomato/watermelon		rin-Tomato
Mammee apple			

Source: Biale and Young (1981).

The physiological behavior of the fruit has importance in the post-harvest biology and technology for extending shelf-life of these commodities. For example, to obtain greater degree of storage and marketing flexibility, climacteric fruit is usually harvested at mature (unripe) green stage since it can ripen normally after harvest. Conversely, non-climacteric fruit must be harvested only when it is fully ripe. Several lines of evidence have demonstrated that ethylene is the crucial phytohormone regulating time of ripening in climacteric fruit (Barry and Giovannoni, 2007; Kader, 2002; Saltveit, 1999). Although non-climacteric fruits typically produce little ethylene after harvesting, many have still been shown to be affected by exogenous ethylene during storage. Stress-induced ethylene may have many physiological effects on commodities. Hence, ethylene control is a target for shelf-life manipulation for both climacteric and non-climacteric fruits (Kader, 2002; Saltveit, 1999; Wills *et al.*, 2007).

Water Loss

Fruits and vegetables typically contain 80-90 per cent water on fresh weight basis (Kader, 2002; Wills *et al.*, 2007). Severed plant organs are highly susceptible to water losses as the water replenishment system is eliminated at harvest. As little as 5 per cent water losses will have adverse effect on appearance, saleble weight, textural quality of many perishable commodities. Therefore, desiccation resulting from water losses is one of the important causes of deterioration of produce during post-harvest period. The major way of water loss in fresh horticultural commodities is through transpiration via stomata (Kader, 2002; Kays and Paull, 2004; Wills *et al.*, 2007). The other paths of water losses are through stem scare, lenticels and cracks resulting from mechanical injury. The transpiration rate is influenced by produce characteristics namely morphological and anatomical characteristics, surface area/volume ratio, surface injury and maturity stage. For example, produce like leafy vegetable with large surface to volume ratio will lose greater percentage of their water far quickly than large spherical fruits. Loose leafy lettuce loses water more rapidly then head lettuce. Besides commodity factors, the rate of post-harvest water losses is primarily dependent on the external vapor pressure deficit (VPD), which in-turn is governed by other environmental factors.

Physiological Disorders

Physiological disorders of fruit and vegetables are those problems resulting from the influence of environmental, cultural factors and mineral imbalance on plant development. A number of physiological disorders arise in controlled environment facilities because environmental and cultural conditions in these facilities often are significantly different from those encounters by plants in the natural environment. Physiological disorder resulting from low temperature effect, respiration related problem and nutritional imbalance are particularly problematic.

The improper temperature may lead to disturbance in the normal metabolism of the harvested commodities. Many vegetables and fruits store the best at temperatures just above freezing, while others are injured by low temperatures and will store the best at 45 to 55°F. Both time and temperature are involved in chilling injury. Damage may occur in a short time if temperatures are considerably below the danger threshold, but some crops can withstand temperatures a few degrees into the danger zone for a longer time. The effects of chilling injury are cumulative in some crops. Low temperatures in transit, or even in the field shortly before harvest, add to the total effects of chilling that might occur in storage (Anon, 1992). Crops such as basil, cucumbers, eggplants, pumpkins, summer squash, okra and sweet potatoes are highly sensitive to chilling injury. Moderately sensitive crops are snap beans, muskmelons, peppers, winter squash, tomatoes and watermelons (Howell, 1993). These crops may look sound when removed from low temperature storage, but after a few days of warmer temperatures, chilling symptoms become evident: pitting or other skin blemishes, internal discoloration, or failure to ripen. Tomatoes, squash and peppers that have been over-chilled may be particularly susceptible to decay such as *Alternaria* rot. Freezing injury, on the other hand results from keeping the commodities below their freezing temperature. The ice crystal so formed cause damage to the tissues which collapse immediately resulting in loss of commodity.

Nutritional disorders which have their origin at pre-harvest nutritional imbalance sometimes appear only after harvest of the produce. In majority of the cases, calcium has been found to associate with post-harvest related deficiency disorder than any other minerals. For example, bitter pit of apple and blossom end rot of tomato are well known calcium deficiency disorders in horticultural crops (Kader, 2002; Wills *et al.*, 2007). Respiratory disorder arises from exposure to very low O_2 (<1 per cent) and/or high CO_2 concentration during storage and packaging.

Mechanical Damage

The high moisture content and soft texture of fruit, vegetables and root crops make them susceptible to mechanical injury, which can occur at any stage from production to retail marketing because of:

- ☆ Poor harvesting practices;
- ☆ Unsuitable field or marketing containers and crates, which may have splintered wood, sharp edges, poor nailing or stapling;
- ☆ Overpacking or underpacking of field or marketing containers;
- ☆ Careless handling, such as dropping or throwing or walking on produce and packed containers during the process of grading, transport or marketing.

It not only directly affects appearance attributes (skin, fresh lesion and browning) but also creates sites for pathogenic infection and water loss. Furthermore, physical injury stimulates ethylene production and respiration in plant tissues which can lead to accelerated senescence (Kader, 2002; Kays and Paull, 2004; Wills *et al.*, 2007).

Pathological Decay

Fruits and vegetables are characterized by the content of wide range of organic substrates and high water activity and thus they are good substrate for microbial spoilage. A significant part of losses of fresh produce during post-harvest is attributed to the diseases caused by fungi and bacteria. The acidic nature of fruit tissues results in their spoilage predominantly by fungi, while vegetables having pH above 4.5 are commonly attacked by bacteria and fungi. The most common pathogen causing decay in fruits and vegetables are species of fungi, *Alternaria, Botrytis, Botryosphaeria, Colletotrichum, Diplodia, Monilinia, Penicillium, Phomopsis, Rhizophus, Sclerotinia*, and of the bacteria, *Erwinia* and *Pseudomonas* (Wills *et al.,* 2007). Generally, fruits and vegetables possess certain degree of resistance to potential pathogens during most of their post-harvest life. But, senescence, ripening or stresses like mechanical damage, chilling injury may render them susceptible to infection by pathogens (Kader, 2002; Kays and Paull, 2004; Wills *et al.,* 2007). Majority of pathogens rely on physical injury or physiological breakdown of the commodity to invade the host tissue, but a few of them like *Colletotrichum* are capable to actively penetrate the skin of healthy produce (Kader, 2002; Wills *et al.,* 2007). This pathogenic infection can occur either before or after harvest. When the product is infected prior to harvest with no obvious symptoms developing until the pathogens are reactivated by onset of congenial conditions such as fruit ripening or favourable climatic condition, it is called 'latent infection' or 'quiescent infection'.This latent infection with pathogen like *Colletotrichum gloeosporioides* causing anthracnose disease of tropical fruits often results rapid and severe post-harvest decay as the infected produce can not be sorted out easily before storage (Kays and Paull, 2004; Wills *et al.,* 2007). Further, the degree of microbial infection at the time of harvest *i.e.* the primary inoculum load plays a significant role in determining the storage life. There exists an inverse relationship between the primary inoculum load of commodities at harvest and the storage life.

Besides, several other physiological changes contribute to a great extent in limiting storage life of fruits and vegetables. Sprouting in onion, ginger, garlic and potatoes is one of the most serious causes of deterioration. Rooting is initiated by a condition of elevated humidity, which may result in rapid decay, shriveling and exhaustion of food reserves in roots and tubers. *Solanine*, a toxic alkaloid is found in green tissues of potatoes which is formed owing to exposure of potatoes to light during storage. Green beans and sweet corn may get tough during prolonged storage due to development of spongy tissue. Similarly, a large number of biochemical changes are associated with the changes in colour, texture, flavour and other sensory quality during storage.

Potential Impact of Climate Change on Post-harvest Quality of Fruits and Vegetables

Our earth has experienced changes in climate many times during its existence, ranging from the ice ages to periods of warmth. During the last several decades increases in average air temperatures have been reported and associated effects on climate have been debated worldwide in a variety of forums. According to studies carried out by the Intergovernmental Panel on Climate Change (IPCC), average air temperatures is expected to increase between 1.4 and 5.8°C by the end of this century, based upon modeling techniques that incorporated data from ocean and atmospheric behavior (IPCC. Climate change, 2001). Exposure to elevated temperatures can cause morphological, anatomical, physiological, and ultimately, biochemical changes in plant tissues and, as a consequence, can have effect on growth and development of different plant organs. These events can cause drastic reductions in commercial yield.

However, by understanding plant tissues physiological responses to high temperatures, mechanisms of heat tolerances and possible strategies to improve yield, it is possible to predict reactions that will take place in the different steps of fruit and vegetable crops production, harvest and post-harvest (Kays, 1997). Besides increase in temperature and its associated effects, climate changes are also a consequence of alterations in the composition of gaseous constituents in the atmosphere. Carbon dioxide (CO_2), also known as the most important greenhouse gas, and ozone (O_3) concentrations in the atmosphere are increasing during the last decade and are affecting many aspects of fruit and vegetable crops production around the globe (Felzer *et al.,* 2007; Lloyd and Farquhar, 2008). The current atmospheric CO_2 concentration is higher than at any time in the past 420,000 years (Petit *et al.,* 1999). Further increases due to anthropogenic activities have been predicted. Carbon dioxide concentrations are expected to be 100 per cent higher in 2100 than the one observed during the pre-industrial era (IPCC, 2007). Even low-levels of ozone in the vicinities of big cities can cause visible injuries to plant tissues as well as physiological alterations (Felzer *et al.,* 2007). The above mentioned climate changes can potentially cause post-harvest quality alterations in fruit and vegetable crops.

Effect of Temperature

Growth and development of fruits and vegetables are influenced by different environmental factors (Bindi *et al.,* 2001). During their development, high temperatures can affect photosynthesis, respiration, aqueous relations and membrane stability as well as levels of plant hormones, primary and secondary metabolites. Most of the physiological processes take place normally in temperatures ranging from 0°C to 40°C. However, cardinal temperatures for the development of fruit and vegetable crops are much narrower and, depending on the species and ecological origin, it can be pushed towards 0°C for temperate species from cold regions, such as carrots and lettuce. On the other hand, they can reach 40°C in species from tropical regions, such as many cucurbits and cactus species (Went, 1953). A general temperature effect in plants is associated with the ratio between photosynthesis and respiration. For a high yield, not only photosynthesis should be high but also the ratio photosynthesis/ respiration should be much higher than one. At temperatures around 15°C, the above mentioned ratio is usually higher than ten, explaining why many plants tend to grow better in temperate regions than in tropical ones (Went, 1953). Photosynthetic activity is proportional to temperature variations. High temperatures can increase the rate of biochemical reactions catalyzed by different enzymes. However, above a certain temperature threshold, many enzymes lose their function, thereby potentially changing plant tissue tolerance to heat stresses (Bieto and Talon, 1996). Temperature is of paramount importance in the establishment of a harvest index. The higher the temperature during the growing season, the sooner the crop will mature. Hall *et al.* (1996) and Wurr *et al.* (1996) reported that lettuce, celery, cauliflower and kiwi grown under higher temperatures matured early than that the same crops grown under lower temperatures. High temperatures on fruit surface owing to prolonged exposure to sunlight hasten ripening and other associated events. One of the classical examples is that of grapes, where berries exposed to direct sunlight ripened faster than those ripened in shaded areas within the canopy (Kliewer and Lider, 1968). Conversely, ripening of 'Hass' avocados is also affected by exposure to high temperatures during growth and development. For fruits exposed to direct sunlight, pulp temperatures reached 35°C and required 1.5 days longer to ripen than those that grew in the shade (Woolf *et al.,* 1999). Further, 'Fuerte' and 'Hass' fruits exposed to direct sunlight were firmer than fruits of the shaded areas. Cell wall enzyme activity (cellulose and polygalacturonase) was negatively correlated with fruit firmness, indicating that sun exposure, *i.e.,* higher temperatures during growth and development, can delay ripening. However, this delay did not occur via a direct effect on the enzymes associated with cell wall degradation (Chan and Linse, 1989). It is also reported that high temperature stresses inhibit ethylene production and cell wall softening in papaya and tomato fruits (Chan, *et al.,*1981; Picton and Grierson, 1988). On the other hand, cucumber fruits exhibit increased tolerance to high temperature stress (32.5°C) with no change in *in vitro* ACC oxidase activity (Chan and Linse, 1989).The above studies suggest that changes in ripening behavior are likely to occur when fruit and vegetable crops are exposed to higher temperatures prior to harvest.

A considerable line of work has been carried out for more than three decades focusing on the quality properties of fruit and vegetable crops exposed to high temperatures during growth and development. Flavor is affected by high temperatures. Apple fruits exposed to direct sunlight had a higher sugar content compared to those fruits grown on shaded sides (Brooks and Fisher, 1926). Grapes also had higher sugar content and lower levels of tartaric acid when grown under high temperatures (Kliewer and Lider, 1968; Kliewer and Lider, 1970). Coombe (1987) observed that a 10°C increase in growth temperature caused a 50 per cent reduction in tartaric acid content. But the synthesis of malic acid was more sensitive to high temperature exposure during growth than was the synthesis of tartaric acid (Kliewer and Lider, 1970; Lakso and Kliewer, 1975). In avocados, dry matter content is used as a harvest indicator due to its direct correlation with oil content, a key quality component (Lee *et al.,* 1983). Higher temperature influenced oil composition in 'Hass' avocados, where the concentration of certain specific fatty acids increased (*e.g.,* palmitic acid by 30 per cent) whereas others did not (*e.g.* oleic acid). Further, avocados with higher dry matter content take longer period to ripen which could pose a serious problem for growers planning to market their fruits immediately after harvest (Woolf and Ferguson, 2000; Woolf *et al.,* 1999). Thus fruit and vegetable growers, packers and shippers must pay close attention to ambient temperatures during growth and development as well as maturity indices to assure harvest at the appropriate time. Mineral accumulation was also reported to be influenced by high temperatures and/or direct sunlight. 'Hass' avocado fruits exposed to direct sunlight had higher calcium (100 per cent), magnesium (51 per cent) and potassium (60 per cent) contents compared to fruits grown under shaded conditions (Woolf *et al.,* 1999). The authors suggested that these changes might be related to water movement through the fruit. Antioxidants in fruit and vegetable crops can also be altered by exposure to high temperatures during the growing season. Wang and Zheng (2001) observed that 'Kent' strawberries grown in warmer nights (18–22°C) and warmer days (25°C) had a higher antioxidant activity than berries grown under cooler (12°C) days. The investigators also observed that high temperature conditions significantly increased the levels of flavonoids and, consequently, antioxidant capacity. It was further corroborated that higher day and night temperatures had a direct influence in strawberry fruit color. Berries grown under those conditions were redder and darker (Galletta and Bringhurst, 1990). Exposure of fruit and vegetable crops to high temperatures can result in physiological disorders also and other associated internal and external symptoms (Table 13.4).

Table 13.4: Symptoms of Heat and Solar Injury of Fruit and Vegetable Crops

Crops	Symptoms
Snap bean	Brown and reddish spots on the pod; spots can coalesce to form a water-soaked area
Cabbage	Outer leaves showing a bleached, papery appearance; damaged leaves are more susceptible to decay
Lettuce	Damaged leaves assume papery aspect; affected areas are more susceptible to decay; tip burn is a disorder normally associated with high temperatures in the field; it can cause soft rot development during post-harvest
Muskmelon	Characteristic sunburn symptoms: dry and sunken areas; green color and brown spots are also observed on rind
Bell pepper	Sunburn: yellowing and, in some cases, a slight wilting
Potato	Black heart: occur during excessively hot weather in saturated soil; symptoms usually occur in the center of the tuber as dark-gray to black discoloration
Tomato	Sunburn: disruption of lycopene synthesis; appearance of yellow areas in the affected tissues
Apple	Skin discoloration, pigment breakdown and water-soaked areas
Avocado	Skin and flesh browning; increased decay susceptibility
Lime	Juice vesicle rupture; formation of brown spots on fruit surface
Pineapple	Flesh with scattered water-soaked areas; translucent fruit flesh

Source: Kader, Lyons and Morris (1974), Sargent and Moretti (2002), Wolf and Ferguson (2000).

Effect of Carbon Dioxide Exposure

The Earth's atmosphere consists basically of nitrogen (78.1 per cent) and oxygen (20.9 per cent), with argon (0.93 per cent) and carbon dioxide (0.031 per cent) comprising next most abundant gases (Lide, 2009). Nitrogen and oxygen are not considered to have significant role in causing global warming as both these gases are virtually transparent to terrestrial radiation. The greenhouse effect is primarily a combination of the effects of water vapor, CO_2 and minute amounts of other gases (methane, nitrous oxide, and ozone) that absorb the radiation leaving the Earth's surface (IPCC, 2001). The warming effect is explained by the fact that CO_2 and other gases absorb the Earth's infrared radiation, trapping heat. Since a significant part of all the energy emanated from Earth occurs in the form of infrared radiation, increased CO_2 concentrations mean that more energy will be retained in the atmosphere, contributing to global warming (Lloyd and Farquhar, 2008). Carbon dioxide concentrations in the atmosphere have increased approximately 35 per cent from pre-industrial times to 2005 (IPCC, 2007). Changes in CO_2 concentration in the atmosphere can alter plant tissues in terms of growth and physiological behavior.

Many of these effects have been studied in detail for some vegetable crops (Bazzaz, 1990; Cure and Acock, 1986; Idso and Idso, 1994). These studies indicated that increased atmospheric CO_2 alters net photosynthesis, biomass production, sugars and organic acids contents, stomatal conductance, firmness, seed yield, light, water, and nutrient use efficiency and plant water potential (Table 13.5). It was evidenced that increases in atmospheric CO_2 (50 per cent higher) increased tuber malformation in approximately 63 per cent, resulting in poor processing quality, and a trend towards lower tuber greening (around 12 per cent) (Högy and Fangmeier, 2009). Higher CO_2 levels (550µmol CO_2/mol) also increased the occurrence of common scab by 134 per cent but no significant changes in dry matter content and specific gravity were noticed. Moreover, higher (550 µmol CO_2/mol) concentrations of CO_2 enhanced glucose (22 per cent), fructose (21 per cent) and reducing sugars (23 per cent) concentrations, reducing tubers quality owing to increased browning and acryl amide formation in French fries. It was also observed that proteins, potassium and calcium content got reduced in tubers exposed to high CO_2 concentrations, indicating loss of nutritional and sensory quality (Table 13.5).

Bindi *et al.* (2001) studied the effects of high atmospheric CO_2 during growth on the quality of wines. They observed that elevated atmospheric CO_2 levels increased fruit dry weight, with increases ranging from 40 to 45 per cent in the 550 mmol CO_2/mol treatment and from 45 to 50 per cent in the 700 mmol CO_2/mol treatment. There was increase in tartaric acid and total sugar contents by around 8 per cent and 14 per cent, respectively with rising in CO_2 levels up to a maximum in the middle of the ripening season but the effect of CO_2 both on the quality parameters got vanished at maturity stage. Overall wine quality was not significantly affected by elevated CO_2. The researchers concluded that the expected rise in CO_2 concentrations may strongly stimulate grapevine production without causing negative repercussions on quality of grapes and wine.

Effect of Ozone Exposure

Ozone in the troposphere is formed as a result of series of photochemical reactions involving carbon monoxide (CO), methane (CH_4) and other hydrocarbons in the presence of nitrogen species ($NO + NO_2$) (Schlesinger, 1991) during the periods of high temperature and solar irradiation, normally during summer seasons (Mauzerall and Wang, 2001). It is also formed, naturally during other seasons, but peak production takes place in the spring (Singh *et al.*, 1978). However, higher concentrations of atmospheric ozone are found during summer due to increase in nitrogen and emission of volatile organic compounds (Mauzerall and Wang, 2001). Concentrations are at maximum in the late afternoon and at minimum in the early morning hours, notably in industrialized cities and their vicinities. The opposite phenomenon occurs at high latitude sites (Oltmans and Levy, 1994). Another potential source for increased levels of ozone in a certain region is via movement by local winds or downdrafts from the stratosphere.

Ozone enters plant tissues through the stomata, causing direct cellular damage, especially in the palisade cells (Mauzerall and Wang, 2001). The damage is probably due to changes in membrane permeability and may or may not result in visible injury,

Table 13.5: Physiological and Quality Parameters of Fruit and Vegetable Crops Affected by Exposure to Increased CO_2 Levels

Physiological or Quality Parameters	Effect of High CO_2	Products	References
Photosynthesis	↑	Potato; spinach	Katnya *et al.* (2005), Jain *et al.* (2007)
Respiration	↓	Asparagus; broccoli; mungbean sprouts; blueberries; tomatoes; pears	Beaudry (1993), Peppelenbos and Leven (1996)
	↑	Potatoes; lettuce, eggplants; lemons; cucumbers; mango	Pal and Buescher (1993), Fonseca *et al.* (2002), Bender *et al.* (1994)
	–	Apples	Peppelenbos and Leven (1996)
Ripening	↓	Tomato	Klieber *et al.* (1996)
Stomatal conductance	↓	Spinach	Leakey *et al.* (2006), Jain *et al.*(2007)
Firmness	–	Tomato	Klieber *et al.* (1996)
	↑	Strawberry; raspberry	Siriphanich (1998), Haffner *et al.* (2002)
Color intensity	↑	Grape	Bindi *et al.* (2001)
Dry matter	↑	Potato	Vorne *et al.* (2002)
Starch	↑	Potato	Vorne *et al.* (2002)
Alcohol	↑	Potato	Vorne *et al.* (2002)
Titratable Acidity	–	Grape	Bindi *et al.* (2001)
Citric Acid	↓	Potato; tomato	Donnelly *et al.* (2001), Islam *et al.* (1996)
Malic Acid	↓	Potato; tomato	Vorne *et al.* (2002), Islam *et al.* (1996)
Ascorbic acid	↑	Potato; strawberry; orange; tomato	Vorne *et al.* (2002), Wang *et al.* (2003), Idso *et al.* (2002), Islam *et al.* (1996)
Reducing sugars	↑	Potato	Vorne *et al.* (2002); Högy and Fangmeier (2009)
	↑	Tomato	Islam *et al.* (1996)
Total phenolics	↑	Grape; strawberry	Bindi *et al.* (2001); Wang *et al.* (2003)
Flavonoids	↑	Grape	Bindi *et al.* (2001)
Anthocyanins	↑	Grape; strawberry	Bindi *et al.* (2001), Wang *et al.* (2003)
Glycoalkaloids	↓	Potato	Vorne *et al.* (2002)
pH	–	Grape	Bindi *et al.*, 2001
Nitrate	↓	Potato; celery; leaf lettuce; Chinese cabbage	Vorne *et al.* (2002), Jin *et al.* (2009)
Volatile compounds	↓	Mango	Lalel *et al.* (2003)
Antioxidant capacity	↓	Scallion; strawberry	Levinea and Paré (2009), Shin *et al.* (2008)

reduced growth but ultimately, cause reduction in yield (Krupa and Manning, 1988). Visible injury symptoms of exposure to low ozone concentrations include changes in pigmentation, also known as bronzing, leaf chlorosis and premature senescence

(Felzer *et al.*, 2007). Since leafy vegetable crops are often grown in the vicinity of large metropolitan areas, it can be expected that increasing concentrations of ozone will result in increased yellowing of leaves. Leaf tissue stressed in this manner could affect the photosynthetic rate, production of biomass and, ultimately, post-harvest quality in terms of overall appearance, color and flavor compounds (Table 13.6). Additionally, it was observed that ozone exposure causes reduction in photosynthesis and increased turnover of antioxidant systems (Percy *et al.*, 2003). It is reported by several workers (Grulke and Miller, 1994; Tjoelker *et al.*, 1995) that higher ozone concentrations can affect both the photosynthetic and respiratory processes (Table 13.6). They verified that branches within the upper canopy of sugar maple (*Acer saceharum* Marsh.) submitted to ozone concentrations of 95 nmol O_3/mol (twice-ambient concentrations) showed reduced light-saturated rates of net photosynthesis by 56 per cent and increased dark respiration by 40 per cent. These researchers also observed that ozone reduced net photosynthesis and impaired

Table 13.6: Physiological and Quality Parameters of Fruit and Vegetable Crops Affected by Exposure to Increased O_3 Levels.

Physiological or Quality Parameter	O_3 Effect	Product	References
Respiration	↑	Blueberry; broccoli; carrot	Song *et al.* (2001)
Dark respiration	↑	Sugar maple	Tjoelker *et al.* (1993)
Photosynthesis	↓	Strawberry; conifers; hardwoods	Amthor and Cumming (1988), Reich (1987)
Visible injury	↑	Black cherry	Chappelka *et al.* (1997)
Viscosity	↑	Potato	Donnelly *et al.* (2001)
Starch	–	Potato	Vorne *et al.* (2002)
Reducing sugars	↓	Potato	Vorne *et al.* (2002)
Citric acid	↑	Potato	Piikki *et al.* (2003)
	–	Tomato	Tzortzakisa *et al.* (2007)
Electrolyte leakage	↑	Persimmon	Salvador *et al.* (2006)
Malic acid	↓	Potato	Piikki *et al.* (2003)
Ascorbic acid	↓	Potato	Vorne *et al.* (2002)
	–	Tomato	Tzortzakisa *et al.* (2007)
N, P	–	Potato	Heagle *et al.* (2003)
K, Mg	↑	Potato	Piikki *et al.* (2003)
Nitrate	↑	Potato	Vorne *et al.* (2002)
Color (L)	↑	Broccoli;	Vorne *et al.* (2002)
	↓	Cucumber, Mushroom	Skog and Chu (2001)
Browning index	↑	Mushroom	Skog and Chu (2001)
Firmness	↑	Cucumber	Skog and Chu (2001)
Isocumarin	↑	Carrot	Hildebrand *et al.* (2008)
Sucrose	–	Potato; carrot	Hildebrand *et al.* (2008)

stomatal function, with these effects depending on the irradiance environment of the canopy leaves.

The review of the pertinent literature related to plant responses to ozone exposure reveals that there exists considerable variation in species response. Greatest impacts in fruit and vegetable crops may occur from changes in carbon transport. Underground storage organs (*e.g.*, roots, tubers, bulbs) normally accumulate carbon in the form of starch and sugars, both of which are important quality parameters for both fresh and processed crops. If carbon transport to these structures is restricted, there is great potential to lower quality in such important crops as potatoes, sweet potatoes, carrots, onions and garlic (Table 13.6). Exposure of other crops to elevated concentrations of atmospheric ozone can induce external and internal disorders, which can occur simultaneously or independently. These physiological disorders can lower down the post-harvest quality of fruit and vegetable crops destined for both fresh market and processing by causing such symptoms as yellowing (chlorosis) in leafy vegetables, alterations in starch and sugar contents of fruits and in underground storage organs. Decreased biomass production directly affects the size, appearance and other important visual quality parameters. Furthermore, impaired stomatal conductance due to ozone exposure can reduce root growth, affecting crops such as carrots, sweet potatoes and beet roots (Felzer *et al.*, 2007) (Table 13.6).

Skog and Chu (2001) carried out a set of experiments to determine the effectiveness of ozone in preventing ethylene-mediated deterioration and post-harvest decay in both ethylene-sensitive and ethylene-producing commodities, when stored at optimal and sub-optimal temperatures. On mushrooms, which have no known site of ethylene activity, effects from ozone exposure would be antimicrobial only. Ozone at the concentration of 0.04 µL/L appeared to have potential for extending the storage life of broccoli and seedless cucumbers, when both were stored at 3°C. Ozone also showed the capability of removing ethylene from the environment, inside cold rooms. At concentrations of 0.4 µL/L, ozone was effective in removing ethylene (1.5–2.0 µL/L) from an apple and pear storage room. Apples and pears subjected to ozone-enriched atmospheres showed no difference in fruits quality. Strawberries *cv.* Camarosa stored for three days under refrigerated storage (2°C) in a ozone-enriched atmosphere (0.35 µL/L) showed a 3-fold increase in vitamin C content when compared to berries stored at the same temperature under normal atmosphere as well as a 40 per cent reduction in emissions of volatile esters in ozonized fruits (Perez *et al.*, 1999). A transient increase in ß-carotene, lutein and lycopene content was observed in ozone-treated tomato fruit, though the effect was not consistent. But sensory evaluation revealed a significant preference for fruits subjected to low-level ozone-enrichment (0.15 µmol/mol) (Tzortzakisa *et al.*, 2007).

The quality of persimmon (*Diospyros kaki* L. F.) fruits (*cv.* Fuyu) harvested at two different harvest dates was evaluated after ozone exposure. Fruits were exposed to 0.15 µmol/mol (vol/vol) of ozone for 30 days at 15°C and 90 per cent relative humidity (RH). Astringency removal treatment (24 h at 20°C, 98 per cent CO_2) was performed and fruits were then stored for 7 days at 20°C (90 per cent RH), imitating commercial conditions. Flesh softening was the most important disorder that appeared when fruits were transferred from 15°C to commercial conditions. Ozone exposure was capable to maintain firmness of second harvested fruits, which were naturally softer than the first harvested fruits, over commercial limits even after 30 days at 15°C plus shelf-life. Ozone-treated fruit showed the highest values of weight loss and maximum electrolyte leakage. However, ozone exposure had no significant effect on color, ethanol, soluble solids and pH. Furthermore, ozone-treated fruits showed no signs of phytotoxic injuries (Salvador *et al.,* 2006).

Therefore, an integrated approach towards post harvest management with appropriate backward linkage at the farm level will go a long way in quality assurance of harvested horticultural produce during subsequent storage and marketing.

Influence of Pre-harvest Factors on Post-harvest Quality of Horticultural Produces

A wide range of biotic and abiotic factors can have effect on the appearance of fruits and vegetables prior to harvest. Poor management decision made during crop production may lead to undesirable texture of the product which may not be liked by the consumers. Many pre-harvest factors have profound impact on flavor by influencing plant growth and development but the influence may be insignificant if the produce is harvested too early or the cultivar is not genetically capable of developing desirable flavor (Mattheis and Fellman, 1999). Pre-harvest factors which may predispose fruits and vegetables for subsequent disorder development are dominated by position of the fruit on the trees, characteristics of fruiting sites, crop load, mineral and carbohydrate nutrition of developing fruit, water relation and response to temperature (Ferguson *et al.,* 1999). Thus, research findings on those factors which are under the control of human beings will offer great opportunities for improving post-harvest quality of fruits and vegetables through adoption of better management practices.

Genetic Material

Growers have the choice of selecting preferred cultivars prior to planting crops. This choice may be limited by availability of planting materials depending on the crop. They should keep in mind, however, the eating needs and desires of the ultimate consumers. In some crops, a great deal of plant breeding has been done to provide a wide range of varieties with different quality attributes. This can be seen in the wide range of commercial fruit and vegetable varieties available to growers for planting. Shapes, sizes, colors, productivity levels, dry matter and taste attributes vary, as well as the ripening times and rates and post-harvest longevity. For example, tomatoes vary in size, shape, sugar content, acidity, dry matter, resistance to pests and diseases, susceptibility to handling damage and rate of post-harvest ripening both on and off the vine. During the 1990s, so-called 'long life' tomato varieties have been made available. These varieties became better due to naturally occurring single gene mutation (known as the *rin,* or ripening inhibitor mutation based on ethylene response). A number of such single-gene ripening mutants of tomato that affect the ripening process, including *Nr* (never ripe), *rin* (ripening inhibitor) and *nor* (non ripening), have been studied. While commercial varieties derived from such materials express very long shelf life, they are generally characterized by, or associated with, decreased fruit quality. There is real potential for combining these qualities and shelf life attributes using genetic engineering. Commercialization, however, will have to wait until society is ready to accept such crops. Environmental conditions (such as light intensity and duration, temperature, water availability, nutrition) modify fruit quality. Different varieties, however, respond relatively similarly to changes in these conditions. Providing optimum conditions for cropping, timing of harvest, storage conditions, post-harvest and marketing methods are also important in determining final product quality at the consumer level (Hewett, 2006).

Cultural Practices

Taste and flavor of horticultural produces are influenced by the environment, agrichemicals, nutrition and management systems that can have impact on flavor through effects on plant development (Mattheis and Fellman, 1999). Appropriate light (intensity and quality) and temperature do influence post-harvest eating quality (Kays, 1999). Both are required for optimal plant productivity and harvest index (the edible dry weight – dry matter or DM – of the crop harvested as a proportion of the total dry weight of the plant). In perennial crops, light utilization is a key determinant of productivity and quality (Snelgar *et al.*, 1998; Tustin *et al.*, 2001) as the leaf area exposed to sun during the day must be adequate to provide the carbon needed for both fruit and vegetative growth. For some tree crops, there is a good understanding of the carbohydrate assimilation, distribution and accumulation within the plant (apples and grapes). However, for others (such as kiwifruit and apricots) there remain gaps in understanding of the factors that influence partitioning of carbon into the different plant parts. Seasonal variations do occur in dry matter accumulation. The best possible example of this is with wine grapes. The best vintages (in terms of wine quality) are those where vine yields are relatively low. However, the radiation, temperature and possibly rainfall conditions during the growing season are such that maximal partitioning of sugars (and other components that contribute to the important quality attributes of wine) are accumulated in the fruit, instead of the shoots, prior to harvest. There is a real need to develop more robust physiological models of carbon assimilation, distribution and accumulation on different plant organs of commercially important crops. This will provide growers with management tools to minimize the proportion of low dry matter fruit that have the potential to provide a bad taste experience for consumers.

Tree geometry will influence the amount of light perceived and captured by fruit trees. Traditional horticultural technologies, using size controlling rootstocks and adaptive pruning and thinning techniques, are good examples of enabling technologies to optimize productivity (crop load) and quality of fruit. These technologies do so by influencing size and form of the tree and hence the volume occupied in an orchard. Conceptual advances in understanding the genetic plasticity of trees, or the extent to which the external or internal environment can influence tree architecture, should provide further insights in factors that control carbon assimilation and distribution within fruit trees (Seleznyova *et al.*, 2003). Numerous studies have shown that improving light penetration into the canopy improve the fruit composition of grapes, raspberry, apple, peach and plums. Judicious pruning generally increases soluble solid content (SSC), anthocyanins, total soluble phenols and reduces titrable acidity (malic acid) and potassium content in fruits. Conversely, shading results in significant differences in the aroma of fruits. High light interception by individual fruit and its surrounding leaves results in good fruit quality traits including fruit colour, size and flavour. Peach fruits in the upper position of the canopy usually have better quality than fruits in the lower, shaded part of canopy. The removal of interior water sprouts can significantly increase light penetration and improve fruit size, colour and SSC of lower position fruits. However, heavy pruning shows the opposite effect due to restricted carbohydrate supply to fruits. In the same way, the practice of removing (pulling) leaves around fruit to increase fruit colour may decrease fruit size and SSC. The ratio of fruit (F) to leaf(L) also influences the fruit quality in peach, cherry, apple and grape. High F/L ratio leads to delayed ripening, lower SSC and smaller fruit size. Fruit thinning increases fruit size but it reduces the total yield. Hence, a balance between yield and fruit size must be achieved for maximum profit. Again, leaving higher number of fruits on a tree not only reduces fruit size but also decreases their SSC. The balance between vegetative and fruit growth can be altered by girdling. Girdling has been found beneficial in grape, jamun and jack fruit. It has been observed that grape vines having extra vigor, produce smaller bunch size with poor berry.

The use of plant growth regulators can have influence on fruit quality and shelf life. Application of gibberallic acid (GA_3) is commercially practiced in grape, strawberry and citrus fruits to improve their appearance and shelf life. Auxins are also used in various forms to alter maturity and improve quality of mandarins, berries and some temperate nuts. Pactobutrazole, Dormax, CIPC etc. are also commercially being utilized by the growers as a pre-harvest treatment in fruit crops to enhance their cosmetic appeal and shelf life during storage. For example, pre-harvest application of putrescine (polyamines) in mango exhibited higher firmness, TSS and lower fruit rot during storage. It also reduces ethylene production of harvested peach fruit and delayed firmness loss. Pre-harvest foliar application of Daminozide (Alar or B9) (@2500 mg/l) leads to colour development in apples and also enhances fruit firmness. It is also been reported that pre-harvest applications of pactobutrazol (@2000 ppm) is effective in minimizing spongy tissue disorder of Alphanso mango. The use of naphthalene acetic acid (NAA) as foliar spray (@300ppm) in pineapple 30 days before harvesting increased the potassium ion concentration in mature fruits and consequently, minimized internal pulp browning (65 per cent) during cold storage.

Mineral Nutrition

Optimum plant performance depends on a balanced and timely availability of mineral nutrients that may be limiting in many soils around the world. Inorganic mineral nutrients can influence the quality of horticultural crops in many ways but particularly in physiological fruit disorders (Ferguson and Boyd, 2002). Much effort has been expended to develop protocols for estimating critical threshold levels of both macro- and microelements for many crops. Management practices have been developed to apply appropriate fertilizers to the crop at times when benefits of yield or quality can be achieved. This may be done with soil or foliar applications, or through irrigation systems (trickle) in the field. It may also be done in Nutrient Film Techniques (NFT) in protected cultivation where very precise formulations can be added at particular stages in the crop phenology to obtain greatest benefits.

Some specific post-harvest quality disorders of fruit and fruit vegetables result from nutritional imbalances (deficiency or excess) of certain minerals elements (Kays, 1999). Excess nitrogen may result in reduced firmness and enhanced susceptibility to post-harvest decay. Of particular importance is calcium, a deficiency of which may induce a range of post-harvest disorders in many fruits and vegetables (Shear, 1975). Some of these are bitter pit in apples and blossom end rot in tomatoes and capsicums (Table 13.7). In papayas, low mesocarp calcium concentrations have been linked with fruit softening (Qiu *et al.,* 1995) and in tomatoes blossom end rot has been associated with low calcium and high potassium fertilization of tomato plants (Ho *et al.,* 1993). In strawberry, Ca deficiency results in small fruits with hard and seedy patches. However, excess level of Ca reduces fruit sweetness in strawberry and cherries. One reason for the development of calcium deficiency symptoms in harvested products is because of the way that calcium is transported around the plant (in the xylem only and not the phloem) and the time at which it is available to be imported into fruit (only early in the development and not during maturation) (Ferguson, 1980; Ferguson and Watkins, 1989; Ferguson *et al.,* 1999; Hewett, 1997). Calcium deficiency is normally overcome by spraying with calcium salts during fruit development or by post-harvest calcium dip/drench treatments of the fruit (Hewett and Watkins, 1991).

Table 13.7: Physiological Disorders Related to Calcium Deficiency in some Horticultural Crops

Crops	Disorders
Apple	Bitter pit, cork spot, cracking, internal browning, Jonathan spot, lenticel blotch/breakdown, low temperature breakdown, senescent breakdown, water core
Avocado	End spot, malformation
Beans	Hypocotyl necrosis
Brussels sprouts	Internal browning
Cabbage	Internal tip burn
Carrots	Cavity spot, cracking
Celery	Blackheart
Cherries	Cracking
Lettuce	Tip burn
Mango	Soft nose
Parsnip	Cavity spot
Pears	Cork spot, Alfalfa greening
Peppers	Blossom end rot
Strawberry	Tip burn
Tomatoes	Blossom end rot

Source. Shear (1975).

Phosphorus level in soil and plant does not have much effect on internal fruit quality but it certainly affects the fruit appearance. Research findings show that there is a relationship between pH regulation, organic acids levels and fruit K content. Optimum level of K exerts a favorable influence on vitamin C content in fruit but K deficiency causes poor color development in peach and small fruit size in oranges. Iron and zinc deficiencies also result in the reduction of fruit size in citrus and poor color development and size reduction in peach. Further, fruit size reduction in strawberry and external corking in apple have been reported to have link with boron deficiency (Sharma *et al.,* 2006). Imbalance in certain nutrients can also have a pronounced impact on fruit shape *e.g.* Zn deficiency alters the shape of peach and cherry fruits. Copper deficiency affects the shape of citrus fruit (misshape) and kernel filling in walnut. Similarly, misshaping of strawberry is associated with molybdenum deficiency.

Water Availability

Water availability is an important factor influencing fruit nutrition and quality. In general, growers adopt water management strategies to minimize moisture stress so as to allow optimal photosynthesis, plant growth and harvestable yield. Irrigation systems vary. However, they apply water on a regular basis (determined by evapo-transpiration demand) well before serious stress conditions occur. However, maximum yields are not always a prerequisite for optimal quality. In some situations with some crops, careful manipulation of water supply may well decrease water usage and improve crop quality without compromising sustainable plant growth. Moisture deficits in soil have been shown to influence ascorbic acid content in fruits (Lee and Kader, 2000). Further, drought stress and use of saline water for irrigation up to certain threshold level improve fruit quality such as TSS, sugars and flavor and shelf-life of the produce. Having very high yields can compromise wine quality. The best wine quality vintages come from those years where environmental conditions often impose stress on the vines (high temperatures, low rainfall). It was further reported that the store fruit of peaches under moderate water stress (50 per cent ET a month prior harvest) did not develop dry and mealy texture (internal break down symptoms) but showed longer shelf life (Crisosto *et al.,* 1994). High crop loads can reduce dry matter in some fruit, such as kiwifruit, and thus may affect the taste and flavor as experienced by the consumer. The use of Regulated Deficit Irrigation (RDI) to minimize water applied without affecting plant performance, and sometimes increasing fruit quality, is a system that has been used successfully for a range of fruit crops (Behboudian and Mills, 1997).

Harvesting and Picking Methods

Deciding when to harvest a crop is often one of the most difficult decisions that a grower has to make. There are certain guiding principles to be followed before selection of fruits or vegetables for harvest. Maturity at harvest has a very important influence on subsequent storage life and eating quality. Horticultural maturity is that stage of development at which a plant or plant part is ready for use by consumers for a particular purpose. This use can occur at any stage of development depending on the commodity (Watada *et al.,* 1984). There are many different ways for determining maturity. Generally maturity is assessed by subjective evaluation of commodities *e.g.* sight- color, size and shape; smell- odour or aroma; touch- texture, hardness or softness; taste- sweetness, sourness, bitterness; morphological changes and resonance- sound when tapped. Different maturity or harvest indices have been devised (see Reid, 2002 for examples). However, combination of both subjective evaluation and objective measurement of appropriate parameters (TSS, acidity, sugar, fat/oil, texture, specific gravity, color, rate of respiration, internal ethylene evolution etc) with respect to particular crop may be more significant for predicting optimum maturity. For example, the kiwifruit industry in New Zealand has used 6.2 per cent soluble solids as a minimum harvest index for many years. However, to obtain optimal flavor when the fruit is eaten ripe, cognisance needs to be taken of the dry matter in the fruit. If the dry matter (DM) is less than 14.5 per cent at harvest, then it is unlikely that the fruit soluble solids at eating will exceed 12.5 per cent, which is the minimum threshold for consumer acceptability (Crisosto and Crisosto, 2001). Fruit acid content at eating ripe can influence perception of taste and flavor and thus acceptability. Unfortunately, DM is not a good harvest index as there is no change in the rate of carbon accumulation (in contrast with soluble solids at about 6.2 per cent when starch commences its conversion to soluble sugars) that can be used by growers to indicate a critical minimum stage of development resulting in fruit of an acceptable eating quality. This apparent relationship between DM and eventual eating quality probably occurs with many fruits as does the interaction with acid concentration. Thus, it is likely that a combination of components need to be considered to decide the optimum time of harvest to allow long storage and shelf life while maintaining excellent eating quality.

Sorting and Grading

Fruits and vegetables show considerable variations in quality due to genetic, environmental and agronomic factors. Grades are made based on size, weight, color, shape, firmness, cleanliness, maturity and visible symptoms of diseases, insect damage and mechanical injury. Deformed fruits and those unmarketable fruits with splits, punctures and rotting along with foreign matter like plant debris, soil or stones are removed during sorting. Even after the produce is sorted and graded in the farm, there may be a further selection before it is packed for quality and size.

Technologies to Improve Post-harvest Quality

The main objectives of post-harvest technology are to prevent deterioration of produce as much as possible along the post-harvest chain and to ensure that maximum market value for the produce is obtained. The technologies involved in post-harvest handling of fruits and vegetables are enormously complicated because of divergent structural origin, developmental stage, physiological status and perishability of the products (Kays and Paull, 2004; Wills *et al.,* 2007). However, to protect the harvested produce by proper packaging, minimize their respiration rate developmental events such as growth or ripening by low temperature storage or manipulation their physiology, eliminating or suppressing pathogenic activity of microorganisms are the bases of all post-harvest techniques.

Temperature Management

Temperature management is the most effective tool for maintaining quality and improving shelf-life of fresh fruits and vegetables as temperature affects the rate of most of the biochemical, physiological, physical and microbiological processes leading to post-harvest deterioration (Kader, 2002; Kays and Paull, 2004; Wills *et al.*, 2007). So the main strategy for preserving harvested horticultural products is to store at low temperature. The major effect of low temperature in preservation of horticultural produce from harvest to end use involves reduction of produce metabolism and consequent delaying of processes leading to quality deterioration and senescence. The typical Q10 values within the physiological temperature range for quality deterioration of most of the horticultural products vary approximately 2 to 3, implying that storability would be double or triple for every 10oC reduction in temperature. It is necessary not only to chill the harvested product but to cool it as quickly as possible in order to maintain the commodity as close to its condition at harvest. **Pre-cooling** is the process of "rapid" removal of field heat/respiratory heat usually practiced for fresh fruits, vegetables and flowers immediately after harvest, before shipment, before storage or before processing depending on the commodity. This is the first step of good temperature management. The primary advantages of pre-cooling are: (a) Inhibition of the growth of decay causing organism, (b) restriction of the enzyme activity, (c) reduction of water loss, (d) reduction in rate of respiration and C_2H_4 liberation and (e) rapid wound healing. The beneficial effect of pre-cooling on prolonging shelf-life is pronounced in metabolically active and highly perishable products such as small berry fruits, and stem vegetables. The production and action of ethylene from harvested fruits and vegetables are temperature dependent. Harvested produce kept at 25°C with 30 per cent RH shows a tendency of 36 times more water loss as compared to that stored at 0°C with 90 per cent RH. Hence pre-cooling serves as an essential practice in any successful cool chain management of horticultural produce. There are basically four methods used for horticultural commodities. These are

1. *Room/air cooling:* Produce is placed in an insulated room equipped with refrigeration units. This method can be used with most commodities, but is slow compared with other options;

2. *Water/hydro cooling:* Dumping produce into cold water, or running cold water over produce, is an efficient way to remove heat, and can serve as a means of cleaning at the same time. In addition, hydro-cooling reduces water loss and wilting. Use of a disinfectant in the water is recommended to reduce the spread of diseases. Hydro-cooling is not appropriate for berries, potatoes to be stored, sweet potatoes, bulb onions, garlic, or other commodities that cannot tolerate wetting;

3. *Forced air cooling:* Fans are used in conjunction with a cooling room to pull cool air through packages of produce. Although the cooling rate depends on the air temperature and the rate of air flow, this method is usually 75–90 per cent faster than room cooling;

4. *Vacuum cooling*: Produce is enclosed in a chamber in which a vacuum is created. As the vacuum pressure increases, water within the plant evaporates and removes heat from the tissues. This system works the best for leafy crops, such as lettuce, which have a high surface-to-volume ratio. To reduce water loss, water is sometimes sprayed on the produce prior to placing it in the chamber. This process is called *hydrovac* cooling; and

5. *Package icing*: Icing is particularly effective on dense products and palletized packages that are difficult to cool with forced air. In top icing, crushed ice is added to the container over the top of the produce by hand or machine. For liquid icing, a slurry of water and ice is injected into produce packages through vents or handholds without removing the packages from pallets and opening their tops. Icing methods work well with high-respiration commodities such as sweet corn and broccoli.

After pre-cooling, the produce should be directly transferred to storage at optimum temperature (Table 13.8) which is usually just above the temperature causing chilling or freezing injury. Refrigeration or low temperature storage is considered the most efficient method to retain the quality of most fruits and vegetables owing to its effect on reduction of respiratory rate, water loss, ethylene emission, senescence and microbial spoilage. However, its use is limited by chilling sensitivity of many products and occasionally by its costs.

Table 13.8: Recommended Storage Temperature for a Selection of Fruits and Vegetables

1–4 °C	5–9 °C	> 10 °C
Apple	Avocado (temperate origin)	Avocado (sub-tropical)
Asparagus	Zucchini	Pawpaw (Papaya)
Berry fruits	French Bean	Grape fruit and Lemon
Broccoli	Passion fruit	Mangoes
Peach and Plum	Egg plant	Banana
Cherry	Capsicum	Pineapple
Grapes	Cucumber	Sweet potato
Lettuce	Mandarin orange	Tomato
Mushroom	Potato	Pumpkin
Carnation	Protea	Ginger

Pusa Zero Energy Cool Chamber is an on-farm cuboidal storage chamber developed at IARI, New Delhi which works on the principle of evaporative cooling. It can be constructed easily anywhere with locally available materials like brick, sand, bamboo, *khaskhas* or straw, gunny bags etc. and its operation needs a steady source of water. It consists of a double-walled bricks having cavity in between that is filled up with fine riverbed sand on all four sides. The bricks are porous enough to allow seepage of water. The water seeped through the walls and sand matrix gets evaporated, and consequently, reduce the temperature of the cool chamber. Water seeping through inner wall provides necessary moisture in the enclosure, and consequently, increases the humidity.

The Major benefit of Pusa zero energy cool chamber could be derived for extension of growing period of button mushrooms with 24 per cent higher yield than conventional growing; orderly marketing of potatoes; quality assurance through better appeal and high retention of essential vitamins in fruits and vegetables; uniform ripening of tomatoes and banana even in peak summer months. It is very effective in storing kinnow-mandarin upto one month with prime quality attributes. The commercial size (6-8 tones) chamber could be successfully utilized for onion storage during rainy season after withholding the water supply.

Temperature management should be initiated from the time of harvest. It is often considered good practice to harvest during the coolest part of the day to reduce product warming. Harvested produce should be protected from the exposure of direct sun-light when heaping fruits or vegetables in the field and then rapidly delivered them to packaging house for pre-cooling. An unbroken cold chain throughout the post-harvest handling is essential to extend and ensure the shelf-life of horticultural produces.

Preventing of Water Losses

The basic principle of minimizing water loss from fruits and vegetables during post-harvest period is to bring down the additional moisture holding capacity of surrounding air (*Wills et al.,* 2007). This can be achieved by commodity treatments such as surface waxing or coating and plastic film wrapping or by environmental manipulation such as lowering down of vapor pressure deficit (VPD) between the products and surrounding air through lowering temperature and/or raising relative humidity or control of air movement. Edible coatings, thin layers of edible material are applied directly on the surface of fruit by dipping, spraying or brushing in addition to or as a replacement for natural protective waxy coatings and it provide a barrier to moisture, oxygen and solute movement. Surface coatings are used extensively on fruits (Table 13.9) to delay ripening, reduce the water loss, improve the finish of the skin and thus enhancing shelf life. These waxes may be of animal origin, vegetable origin or may be produced synthetically. Despite the fact that most fruit and vegetable crops retain better quality at high relative humidity (80 to 95 per cent), but at this humidity, disease growth is encouraged. The cool temperatures in storage rooms help to reduce disease growth, but sanitation and other preventative methods are also required. In general, it is recommended that 90 per cent and 98-100 per cent RH are the optimal compromise conditions for fruits and leafy vegetables storage respectively (Kader, 2002; Wills *et al.,* 2007).

Preventing of Post-harvest Diseases

Many types of post-harvest infectious diseases affect fresh fruits and vegetables. The control of biotic post-harvest diseases depends on understanding the nature of disease organisms, the conditions which promote their occurrence, and the factors that affect their capacity to cause losses. **Sanitation** is of great concern to produce handlers, not only to protect produce against post-harvest diseases, but also to protect

Table 13.9: Specific Coating Applications for different Fruits

Coating Materials	Fruits
Prolong	Banana
Semperfresh	Banana
Semperfresh with organic acid	Banana
Ban-seel	Banana and plantains
Tal prolong, Semperfresh and applewax	Apple
Nutri-save	Golden delicious apple
Semperfresh	Granny smith apple
Brilloshine	Apple, avocado, melons and citrus fruits
Nu-coatflo, Brilloshine and Citrashine	Citrus fruits
Semperfresh	Guava
Palm oil	Guava
Vapor gard	Mango
Chitosan	Strawberry and raspberry
N,O-Carboxymethyl chitosan	Fruits

consumers from foodborne illnesses. *E. coli* 0157:H7, *Salmonella, Chryptosporidium, Hepatitis* and *Cyclospera* are among the disease-causing organisms that have been transferred via fresh fruits and vegetables (Suslow, 1997; Melnick, 1998). Chlorine in the form of sodium hypochlorite solution (for example, Clorox™) or as a dry, powdered calcium hypochlorite can be used in hydro-cooling or wash water as a disinfectant. For the majority of vegetables, chlorine in wash water should be maintained in the range of 75–150 ppm (parts per million.). **Ozonation** is another technology that can be used to sanitize produce. It is the most effective natural bactericide of all the disinfecting agents. Ozone has long been used to sanitize drinking water, swimming pools and industrial wastewater. Fruit and vegetable growers have begun using it in dump tanks as well, where it can be thousands of times more effective than chlorine. Ozone not only kills whatever foodborne pathogens might be present, it also destroys microbes responsible for spoilage. **Post-harvest biological control** is another relatively new approach and offers several advantages over conventional biological control. Several biological control agents have been developed in recent years, and a few have actually been registered for use on fruit crops. The first biological control agent developed for post-harvest use was a strain of *Bacillus subtilis* (Pusey and Wilson, 1984). It controlled peach brown rot, but when a commercial formulation of the bacterium was made, adequate disease control was not obtained (Pusey, 1989). More recently, a strain of *Pseudomonas syringae* van Hall was found that controlled both Blue and Gray Mold of pome fruit (Janisiewicz and Marchi, 1992). It was subsequently registered, and is now sold commercially for postharvest disease control (Janisiewicz and Jeffers, 1997). Other bacterial microorganisms are in the process of development for postharvest disease control. For example, strains of *Bacillus pumilus* and *Pseudomonas fluorescens* have been identified that exhibit successful control of *B. cinerea* in field trials of strawberry (Swalding and Jeffries, 1998). Yeasts such as *Pichia guilliermondii* (Wisniewski *et al.*, 1991) and *Cryptoccocus laurentii*, a yeast that occurs naturally on apple leaves, buds, and fruit (Roberts, 1990) were the first to be applied for control of postharvest decay on fruit. The yeast, *Candida oleophilia* is registered for control of postharvest decay on fruit crops. The yeasts, *Cryptococcus infirmo-minutus* and *Candida sake* successfully control brown rot and blue mold on sweet cherry (Spotts *et al.*, 1998), and three diseases of apple (Vinas *et al.*, 1998), respectively, and may be developed commercially.

Although there is no doubt that biocontrols are effective, they do not always give consistent results. Compatibility with chemicals used during handling is also important. Indications are that biological control agents must be combined with other disease control strategies if they are to provide acceptable control. **Irradiation,** although ultraviolet light has a lethal effect on bacteria and fungi that are exposed to the direct rays, there is no evidence that it reduces decay of packaged fruits and vegetables (Hardenburg *et al.*, 1986). More recently, low doses of ultraviolet light (254 nm UV-C) irradiation reduced post-harvest brown rot of peaches (Stevens *et al.*, 1998). In this case, the low dose ultraviolet light treatments had two effects on brown rot development; reduction in the inoculum of the pathogen and induced resistance in the host. However, it has not become a practical post-harvest treatment as yet and requires more research.

Gamma radiation has been studied for controlling decay, disinfestation and extending the storage and shelf-life of fresh fruits and vegetables. Dosages of 1.5 to 2 kilogray (kGy), and some cases 3.0 kGy (300 krad), have been found effective in controlling decay in several products (Hardenburg *et al.*, 1986). Commercial application of gamma radiation is limited due to the cost and size of equipment needed for the treatment and to uncertainty about the acceptability of irradiated foods to the consumer (Hardenburg *et al.*, 1986). Gamma irradiation may be used more in the future once methyl bromide is no longer available to control insect infestation in stored products. All uses of methyl bromide are being phased out to avoid any further damage to the protective layer of ozone surrounding the earth.

Atmosphere Modification

Changes in the concentrations of gases around the horticultural products can significantly prolong their storage life owing to reduction in respiration rate of produce, retardation of senescence and inhibition of many spoilage causing microorganisms. Atmosphere at ambient conditions comprises of 78.08 per cent N_2, 20.98 per cent O_2 and 0.03 per cent CO_2 under normal conditions. Any deviation from this normal atmosphere composition *e.g.* elevated level of CO_2 reduced level of O_2, N_2 or any other combination is known as '**Modified Atmosphere**'. When this deviated normal atmosphere is precisely kept under control then it is termed as "**Controlled Atmosphere**". This control can be done in package (Controlled Atmosphere Packaging) or in the storage chamber (Controlled Atmosphere storage). Similar is the case for Modified Atmosphere Storage (MA-storage) and Modified Atmosphere Packaging (MAP). Generally, O_2 below 8 per cent and CO_2 above 1 per cent are used in CA-storage. Atmospheric modification is a supplementary practice to temperature management in preserving quality and safety of fresh fruits, vegetables, ornamentals and their products throughout post-harvest handling.

Modified atmosphere packaging consisting of decreased O_2 and elevated CO_2 concentration compared to air which is created either by rapidly flushing the head-space of the package with the desired gas mixture or by allowing the produce to respire inside the package so that an equilibrium is slowly attained. Extensive continued use of MAP in preservation of horticultural produce is anticipated for the future. One of the newest trends in MAP is the shrink-wrapping of individual produce items. Shrink-wrapping has been used successfully to package apples, mangoes and a variety of tropical fruits. Shrink-wrapping with an engineered plastic wrap can reduce shrinkage, protect the produce from disease, reduce mechanical damage and provide a good surface for stick-on labels. Essentiality of CA/MA technology should be justified only if (a) the commodities are having high market value (b) it significantly enhances storage life (c) it retains significantly better quality (d) it fetches better price compared to conventional cool stored produce. Retardation of ripening, reduction in decay, prevention of specific disorders and maintenance of product texture are some of the potential advantages of CA/MA storage.

Beneficial Effects of CA (In optimum commodity specific composition)

☆ Retardation of senescence (including ripening) and associated biochemical and physiological changes *i.e.* slowing down rates of respiration, ethylene production, softening and compositional changes.

☆ Reduction of sensitivity to ethylene action at O_2 levels < 8 per cent and/or CO_2 levels > 1 per cent.

☆ Alleviation of certain physiological disorders such as chilling injury of avocado and some storage disorders, including scald of apples.

☆ CA can have a direct or indirect effect on postharvest pathogens (bacteria and fungi) and consequently decay incidence and severity. For example, CO_2 at 10 to 15 per cent significantly inhibit development of *Botrytis* rot on strawberries, cherries and other perishables.

☆ Low O_2 (< 1 per cent) and/or elevated CO_2 (40 to 60 per cent) can be a useful tool for insect control during storage of dried products from fruits, vegetables, flowers, nuts and grains.

Detrimental Effects of CA (Above or below optimum composition for the commodity)

☆ Initiation and/or aggravation of certain physiological disorders such as internal browning in apples and pears, brown stain of lettuce, and chilling injury of some commodities.

☆ Irregular ripening of fruits, such as banana, mango, pear, and tomato can result from exposure to O_2 levels below 2 per cent and/or CO_2 levels above 5 per cent.

☆ Development of off-flavors and off-odors at very low O_2 concentrations and very high CO_2 (as a result of anaerobic respiration and fermentative metabolism)

☆ Increased susceptibility to decay when the fruit is physiologically injured by too-low O_2 or too-high CO_2 concentrations.

Table 13.10: Classification of Fruits and Vegetables according to their Tolerance to Low O_2 Concentrations

Minimum O_2 Concentration Tolerated (per cent)	Commodities
1.0	Some cultivars of apples and pears, broccoli, mushroom, garlic, onion
2.0	Most cultivars of apples and pears, kiwifruits, apricot, cherry, nectarine, peach, plum, strawberry, papaya, pineapple, olive, cantaloupe, green bean, celery, lettuce, cabbage, cauliflower, brussels sprouts
3.0	Avocado, persimmon, tomato, pepper, cucumber, artichoke
5.0	Citrus fruits, green pea, asparagus, potato, sweet potato

Source: Kader *et al.*, 1989.

Table 13.11: Classification of Fresh Fruits and Vegetables according to their Tolerance to Elevated CO_2 Concentrations

Maximum O_2 Concentration Tolerated (per cent)	Commodities
2	Apples (Golden Delicious), Asian and European Pear, apricot, grape, olive, tomato, sweet pepper, lettuce, Chinese cabbage, artichoke, sweet potato
5	Most cultivars of apples and peach, nectarine, plum, orange, avocado, banana, mango, papaya, kiwifruit, pea, chilli, egg plant, cauliflower, cabbage, brussels sprouts, radish, carrot

Source: Kader *et al.*, 1989.

Reducing Damaging Effect of Ethylene

The presence of ethylene in the atmosphere emanating from ripening fruits in the market or storage room or from the exhaust gases of vehicle has been a major concern not only for unripe climacteric fruits but also for non-climacteric fruits and vegetables during the post-harvest handling as it hastens ripening, loss of chlorophyll, also increases the susceptibility of the product to rots. Preventing ethylene build up around the product is often the simplest method of reducing the damaging effects of ethylene on shelf-life. Therefore, fast removal of emitted ethylene from the surrounding atmosphere of produce, creation of storage environment unfavorable for ethylene action and technologies to ethylene biosynthesis of tissue are being widely utilized in commercial practices (Wills *et al.*, 2007). A simple physical method to prevent ethylene accumulation is to ensure good air circulation inside the storage room and ventilation with external fresh air if needed. The traditional method had been to use potassium permanganate or 'Purafil' which reacts with ethylene, resulting in production of CO_2 and water. In order to scrub the air efficiently it is the best to spread the potassium permanganate over as larger a surface area as possible either in trays or within highly permeable bags. There is another relatively new compound called 1-methylcyclopropene (1-MCP) which opened a new era of minimizing ethylene damage on marketing quality and storage life of fruits and vegetables after harvest (Blankenship and Dole, 2003; Reid and Staby, 2008; Wills *et al.*, 2007). 1-MCP is an antagonist of ethylene responses and acts by occupying ethylene receptor such that ethylene can not bind and hence its signaling is blocked (Blankenship and Dole, 2003; Wills *et al.*, 2007). It has a non-toxic mode of action, negligible residue and is active at very low concentration. Polyamines (PAs), low molecular weight small aliphatic amines, have been proposed to be a new category of plant growth regulator and involved in large spectrum of physiological and biological processes of fruits. PAs in their free forms have been reported as anti-senescent agent, and hence increase shelf life from both endogenous and exogenous applications (Valaro *et al.*, 1998). The main effects of PAs in post-harvest management of fruits and vegetables are inhibition of ethylene biosynthesis, enhancement of fruit firmness, reduction of respiration rate, chilling injury and mechanical damage, and prevention of colour changes. The most common polyamines are putrescine, spermidine and spermine found in every plant cell. These PAs inhibit biosynthesis of ethylene as polyamine and ethylene biosynthesis are linked through the common precursor S-adenosylmethionine (SAM) but these two molecules exhibit opposite effects in relation to senescence. Ethylene production is associated with the biosynthesis of ACC. Most of the observations indicate that various PAs can delay senescence by inhibiting ACC synthesis (Lee *et al.*, 1997). It was observed that apricot fruit treated with putrescine (1 mM) by low pressure infiltration and plum fruit treated by immersion method showed significant reduction in ethylene levels as compared to untreated fruits (Romero *et al.*, 2002). Further, exogenous application of PAs (1 mM putrescine and 1 mM spermidine) have been reported to reduce chilling injury of pomegranate when stored at 2°C and 90 per cent RH by protecting the membrane lipid from being conversed in physical state (liquid-crystalline to solid-gel state) and lipid peroxidation due to its antioxidant property. PAs also maintain firmness of fruit due to their cross-linkage to the carboxyl group of the pectic substances in the cell wall, resulting in rigidification, thus blocks the access of degrading enzymes (Polygalacturonase, pectin methyl esterase, pectin esterase and cellulase) and reduce rate of softening subsequently the produce becomes less prone to mechanical damage. The firmness of mango fruits extended up to 28 days after storage at 13°C when treated with putrescine (1 mM), spermidine (0.5 mM) and spermine (0.01 mM) compared to untreated fruits (Malik and Singh, 2005).

Thermal Treatments

The consumers now-a-days are more aware of the fact that chemical treatments of fruits and vegetables to control pests, pathogenic organisms and physiological disorders are potentially harmful not only to their health but also to the environment. Thus, there is an increasing trend towards the use of natural compounds or physical treatments to control insects and diseases in fresh horticultural produce (Kader, 2002; Lurei, 1998; Wills *et al.*, 2007). Thermal treatments have actually been considered as environment friendly methods of deterioration control either alone of in combination with other methods. The most commonly used thermal treatments are hot water immersion, forced hot air treatment and vapor heat treatment. Hot water immersion and hot water rinsing and brushing (HWRB) are the most commercially followed techniques of thermal treatment. Thermal treatments not only remove soil and dust, but also fungal spores from the fruit surface. It is also effective in eradicating quiescent infections of fungi which otherwise get established on and beneath the cuticle and within the pedicel under favorable atmospheric condition. Additionally it is used to control insect infestation on fruits. Thermal treatment causes recrystallization or melting of the waxy layer on the fruit surface which seal barely visible cracks in the cuticle through which water could escape. This sealing of cracks or natural openings significantly limits the sites of fungal penetration and reduces weight loss, thus maintaining fruit firmness after prolonged storage. Thermal treatment also stimulates infected fruit to build up resistance response by the production of lignin-like material at the infection site, followed by accumulation of the phytoalexins and scoparone. While, HWRB treatment reduces respiration rate, ethylene evolution rate and enzymatic activity. This treatment inhibits activity of cell wall degrading enzyme and ethylene-forming enzyme (EFE) and results in lowering of fruit softening which prolong the shelf life of fruits. Many fruits and vegetables tolerate exposure to water temperatures of 50 to 60°C for upto 10 minutes but shorter exposure at these temperatures can control many post-harvest plant pathogens (Barkai-Golan and Phillips, 1991). In contrast, hot water dips for fruit require 90 minutes exposure to 46°C. Fungicides effectiveness can be enhanced by applying the fungicides in a hot water bath, thus allowing more effective fungal control with a reduction in chemicals. This has been particularly effective on citrus with fungicides thiabendazole and imazalil (McDonald *et al.,* 1991; Wild, 1993; Schirra and Mulas, 1995 a,b). In addition, generally recognized as safe (GRAS) compounds have been applied in hot water to improve the efficiency of their anti-fungal action. Heated solution (45°C) of sulfur-di-oxide, ethanol or sodium carbonate are used to control green mould (*Penicillium digitatum*) on citrus fruits (Smilanick *et al.,* 1995, 1997). These compounds were as effective as an imazalil dip at 25°C in controlling artificial inoculation of the fungus (Smilanick *et al.,* 1995).

Vapor heat treatment is a method of heating fruits with air saturated with water vapor at temperatures of 40-50°C to kill insect eggs and larvae as a quarantine treatment before fresh market shipment (Animal and Plant Health Inspection Service, 1985). Heat transfer is by condensation of water vapor on the cooler fruit surface. In modern facilities, the vapor heat includes forced air which circulates through the pallets and heats the commodity more quickly than vapor heat without forced air. Commercial facilities operate in many countries, mainly for use on sub-tropical fruits particularly mango and papaya (Paull, 1994). In addition, studies have been conducted for using vapor or moist forced air to disinfest many fruits and vegetables from various insect pests (Shellie and Mangan, 1994). The treatment consists of a period of warming (approach time) which can be faster or slower depending on a commodity's sensitivity to high temperatures. Then there is a holding period when the interior temperature of the produce reaches the desired temperature for the length of time required to kill the insect. The last part is the cooling down period which can be air-cooling (slow) or hydro-cooling (fast). Thus, there are a number of components of the treatment which can be manipulated to find the best combination for elimination of the insect pest without damaging the commodity. A recommended treatment for citrus, mangoes, papaya and pineapple is 43°C in saturated air for 8 hours and then holding the temperature for further 6 hours. Moreover VHT is mandatory for export of mangoes.

Conclusion

Today fresh produce consumer not only look for traditional quality attributes such as appearance, firmness and flavor but also value other parameters including nutrients, bioactive compounds availability, antioxidants and aroma. There is growing concern about food safety and environmental issue. Management tools exist for growers to minimize the effects of extreme environmental events and to manipulate timing of harvest. A number of enabling technologies are available for optimizing product quality through manipulation of nutrition, water and light (plant architecture) to minimize post-harvest disorder and quality deterioration as well as to optimize carbon assimilation, distribution and accumulation in harvested organs. Achieving high dry matter within many harvested products seems to be a prerequisite to enhance sugar and flavor components especially in perennial tree crops. Innovative technologies for maintaining optimal temperature and relative humidity, delaying losses of flavor and nutritional quality by supplemental treatments and ensuring safety will continue to be developed through collaboration between public and private organizations.

References

1. Amthor, J. S., and Cumming, J. R. (1988). Low levels of ozone increase bean leaf maintenance respiration. *Canadian Journal of Botany*, 66: 724–726.

2. Animal and Plant Health Inspection Service (1985). Section III,9 and section VI-T106. Plant protection and quarantine manual. US Dept. Agric., Wash. DC.

3. Anon (1992) Put it on ice. American Vegetable Grower. June. pp. 17–18.

4. Barkai-Golan, R. and Phillips, D.J. (1991). Post-harvest heat treatment of fresh fruits and vegetables for decay control. *Plant Dis.*, 75: 1085–1089.

5. Barry, C.S. and Giovannoni, J.J. (2007). Ethylene and fruit ripening. *Journal of Plant Growth Regulation*, 26: 143-159.

6. Bazzaz, F. A. (1990). The response of natural ecosystems to the rising global CO_2 levels. *Annual Review of Ecology and Systematics*, 21: 167–196.

7. Beaudry, R. M. (1993). Effect of carbon dioxide partial pressure on blueberry fruit respiration and respiratory quotient. *Postharvest Biology and Technology*, 3: 249–258.

8. Behboudian, M.H. and Mills, T.M. (1997). Deficit irrigation in decidious orchards. *Hort. Rev.*, 21: 105-131.

9. Bender, R. J., Brecht, J. K., and Campbell, C. A. (1994). Responses of 'Kent' and 'Tommy Atkins' mangoes to reduced O_2 and elevated CO_2. *Procedures of the Florida State Horticultural Society*, 107: 274–277.

10. Bide, J. B. and Young, R. E. (1981) "Respiration and Ripening in Fruits-Retrospect and Prospect." In: Friend, J. and Rhodes,M. J. C. (Eds.), "Recent Advance in the Biochemistry of Fruits and Vegetables.", Academic Press, New York, pp.1.

11. Bieto, J. A., and Talon, M. (1996). Fisiologia y bioquimica vegetal. Madrid: Interamericana, McGraw-Hill., pp.581.

12. Bindi, M., Fibbi, L., and Miglietta, F. (2001). Free air CO_2 Enrichment (FACE) of grapevine (*Vitis vinifera* L.): II. Growth and quality of grape and wine in response to elevated CO_2 concentrations. *European Journal of Agronomy*, 14: 145–155.

13. Blankenship, S.M. and Dole, J.M. (2003). 1-Methylcyclopropene: a review. *Postharvest Biology and Technology*, 28: 1-25.

14. Brooks, C., and Fisher, D. F. (1926). Some high temperature effects in apples: Contrasts in the two sides of an apple. *Journal of Agricultural Research*, 23: 1–16.

15. Chan, H. T., and Linse, E. (1989). Conditioning cucumbers to increase heat resistance in the EFE system. *Journal of Food Science*, 54: 1375–1376.

16. Chan, H. T., Tam, S. Y. T., and Seo, S. T. (1981). Papaya polygalacturonase and its role in thermally injured ripening fruit. *Journal of Food Science*, 46: 190–197.

17. Chappelka, A., Renfrob, J., Somers, G., and Nash, B. (1997). Evaluation of ozone injury on foliage of black cherry (*Prunus serotina*) and tall milkweed (*Asclepias exaltata*) in Great Smoky Mountains National Park. *Environmental Pollution*, 95(1): 13–18.

18. Coombe, B. G. (1987). Influence of temperature on composition and quality of grapes. *Acta Hortic*, 206: 23–35.

19. Crisosto, C.H. and Crisosto, G.M. (2001). Understanding consumer acceptance of early harvested "Hayward" kiwifruit. *Postharvest Biology and Technology*, 22: 205–213.

20. Crisosto, C.H., Johnson, R.S., Luza, J.G. and Crisosto, G.M. (1994). Irrigation regimes affect fruit soluble solids content and the rate of water loss of 'O'Henry' peaches. *HortScience*, 29: 1169-1171.

21. Cure, J. D., and Acock, B. (1986). Crop responses to carbon dioxide doubling: A literature survey. *Agricultural Forest and Meteorology*, 38(1/3): 127–145.

22. Donnelly, A., Lawson, T., Craigon, J., Black, C. R., Colls, J. J., and Landon, G. (2001). Effects of elevated CO_2 and O_3 on tuber quality in potato (*Solanum tuberosum* L.). *Agriculture, Ecosystems and Environment*, 87(3): 273–285.

23. Felzer, B. S., Cronin, T., Reilly, J. M., Melillo, J. M., and Wang, X. (2007) Impacts of ozone on trees and crops. *Compters Rendus Geosci.*, 339: 784–798.

24. Ferguson, I., Volz, R. and Woolf, A. (1999). Preharvest factors affecting physiological disorders of fruit. *Postharvest Biology and Technology*, 15: 255–262.

25. Ferguson, I.B. (1980). Movement of mineral nutrients into the developing fruit of the kiwi fruit *Actinidia chinensis* Planch. *New Zealand Journal of Agricultural Research*, 23: 349–353.

26. Ferguson, I.B. and Boyd, L.M. (2002). Inorganic nutrients and fruit quality. In: Knee, M. (Ed.), Fruit Quality and its Biological Basis, England, Sheffield Academic Press, pp.15–45.

27. Ferguson, I.B. and Watkins, C.B. (1989). Bitter pit in apple fruit. *Horticultural Reviews*, 11: 289–355.

28. Fonseca, S. C., Oliveira, F. A. R., and Brecht, J. K. (2002). Modelling respiration of fresh fruits and vegetables for modified atmosphere packages: A review. *Journal of Food Engineering*, 52: 99–119.

29. Galletta, G. J., and Bringhurst, R. S. (1990). Strawberry management. In: Galletta, G. J. and. Bringhurst, R. S. (Eds.), Small fruit crop management, Prentice-Hall: Englewood Cliffs. pp. 83–156.

31. Grulke, N. E., and Miller, P. R. (1994). Changes in gas exchange characteristics during the lifespan of giant sequoia: Implications for response to current and future concentrations of atmospheric ozone. *Tree Physiology*, 14: 659–668.

32. Haard, N. F. (1983) "Edible Plant Tissues". (In) Fennema, O. (Ed.), "Principles of FMD Science. Part l-Food Chemistry," 2nd **ed.,** Marcel Dekker. New York, chap. 16.

33. Haffner, K., Rosenfeld, H. j., Skrede, G., and Wang, L. (2002). Quality of red raspberry, *Rubus idaeus* L. cultivars after storage in controlled and normal atmospheres. *Postharvest Biology and Technology*, 24(3): 279–289.

34. Hall, A. J., McPherson, H. G., Crawford, R. A., and Seager, N. G. (1996) Using early season measurements to estimate fruit volume at harvest in kiwifruit. New Zealand *Journal of Crop and Horticultural Science*, 24: 379–391.

35. Hardenburg, R.E., Watada, A.E. and Wang, C.Y.(1986). The commercial storage of fruits and vegetables, and florist and nursery stocks. *USDA Agric. Handbook* No. 66.

36. Heagle, A. S., Miller, J. A., and Pursley, W. A. (2003). Atmospheric pollutants and trace gases growth and yield responses of potato to mixtures of carbon dioxide and ozone. *Journal of Environmental Quality*, 32: 1603–1610.

37. Hewett, E.W. (1997). Fruit quality and tree nutrition. In: Currie, L.D. and Loganathan, P. (Eds.), Proceedings of Workshop Nutritional Requirements of Horticultural Crops, Massey University Fertiliser and Lime Research Centre, Occasional Report No. 10, pp.159–171.

38. Hewett, E.W. (2006). An overview of preharvest factors influencing postharvest quality of horticultural products. *Int. J. Postharvest Technology and Innovation*, 1 (1): 4–15.

39. Hewett, E.W. and Watkins, C.B. (1991). Bitter pit control by sprays and vacuum infiltration of calcium in Cox's Orange Pippin apples. *HortScience*, 26: 284–286.

40. Hildebrand, P. D., Forney, C. F., Jun, S., Lihua, F., and McRae, K. B. (2008). Effect of a continuous low ozone exposure (50 nL L^{-1}) on decay and quality of stored carrots. *Postharvest Biology and Technology*, 49(3): 397–402.

41. HO, L.C., Belda, R., Brown, M., Andrews, J. and Adams, P. (1993). Uptake and transport of calcium and the possible cause of blossom-end rot in tomato. *Journal of Experimental Botany*, 44: 509-518.

42. Högy, P., and Fangmeier, A. (2009). Atmospheric CO_2 enrichment affects potatoes: 2 tuber quality traits. *European Journal of Agronomy*, 30: 85–94.

43. Howell, J. C. (Ed.) (1993) Postharvest handling. Vegetable Notes: Growing and Marketing Information for Massachusetts Commercial Growers. pp. 1–5.

44. Idso, K. E., and Idso, S. B. (1994). Plant responses to atmospheric CO_2 enrichment in the face of environmental constraints: A review of the past 10 years' research. *Agricultural and Forest Meteorology*, 69: 153–203.

45. Idso, S. B., Kimball, B. A., Shaw, P. E., Widmer, W., Vanderslice, J. T., Higgs, D. *et al.* (2002). The effect of elevated atmospheric CO_2 on the vitamin C concentration of (sour) orange juice. *Agriculture, Ecosystems and Environment*, 90: 1–7.

46. IPCC. Climate change (2001): Working group II: Impacts, adaptations and vulnerability. <http://www.grida.no/climate/ipcc_tar/wg2/005.html> Accessed 13.03.09.

47. IPCC. Climate change (2007). (In) Solomon, S. Qin, D. Manning, M. Chen, Z. Marquis, M. Averyt, K. B. Tignor, M. Miller, H. L. (Eds.), The physical science basis contribution of working group I to the fourth assessment report of the intergovernmental panel on climate change. Cambridge, United Kingdom: Cambridge University Press, pp. 996.

48. Islam, M. S., Matsui, T., and Yoshida, Y. (1996). Effect of carbon dioxide enrichment on physic-chemical and enzymatic changes in tomato fruits at various stages of maturity. *Scientia Horticulturae*, 65: 137–149.

49. Jain, V., PAL, M., RAJ, A., and Khetarpal, S. (2007). Photosynthesis and nutrient composition of spinach and fenugreek grown under elevated carbon dioxide concentration. *Biologia Plantarum*, 51(3): 559–562.

50. Janisiewicz, W.J. and Jeffers, S.N. (1997). Efficacy of commercial formulation of two biofungicides for control of blue mold and gray mold of apples in cold storage. *Crop Prot.*, 16: 629-633.

51. Janisiewicz, W.J. and Marchi, A. (1992). Control of storage rots on various pear cultivars with a saprophytic strain of *Pseudomonas syringae*. *Plant Dis.*, 76: 555-560.

52. Jin, C., Du, S., Wang, Y., Condon, J., Lin, X., and Zhang, Y. (2009). Carbon dioxide enrichment by composting in greenhouses and its effect on vegetable production. *Journal of Plant Nutrition and Soil Science*, 172: 418–424.

53. Kader, A.A. (2002). Postharvest biology and technology: an overview. (In) Kader, A.A. (Ed.), Postharvest Technology of Horticultural Crops (3rd ed.), University of California, Agriculture and Natural Resources, Publication 3311, pp. 39-47.

54. Kader, A.A. (2005) Increasing food availability by reducing postharvest losses of fresh produce. *Acta Hort.* 682: 2169-2175.

55. Kader, A. A., Lyons, J. M., and Morris, L. L. (1974). Postharvest responses of vegetables to preharvest field temperature. *HortScience*, 9(6): 523–527.

56. Kader, A.A. and Saltveit, M.E. (2003). Atmosphere modification. (In) Bartz, J.A. and Brecht, J.K. (Eds.), Postharvest physiology and pathology of vegetables (2nd ed.), Marcel Dekker Inc., New York, USA.

57. Kader, A.A., Zagory, D.and Kerbel, E.L. (1989). Modified atmosphere packaging of fruits and vegetables. *Crit. Rev. Food Sci. Nutr.*, 28: 1–30.

58. Katnya, M. A. C., Hoffmann-Thoma, G., Schriera, A. A., Fangmeierd, A., Jägerb, H. J., and van Bel, A. J. E. (2005). Increase of photosynthesis and starch in potato under elevated CO_2 is dependent on leaf age. *Journal of Plant Physiology*, 162(4): 429–438.

59. Kays, J.J. and Paull, R.E. (2004). Postharvest Biology. Exon Press, USA, pp.568.

60. Kays, S. J. (1997). Postharvest physiology of perishable plant products. Athens: AVI., pp. 532.

61. Kays, S.J. (1999). Preharvest factors affecting quality'. *Postharvest Biology and Technology*, 15: 233–247.

62. Klieber, A., Ratanachinakorn, B., and Simons, D. H. (1996). Effects of low oxygen and high carbon dioxide on tomato cultivar 'Bermuda' fruit physiology and composition. *Scientia Horticulturae*, 65(4): 251–261.

63. Kliewer, M. W., and Lider, L. A. (1968). Influence of cluster exposure to the sun on the composition of Thompson seedless fruit. *American Journal of Enology and Viticulture*, 19: 175–184.

64. Kliewer, M. W., and Lider, L. A. (1970). Effects of day temperature and light intensity on growth and composition of *Vitis vinifera* L. fruits. *Journal of the American Society for the Horticultural Science*, 95: 766–769.

65. Krupa, S. V., and Manning, W. J. (1988). Atmospheric ozone: Formation and effects on vegetation. *Environmental Pollution*, 50: 101–137.

66. Lakso, A. N., and Kliewer, W. M. (1975). The influences of temperature on malic acid metabolism in grape berries I. Enzyme responses. *Plant Physiology*, 56: 370–372.

67. Lalel, H. J. D., Singh, Z., and Tan, S. C. (2003). Distribution of aroma volatile compounds in different parts of mango fruit. *Journal of Horticultural Science and Biotechnology*, 78: 131–138.

68. Leakey, A. D., Bernacchi, C. J., and Long, R. A. (2006). Long-term growth of soybean at elevated $[CO_2]$ does not cause acclimation of stomatal conductance under fully open-air conditions. *Plant Cell and Environment*, 29(9): 1794–1800.

69. Lee, M. M., Lee, S. H. and Park, K. Y. (1997). Effects of spermine on ethylene biosynthesis in cut carnation (*Dianthus caryophyllus*) flowers during senescence. *J. Plant Physiol.*, 151: 68-73.

70. Lee, S. K., Young, R. E., Shiffman, P. M., and Coggins, C. W. Jr. (1983). Maturity studies of avocado fruit based on picking date and dry weight. *Journal of the American Society for Horticultural Science*, 108: 390–394.

71. Lee, S.K. and Kader, A.A. (2000). Preharvest and postharvest factors influencing vitamin C content of horticultural crops. *Postharvest Biology and Technology* 20: 207-220.

72. Levinea, L. H., and Paré, P. W. (2009). Antioxidant capacity reduced in scallions grown under elevated CO2 independent of assayed light intensity. <http: // www. sciencedirect.com/science?_ob=ArticleURL and _udi=B6V3S-4WMDHPN-7641 and _user=10 and _rdoc=1 and _fmt= and _orig=search and _sort=d and _docanchor765= and view=c and _searchStrId=990842389 and _rerunOrigin=google and _acct=C000050221 and _ vers766ion=1 and _urlVersion=0 and _userid=10 and md5=ee834de588a43459bc1fe2f62 060e919> Accessed 28.07.09.

73. Lide, D. R. (2009). CRC handbook of chemistry and physics (90th ed.). Boca Raton: CRC Press. pp. 2804.

74. Lin, Z., Zhong, S. and Grierson, D. (2009). Recent advances in ethylene research. *Journal of Experimental Botany*, 60: 3311-3336.

75. Lloyd, J., and Farquhar, G. D. (2008) Effects of rising temperatures and [CO_2] on the physiology of tropical forest trees. *Philosophical Transactions of the Royal Society of Biological Sciences*, 363: 1811–1817.

76. Lurei, S. (1998). Postharvest heat treatments. *Postharvest Biology and Technology*, 14: 257-269.

77. Malik, A. U. and Singh, Z. (2005). Pre-storage application of polyamines improves shelf-life and fruit quality of mango. *J. Hortic. Sci. Biotechnol.*, 80(3): 363-369.

78. Mattheis, J.P. and Fellman, J.K. (1999). Preharvest factors influencing flavor of fresh fruit and vegetables. *Postharvest Biology and Technology*, 15: 227–232.

79. Mauzerall, D. L., and Wang, X. (2001). Protecting agricultural crops from the effects of tropospheric ozone exposure: Reconciling science and standard setting in the United States, Europe, and Asia. *Annual Review of Energy and the Environment*, 26: 237–268.

80. McDonald, R.E., Miller, W.R., McCollum, T.G. and Brown, G.E. (1991). Thiabendazole and imazalil applied at 53°C reduce chilling injury and decay of grapefruit. *HortScience*, 26: 397–399.

81. Melnick, R. (1998). Safety sets the table. *American Vegetable Grower*, February. pp. 9– 11, 13, 15.

82. Oltmans, S. J., and Levy, H. (1994). Surface ozone measurements from a global network. *Atmospheric Environment*, 28: 9–24.

83. Pal, R. K., and Buescher, R. W. (1993). Respiration and ethylene evolution of certain fruits and vegetables in response to carbon dioxide in controlled atmosphere storage. *Journal of Food Science and Technology*, 30: 29–32.

84. Paull, R.E. (1994). Response of tropical horticultural commodities to insect disinfestation treatments. *HortScience,* 29: 988–996.

85. Peppelenbos, H. W., and van't Leven, J. (1996). Evaluation of four types of inhibition for modelling the influence of carbon dioxide on oxygen consumption of fruits and vegetables. *Postharvest Biology and Technology,* 7: 27–40.

86. Percy, K. E., Legge, A. H., and Krupa, S. V. (2003). Troposphere ozone: A continuing threat to global forests? In: Karnosky, D. F. E. A. (Ed.), Air pollution. Elsevier Ltd.: Global Change and Forests in the New Millenium. pp. 85–118.

87. Petit, J. R., Jouzel, J., Raynaud, D., Barkov, N. I., Barnola, J. M., Basile, I., Bender, M. Chappellaz, J., Davisk, M., Delaygue, G. Delmotte, M. Kotlyakov, V. M. Legrand, M. Lipenkov, V. Y., Lorius, C., Pepin, L., Ritz, C. Saltzmank, E. and Stievenard, M. (1999). Climate and atmospheric history of the past 420, 000 years from the Vostok ice core, Antarctica. *Nature,* 399: 429–436.

88. Picton, S., and Grierson, D. (1988). Inhibition of expression of tomato-ripening genes at high temperature. *Plant Cell and Environment,* 11: 265–272.

89. Piikki, K., Vorne, V., Ojanperä, K., and Pleijel, H. (2003). Potato tuber surgars, starch and organic acids in relation to ozone exposure. *Potato Research,* 46(1–2): 67–79.

90. Pusey, P.L. (1989). Use of Bacillus subtilis and related organisms as biofungicides. *Pesticide Sci.,* 27: 133-140.

91. Pusey, P.L. and C.L. Wilson. (1984). Postharvest biological control of stone fruit brown rot by *Bacillus subtilis. Plant Dis.,* 68: 753-756.

92. Qiu, Y., M.S. Nishina and R.E. Paull. (1995). Papaya Fruit Growth, Calcium Uptake, and Fruit Ripening. *J. Amer. Soc. Hort. Sci.* 120(2): 246-253.

93. Reich, P. B. (1987). Quantifying plant response to ozone: A unifying theory. *Tree Physiology,* 3: 63–91.

94. Reid, M.S. and Staby, G.L. (2008). A brief history of 1-methylcyclopropene. *HortScience,* 43: 83-85.

95. Roberts, R.G. (1990). Biological control of gray mold of apple by *Cryptoccus laurentii. Phytopathology,* 80: 526-530.

96. Romero, D. M., Serrano, M., Carbonell, A., Burgos, L., Riquelme, F. and Valero, D. (2002). Effects of postharvest putrescine treatment on extending shelf life and reducing mechanical damage in apricot. *Food Chem. Toxicol.,* 67(5): 1706-1712.

97. Saltveit, M.E. (1999). Effect of ethylene on quality of fresh fruits and vegetables. *Postharvest Biology and Technology,* 15: 279-292.

98. Salvador, A., Abad, I., Arnal, L., and Martinez-Javegam, J. M. (2006). Effect of ozone on postharvest quality of persimmon. *Journal of Food Science,* 71(6): 443–446.

99. Sargent, S. A., and Moretti, C. L. (2002). Tomato. <http://www.ba.ars.usda.gov/hb66/138tomato.pdf> Accessed 22.03.09.

100. Schirra, M. and Mulas, M. (1995a). Influence of postharvest hot-water dip and imazalil-fungicide treatments on cold-stored 'Di Massa' lemons. *Adv. Hort. Sci.,* 1: 43–46.

101. Schirra, M. and Mulas, M. (1995b). Improving storability of 'Tarocco' oranges by postharvest hot-dip fungicide treatments. *Postharv. Biol. Technol.* 6: 129–138.

102. Schlesinger, W. H. (1991). Biogeochemistry: An analysis of global change. New York: Academic Press. pp. 443.

103. Seleznyova, A.N., Thorp, T.G., White, M., Tustin, S. and Costes, E. (2003). Application of architectural analysis and AMAPmod methodology to study dwarfing phenomenon: the branch structure of "Royal Gala" apple grafted on dwarfing and non-dwarfing rootstock/interstock combinations. *Annals of Botany,* 91: 665–672.

104. Sharma, R. R., Patel, V. B. and Krishna, H. (2006). Relationship between light, fruit and leaf mineral content with albinism incidence in strawberry (*Fragaria x ananassa* Duch.). *Sci. Hortic.,* 109: 66-70.

105. Shear, C.B. (1975). Calcium-related disorders of fruits and vegetables. *HortScience,* 10: 361–365.

106. Shellie, K.C. and Mangan, R.L. (1994). Postharvest quality of 'Valencia' orange after exposure to hot, moist forced air for fruit y disinfestation. *HortScience,* 29: 1524–1527.

107. Shin, Y., Ryu, J. A., Liu, R. H., Nock, J. F., Polar-Cabrera, K., and Watkins, C. B. (2008). Fruit Quality, Antioxidant Contents and Activity, and Antiproliferative Activity of Strawberry Fruit Stored in Elevated CO_2 Atmospheres. *Journal of Food Science,* 73(6): 339–344.

108. Singh, H. B., Ludwig, F. L., and Johnson, W. B. (1978). Tropospheric ozone: Concentrations and variabilities in clear remote atmospheres. *Atmospheric Environment,* 12: 2185–2196.

109. Siriphanich, J. (1998). High CO_2 atmosphere enhances fruit firmness during storage. *Journal of the Japan Society of Horticultural Science,* 6: 1167–1170.

110. Skog, L. J., and Chu, C. L. (2001). Effect of ozone on qualities of fruits and vegetables in cold storage. *Canadian Journal of Plant Science,* 81: 773–778.

111. Smilanick, J.L., Mackey, B.E., Reese, R., Usall, J. and Margosan, D.A. (1997). Inuence of concentration of soda ash, temperature, and immersion period on the control of postharvest geen mold of oranges. *Plant Dis.,* 80: 79–83.

112. Smilanick, J.L., Margosan, D.A. and Henson, D.J. (1995). Evaluation of heated solutions of sulfur dioxide, ethanol, and hydrogen peroxide to control postharvest green mold of lemons. *Plant Dis.,* 79: 742–747.

113. Snelgar, W.P., Hopkirk, G., Seelye, R.J., Martin, P.J. and Manson, P. (1998). Relationship between canopy density and fruit quality in kiwi fruit. *New Zealand Journal of Crop and Horticultural Science,* 26: 223–232.

114. Song, J., Fan, L., Forney, C. F., Jordan, M. A., Hildebrand, P. D., Kalt, W., *et al.* (2001). Effect of ozone treatment and controlled atmosphere storage on quality and phytochemicals in high blueberries. *Acta Horticulturae*, 600: 417–424.

115. Spotts, R.A., Cervantes, L.A., Facteau, T.J. and Chand-Goyal, T. (1998). Control of brown rot and blue mold of sweet cherry with preharvest iprodione, postharvest *Cryptococcus infirmo-miniatus*, and modified packaging. *Plant Dis.*, 82: 1158-1160.

116. Stevens, C., Khan, V.A., Lu, J.Y., Wilson, C.L., Pusey, P.L., Kabwe, M.K., Igwegbe, E.C.K., Chalutz E. and Droby, S. (1998). The germicidal and hormetic effects of UV-C light on reducing brown rot disease and yeast microflora of peaches. *Crop Protection,* 17: 75-84.

117. Suslow, T. (1997). Microbial food safety: an emerging challenge for small-scale growers. *Small Farm News,* June–July. pp. 7–10.

118. Swalding, I.R. and Jeffries, P.J. (1998). Antagonistic properties of two bacterial biocontrol agents of grey mould disease. *Biocontrol Sci. Technol.* 8: 439-448.

119. Tjoelker, M. G., Volin, J. C., Oleksyn, J., and Reich, P. B. (1995). Interaction of ozone pollution and light effects on photosynthesis in a forest canopy experiment. *Plant, Cell and Environment*, 18: 895–905.

120. Tjoelker, M. G., Volin, J. C., Oleksyn, J., and Reich, P. B. (1993). Light environment alters response to ozone stress in seedlings of *Acer saccharum* Marsh and hybrid Populus L. I. *In situ* net photosynthesis, dark respiration and growth. *New Phytologist*, 124: 627–636.

121. Tustin, D.S., Cashmore, W.M. and Bensley, R.B. (2001). Pomological and physiological characteristics of Slender Pyramid centre leader apple (*Malus domestica*) planting systems grown on intermediate vigour, semi-dwarfing and dwarfing rootstocks. *New Zealand Journal of Crop and Horticultural Science*, 29: 195–208.

122. Tzortzakisa, N., Borlanda, A., Singletona, I., and Barnes, J. (2007). Impact of atmospheric ozone-enrichment on quality-related attributes of tomato fruit. *Postharvest Biology and Technology*, 45(3): 317–325.

123. Valero, D., Martinez, D., Riquelme, F. and Serrano, M. (1998). Polyamine response to external mechanical bruising in two mandarin cultivars. *Hort Science.*, 33(7): 1220-1223.

124. Vinas, I., Usall, J., Texido, N. and Sanchis, V. (1998). Biological control of major postharvest pathogens in apple with Candida sake. *Int. J. Food Micro.*, 40: 9-16.

125. Vorne, V., Ojanperä, K., De Temmerman, L., Bindi, M., Högy, P., Jones, M., *et al.* (2002). Effects of elevated carbon dioxide and ozone on potato tuber quality in the European multiple-site experiment 'CHIP-project'. *European Journal of Agronomy*, 17(4): 369–381.

126. Wang, S. Y., and Zheng, W. (2001). Effect of plant growth temperature on antioxidant capacity in strawberry. *Journal of Agricultural Food Chemistry*, 49: 4977–4982.

127. Wang, S. Y., Bunce, J. A., and Maas, J. (2003). Elevated carbon dioxide increases contents of antioxidant compounds in field-grown strawberries. *Journal of Agricultural and Food Chemistry*, 51: 4315–4320.

128. Watada, A.E., Herner, R.C. and Kader, A.A. (1984). Terminology for the description of developmental stages for horticultural crops. *HortScience*, 19: 20–21.

129. Went, F. W. (1953). The Effect of Temperature on Plant Growth. *Annual Review of Plant Physiology*, 4: 347–362.

130. Wild, B.L. (1993). Reduction of chilling injury in grapefruit and oranges stored at 1°C by prestorage hot dip treatments, curing, and wax application. *Aust. J. Exp. Agric.*, 33: 495–498.

131. Wills, R.B.H., McGlasson, W.B., Graham, D. and Joyce, D.C. (2007). Postharvest-An introduction to the physiology and handling of fruits, vegetables and ornamentals (5th ed.). CAB International, Oxfordshire, UK, pp.227.

132. Wisniewski, M., Biles, C., Droby, S., McLaughlin, R., Wilson, C.and Chalutz, E.(1991). Mode of action of the postharvest biocontrol yeast Pichia guilliermondii. 1. Characterization of attachment to *Botrytis cinerea. Physiol. Molec. Plant Pathol.* 39: 245-258.

133. Woolf, A. B., and Ferguson, I. B. (2000). Postharvest responses to high fruit temperatures in the field. *Postharvest Biology and Technology*, 21: 7–20.

134. Woolf, A. B., Ferguson, I. B., Requejo-Tapia, L. C., Boyd, L., Laing, W. A., and White, A. (1999). 'Impact of sun exposure on harvest quality of 'Hass' avocado fruit. *Revista Chaingo Serie Horticultura*, 5: 352–358.

135. Wurr, D. C. E., Fellows, J. R. and Phelps, K. (1996). Investigating trends in vegetable crop response to increasing temperature associated with climate change. *Scientia Horticulturae*, 66: 255–263.

2015, Horticulture for Nutrition Security
Editor: **Prof. K.V. Peter**
Published by: **DAYA PUBLISHING HOUSE, NEW DELHI**

Pages 301–329

Chapter 14

Physiology of Spoilage of Temperate Fruits

Parshant Bakshi, Ganganpreet Kour
and F.A. Masoodi

Fruits are very important food commodities not only in India but all over the world. India, which is the second most populated country of the world, is still struggling to achieve self-sufficiency to feed about 800 million people. For this purpose, fruits have got their specific importance to provide a balanced and healthy diet to the people. India is the second largest producer of fruits in the world. Though, we are producing adequate quantities of fruits, yet on account of losses in the field as well as in storage, they become inadequate. Generally, 20-25 per cent of fresh harvested produce is rendered unfit for consumption due to spoilage after harvesting. Post harvest loss of fruits is defined as "that weight of wholesome edible product (exclusive of moisture content) which is normally consumed by human and that has been separated from the medium and sites of its immediate growth and production by deliberate human action with the intention of using it for human feeding but which for any reasons fails to be consumed by human". Not only quantity and quality but even the appearance of fruits are affected and their market value is reduced. Fresh fruits are living dynamic systems. They respire and transpire even after harvest. Respiration is the process of O_2 uptake and oxidation of energy-rich cellular organic substances such as starch, sugar and organic acids to produce CO_2 and H_2O and energy. Before harvesting, when fruits are attached to parent plant, losses due to respiration and transpiration are replaced by water, photosynthates and minerals from the plants. After harvest, when they are removed from the plant, losses of respirable substrates and moisture are not replaced and deterioration occurs. For instance, plum fruit is highly delicate and perishable and demands immediate disposal and utilization. After harvesting, biochemical changes in fruits are continuous which lead to fruit softening and spoilage. The acidity of the fruit increased with days due to spoilage. Consequently, the taste of spoilt fruit ranges from loss of good characteristic tastes to the development of objectionable tastes. Thus, a spoilt fruit develops acidic taste which is often bitter (Ihekeronye and Ngoddy, 1985). If these changes are reduced, the storage life of fresh fruits can effectively be increased and spoilage can be reduced. Harvested fruits are highly perishable agricultural commodities, and their damaged tissues themselves function as an excellent substrate for the growth of spoilage and pathogenic micro-organisms. These micro-organisms cause the decay of fresh fruits or serve as medium of human disease (Beuchat, 1998; Sharma *et al.,* 2009).

What is Fruit Spoilage?

All the changes in fruit after harvest which cause it to lose its desired quality and eventually become inedible are called fruit spoilage or rotting. As long as they are not harvested, their quality remains relatively stable if they are not damaged by disease or eaten by insects or other animals. Once the fruits are cut off from their natural nutrient supply, their quality begins to diminish. This is due to a natural process that starts as soon as the biological cycle is broken by harvesting. As soon as it is harvested, the fruit then begins to spoil or rot.

Types of Spoilage

The peel of a fruit provides natural protection against micro-organisms. However, if this shield is damaged by falling, crushing, cutting, peeling or cooking, the chance of spoilage increases considerably. The various types of spoilage in temperate fruits are:

☆ Physical spoilage

☆ Physiological ageing

☆ Spoilage due to insects or rodents

☆ Mechanical damage

☆ Chemical and enzyme spoilage

☆ Microbial spoilage

Physical spoilage is caused by dehydration. Physiological ageing occurs as soon as the biological cycle is broken through harvesting. Neither of the process can be prevented, but they can be delayed by storing the fruits in a dry and draft-free area at as low a temperature as possible. Insects and rodents can cause a lot of damage. Not only by eating the products, but also by passing on micro-organisms through their hair and droppings. The affected parts of the plants are susceptible to diseases. Chemical and enzyme spoilage occur especially when fruits are damaged by falling or breaking. Such damage can release enzymes that trigger chemical reactions. For example apples and other types of fruit turn brown. The fruit can also become rancid.

The same processes can also be triggered by insects: the fruit becomes damaged, which causes enzymes to be released.

Chemical Changes Responsible for Spoilage of Fruits

Enzymes

Enzymatic activities in fruits can cause browning and softening of tissues. Typically, these reactions are catalysed by phenoloxidase enzymes, which react with phenol compounds and oxygen to form undesirable brown pigments. Another form of browning which happens due to non-enzymatic activity is Millard Browning. This non-enzymatic browning occurs due to reaction between proteins (amino-acids) and reducing sugars. This is associated with loss in nutritional value along with the browning and change in the texture of food products. The essential amino acid lysine, which readily reacts with reducing sugars, is quickly lost. Gong *et al.* (2001) reported that the incidence of browning disorder was related to the harvest time of fruit and the activity of superoxide dismutase. Superoxide dismutase activity was lower in late harvested 'Braeburn' apple fruit that had a higher disorder incidence. The risk of browning disorder was higher after growing seasons with lower cumulative temperature and lower activities of superoxide dismutase and catalase in fruit. The fruit harvested from the orchard located in a colder area had lower catalase activity, lower lipid soluble antioxidant levels, and a higher risk of browning disorder.

Phenolics and Tissue Browning

Tissue browning due to oxidation of phenolic compounds by polyphenol oxidase results from loss of compartmentalization within the cells when exposed to physical or physiological stresses. Lu and Toivonen (2000) observed a slower browning rate during storage at 1°C for 2 weeks of slices made from 'Spartan' apples that had been kept in 100 kPa O_2 for 12 days at 1°C prior to cutting compared with those kept in air. Exposure of 'Bartlett' pear slices to 40, 60, or 80 kPa O_2 for 4 days at 10°C did not influence their browning rate compared with slices kept in air. Ding *et al.* (2009) studied the biological basis of browning in peach fruit during storage at low temperatures and they found decrease in H_2O_2 content was correlated with browning, whereas phenol content and activities of PPO and POD were not correlated with the change in H_2O_2 content. Moreover, H_2O_2 content was influenced by different responses of antioxidants at different storage conditions. The main effect of H_2O_2 on browning was to regulate its appearance and development as a signal molecule, and lower H_2O_2 content was beneficial to browning. The polyphenols responsible for browning of temperate fruits are enlisted in Table 14.1.

Physiological Spoilage

Solomos *et al.* (1997) reported that 'Gala' and 'Granny Smith' apples exposed to 100 kPa O_2 developed extensive skin injury to that which occurs under 1 kPa or lower O_2 atmospheres. The activity of cis-aconitase in the fruit was inhibited by 100 kPa O_2, thereby disrupting the TCA cycle. Whitaker *et al.* (1998) reported that production of farnesene and trienol, related to development of storage scald, increased in apples kept in 100 kPa O_2 atmospheres at 0°C for up to 3 months. 'Granny Smith' apples

Table 14.1: Polyphenols Involved in Browning of Temperate Fruits

Source	Phenolic Substrates
Apple	Chlorogenic acid (flesh), catechol, catechin (peel), caffeic acid, 3,4-dihydroxy-phenylalanine (DOPA), 3,4-dihydroxy benzoic acid, *p*-cresol, 4-methyl catechol, leucocyanidin, *p*-coumaric acid, flavonol glycosides
Apricot	Isochlorogenic acid, caffeic acid, 4-methyl catechol, chlorogenic acid, catechin, epicatechin, pyrogallol, catechol, flavonols, *p*-coumaric acid derivatives
Peach	Chlorogenic acid, pyrogallol, 4-methyl catechol, catechol, caffeic acid, gallic acid, catechin, dopamine
Pear	Chlorogenic acid, catechol, catechin, caffeic acid, DOPA, 3,4-dihydroxy benzoic acid, *p*-cresol
Plum	Chlorogenic acid, catechin, caffeic acid, catechol, DOPA

stored under 100 kPa O_2 were completely 'bronzed' after 3 months and contained high ethanol concentrations. Ripening of mature-green, climacteric fruits was slightly enhanced by exposure to 30-80 kPa O_2, but levels above 80 kPa retarded their ripening and caused O_2 toxicity disorders on some fruits. High O_2 concentrations enhance some of the effects of ethylene on fresh fruits, including ripening, senescence, and ethylene-induced physiological disorders (Kader and Ben-Yehoshua, 2000). Apple scald causes more losses than all the other physiological storage diseases combined, being particularly severe on York Imperial, Grimes, Black Twig, Arkansas Black, Rome, and Stayman. It can be distinguished from all other apple diseases by its preference for the greener side of the fruit, the flesh of which is sometimes decayed to a depth of half an inch. In general, apples held at 60° to 70° F scald three to four weeks earlier than those held at 50°; those at 50° about four weeks earlier than those at 40°; and those at 40° about three weeks earlier than those at 32° F. The higher temperatures are frequently encountered in cases of delayed storage. The time immediately following picking is a critical period during which refrigeration is urgently required. There are, however, other factors besides temperature to be considered. The green portion of the skin is the most susceptible to scald, and measures to secure proper colouring of the fruit are desirable. Castro *et al.* (2008) observed flesh browning (FB) disorder in stored Pink Lady apples in air or controlled atmosphere (CA) with 1.5 kPa O_2 and 5 kPa O_2 at 0.5°C for 2 and 4 months and reported that both brown and surrounding healthy tissues in apples with FB showed a decrease in ascorbic acid and an increase in dehydroascorbic acid during the first 2 months of storage in CA. Undamaged, CA-stored apples retained a higher concentration of ascorbic acid after 2 months in storage. The level of hydrogen peroxide (H_2O_2) increased more in the flesh of CA-stored apples than in air-stored apples, an indication of tissue stress. Further, polyphenoloxidase (PPO) activity was similar for apples kept in air or CA storage and between undamaged and damaged fruit.

Biotic Spoilage

Damaged fruits, which are usually somewhat acidic, are very susceptible to the growth of bacteria (*e.g. Lactobacillus*), yeasts (*e.g. Saccharomyces*), and molds (*e.g. Rhizopus*). Though, not visible to the naked eye, bacteria, yeast and molds can be present in large numbers (Figure 14.1). Spoilage caused by yeasts, moulds and bacteria develops slowly and is not always noticeable. The most important sources of microbial contaminations are sand, water, air, and pests such as insects and rodents. Micro-organisms are everywhere around us. The occurrence of spoilage in fruits by micro-organisms depends on the types of organisms present and whether the fruit under its existing condition of storage can support the growth of any or all of them. Only certain species out of all the organisms present in a fruit will be able to thrive well and numerous cell wall degrading enzymes can be secreted by these pathogens to breach and use the plant cell walls as nutrient sources that reduced post-harvest life and finally lead to develop inedible, undesirable quality and soft rot spoilage (Raviyan *et al.*, 2005; Tomassini *et al.*, 2009; Al-Hindi *et al.*, 2011). Spoilage by micro-organisms may be influenced by some qualities such as water content, pH, temperature, texture and nutrient composition of the fruit (Lloyd, 1993). Also, transporting equipment such as sack, basins, baskets etc. after harvesting can serve as prime source of pathogen transmission. Water used in washing the fruits may be contaminated with large number of microbes and as such transfer them directly to the fruit and due to the high nutritional constituents of the fruits they serve as a fertile ground for microbes especially fungi (Salunkhe, 1974). Charley (1999) observed that fresh fruits carry out the physiological function of respiration, thereby absorbing and releasing gases and other materials from and to environment. This condition makes them susceptible to microbial infection under condition of high ambient temperature and relative humidity which lead to their deterioration during post harvest period. Physiological stability of fruit varies and the damage to the quality of the fruit does not occur until the micro-organisms have gained access to the living tissues through the skin. Fruits come in contact with the disease causing organisms through different sources such as field infections or during harvesting if it comes in contact with the soil which is an important factor that aids spoilage of stored fruits. Air is another factor responsible for spoilage of fruits during their post-harvest period, fungal spores are always in the

Figure 14.1: *Penicillium expansum*   *Monilinia* sp.

air, these spores and other particulate ones constitute the air spora (Holtmeyer and Wallin, 1981).

Fungus

Most of the fungal species responsible for food spoilage spend most of their time in soil. The micro-organisms responsible for fruit spoilage are almost always present in the fruits long before the fruit actually spoils. Spoilage fungi are considered toxigenic or pathogenic. Toxigenic fungi have been isolated from spoiling fruits (Tournas and Stack, 2001). Pathogenic fungi, on the other hand, could cause infections or allergies (Monso, 2004). These micro-organisms can make their way into fruit "on the seed itself, during crop growth in the field, during harvesting and post-harvest handling or during storage and distribution" via such vectors as "soil particles, airborne spores, and irrigation water". Some of the most common fungal species responsible for fruit spoilage include species of the *Penicillium, Geotrichum, Fusarium, Botrytis, Colletotrichum, Micor, Monilinia, Rhizopus* and *Phyophthora*. Each of these groups contains numerous individual species, and these fungi are known to infect the most commonly eaten fruits in the country, including apples, berries, peaches and pears. Apple decay in the cooler regions was caused mainly by *Colletotrichum* spp. which is one of the widespread pathogenic micro-organisms during storage in chambers without modified atmosphere (Osterloh, 1994). Fruits contain high levels of sugar and nutrient element and their low pH values make them particularly desirable to fungal decayed (Singh and Sharma, 2007). Tancinova *et al.* (2013) isolated and identified 8 genera (*Penicillium, Monilinia, Botrytis, Aspergillus, Cladosporium, Epicoccum, Fusarium* and *Geotrichum*) of filamentous fungi from 30 apples with rotting and reported that *Penicillium expansum* was mainly responsible for rotting of 40 per cent apples during storage (Table 14.2).

Table 14.2: Microscopic Fungi Causing Rot of Apples

Sl.No.	Variety	Determined Pathogens
1.	Braeburn	*Botrytis cinerea; Penicillium expansum;* Yeast
2.	Without variety determination, sold as "red loose"	*Penicillium expansum; Penicillium solitum; Cladosporium* sp.; Yeast
3.	Jonagold	*Monilinia* sp.; Yeast
4.	Red Chief	*Penicillium expansum*
5.	Idared	*Penicillium expansum, Penicillium funiculosum, Penicillium citrinum*
6.	Rubín	*Penicillium solitum,* Yeast
7.	Golden Delicious	*Penicillium expansum, Cladosporium* sp.
8.	Red Delicious	*Penicillium expansum, Penicillium roqueforti*
9.	Galla	*Penicillium expansum, Penicillium citrinum, Fusarium* sp.
10.	Fugi	*Alternaria* sp.; *Penicillium expansum, Geotrichum* sp.; *Monilinia* sp.; *Aspergillus versicolor,* Yeast
11.	Glosten	*Penicillium expansum*
12.	Granny Smith	No fungus
13.	Glosten	*Epicoccum nigrum, Cladosporium* sp.

Cell Wall Degradation

The process of infection in case of fungal invasion involves the development of fungal penetrating structure and colonization of fungi. It is a critical phase in the microbial spoilage of post harvested fruits involving the ability of the fungi to establish itself within the host and the magnitude of symptoms of the induced disease is a reflection of the extent of colonization (Chuku *et al.*, 2008). Spoilage fungi exploit the fruit using extracellular lytic enzymes that degrade the cell wall of fruit to release water and other intercellular constituents for using as nutrients for their growth. Therefore, the 445 isolated fungi *F. oxysporum, A. oryzae, A. awamori, A. phoenicis, A. tubingensis, A. niger, A. flavus, A. japonicus, A. foetidus* and *R. stolonifer* were cultured on their spoilage fruits of banana, orange, lemon, tomato and peach, apple, grape, date, mango, pokhara, apricot and kiwi peels in comparison with potato dextrose broth (PDB) medium. Xylanase, polygalacturonase, cellulase and amylase were detected in the cell-free broth of all tested fungi. Xylanase and polygalacturonase had the highest level of contents as compared to the cellulase and amylase. This is consistent with the results of several researchers who suggested that polygalacturonases and xylanases are important pathogenecity factors for spoilage fungi (Dimatteo *et al.*, 2006). Recently, Niturea *et al.* (2008) reported that in both acidic and alkaline conditions, the organism produced significant levels of inducible xylanase and amylase enzymes and the production of cellulase was lower compared with other enzymes. The secretion of pectin degrading enzymes during infection to the plants has been reported from various plant pathogenic fungi such as *F. oxysporum, Botrytis cinerea, Sclerotinia sclerotiorum*. In the present study, xylanase had the highest level in *A. tubingensis* (2786 ± 55 units/100 ml), and *A. awamori* (1713 ±85 units/100 ml) grown on peach and lemon with agitation, respectively, and *F. oxysporum* (3535 ± 176 units/100 ml), *A. niger* (1289 ± 77 units/100 ml) grown on banana and apple with stationary, respectively. Low level of xylanase activity (> 500 units/100 ml and < 1000 units/100 ml) was detected for some tested fungi grown on fruit peels and PDB with agitation and stationary phases. Comparing the polygalacturonase, the highest level of activity was detected in the *A. japonicus* (7433 ± 327 units/100 ml), *R. stolonifer* (4547 ± 227 units/100 ml), *A. niger* (4197 ± 209 units/100 ml) grown on PDB with stationary and *A. oryzae* (3124 ± 62 units/100 ml) and *A. niger* (4416 ± 44 units/ 100 ml) grown on orange peel and PDB with agitation, respectively. Several tested fungi grown on fruit peels and PDB with agitation and stationary phases had moderate (> 500 units/100 ml and < 2000 units/100 ml) and low (< 500 units/100 ml) polygalacturonase activity levels. The levels of xylanase activity were very low in all tested fungi grown on PDB as compared to fruit peels. Previous studies reported that the same tested fungi have been produced from several plant cell wall degrading enzymes. Di Pietro *et al.* (2003) stated that *Fusarium* was able to secrete several cell wall degrading enzymes such as cellulase, xylanase, amylase and pectinase. *F. oxysporum* produced high level of xylanase (Simoes *et al.*, 2009). Tissues infected by *F. oxysporum* produced the highest pectolytic enzyme activity among the fungi studied (Bahkali *et al.*, 1997). Filamentous fungi, *Aspergillus* spp. are widely distributed among the spoilage fruit fungi and also secreted several plant cell wall degrading enzymes. Induction of polygalacturonases from *A. oryzae* by pectin was significantly higher than when rinds of citrus fruits were used as inducer. *A. oryzae* produced xylanase and polygalacturonase in solid-state and submerged cultures (Oda *et al.,* 2006). Botrytis fruit rot is a universal problem in strawberry growing areas. The minimum growth temperature for the pathogen is about -2°C (Sommer *et al.,* 1973). Development continues on cold-stored fruit although progress is most rapid at room temperature, especially if free moisture is present (Ryall and Pentzer, 1982). Infected fruits show brown discoloration, but even when completely decayed, fruit may preserve their integrity with very little juice exudation. Afterward, the mycelium grows, changing colour from white to gray as the sporulation begins (Dennis, 1983; Maas, 1992). The 'box rot' of dried French prunes, which is soft, sticky, macerated areas on the fruit and slippage of the skin under slight pressure due to the activity of pectinolytic enzymes produced by these fungi (Sholberg and Ogawa, 1983).

Wound Infection

Many pathogens associated with quiescent infections penetrate host surfaces through wounds or natural infections, whereas others are restricted to wounds for gaining entry. Wounds in apple fruit are primary infection sites for *Botrytis cinerea* Pers. Fr and *Penicillium expansum* Link, fungi that cause gray mold and blue mold, respectively (Rosenberger 1990; Prins *et al.,* 2000). These diseases are particular problems in the post-harvest storage of apple fruits. It is generally accepted that as plant wounds age, they undergo healing, which is accompanied by the development of resistance in wounds to pathogen infection (Bostock and Stermer, 1989). Wound healing in apple fruits increased resistance in puncture wounds to infection by *B. cinerea* and *P. expansum* (Lakshminarayana *et al.,* 1987). Wounds that were inoculated within a few minutes after infliction had greater gray mold or blue mold incidence after incubation compared to wounds that were four days old prior to inoculation. Diameters and incidences of gray mold and blue mold at 22°C in fruits inoculated with 50 conidia per wound were the greatest in fresh puncture wounds than in 1 day-old and 3 day-old puncture wounds. Conversely, both fungal decays were the lowest in slice wounds compared to greater decay levels in older wounds. Wounds inoculated with 500 conidia exhibited similar decay reactions. Decay diameter and incidence caused by either pathogen in puncture wounds were the greatest in fresh wounds and least in 3 day-old wounds. The converse was found in slice wounds. Decay diameter and incidence in wounds of fruits stored at 3°C for 30 days were the greatest in fresh wounds and least in 3 day-old puncture wounds. The opposite effect was observed with slice wounds (Filonow, 2005). Two wound pathogens, *Penicillium expansum* and *Botrytis cinerea*, if not scrupulously cleaned from fruits prior to storage or if fruits with infected wounds have not thoroughly been culled from the lot, can cause significant crop loss as these spoilage fungi eventually degrade the wound sites, create lesions, and cross-contaminate adjacent fruits. If fruits receive improper pre-harvest fungicide application, poor washing, and/or inadequate culling, an expanding infestation of spoilage micro-organisms can destroy a substantial portion of a stored lot of fruits. *P. expansum* and *B. cinerea* are pathogens of apples, pears and a number of other pectin-rich fruits. *B. cinerea* is an especially sophisticated and selective plant pathogen which possesses multiple cutinases and lipases that are capable of degrading plants rich in pectin (Van Kan, 2006). Colonization and lesion development more typically and more rapidly occur within the damaged or otherwise compromised plant tissue. External damages such as bruising, cracks and punctures create sites for establishment and outgrowth of the spoilage microbes. Lesion development can relatively be rapid, occurring within days or weeks. This presents the risk that rapidly reproducing spoilage micro-organisms will arrive within open wound sites at the packing facility, and thereby, through shedding from the asymptomatic wound, present the potential for cross contamination within the facility during handling, culling, washing, sorting and packing before storage. Such cross-contamination to some degree is inevitable and, if not carefully managed with a robust facility sanitation program, could lead to the establishment of a population of spoilage microbes endemic to the facility that may be difficult to eradicate. A further and potentially more serious complication is the introduction of spoilage micro-organisms into the cold storage facility already established in wound sites on product, whether the product is in bins or boxed and palletized. Depending upon storage conditions and storage time (greater than 12 months for certain robust crops), and if not carefully managed, these "primed" spoilage micro-organisms can have a devastating impact on the stored product. Apples, for example, are stored in very large, controlled atmosphere storage rooms, either in wooden bins or boxed and ready for distribution (Watkins *et al.,* 2004). Sever *et al.* (2012) identified 32 *Fusarium* isolates from rotten apple fruit of cultivars Golden Delicious, Jonagold, Idared and Pink Lady, stored in Ultra Low Oxygen (ULO) conditions. *Fusarium* rot was detected in 9.4 per cent to 33.2 per cent of naturally infected apples, depending on the cultivar. *F. pseudograminearum, F. semitectum, F. crookwellense,* and *F. compactum* were identified by morphological characteristics. *F. avenaceum* can produce several mycotoxins and its dominance in *Fusarium* rot points to the risk of mycotoxin contamination of apple fruit juices and other products for human consumption. Pathogenicity tests showed typical symptoms of *Fusarium* rot in most of the inoculated wounded apple fruits. In this respect *Fusarium avenaceum,* as the dominant cause of *Fusarium* rot in stored apple fruits is a typical wound parasite.

Pathogen Attacks

The surfaces of fruit have evolved to prevent as impervious a barrier as feasible with multiple layers of chemical and mechanical defences. Thus, while the first step in spoilage is getting the pathogen onto the fruit, the second, more important step in spoilage is getting the pathogen into the produce. Obligate phytopathogens have evolved a suite of tools that allows them to penetrate the plant defences. These may take the form of specialised physical structures such as fungal appressoria in addition to specialized enzymes that can degrade the cellulose, hemicellulose, pectin and cutin of the epidermal and endodermal tissues. Cellulase is of notable importance with regard to field diseases they are considered to be relatively less significant that pectinases with regard to storage diseases (Brackett, 1997). Many damaging forms of spoilage are caused by organisms that by themselves are unable to penetrate the epidermis of the produce. These organisms, often referred to as secondary or opportunistic pathogens, rely on a breach of the surface integrity to gain entry to the inner tissues of the fruit. This opening may be a naturally occurring anatomical structure opening such as a stomata, hydathode, stem end scar, etc., or it may be a breach caused by an obligate pathogen, a pest or some other biotic agent. Frequently, the breach is of biotic origin, such as a puncture, fracture, abrasion or such other wound resulting from mishandling at some stage of the production cycle. Poorly designed or maintained equipment can cause these wounds at a variety of stage during the growth, harvest, washing, packing or shipping of the produce. *P. expansum* is normally incapable of penetrating the unbroken skin of the fruit and does little harm to the crop on the trees, although it has been observed to do so following injury by codling moth. In commercial storage and in transit, it probably causes 80 to 95 per cent of the total rots, while in the local markets and home storage the losses are estimated to exceed 10 per cent. The fungus enters through stem punctures, but sometimes through finger-nail scratches by pickers, insect injuries, scab spots, bruises and all kinds of wounds. The disease may spread from one apple to another by the dissemination of the spores or by actual contact. Low temperatures greatly delay the development of the mould, more particularly at the inception of decay than during its later development. The losses from the disease may be minimized by careful handling, early cooling of the apples to 32° F and securing sanitary conditions in the packing houses (Fisher, 1922).

Molds

Molds are fungi which cover surfaces as fluffy mycelia and usually produce masses of asexual, or sometimes sexual spores. Some spoilage molds are toxigenic while others are not (Pitt and Hocking, 1997). Spoilage molds can be categorized into four main groups:

Zygomycetes are considered relatively primitive fungi but are widespread in nature, growing rapidly on simple carbon sources in soil and plant debris, and their spores are commonly present in indoor air. Generally they require high water activities for growth and are notorious for causing rots in a variety of stored fruits including strawberries. Some common bread molds also are *zygomycetes*. Some *zygomycetes* are also utilized for production of fermented soy products, enzymes and organic chemicals. The most common spoilage species are *Mucor* and *Rhizopus*. *Rhizopus* is strictly wound invading since the initial infection occurs through breaks, bruises or abrasions on the fruit surface such as peach and strawberry (Maas, 1992). Sporulation occurs only on fruit lesions. The pathogen appears unable to colonize and then sporulate on plant debris in the field (Ryall and Pentzer, 1982; Dennis, 1983). Spores form freely at favourable temperatures and readily infect injuries on other fruits. Diseased fruits usually disintegrate, which release abundant cell sap, hence the disease descriptor, leak. *Zygomycetes* are not known for producing mycotoxins but there are some reports of toxic compounds produced by a few species.

Penicillium and related genera are present in soils and plant debris from both tropical and Antarctic conditions but tend to dominate spoilage in temperate regions. They are distinguished by their reproductive structures that produce chains of conidia. Although they can be useful to humans in producing antibiotics and blue cheese, many species are important spoilage organisms, and some produce potent mycotoxins (*patulin, ochratoxin, citreoviridin, penitrem*). *Penicillium* spp. cause visible rots on citrus, pear and apple fruits and cause enormous losses in these crops. Some species can attack refrigerated and processed products such as jams. A related genus,

Byssochlamys, is the most important organism causing spoilage of pasteurized juices because of the high heat resistance of its spores.

Aspergillus and related molds generally grow faster and are more resistant to high temperatures and low water activity than *Penicillium* spp. and tend to dominate spoilage in warmer climates. Many *aspergilli* produce mycotoxins *viz: aflatoxins, ochratoxin, territrems*, cyclopiazonic acid.

Other molds, belonging to several genera, have been isolated from spoiled fruits. These are not the major causes of spoilage but can be a problem for some processed products. *Fusarium* spp. cause plant diseases and produce several important mycotoxins but are not important spoilage organisms. However, their mycotoxins may be present in harvested produce and pose a health risk. During refrigeration, some molds may produce mycotoxin compounds which are capable of inducing mycotoxicoses in man following ingestion or inhalation (Effiuvwevwere, 2000; Tournas and Stack, 2001). Many of the molds secrete toxic compounds like *aflatoxins, patulin* etc, beside production of *alfatoxins*, and the molds like *A. parasiticus* also produce compositional changes in the fruits (Sinha and Singh, 1987). The highest microbial diversity on the apple fruit surface was found when stored in a cooler conventional atmosphere, suggesting that these conditions were not suitable for apple fruit storage. A modified atmosphere reduced the development of bacteria, yeasts and molds. The total amount of bacteria was reduced by 50 per cent in the used modified atmosphere, as compared to samples stored in a cooler. The proportion of molds increased three-fold in control samples, but their number in the modified atmosphere did not change significantly. Storage of apple fruits in an Ultra Low Oxygen chamber with 1.5 per cent O_2 and 2.5 per cent CO_2 appeared to be most prospective for several commercial apple cultivars. The cultivars 'Auksis' and 'Orlik' showed the least proportion of damaged fruit after storage in these conditions (Juhnevica *et al.,* 2011). Microscopic fungi are the most important spoilage factor of stored apples. Only from one rot apple was not isolated microscopic fungus. From 83 per cent rotten apples molds were isolated. The most important spoilage species *Penicillium expansum* was isolated from 43 per cent rot apples (Table 14.3). All *in vitro* tested isolates of *Penicillium expansum* were detected as producers of mycotoxin patulin (Tancinova *et al.,* 2013).

Table 14.3: *In vitro* Productions of Mycotoxins by *Aspergilli* and *Penicillia* Isolated from Apples Tested by Means of Thin Layer Chromatography

Species	Number of Tested Isolates	Detected Toxin	Evaluation	
			+	−
Aspergillus versicolor	1	Sterigmatocystin	1	0
Penicillium citrinum	1	Citrinin	1	0
Penicillium expansum	12	Patulin	12	0
	12	Citrinin	12	0
Penicillium roqueforti	1	Roquefortin C	1	0

+: Confirmed production of mycotoxin; −: Not detected production of mycotoxins.

Like yeasts, mold populations are reported in various types of fresh-cut fruits (Nguyen-the and Carlin, 1994; Hagenmaier and Baker, 1998) and visible molds have resulted in inedible fresh-cut fruits such as strawberry, honeydew and cantaloupe (O'Connor-Shaw *et al.,* 1994). Since molds are usually detected and enumerated using the same plating media as yeasts and reported in the same category, their species most often are not identified and reported for contamination of fresh-cut produce. High acid (*i.e.* low pH): Spoilage is mainly by molds in tree fruits (*e.g.* apple, peach) and soft fruits (*e.g.* strawberries, raspberries).

Toxigenic Fungi in Dried Fruits

Dried fruit is fruit that is preserved by removing the original water content naturally, through sun drying or artificially, by the use of specialized dryers or dehydrators. Traditional dried fruits such as raisins, figs, dates, apricots and prunes have been a staple of Mediterranean diets for millennia. The Mediterranean region is very favourable for production of dried fruits, not only with its climatic conditions, but also its exceptional fertile lands. Nearly half of the dried fruits sold throughout the world are raisins, followed by dates, prunes (dried plums), figs, apricots, peaches, apples and pears. Dates, prunes, apricots, figs and raisins are the major dried fruits produced in the Mediterranean area. Dried fruits are not perishable but can support mold growth, some of which can produce mycotoxins. Although the most important mycotoxins occurring in Mediterranean crops are aflatoxins (B_1, B_2, G_1 and G_2) and ochratoxin A. The type and level of mycotoxins and toxigenic molds vary by crop and also by country and in some cases geographic location within a country (Ozer *et al.,* 2012). Dried fruits are susceptible to mold growth and mycotoxin formation because of their high sugar content, method of harvest and drying conditions (Trucksess and Scott, 2008). The main problems related to sun drying of fruits are contact with the soil and infection risk by attack of insects and pathogens during outside drying. Moreover, mold growth is related directly with the moisture content of dried fruits (Piga *et al.,* 2004). Toxigenic group of fungi which are commonly found in dried apricots, dates and prunes is *Aspergillus* section Nigri (called black *Aspergilli*) which includes *Aspergillus carbonarius* and the members of *Aspergillus niger* aggregate (Figure 14.2). The black *Aspergilli* are not only most common fungi responsible for spoilage

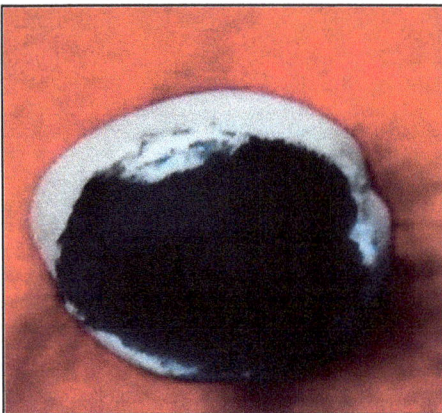

Figure 14.2: *Rhizopus* Rot   *Aspergillus* Rot.

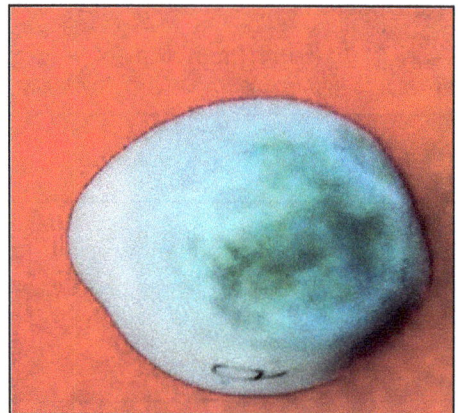

and biodeterioration of materials, but are also extensively used for various biotechnological processes including production of various enzymes and organic acids. In addition, some black *aspergilli* can produce ochratoxin A (OTA) in various food commodities (Schuster *et al.,* 2002). OTA is a mycotoxin produced by several fungal species belonging to *Aspergillus* sections *Circumdati, Flavi* and *Nigri* and by *Penicillium verrucosum* and *P. nordicum*. These species can frequently be found in a variety of beverages, therefore OTA contamination may occur in diverse foodstuffs including coffee, cocoa, beer, wine, grape juice and dried fruits as well as their products (Desphande, 2002). OTA is a potent nephrotoxic and hepatocarcinogenic mycotoxin that primarily affects the kidneys in animals and has been associated with Balkan Endemic Nephropathy and urethelial tumors in humans. OTA is classified as a possible human carcinogen (group 2B) by the International Agency for Research on Cancer (IARC, 1993). Iamanaka *et al.* (2005) analysed 14 dried apricots, 10 dates and 21 prunes (Table 14.4) for the presence of the toxigenic fungi. Dried fruit samples originated from Turkey, Spain, Mexico, Tunisia, USA, Argentina and Chile were analysed. 1.5 per cent of date samples originated mostly from mediterranean countries was mainly contaminated with *A. niger*, while none of the apricot samples were contaminated by any fungi. On the other hand, 8 per cent and 0.5 per cent of the prune samples were contaminated with *A. niger* and *A. ochraceus*, respectively. 15 per cent of *A. niger* strains were found to be ochratoxigenic while, 87 per cent of *A. ochraceus* strains have the capacity of ochratoxin production among all samples. However, it was reported that most of the strains identified previously as *A. ochraceus* should be *A. westerdijkiae*, a new species recently described as very similar and morphologically indistinguishable from *A. ochraceus*. So, most ochratoxigenic isolates which have been previously identified as *A. ochraceus* are now recognized as *A. westerdijkiae* (Frisvad *et al.,* 2004).

Table 14.4: Mycotoxigenic Fungi in Dried Apricots, Prunes and Dates in Mediterranean Crops

Sample	No. of Samples	Fungi	Contaminated Samples[a]
Dried apricot	14	None	0
Direct apricot	1	*P. chrysogenum*	100
Date	10	*A. niger*	NI
Prune	21	*A. niger, A. ochraceus*	80.5
Prune	1	*P. chrysogenum*	100

[a] NI = No information.

Yeast

Yeasts are a subset of a large group of organisms called fungi that also include molds and mushrooms. They are generally single-celled organisms adapted for life in specialized, usually liquid, environments and, unlike some molds and mushrooms, do not produce toxic secondary metabolites. Yeasts can grow with or without oxygen (facultative), since yeasts are anaerobic and can develop in environment without oxygen (Nikolaeva, 2007). The amount of yeasts on apples stored in ultra low oxygen chambers was relatively high. The air contamination level in the cooler was five times higher than that in a controlled modified atmosphere. In addition, the gas composition in the cooler was beneficial for the development of micro-organisms, and it is possible that secondary contamination occurred. Yeast is widely found in the environment including the surface of the fruits, leaves and flowers etc. The yeast growth largely depends on nature of fruit product. Yeast degrades starch, pectin, sugars and changes the composition of a particular processed fruit product and finally leading to its degradation. *Zygosaccharomyces* and related genera tolerate high sugar and high salt concentrations and are the usual spoilage organisms in foods like dried fruit and jams. They usually grow slowly, producing off-odours and flavours and carbon dioxide that may cause food containers to swell and burst. *Debaryomyces hansenii* can grow at salt concentrations as high as 24 per cent, accounting for its frequent isolation from salt brines used for olives.

Spoilage yeasts and molds can grow on raw and processed foods where the environmental conditions for most bacteria are unfavourable (low pH, low water activity (a_w)). The nutrients and oxygen available in the food are the main factors determining the kind of fungal spoilage. Molds require oxygen for their growth, but dissolved oxygen in the foodstuffs is more important here than atmospheric oxygen tension. Fermentative yeasts are able to grow without oxygen. Molds produce a vast number of enzymes: lipases, proteases, carbohydrases for the degradation of complex molecules, and can utilize nitrogen and carbon sources in many forms from nitrates to proteins and from simple sugars to complex carbohydrates. On the contrary, many types of yeast are unable to assimilate nitrate or complex carbohydrates such as starch, and require vitamins for their growth (Pitt and Hocking, 2009). Ethanol fermenting yeasts, *Saccharomyces*, *Schizosaccharomyces*, *Zygosaccharomyces* strains, cause the deterioration of fruit juices, soft drinks and fruit purees. The film-forming yeast *Pichia anomala* is reported to cause spoilage in fruit juices. Mold growth on raw or processed product leads to textural and sensorial changes: softening, off-odours and off-flavours. The most important aspect is, however, the formation of mycotoxins. Mycotoxins are secondary fungal metabolites and are toxic to humans, causing severe disorders like cancer, and immune suppression. Since mycotoxins are very stable and mainly resistant against heat treatment and acidic environment, they remain in the food during processing and storage, causing a serious food safety problem (Filtenborg *et al.*, 1996). Mycotoxins found in juices made from pomaceous or stone fruits are patulin and citrinin. Patulin is a strong antibiotic but it is toxic to humans. Yeasts and molds are able to degrade sorbic acid to 1, 3-pentadiene, causing a kerosene-like off-odour (Filtenborg *et al.*, 1996; Stratford, 2007). Higashihara *et al.* (2009) reported that to prevent fungal spoilage and patulin production it was demonstrated that apples needed to be stored just below 0°C. It was also shown that once fungal decay had commenced, it spread relatively quickly and that patulin was produced, even when apples were stored at +1°C. However, when apples were stored at +5°C or below there was some time lag before decay and the onset of patulin formation. It is very important to store apples at +5°C or less as soon as possible after harvest. Tournas and Memon (2009) analysed a total of 424 apple samples comprised of six varieties (Gala, Red Delicious, Golden Delicious, Fuji, Granny Smith and

Braeburn) for internal fungal contamination and found twelve per cent of the intact apples showed visible growth after 2-4 weeks of incubation at room temperature. *Penicillia* (including the patulin producer, *Penicillium expansum*) were the most frequent, found in 8 per cent of the samples followed by *Fusarium* and *Alternaria* spp. (each found in 3 per cent of the samples tested). The highest mold incidence was observed in the Red Delicious and Fuji and the lowest in the Granny Smith. A variety of microfungi including members of the toxigenic genera *Alternaria*, *Penicillium* and *Fusarium* were isolated from the apple cores. The predominant molds were *Alternaria*, *Cladosporium*, *Penicillium* and *Fusarium* spp. recovered from 50, 22, 33 and 23 per cent of the analysed samples, respectively. Less common were *Ulocladium* spp., *Botrytis cinerea* and *Aureobasidium pullulans* found in less than 4 per cent of the samples. Yeasts were found only in 2 per cent of the samples.

Bacteria

Bacteria exist on plants as part of complex microbial ecosystems. Plant-associated microbial ecosystems are dynamic over time (Upper *et al.*, 1989) becoming larger and more diverse in wet weather and smaller in dry weather. The plant is also dynamic, undergoing growth, maturation, senescence and death. Phyllosphere ecosystem may be profoundly disturbed by severe storms featuring wind, hail or heavy rainfall; insects that create wounds; or pathogens that infect the plant. *Erwinia amylovora* is a native pathogen of wild, rosaceous hosts in eastern North America. It was the first bacterium proven to be a pathogen of plants. Today, fire blight is an important disease of apples and pears in many parts of the world. Indeterminate, water-soaked lesions form on fruit surface and later turn brown to black. Droplets of bacterial ooze may form on lesions, usually in association with lenticels. Severely diseased fruits blacken completely and shrivel (Johnson, 2000). Two species of *Pseudomonas* (*P. fluorescens* and *P. viridiflava*) comprise up to 40 per cent of the naturally occurring bacteria on the surface of fruits and cause nearly half of post-harvest rot of fresh produce stored at cold temperatures (Doyle, 2007). The bacterium *Erwinia carotovora* is a highly effective spoilage microbe that causes soft rot across a broad host range of some fruits (Lund, 1983). Soft rot is a form of decay characterized by a watery transparency in infected leafy plant parts and watery disintegration of non leafy plant materials. "Soft-rot erwinia" tends to initiate infection and decay at wound sites and, once established, can quickly advance to total destruction of the product. Soft-rot erwinia expresses four pectin-degrading extracellular enzymes: pectin lyase, polygalacturonase, pectin methylesterase, and pectate lyase. Of these enzymes, pectate lyase is primarily responsible for extensive decay. *E. carotovora* has built-in redundancy for this apparently critical pathogenicity factor, expressing four distinct extracellular pectate lyase isozymes (Barras *et al.*, 1994). Soft-rot erwinia is active only at temperatures of 20°C and above, which reinforces the need to maintain a continuous cold chain from immediately post harvest to retail to successfully manage this ubiquitous spoilage bacterium. Another group of soft-rotting bacteria, the fluorescent pseudomonads (*i.e.*, *Pseudomonas fluorescens* and *Pseudomonas viridiflava*), can decay plant tissue at temperatures at or below 4°C.

Pathogen Attack

Bacterial and fungal phytopathogens can be introduced to fruits at any point during the growth, maturation, harvest, storage, or shipment of the produce by exposure to contaminated water, dust, mechanical equipment etc. (Beuchat, 1995). Two of the most commonly destructive bacteria of stored produce are *E. carotovora*, *Pv. Carotovora* and *Ps. Fluosrescens* (Agrios, 1997). These are responsible for the soft rot of a wide variety of fruits in storage. Stored produce that is clean from the field may still become infected by contact with contaminated surfaces and via infiltration or absorption of contaminated wash water. Wounds provide common entry points, and the bacteria multiply in intercellular spaces. Produce that is under attack by soft-rotting bacteria quickly degrade, such that by the time a soft rot infection is apparent by appearance or by a characteristics smell the shelf life of the produce can be counted in days, if not hours. In the initial phase of infection by *Erwinia* the bacteria reproduce on the surface and internally without digesting the polysaccharide matrix of the cell wall when the population density reaches a critical level, a quorum sensing pathway is initiated Acyl-homoserine lactone (AHL) is a signal molecule that initiates the production of pectolytic enzymes that release oligo and monosaccharides on which the bacteria feed, degrading the plant tissue in the process (Dong *et al.*, 2001; Leadbetter, 2001). The bacteria multiply rapidly in the sugar and nutrient rich medium of the degraded tissue and because the bacteria are already at elevated population levels before the tissue degradation triggers plant defense responses. Sokolowska (2013) reported that two *A. acidoterrestris* strains caused spoilage after sixteen days of incubation at 25°C and tested strains produced guaiacol in apple juice enriched with vanillic acid, after two days of incubation at 45°C. Also, the Growth of two above strains was observed after 10 and 34 days of incubation, at 25°C, apple juice enriched with vanillic acid was observed, whereas guaiacol was produced at this condition only by one of tested strains.

Abiotic Spoilage

It is due to different physical and chemical changes in the product *viz.*, hydrolytic action of enzymes, oxidation of fats, putrefaction of proteins, browning reaction between proteins and sugars and physical changes of wilting control, melting etc. Temperature control is the major factor to provide longevity in shelf life to the fruits and vegetables.

Storage of Fruits Under Low Temperature

In the commercial storage of fruits, it is essential that they should be kept in such a condition that the group of chemical and physical processes usually associated with living organisms and characterized for want of a better name as life processes or vital activities can proceed without interruption. At the same time, it is essential that these processes be slowed down as much as possible, so that this portion of the life cycle of the organism will not be completed too quickly and the product becomes unfit for consumption through autolytic decomposition or be broken down by the action of micro-organisms. The physiological problems of commercial cold storage of plant products are, of course, somewhat complex; as in most problems in plant physiology the method of attack has been essentially from a chemical standpoint.

The fundamental principle which apparently governs the relation of many of the processes which go on in plants, or which at least seems to explain their relation to temperature, was suggested by Van't Hoff (1896). This rule is that the rate of chemical reaction doubles for every rise of 10°C in temperature. It has been found that many of the processes which go on in plants take place in accordance with this rule; that is, within limits there is an acceleration of the processes if the temperature is raised and retardation with the lowering of the temperature. For instance the storage temperature and warming experiments provide clear evidence that superficial scald can induce by chilling. 'Granny Smith' apples were stored at temperatures ranging from 0° to 20°C. Fruit stored at 0° or 4°C developed superficial scald. At 10°C, surface defects occurred but they were not typical symptoms of scald, and at 15° or 20°C no symptoms developed. Accumulation of faroesene and conjugated trienes in fruit peel correlated with increasing ethylene production, which was greater at higher temperatures. However, concentrations of conjugated trienes were highest at 0° and 4°C. When fruit were kept at 10°C for 5 or 10 days before storage, scald development after storage was not reduced. An interruption of 0°C storage with a single warming period at 10° or 20°C reduced scald development after 25 weeks of storage, maximum reduction occurring when fruit were warmed for 3 to 5 days at 20°C after 1 to 4 weeks at 0°C. Amelioration of scald declined as time at 0°C before warming increased. Diphenylamine application after conjugated trienes increased during warming, but at the end of storage (when scald was developing) the conjugated triene concentrations in peel were reduced in fruit that had been warmed. Warming slightly increased yellowing, softening and greasiness of fruit after storage (Watkins *et al.*, 1995). Magness (1920) indicates that at high temperatures there is an accumulation of CO_2 in fleshy fruits and a very low pressure of O_2, while at low temperatures the ratio O_2/CO_2 within the fruit is considerably higher. This might well influence the kind of respiration, and the energy changes might at high temperatures be by intramolecular rearrangement and partial splitting rather than by a breaking down of the molecule to CO_2 and H_2O, while at low temperatures in the presence of larger amounts of oxygen the molecule might be more completely decomposed. Experiments at constant temperatures carried out for long periods will undoubtedly furnish some idea of the amount of CO_2 given off at these temperatures under the conditions existing in the tissue. Crisosto *et al.* (1999) observed chilling injury in peach, nectarine and plum cultivars when fruit was stored at both storage temperatures (0°C and 5°C) within 5 weeks of storage. Woolliness is a chilling injury phenomenon occurring in nectarines held at low temperatures for extended periods. It is a disorder marked by altered cell wall metabolism during ripening leading to a dry, woolly texture in the fruit. One was holding the fruits for 2 days at 20°C before 0°C storage and the second were having ethylene present during cold storage (ethylene). Immediately stored fruit (control) had 88 per cent woolliness while 7 per cent of delayed storage and 15 per cent of ethylene treated fruit showed woolliness. The severity of the injury in individual fruits was closely related to inhibition of ethylene evolution. Woolly fruit had higher levels of 1-aminocyclopropane-1-carboxylic acid (ACC) and less 1-aminocyclopropane-1-carboxylic acid oxidase activity than healthy fruit. It is suggested that ethylene is essential for promoting the proper sequence of cell wall hydrolysis necessary for normal fruit softening. This is in contrast to chilling injury in other fruits, whereby ethylene is often a sign of incipient damage. Respiration was also found to be associated with chilling injury, in that fruit with woolliness had a depressed respiration (Zhou *et al.*, 2001).

Changes in the chemical composition of the product, while not always of value in indicating the constituents used in respiration, are of interest in themselves. The effect of low temperatures on changes which take place in carbohydrates in plants is of especial interest. The three strawberry cultivars (Dover, Campineiro, and Oso Grande) were stored at 6, 16 and 25°C, for 6 days. Low temperature negatively affected anthocyanin and vitamin C accumulation, and positively affected soluble sugars, while flavonols, ellagic acid and total phenolic contents remained almost the same or even decreased at all temperatures. Despite differences in anthocyanin content between varieties and its increase during storage (higher with increasing temperature), there was no difference in the antioxidant activity between cultivars, which decreased after harvesting, independently of the temperature of storage. Variations in the proportion of dehydroascorbic acid/ascorbic acid showed that there were differences between cultivars concerning adaptation of the fruit to low temperatures. It was observed that cold storage is an effective way to maintain strawberry quality, but a compromise between sensorial and nutritional values can be achieved at 16°C, for all the cultivars (Cordenunsi *et al.*, 2005). Skin and flesh of 'Selva' strawberries (*Fragaria* x *ananassa* Duch.) stored at 5°C in air or 2 kPa O_2 became darker red and accumulated anthocyanin levels, but these changes were reduced in fruit stored in air +20 kPa CO_2, 2 kPaO_2 + 20 kPa CO_2, 0.5 kPa O_2, and 0.5 kPa O_2 + 20 kPa CO_2. Increasing pH and decreasing titratable acidity in tissues during storage, especially in the internal tissues, were more marked in fruit stored in high CO_2 atmospheres. Since pH affects colour expression of the anthocyanin pigment, these changes may contribute to the observed changes in colour. Combined citric and malic acid concentrations were higher in the external tissues (10.6 mg g^{-1}) than internal tissues (5.5 mg g^{-1}), and decreases in concentration of both acids were greater in fruit kept in high CO_2 atmospheres. Succinic acid in fruit tissues was present in low concentrations, but usually increased in high CO_2 atmospheres (Holcraft and Kader, 1999). The pH of the internal tissues was higher and tended to increase with CO_2 treatment. This combined with the reduced synthesis of anthocyanin under elevated CO_2 atmospheres (Holcroft, 1998) may explain the pale internal flesh colour of strawberries stored under conditions. With stored fruits which contain considerable quantities of acids, the acids seem to be utilized in respiration, especially at the low temperatures, though there may be some decrease in carbohydrate content as well. The acids as a rule break down more rapidly at high temperatures than at low temperature.

At low temperature, the water which is the main constituent of fruits, freezes, and most of these reactions cease. A few reactions may proceed slowly after the fruit is solidly frozen, but most of them will take place only in liquid solution. When the organism is thawed and the water or solute is again liquid, many of the chemical reactions which were going on before freezing are resumed. In the case of most plant parts, however, there is a disorganization of the chemical and physical equilibria, and the processes are not checked and balanced as in the living plant. This, of course, leads to a breaking down and a decomposition of the tissue. Many fruits, however, can be cooled to temperatures below their freezing points and have the crystallization of the water which actually take place within the tissues without apparent injury, and the occurrence of local injuries, that is, injuries to certain portions of the tissue, due to freezing, is quite common. Apples may be frozen lightly and thawed without apparent injury. Hard, ripe Bartlett pears have been frozen solid, removed from the freezing room, and ripened normally without any indications that the tissues of the fruit had been injured by the solidification of the water. Ripe pears or immature pears of this variety may be seriously injured by freezing. However in apples, slight freezing sometimes causes local discolorations around the fibro vascular bundles or dark-colour areas in the tissue. In freezing fruits, they may frequently be under cooled far below the freezing point without the formation of ice or without any apparent injury to the tissues.

Chilling Injury

There are many symptoms of chilling injury in different species and tissues and they occur after varying periods of exposure to different chilling temperatures. In general, it is currently thought that the chilling temperature is transduced into a physiological change and also involves a phase change in some cellular membrane or enzyme system (Saltveit, 2000). The resultant alteration in metabolism produces the myriad of symptoms that are grouped together under the term chilling injury. Lyons (1973) proposed that, at chilling temperatures, a portion of the cell membrane of sensitive plants undergoes a phase transition from liquid-crystalline to solid gel. This conformational change would then give rise to the physiological changes associated with Chilling injury (Figure 14.3). It is hypothesized that physical changes in the membranes of chilling-sensitive tissue are reflected in changes in their biological function (Saltveit and Morris, 1990). Two consequential effects on membranes would be changes in membrane permeability and the activity of associated enzymes. Saltveit (2000) stated that the increases in ion leakage that are highly correlated with symptoms of chilling injury increase significantly only after days of chilling for most fruit crops. Permeability changes would result in changes in cytosolic levels of calcium, since the calcium concentration in the cell wall and vacuole are orders of magnitude greater than in the cytosol. Calcium is involved in many important regulatory mechanisms in the plant cell, and an unregulated influx of calcium could cause serious injury. There are similarities between the effect of treatments that raise the cytosolic levels of calcium and the effects of chilling injury in chilling-sensitive plants For example, chilling causes microtubule depolymerization, as does artificially raising the cytosolic calcium concentration. The membrane-phase change theory has continually undergone revisions to accommodate new information concerning how low temperatures might realistically be affecting membranes. In response to these observations, it has been argued that minor compositional components of the membrane, and not the bulk lipids, may dictate the level of chilling sensitivity. The chilling sensitivity of small discs or segments of tissue excised from chilling sensitive species was significantly altered by prior temperature exposure subsequent to holding the tissue at chilling temperatures as measured by a number of physiological processes sensitive to chilling. Exposure to 32° and 12°C had no effect on the rate of ion leakage

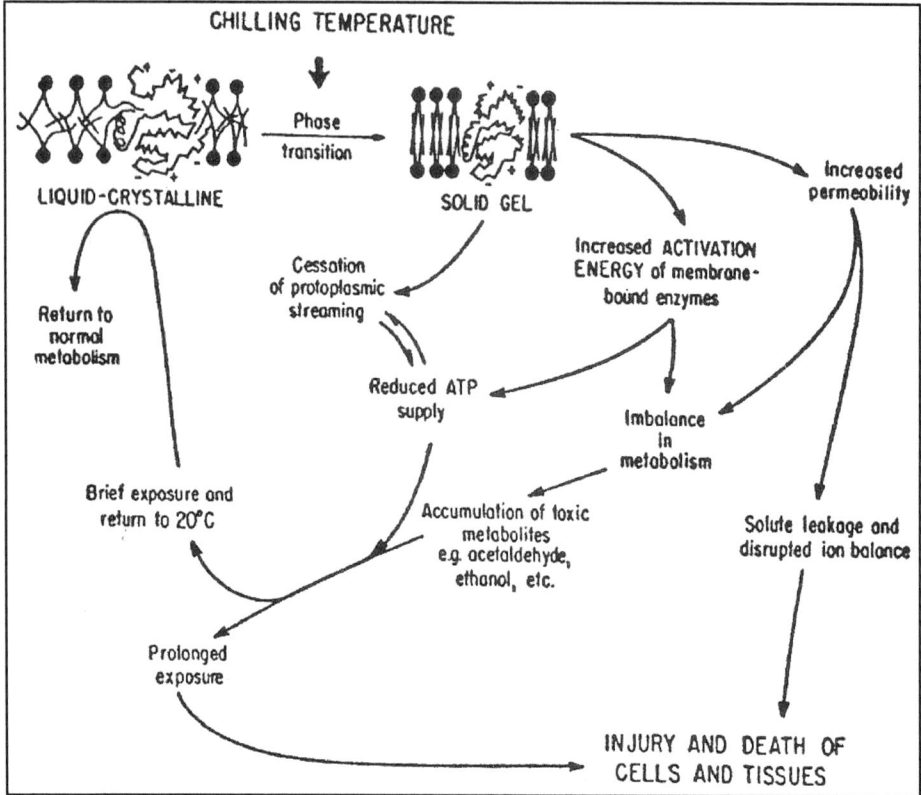

Figure 14.3: Schematic Pathway Leading to Chilling Injury in Sensitive Plant Tissue. (*Source:* Maas J L. 1992).

from fruit tissue of chilling tolerant species (*Malus domestica* Borkh cv. Golden Delicious, *Pyrus communis* L cv. Bartlett) (Saltveit,1991).

Phosphofructokinase (PFK), a key enzyme in glycolysis, exists in an active tetrameric form that is dissociated into inactive dimers at low temperatures (Dixon *et al.*, 1981). Certain enzymes have thresholds for activation or synthesis near the chilling threshold temperature. In addition to the effect of low temperatures on enzyme induction, low temperatures are also known to slow the decay of enzymes. Thermal displacement of equilibrium can lead to a shift in metabolism. The rate of enzymatic activity usually declines with declining temperature. The extent of the decline differs for each enzyme. In biological systems, there may be several series of reactions competing for the same substrate. If these reactions have different activation energies or Q_{10} values, a change in temperature could markedly shift the proportion of substrate being converted into the various products, leaving the cell devoid of sufficient product from one series of reactions or producing toxic levels of product by another pathway. As a result of these rate imbalances induced by chilling temperatures, there may also be a change in the pool size of metabolites that serve as metabolic regulators or protein protectants. Loss of feedback inhibition could accelerate breakdown of metabolites and enzymes. It has frequently been noted that many enzymes are more stable in the presence of their substrates. Unregulated alterations in substrate concentration could also affect metabolism by altering the rate of enzyme stability. Flesh woolliness in peach has been mainly attributed to the imbalance between polygalacturonase and pectin esterase (Zhou *et al.*, 2000).

Ethylene is an important plant hormone, the synthesis of which is actively regulated by the plant (Abeles *et al.*, 1992). Many post harvest processes are influenced by the biological activity of ethylene. Peaches become mealy if improperly treated with C_2H_4. Kiwifruit is very sensitive to C_2H_4, and exposure to 30 ppb can cause unacceptable softening in storage. During chilling, the synthesis of the immediate precursor of ethylene (ACC) is often stimulated, while the ability of the plant to convert ACC to ethylene is progressively diminished (Wang and Adams, 1982). Intermittent warming refers to the periodic interruption of an injurious chilling exposure with periods at non-chilling temperatures. Enhanced respiration and ethylene production occur during intermittent warming and are thought to reflect the dissipation of toxic products in the chilled cells, so that they do not reach injurious levels (Cabrera and Saltveit, 1990). Bakshi and Masoodi (2005) stated that intermittent warming coupled with modified atmospheric storage was found very effective in preventing chilling injury and prolong storage life of fruit.

Freezing Injury

Ice nucleation will first appear in the purest water, it usually occurs extracellularly in the xylem, intercellular spaces, or cell wall. Supercooling denotes the capacity of water to be cooled below its freezing point before the formation of ice crystals occurs. In undisturbed tissue, supercooling can be a few degrees below the freezing point, but agitation during transport usually precludes supercooling. The amount of energy that must be removed to freeze water (80 kcal kg^{-1}) is great in comparison with that needed to cool water ($\sim$ 1.0 kcal kg^{-1}°C^{-1}) or ice ($\sim$0.5 kcal kg^{-1} °C^{-1}). If energy is being removed from a commodity at a constant rate, its temperature will steadily fall until freezing is initiated, at which point its temperature will remain constant until all the available water is frozen. This retarding of the temperature drop under moderate freezing conditions of a few degrees below zero can protect sensitive tissues from freezing injury. If supercooling has occurred, the temperature of the freezing commodity will actually increase as it freezes (*e.g.*, going from -2° to 0°C).

Ice crystals that form during very rapid freezing are too small to cause mechanical damage to sub cellular structure. When tissue is cooled under natural conditions, ice usually forms first within the cell walls. This ice formation is not lethal, and the tissue fully recovers if warmed. However, when plants are exposed to freezing temperatures for an extended period, the crystals in the walls continue to grow and extend into the protoplast, causing lethal damage. Freeze resistant species somehow limit the growth of crystals to the cell walls and intercellular spaces. During rapid freezing, the protoplast, including the vacuole, super cools, that is the cellular water remains liquid even at temperatures several degrees below its freezing point. Several hundred molecules of ice are needed for an ice crystal to begin forming. The process whereby these hundreds of ice molecules start to form a stable ice crystal is called ice nucleation. Some large polysaccharides and proteins facilitate ice crystal formation are called ice nucleates in plant cell. As ice crystal begins to grow from endogenous ice nucleates and the resulting relatively large intercellular ice crystal causes extensive damage to the cell. If cooling is quite rapid, water movement from the cell may be insufficient to maintain extracellular ice formation and the temperature of the tissue will drop to the point where intracellular ice crystals will form. This is usually lethal to the cell because of protein denaturation and membrane disruption due to the removal of water and the concentration of the resultant solution. Freezing injury is not the result of ice crystal formation themselves, but of the change in water activity in the freezing tissue because of the exclusion of solutes from the ice crystal and the resulting concentration of the remaining aqueous solution. Ice crystals appear sharp and able to puncture delicate cellular membranes. However, an ice crystal forms by the deposition of water molecules on its surface and therefore cannot exert any force to puncture or mechanically disrupt the cell. Although frozen tissue is more prone to mechanical injury, cellular dehydration, not mechanical perturbation, is the cause of most freezing injury.

Methylcyclopropene

In strawberries (*Fragaria* x *anaassa*), methylcyclopropene (1-MCP) concentrations greater than 15 nl l^{-1} are associated with increased decay of fruit (Ku *et al.,* 1999). Jiang *et al.* (2001) also found an increase in disease with high 1-MCP concentrations on strawberries. 1-MCP may inhibit a beneficial metabolic response or stimulate an undesirable characteristic, possibly relating to a natural defense mechanism. Lower phenolic content in 1-MCP-treated strawberries was thought to account for the increased disease incidence (Jiang *et al.,* 2001). Wooliness and reddening of nectarines was more prominent in 1-MCP-treated fruit as compared with untreated. A higher incidence of internal browning occurred when apricots were treated with 1-MCP before storage, but internal browning was not increased if fruits were treated after storage (Dong *et al.,* 2002). While both treated and untreated peaches stored at 5°C had internal browning, the problem was more exaggerated in 1-MCP-treated fruit. Internal browning was not associated with 1-MCP treatment when peaches were stored at 0 or 10°C or later harvested fruit was used (Fan *et al.,* 2002). Fan and Mattheis, (2001) observed that some irradiated gala apple fruit stored at 20°C for 3 weeks developed internal browning, and MCP-treated fruit had more injury than control fruit. Storage at 0°C after irradiation greatly reduced development of internal browning.

At last, it is concluded that post-harvest spoilage of temperate fruits occurs due to improper harvesting, transportation, storage and distribution. The post-harvest life of fruits is governed by water content, respiratory rate, ethylene production, endogenous plant hormones and exogenous factors such as microbial growth, temperature, relative humidity and atmospheric compositions. The post-harvest losses of temperate fruits can considerably be minimized and their storage life can be increased by careful manipulation of these factors. The loss can be minimized by adopting necessary cultural operations, careful handling and packaging. The use of post-harvest heating of fruits, appropriate chemicals at pre and post-harvest stage, may extend the availability of fruits over a long period by protecting them from microbial as well as environmental agencies of damage. These can also be stored under controlled atmosphere, modified atmosphere, low temperature with appropriate chemical treatment to delay senescence and inhibit microbial decay. Thus, the reduction of post-harvest spoilage of fruits is a complementary means for increasing production. It may not be necessary to considerably step up the production of fruits with the growing demand if the post-harvest loss is reduced to a great extent. The cost of preventing losses after harvest in general is less than producing a similar additional amount of fruit crop of the same quality.

References

1. Abeles F B, Morgan P W and Saltveit, M E. (1992). *Ethylene in plant biology*. Vol 15, 2nd ed. Academic press, San Diego, Califorina.

2. Agrios G.A. (1997). *Plant Pathology*, 4th ed., Academic Press, San Diego, CA.

3. Al-Hindi R R, Al-Najada A R and Mohamed S A. (2011). Isolation and identification of some fruit spoilage fungi: Screening of plant cell wall degrading enzymes. *African Journal of Microbiological Research*, **5** (4): 443-8.

4. Bahkali A R, A1-Khaliel A S and Elkhider K A. (1997). *In vitro* and *in vivo* production of pectolytic enzymes by some phytopathogenic fungi isolated front southwest Saudi Arabia. King Saud University **9**: 125-37.

5. Bakshi P and Masoodi F A. (2005). Use of intermittent warming to control chilling injury in peach during storage. *Acta Horticulturae*, **696**: 523-526.

6. Barras F, Van Gijsegem F and Chatterjee A K. (1994). Extra cellular enzymes and pathogenesis of soft-rot erwinia. *Annual Review Phytopathology*, **32**: 201-34.

7. Beuchat L R. (1995). Pathogenic micro-organisms associated with fresh produce. *Journal of Food Protection*, **59**: 204-16.

8. Beuchat L R. (1998). Surface decontamination of fruits and vegetables eaten raw: A review. Food Safety Unit, World Health Organization, WHO/FSF/98.2, pp. 1-42.

9. Bostock R M and Stermer B A. (1989). Perspectives on wound healing in resistance to pathogens. *Annual Review of Phytopathology*, **27**: 343-71.

10. Brackett R E. (1997). Fruits, vegetables and grains in Food microbiology: Fundamentals and Frontiers, Doyle, M. P., L. R. Beuchat and T. J. Montville, eds., American Society of Microbiology, Washington, DC. pp: 117-28.

11. Cabrera R M and Saltveit J M E. (1990). Physiological response to chilling temperatures of intermittently warmed cucumber fruit. *Journal of American Society of the Horticultural Science*, **115**: 256-61.

12. Castro E D, Barrett D M, Jobling J and Mitchan E J. (2008). Biochemical factors associated with a CO_2 induced flesh browning disorder of Pink lady apples. *Postharvest Biology and Technology*, **48**: 182-91.

13. Charley V L. (1999). The Prevention of microbiological spoilage in fresh fruit. *Journal of Science and Food Agriculture*, **10**: 399-400.

14. Chuku EC, Ogbonna, DN, Onuegbu, BA and Adeleke, MTV. (2008). Comparative studies on the fungi and biochemical characteristics of Snake Gaurd (*Trichosanthes curcumerina* Linn.) and Tomato (*Lycopersicon esculentus* Mill.) in Rivers State, Nigeria. *Journal of Applied Science*, **8**(1): 168-172.

15. Cordenunsi B R, Genovese M I, Nascimento J R O D, Hassimotto N M A, Santos R J D and Lajolo F M. (2005). Effects of temperature on the chemical composition and antioxidant activity of three strawberry cultivars. *Food Chemistry*, **91**: 113-121.

16. Crisosto C H, Mitchell F G and Ju Z. (1999). Susceptibility to chilling injury of peach, nectarine and plum cultivars grown in California. *HortScience*, **34**(6): 1116-1118.

17. Dennis C. (1983). Soft fruits. Post harvest pathology of fruits and vegetables. In: C Dennis (ed.). Academic Press, New York. pp. 23-42.

18. Desphande S S. (2002). Fungal toxins In: Handbook of Food Toxicology, Marcel Dekker, Inc. New York, NY, USA, pp. 413–417.

19. Dimatteo A, Bonivento D, Tsernoglou D, Federici L and Cervone F. (2006). Polygalacturonase-inhibiting protein (PGIP) in plant defence: a structural view. *Phytochem*istry, **67**: 528-533.

20. Di Pietro A, Madrid MP, Caracuel Z, Delgado-Jarana J and Roncero MIG (2003). *Fusarium oxysporum*: exploring the molecular arsenal of a vascular wilt fungus. *Molecular Plant Pathology*, **4**: 315-326.

21. Dixon W L, Franks F and Ap. Rees T. (1981). Cold lability of phosphofructokinase from potato tubers. *Phytochemistry*, **20**: 969-972.

22. Dong L, Lurie S and Zhou H. (2002). Effect of 1-methylcyclopropene on ripening of 'Canino' apricots and 'Royal Zee' plums. *Postharvest Biology and Technology*, **24:** 135-145.

23. Dong Y H, Wang LH, Xu J L, Zhang H B, Zhang X F and Zhang L H. (2001). Quenching quorum-sensing dependent bacterial infection by an N-acyl homoserine lactonase. *Nature*, **411**: 813-817.

24. Doyle M E. (2007). Microbial Food Spoilage- Losses and Control Strategies. A Brief Review of the Literature. pp. 1-16.

25. Effiuvwevwere BJO. (2000). Microbial spoilage agents of tropical and assorted fruits and vegetables (An Illustrated References Book). Paragraphics Publishing Company, Port Harcourt. pp. 1-39.

26. Fan X, Argenta L and Mattheis J P. (2002). Interactive effects of 1-MCP and temperatures on 'Elberta' peach quality. *HortScience*, **37**: 134-138.

27. Fan, X and Mattheis J. (2001). 1- Methylcyclopropene and storage temperature influence responses of gala apple fruit to gamma irradiation. *Post Harvest Biology and Technology*, **23**(2): 143-151.

28. Filonow A B. (2005). Wound type in apple fruits affect wound resistance to decay causing fungi. *Journal of Plant Pathology*, **87**(3): 233-238.

29. Filtenborg O, Frisvad J C and Thrane U. (1996). Molds in Food spoilage. *International Journal of Food Microbiology*, **33**: 85-102.

30. Fisher D F. (1922). Spoilage of Apples after harvest. Rept. Proc. Thirty-second Awn. Convention Brit. Columbia Fruit-Growers' Assoc. held at Victoria, B. C. 18th to 20th Jan1922. pp. 68.

31. Frisvad JC, Frank JM, Houbraken JAMP, Kuijpers, AFA and Samson RA. (2004). New ochratoxin A producing species of *Aspergillus* section Circumdati. *Studies in Mycology*, **50**: 23–43.

32. Hagenmaier R D and Baker R A. (1998). A survey of the microbial population and ethanol content of bagged salad. *Journal of Food Protection*, **61**, 357–359.

33. Higashihara K, Takeuchi M, Bandoh S, Saegusa Y, Miyagawa H and Goto T. (2009). Effect of storage temperature on the growth of *Penicillium expansum* and its production of patulin in apples. *Mycotoxins*, **59** (1): 7-13.

34. Holcroft DM. (1998). Biochemical basis of changes in the colour of strawberry fruit stored in controlled atmospheres. Ph. D. Dissertation, University of California, Davis.

35. Holcraft D M and Kader A A. (1999). Controlled atmosphere induced changes in pH and organic acid metabolism may affect colour of stored strawberry fruit. *Postharvest Biology and Technology*, **17**: 19-32.

36. Holtmeyer M G and Wallin J R. (1981). Incidence and Distribution of airborne spores of *Aspergillus flavus*. In: Plant Disease **6**: 54-60.

36. Iamanaka BT, Taniwaki MH, Vicente E and Menezes HC. (2006). Fungi producing *ochratoxin* in dried fruits. *Advances in Experimental Medicine and Biology*, **571**: 181–188.

37. International Agency for Research on Cancer (IARC) (1993). Some naturally occurring substances: food items and constituents, heterocyclic aromatic amines and mycotoxins. Monograph 56. Lyon International Agency for Research on Cancer, 571 pp.

38. Ihekeronye A I and Ngoddy P O. (1985). Integrated Food Science and Technology for the tropics, Macmillan Publishers Ltd. London pp385.

39. Jiang Y, Joyce DC and Terry LA. (2001). 1-methylcyclcopropene treatment affects strawberry fruit decay. *Postharvest Biology and Technology*, **23**: 227-232.

40. Johnson KB. (2000). Fire blight of apple and pear. *The Plant Health Instructor*.

41. Juhnevica, K., Skudra, G. and Skudra, L. (2011). Evaluation of microbiological contamination of apple fruit stored in a modified atmosphere. *Environmental and Experimental Biology*, **9**: 53-59.

42. Kader AA and Ben-Yehoshua S. (2000). Effects of superatompheric oxygen levels on postharvest physiology and quality of fresh fruits and vegetables. *Postharvest Biology and Technology*, **20**: 1-13.

43. Ku VVV, Wills RBH and Ben-Yehoshua S. (1999). 1-Methylcyclopropene can differentially affect the post harvest life of strawberries exposed to ethylene. *HortScience*, **34**: 119-120.

44. Lakshiminarayana S, Sommer NF, Polito V and Fortlage RJ. (1987). Development of resistance to infection by *Botrytis cinerea* and *Penicillium expansum* in wounds of mature apple fruits. *Phytopathology,* **77**: 1674-1678.

45. Leadbetter J R. (2001). Plant microbiology: quieting the raucous crowd. *Nature*, **411**: 748-749.

46. Lloyd B B. (1993). Fungi in food. An Overview In: Encyclopaedia of Food Science, Food Technology and Nutrition. Academic Press Ltd. London pp. 4327-4337.

47. Lu C and Toivonen PMA. (2000). Effect of 1 and 100 kPa O_2 atmospheric pretreatment of whole 'Spartan' apples on subsequent quality and shelf-life of slices stored in modified atmosphere packages. *Postharvest Biology and Technology*, **18**: 99–107.

48. Lund B M. (1983). Bacterial spoilage. Post-harvest pathology of fruits and vegetables. In C. Dennis (Ed.), London: Academic Press. Pp 218–257.

49. Lyons J M. (1973). Chilling injury in plants. Annual Review. *Plant Physiology*, **24**: 445-466.

50. Maas J L. (1992). Compendium of strawberry diseases. APS Press and Agr. Res. Serv., U. S. Dept. Agr., St. Paul, Minn.

51. Magness J R. (1920). Composition of gases in intercellular spaces. *Botanically Gazette*, **70**: 308-316.

52. Monso EM. (2004). Occupational asthma in greenhouse workers. *Current Opinion Pulmonary Medicine*, **10**: 147-150.

53. Nguyen-the C and Carlin F. (1994). The microbiology of minimally processed fresh fruits and vegetables. *Critical Reviews in Food Science and Nutrition*, **34**, 371–401.

54. Nikolajeva V. (2007). *Food Microbiology*. LU Akademiskaisapgads, Riga, p 130.

55. Niturea SK, Kumarb AR, Parabc PB and Panta A. (2008). Inactivation of polygalacturonase and pectatelyase produced by pH tolerant fungus *Fusarium moniliforme* NCIM 1276 in a liquid medium and in the host tissue. *Microbiological Research*, **163**: 51-62.

56. O'Connor-Shaw R E, Roberts R, Ford A L and Nottingham S M. (1994). Shelf life of minimally processed honeydew melon, kiwifruit, papaya, pineapple and cantaloupe. *Journal of Food Science*, **59**: 1202–1206, 1215.

57. Oda K, Kakizono D, Yamada O, Iefuji H, Akita O and Iwashita K. (2006). Proteomic analysis of extracellular proteins from *Aspergillus oryzae* grown under submerged and solid-state culture conditions. *Applied and Environmental Microbiology*, **72**: 3448-3457.

58. Osterloh D. (1994). Obstlagerung. Deutscher Landwirtschaftsverlag, Berlin, pp. 320–342.

59. Ozer H, Oktay Basegmez H I and Ozay G. (2012). Mycotoxin risks and toxigenic fungi in date, prune and dried apricot among Mediterranean crops. *Phytopathologia Mediterranea*, **51**(1): 148-157.

60. Piga A I, Pinna K B, Ozer M A and Aksoy U. (2004). Hot air dehydration of figs (*Ficus carica* L.): drying kinetics and quality loss. *International Journal of Food Science and Technology*, **39**: 793–799.

61. Pitt J I and Hocking A D. (1997). Fungi and Food Spoilage. Blackie Academic and Professional, New York.

62. Pitt J I and Hocking A D. (2009). Primary keys and miscellaneous fungi. In: Pitt, J. I., Hocking, A. D. eds., Fungi and food spoilage. New York NY: Springer, 1-9: 122-124.

63. Prins TW, Tudzynski P, von Tiedmann A, Tudzynski B, TenHave A, Hansen ME, Tenberge K and Van Kan J A I. (2000). Infection strategies of *Botrytis cinerea* and related necrotrophic pathogens. In: Kondrad J.W. (ed.). Fungal Pathology, pp. 33-64. Kluwer Academic, Boston, MA, USA.

64. Raviyan P, Zhang Z and Feng H. (2005). Ultrasonicationfor tomato pectinmethylesterase inactivation: Effect of cavitation intensity and temperature on inactivation. *Journal of Food Engineering*, **70**: 189–196.

65. Rosenberger D.A. (1990). Postharvest diseases. In: Jones A. L. Aldwinckle H.S. (eds). Compendium of apple and pear diseases, pp. 53-54.APS Press, St. Paul, MN, USA.

66. Ryall A L and Pentzer W T. (1982). Disease and injuries of small fruits during marketing. Handling, transportation and storage of fruits and vegetables. In: A. L. Ryall and W. T. Pentzer (eds.). Fruits and tree nuts, vol 2. 2nd ed. AVI Publication Co., Westport, conn. Pp 519-547.

67. Salunkhe D K. (1974). Development in Technology of storage and handling of fresh fruits and vegetables. *Critical Review of Food Technology*, 3: 15-54.

68. Saltveit J M E. (1991). Prior temperature exposure affects subsequent chilling sensitivity. *Physiologia Plantarum*, **82**: 529-536.

69. Saltveit J M E. (2000). Discovery of chilling injury. In: Kung, S. D. and Yang, S. F. (eds.), Discoveries in plant biology. Vol 3, World Scientific Publishing, Singapore, pp. 423-448.

70. Saltveit J M E. (2002). The rate of ion leakage from chilling-sensitive tissue does not immediately increase upon exposure to chilling temperatures. *Postharvest Biology and Technology*, **26**: 295–304.

71. Saltveit J M E and Morris L L. (1990). Overview of chilling injury of horticultural crops. - In Chilling Injury of Horticultural Crops (Chien Yi Wang, Ed.), CRC Press, Boca Raton, FL, ISBN 0-8493-5736-5, pp, 3-15.

72. Schuster E, Dunn-Coleman N, Frisvad JC and Van Dijck PWM. (2002). on the safety of *Aspergillus niger*-a review. *Applied Microbiolology and Biotechnology*, **59**: 426–435.

73. Sharma R R, Singh D and Singh R. (2009). Biological control of post-harvest diseases of fruits and vegetables by microbial antagonists: A review. *Biological Control*, **50** (3): 205-221.

74. Sholberg P L and Conway W S. (2004). Postharvest pathology. In The commercial storage of fruits, vegetables, and florist and nursery stocks, USDA-ARS Agriculture Handbook Number 66. Draft – revised April 2004.

75. Sholberg PL and Ogawa JM. (1983). Relation of postharvest decay fungi to the slip-skin maceration disorder of dried French prunes. *Phytopathology*, **7**(5): 708 -713.

76. Simoes MLG, Tornisielo SMT and Tapia DMT. (2009). Screening of culture condition for xylanase production by filamentous fungi. *African Journal of Biotechnology*, **8**: 6317-6326.

77. Sinha K K and Singh A. (1987). Chemical changes in apples due to Aflatoxin producing *Aspergilli*. *Journal of Food Science and Technology*, **24**(1): 44.

78. Singh D and Sharma R R. (2007). Postharvest diseases of fruit and vegetables and their management. Sustainable Pest Management. In: Prasad, D. (Ed.). Daya Publishing House, New Delhi, India.

79. Sokolowska B, Skapska S, Sionek B, Niezgoda J and Chotkiewicz M. (2013). *Alicyclobacillus acidoterrestris* - growth and production of odour compounds in apple juice. *Postepynaukii Technologii Przemys³uRolno-Spozywczego*, **68** (1): 19-37.

80. Solomos T, Whitaker B and Lu C. (1997). Deleterious effects of pure oxygen on 'Gala' and 'Granny Smith' apples. *HortScience*, 32: 458 abstract.

81. Sommer N F, Fortlage R J, Mitchell F G and Maxie EC. (1973). Reduction of post harvest losses of strawberry fruits from gray mold. *Journal of American Society Horticulture Science*, **98**: 285-288.

82. Stratford M. (2007). Food and beverages spoilage yeasts. Yeasts in food and beverages. In: Quarol A. and Fleet, G. H. (Eds). The yeast handbook, Berlin: Springer, 335-379.

83. Tancinova D, Barborakova Z, Kacinova J, Maskova Z and Volckova M. (2013). The occurrence of micromycetes in apples and their potential ability to produce mycotoxins. *Journal of Microbiology, Biotechnology and Food Sciences*, **2** (Special issue 1) 1800-1807.

84. Tomassini A, Sella L, Raiola A, D'Ovidio R and Favaron F. (2009). Characteri-zation and expression of *Fusarium graminearum*en do polygalacturonases *in vitro* and during wheat infection. *Plant Pathology*, **58**: 556-564.

85. Tournas VH and Stack ME. (2001). Production of alternariol and alternariol methyl ether by *Alternaria alternata* grown on fruits at various temperatures. *Journal of Food Protection*, **64**: 528-532.

86. Tournas V H and Memon S U. (2009). Internal contamination and spoilage of harvested apples by patulin-producing and other toxigenic fungi. *International Journal of Food Microbiology*, **133** (1/2): 206-209.

87. Trucksess MW and Scott PM. (2008). Mycotoxins in botanicals and dried fruits: a review. *Food Additives and Contaminants*, **25**: 1–12.

88. Upper C D, Cook R J, Atlas R M, Suslow T V and Van Alfen N. (1989). The ecology of plant-associated microorganisms. National Academy Press, Washington, DC.

89. Van Kan J A L. (2006). Licensed to kill: the lifestyle of a necrotrophic plant pathogen. *Trends in Plant Science*, **11**: 247–253.

90. Van't Hoff J H. (1896). Studies in chemical dynamics (English translation). London. Pp 286.

91. Wang C Y and Adams D O. (1982). Chilling induce ethylene production in cucumber (*Cucumis sativas* L.). *Plant Physiology*, **69**(2): 424-427.

92. Watkins C B, Bramlage W J and Cregoe BA. (1995). Superficial scald of 'Granny Smith' apples is expressed as a typical chilling injury. *Journal American Society of Horticultural Science*, **120**(1): 88-94.

93. Watkins C B, Kupferman E, and Rosenberger D A. (2004). Apple. In The commercial storage of fruits, vegetables, and florist and nursery stocks, USDA-ARS Agriculture Handbook Number 66.

94. Whitaker BD, Solomos T and Harrison DJ. (1998). Synthesis and oxidation of a-farnesene during high and low O2 storage of apple cultivars differing in scald susceptibility. *Acta Horticulturae*, **464**: 165–170.

95. Zhou HW, Dong L, Arie R B and Lurie S. (2001). The role of ethylene in the prevention of chilling injury in nectarines. *Journal of Plant Physiology*, **158**: 55-61.

96. Zhou HW, Dong L, Ben-Arie R and Lurie S. (2000). Pectinesterase, polygalacturonase and gel formation in peach pectin fractions. *Phytochemistry*, **55**: 191–195.

2015, Horticulture for Nutrition Security
Editor: **Prof. K.V. Peter**
Published by: **DAYA PUBLISHING HOUSE, NEW DELHI**

Pages 331–343

Chapter 15

Resistance to Abiotic Stress in Vegetables and Spices

K.V. Peter, S. Nirmala Devi,
K. Nirmal Babu and P.G. Sadhankumar

Abiotic stress is the negative impact of non-living factors in the living organism in a specific environment. They are naturally occurring intangible factors which are essentially unavoidable and most harmful The basic stress factors are heat, drought, chilling or freezing temperature, flood, salinity and mineral deficiency and toxicity. Abiotic stress causes average yield loss of about 50 per cent as shared by high temperature (20 per cent), low temperature (7 per cent), salinity (10 per cent), drought (9 per cent) and others (4 per cent). In order to sustain production of vegetable and spices under these challenges,there should be a package to manage abiotic stress which includes use of resistant/tolerant varieties and adoption of appropriate production techniques.

Climate models indicate that warming will take place in next several decades leading to rise in temperature and change in rainfall in terms of quantity and quality. Increase in acidity, alkalinity and salinity can be expected. The increase in temperature by $1°C$ is likely to shift potential vegetable growing areas to nontraditional areas. The growth and development will be more rapid and production timing will change. Change in winter period and chilling duration will be reduced in temperate regions and temperate vegetable production will be affected. Pollination will be affected and flowers may abort. Heat unit will be met in lesser time leading to early maturity and faster ripening. Irrigation requirement will increase due to high temperature and faster growth of plants. Soil temperature will increase leading to advancing the planting time. The identification of mechanism of stress tolerance has provided approaches and methods for managing stress. Soil and climatic requirement of important vegetable crops and spices are given in Table 15.1 (Singh, 2012).

Table 15.1: Soil and Climatic Requirements of Vegetables and Spices

Crops	Soil pH	Temperature (°C)		Rainfall(mm)/Water Requirement (mm/ha)
		Max.	Min.	
Tomato	5.5-7	15	24	330 mm/ha
Brinjal	5.5-6.6	21	27	486 mm/ha
Chilli	6.5	20	35	640 mm/ha
Melons	6-7	30	35	450 mm/ha
Peas	5.5-6.0	10	18	240 mm/ha
French bean	5.5-6.0	15	25	300-350 mm/ha
Onion		20	25	500 mm/ha
Blackpepper	5-6	10	40	1500-3000mm
Cardamom	4.2-6.8	10	35	1500-5000mm
Ginger	5-7	19	28	1500-3000mm
Turmeric	4.3-7.5	18	30	640-4000mm

Drought

On an average, 28 per cent of area in India is vulnerable to drought. The world's water supply is fixed, thus increasing population pressure and competition for water resources will make the effect of successive droughts more severe. Inefficient water usage all over the world and inefficient distribution systems in developing countries further decreases water availability. Severe water stress conditions will affect productivity of vegetables. In combination with elevated temperatures, decreased precipitation could cause reduction of irrigation water availability leading to severe crop water-stress conditions. Vegetables, being succulent products consists of more than 90 per cent water. Thus, water greatly influences the yield and quality of vegetables; drought conditions drastically reduce vegetable productivity. Drought stress causes an increase of solute concentration in the environment (soil), leading to an osmotic flow of water out of plant cells. This leads to an increase of the solute concentration in plant cells, thereby lowering the water potential and disrupting membranes and cell processes such as photosynthesis. The timing, intensity, and duration of drought spells determine the magnitude of the effect of drought. Plants resist water or drought stress in many ways. In slowly developing water deficit, plants may escape drought stress by shortening their life cycle (Chaves and Oliveira 2004). However, the oxidative stress of rapid dehydration is very damaging to the photosynthetic processes (Ort, 2001; Chaves and Oliveira, 2004).

Successful management of water available for agricultural uses depend on better agronomic practices and enhanced understanding of water productivity. Promotion of drip and sprinkler irrigation systems in rainfed areas, furrow irrigated raised bed planting systems, mulching, application of antitranspirants like thiourea and KNO_3 will moderate stress. For mitigating impact of drought in vegetable crops, the genetic resources of under/unexploited vegetable crops should be exploited for consumer acceptability and productivity. The ecological limits and management practises of the promising species, its suitability in different cropping systems, post harvest techniques and marketing opportunities should be studied.

Genetic studies for drought tolerance have been conducted in solanaceous, cucurbitaceous and leguminous vegetables. The Tomato Genetics Resource Center at the University of California, Davis has assembled a set of stress tolerant tomato germplasm including accessions of *S. cheesmanii*, *S. chilense*, *S. lycopersicum*, *S. lycopersicum* var. *cerasiforme*, *S. pennellii*, *S. peruvianum* and *S. pimpinellifolium*. Drought tests show that *S. chilense* is five times more tolerant of wilting than cultivated tomato. *S. pennellii* has the ability to increase its water use efficiency under drought conditions unlike the cultivated tomato. It has thick, round waxy leaves, is known to produce acyl-sugars in its trichomes, and its leaves are able to take up dew (Rick 1973). Transfer and utilization of genes from these drought resistant species will enhance tolerance of tomato cultivars to dry conditions. Although wide crosses with *S. pennellii* produce fertile progenies, *S. chilense* is cross-incompatible with *S. lycopersicum* and embryo rescue through tissue culture is required to produce progeny plants. Field tolerance against mild and severe drought condition was observed in thirty varieties of tomato.In pot culture studies,nine genotypes were found promising against drought stress (IIVR, 2012). The vegetable varieties tolerant to drought is given in Table 15.2.

Table 15.2: Drought Tolerant Varieties in Vegetable Crops

Crops	Variety
Tomato	Arka Vikas, Arka Meghali, Punjab NR7, Mani Khammu, Mani Leima
Brinjal	PKM 1, Azad Kranti
Chilli	Arka Lohit
Cowpea	Arka Garima
Dolichos bean	Arka Jay and Arka Vijay
Long melon	Arka Sheetal
Capsicum sp.	*C. chacoense*, *C. cardenasi*

Temperature

Vegetable crops adapt to different environmental conditions but performs well under optimum temperature, remain dormant at lower temperature due to limited metabolic activity and forcibly mature at high temperature. Beyond these limits, if exposure is for prolonged periods damage will be permanent and plant dies.

High Temperature

In the tropics, high temperature conditions are often prevalent during the growing season and, with changing climate, crops will be subjected to increased temperature stress. High temperature stress disrupts the biochemical reactions fundamental for normal cell function in plants. Significant inhibition of photosynthesis occurs at temperatures above optimum, resulting in considerable loss of potential productivity.

Vegetative and reproductive processes in tomatoes are strongly modified by temperature alone or in conjunction with other environmental factors (Abdalla and Verkerk, 1968). It primarily affects the photosynthetic functions and cause significant losses in tomato productivity due to reduced fruit set, and smaller and lower quality fruits (Stevens and Rudich 1978). Pre-anthesis temperature stress is associated with developmental changes in the anthers, particularly irregularities in the epidermis and endothesium, lack of opening of the stromium, and poor pollen formation (Sato *et al.,* 2002). Hazra *et al.* (2007) summarized the symptoms causing fruit set failure at high temperatures in tomato as bud drop, abnormal flower development, poor pollen production, dehiscence, and viability, ovule abortion and poor viability, reduced carbohydrate availability, and other reproductive abnormalities. Flower drop in peas, cole crops, fenugreek and broad beans was observed as a result of high temperature. Tuber formation and development in potato was affected leading to lower yield. In cucurbits, sex ratio was affected resulting in malformed fruits. Seed production was also affected. Flower drop resulted in lower fruit setting in brinjal and chilli and fruit size was reduced in sweet pepper. In pepper, high temperature exposure at the pre-anthesis stage did not affect pistil or stamen viability, but high post-pollination temperatures inhibited fruit set, suggesting that fertilization is sensitive to high temperature stress (Erickson and Markhart, 2002). In onion and garlic, tip burning takes place under high temperature. Early maturity of bulbs resulted in lower yields. In cole crops quality of produce was deteriorated.

High temperature stress can be managed to some extent by application of frequent irrigation, mulching and use of tolerant varieties.The heat tolerant tomato lines were developed using breeding lines and landraces from the Philippines (*e.g.* VC11-3-1-8, VC 11-2-5, Divisoria-2) and the United States (*e.g.* Tamu Chico III, PI289309) (Opena *et al.,* 1989). However, lower yields in the heat tolerant lines are still a concern. Heat tolerance in CL5915, based on fruit set and fruit number per cluster, is controlled by additive and dominant effects (Hanson *et al.,* 2002). Capsicum variety Pusa Deepthi and potato variety Kufri Surya are tolerant to heat. Leaf amaranth varieties Tainung No.1and TainungNo.2 are tolerant to heat and cold.On the basis of morpho-physiological characters and fruitset, eight genotypes found promising against heat stress in field and seven were found promising for heat tolerance under net house condition. (IIVR,2012) Although field screening is effective for evaluating heat tolerance the accuracy and speed of the process could be improved through the use of molecular markers.

Low Temperature

In India, frost/cold wave occurs in December, January and February in North, West, Central and Peninsular India. Uttar Pradesh, Rajasthan, Haryana Punjab, Himachal Pradesh, northern parts of Madhya Pradesh, Uttarkhand and Jammu and Kashmir are affected by low temperature stress. Formation of ice in inter and intra cellular spaces of cells leads to mechanical damage of protoplasm and plasma membrane. Incidence of blight disease (early and late) in tomato and potato has been reported to be associated with low temperature. Low temperature stress can be managed by light, frequent irrigation, mulching, application of chemicals that enhance resistance to cold stress, mixed cropping and cultivation of cold/frost resistant varieties.Cold tolerance genotypes possess higher level of linolenic acid in several phospholipids than the cold sensitive. Hardening the plants by low temperature treatment results in a higher phospholipid level especially phosphatidyl choline, more unsaturated phospholipid and a lower sterol: phospholipid ratio, all of which contribute to greater membrane fluidity under lower temperature limit.

Salinity and Alkalinity

Salt affected soils are classified into alkali soils and saline soils. Alkali soils are characterised by presence of more than 15 per cent exchangeable Na, pH more than 8.5and EC <4.0dS/m. Saline soils are rich in soluble salts of Na, Ca and Mg and have exchangeable sodium <15 per cent,EC 4.0 and pH<8.5.

Salinity

Vegetable production is threatened by increasing soil salinity particularly in irrigated croplands which provide 40 per cent of the world's food. According to the United States Department of Agriculture (USDA), onions are sensitive to saline soils, while cucumbers, eggplants, peppers, and tomatoes are moderately sensitive. In hot and dry environments, high evapotranspiration results in substantial water loss, thus leaving salt around the plant roots which interferes with the plant's ability to uptake water. Physiologically, salinity imposes an initial water deficit that results from the relatively high solute concentrations in the soil, resulting in ion-specific stresses from altered K^+/Na^+ ratios that leads to a build up in Na^+ and Cl^- concentrations that are detrimental to plants (Yamaguchi and Blumwald 2005). Plant sensitivity to salt stress is reflected in loss of turgor, growth reduction, wilting, leaf curling and epinasty, leaf abscission, decreased photosynthesis, respiratory changes, loss of cellular integrity, tissue necrosis, and potentially death of the plant (Cheeseman 1988). Salinity also affects agriculture in coastal regions which are impacted by low-quality and high-saline irrigation water due to contamination of the groundwater and intrusion of saline water due to natural or man-made events. Salinity fluctuates with season, being generally high in the dry season and low during rainy season. Coastal areas are threatened by specific, saline natural disasters which can make agricultural lands unproductive. Although the seawater rapidly recedes, the groundwater contamination and subsequent osmotic stress causes crop losses and affects soil fertility. Several studies have shown that increased salinity produces vegetables with higher contents of sugars, organic acid and dry matter. Saline stress activates antioxidative response (Coll *et al.,* 2010c).Levels of ascorbic acid increases with salinity due to detoxification of free radicals. Moderate stress enhanced levels of carotenoids.

Salt stress can be managed by ground water recharging during rainy season, addition of gypsum, widespread use of drip irrigation and developing varieties resistant to salt stress using biotechnological and genetic engineering approaches.

Attempts to improve the salt tolerance of crops through conventional breeding programs have very limited success due to the genetic and physiologic complexity of this trait (Flowers 2004). In addition, tolerance to saline conditions is a developmentally regulated, stage-specific phenomenon; tolerance at one stage of plant development does not always correlate with tolerance at other stages (Foolad 2004). Success in breeding for salt tolerance requires effective screening methods, existence of genetic variability, and ability to transfer the genes to the species of interest. Most commercial tomato cultivars are moderately sensitive to increased salinity and only limited variation exists in cultivated species. Genetic variation for salt tolerance during seed germination in tomato has been identified within cultivated and wild species. Potential sources of resistance to salt stress have been observed in *S. cheesmanii, S. peruvianum, S. pennelii, S. pimpinellifolium,* and *S. habrochaites* (Flowers 2004, Foolad 2004, Cuartero *et al.,* 2006). Utkal Pallavi is a salinity tolerant tomato variety.In pepper, salt stress significantly decreases germination, shoot height, root length, fresh and dry weight, and yield. Yildirim and Guvenc (2006) reported that pepper genotypes Demre, Ilica 250, 11-B-14, Bagci Carliston, Mini Aci Sivri, Yalova Carliston, and Yaglik 28 can be used as sources of genes to develop pepper cultivars with improved germination under salt stress.

Flooding

Vegetable production occurs in both dry and wet seasons in the tropics. However, production is often limited during the rainy season due to excessive moisture brought about by heavy rain. Most vegetables are highly sensitive to flooding and genetic variation with respect to this character is limited. In general, damage to vegetables by flooding is due to the reduction of oxygen in the root zone which inhibits aerobic processes. Flooded tomato plants accumulate endogenous ethylene that causes damage to the plants (Drew, 1979). Low oxygen levels stimulate an increased production of an ethylene precursor, 1-aminocyclopropane-1-carboxylic acid (ACC), in the roots. The rapid development of epinastic growth of leaves is a characteristic response of tomatoes to water-logged conditions and the role of ethylene accumulation has been implicated (Kawase, 1981). The severity of flooding symptoms increases with rising temperatures; rapid wilting and death of tomato plants is usually observed following a short period of flooding at high temperatures (Kuo *et al.,* 1982). The impact of water logging can be negated by following ridge and furrow method of cultivation, giving provision for drainage and growing varieties tolerant to submergence.

Heavy Metals

The input of heavy metals originating from fertilisers,certain pesticides and industrial activities result in soil contamination leading to reduced seed germination, root elongation,biomass production,inhibition of chlorophyll synthesis,and disturbances in cellular metabolism and chromosome distortion.In a survey conducted in Japan (Arao *et al.,* 2008), approximately 7 per cent brinjal fruits contain cadmium at concentrations exceeding international limits for fruit vegetables. Leafy vegetables and tuber crops are generally grown in contaminated soils and they are hyper accumulators of heavymetals in their edible parts.The risk for human health is associated with lead, arsenic, cadmium mercury and fluoride.

Grafting

In addition to the advanced crop management techniques and development of resistant/tolerant varieties to mitigate abiotic stress, grafting is also considered as a viable proposition to induce higher tolerance to abiotic stress conditions like drought, salinity, alkalinity, thermal stress, water stress, heavy metals and organic pollutants. Vegetable grafting was first adopted to enhance their ability to cope up with biotic stresses. Lately, this technique has also been proposed as a way to enhance vegetable tolerance to abiotic soil stresses, such as low soil temperature, drought, salinity and flooding (Ahn *et al.,* 1999; AVRDC, 2001).

Grafting cucumber on cold resistant *C. ficifolia* and watermelon, melon, cucumber and summer squash onto low temperature tolerant rootstocks such as interspecific hybrid between *Cucurbita maxima* x *C. moschata* reduced the risk of severe growth inhibition caused by low soil temperature in winter. Yield losses caused by salinity stress in high-yielding genotypes of solanaceae and cucurbitaceae can be avoided or reduced by grafting. The mechanism of salt tolerance in grafted plants is related to the root morphology, salt exclusion in the shoots, induction of hormones-mediated changes in plant growth and induction of anti oxidant defense system (Colla *et al.,* 2010). Grafting brinjal variety Suqiqie on *Solanum torvum* improved growth performance under saline condition (Wei *et al.,* 2009). Grafting cucumber cultivar onto *C. ficifolia* and *Lagenaria siceraria* resulted in higher yields under different salinity levels. Similarly *C.melo* cultivars grafted onto hybrids of *C. maxima* x *C. moschata* exhibited higher yield (Huang *et al.,* 2009).Tomato can be successfully grown even under flooded conditions by grafting on to eggplant rootstocks (EG195 and EG203) as it can survive for days under water. (Black *et al.,* 2003). Sweet peppers can be grown successfully by grafting onto chilli rootstocks (PP0237-7502, 0242-62, and Lee B) (Manuel and Deng-Lin, 2009).Arao *et al.* (2008)obtained reduced cadmium concentration in brinjal fruits by grafting on *S. torvum.* Concentration of cadmium and nickel in cucumber was restricted by grafting on *C. maxima* x *C. moschata* rootstocks (Savvas *et al.,* 2012). Autotoxic potential of watermelon, melon and cucumber due to phenolic acids of root tissue and root exudates could be overcome by grafting on *C. ficifolia.* Transgenics can also be used as rootstocks (TRANSGRAFTS) for imparting resistance to abiotic stress where GMOs are not permitted.

Tolerance of vegetables crops to abiotic stresses could be further enhanced through integration of grafting technique with sustainable control strategies like inoculation with arbuscular mycorrhiza and organic/foliar fertilization. Oztekin *et al.* (2002) reported better performance of AM fungi inoculated tomato rootstocks under saline conditions. In alkaline soils, zucchini squash inoculated with arbuscular mycorrhiza gave higher yield and total biomass(Cardarelli *et al.,* 2010).Root or foliar application of Silicon has improved growth and performance of crops under water stress(Ma *et al.,* 2004) and salinity(Savvas *et al.,* 2009).Exogenous application of osmoprotectant proline is effective in mitigating stresses due to water deficit(Ali *et al.,* 2007) and salt stress(Ashraf and Foolad,2007) in vegetable crops.

Knowledge on structural organisation and functional properties of genetic variation for stress related traits allow gene based selection through molecular markers.These findings are translated to stress tolerant crop varieties by germplasm screening, conventional breeding, marker assisted selection and plant transformation. Molecular marker analysis of stress tolerance in vegetables is limited but efforts are underway to identify QTLs underlying tolerance to stresses. Martin *et al.* (1989) identified three QTLs linked to water use efficiency in *S. pennellii* based on 13C composition. Three independent yield-promoting regions were identified in *S. pennellii* when grown in both wet and dry field conditions in Israel (Gur and Zamir 2004), while Foolad *et al.* (2003) identified four QTLs associated with seed germination drought tolerance, two of which were contributed by *S. pimpinellifolium*. QTL mapping indicates that salt tolerance is quantitatively inherited (Foolad 2004). Lin *et al.* (2006) identified random amplified polymorphic DNA (RAPD) markers linked to heat tolerance in tomato line CL5915. Studies indicate that stress tolerance is quantitatively inherited and in some cases tolerance is dependent on the developmental stage of the plant. Consequently, multiple genes are predicted to be involved with the expression of stress tolerance. Integration of QTL analysis with gene discovery and modeling of genetic networks will facilitate comprehensive understanding of stress tolerance, permit the development of useful and effective markers for marker-assisted selection, and identify candidate genes for genetic engineering.

Approximately 130 drought-responsive genes have been identified using microarrays (Reymond *et al.*, 2000, Seki *et al.*, 2001). These genes are involved with transcription modulation, ion transport, transpiration control and carbohydrate metabolism. *DREB1A* and *CBF*, and *HSF* genes are transcription factors implicated in drought and heat response, respectively (Sakuma *et al.*, 2002,Sung *et al.*, 2003). Cell wall invertase (*INV*) and sucrose synthase (*SUSY*) play key roles in carbohydrate partitioning in plants (Déjardin *et al.*, 1999) and this regulation of carbohydrate metabolism in leaves may represent part of the general cellular response to acclimation and contribute to osmotic adjustment under stress. The *ERECTA* gene regulates plant transpiration efficiency in *Arabidopsis thaliana* (Masle *et al.*, 2005), and the *NHX* and *AVP1* genes are associated with ion transport (Zhang and Blumwald 2001). Although the function of these genes has been elucidated, particularly in *A. thaliana*, only a few genes have contributed to a tolerant phenotype when over-expressed in vegetables (Zhang *et al.*, 2004). Expression of *AVP1*, a vacuolar H^+ pyrophosphatase from *A. thaliana*, in tomato resulted in enhanced performance under soil water deficit (Park *et al.*, 2005). The engineered tomato has a stronger, larger root system that allows the roots to make better use of limited water. The control plants suffered irreversible damage after five days without water as opposed to transgenic tomatoes which began to show water-stress damage only after 13 days but recovered completely as soon as water was supplied. The *CBF/DREB1* genes have been used successfully to engineer drought tolerance in tomato and other crops (Hsieh *et al.*, 2002). Constitutive over-expression of *CBF* genes results in salt, cold, or drought tolerance in several plant species. However, in addition to increased stress tolerance, the transgenic plants were dark-green and were stunted, with higher levels of soluble sugars and proline (Liu *et al.*, 1998). The use of stress-inducible promoters that have a low background expression of *CBF* under normal growth conditions can achieve increased stress tolerance without plant growth retardation (Lee *et al.,* 2003). Maintaining increased expression of the *A. thaliana* tonoplast membrane Na$^+$/H$^+$ antiporter, *AtNHX1*, under a strong constitutive promoter, was reported to result in salt-tolerant tomatoes (Zhang and Blumwald 2001). The transgenic tomato plants grown in the presence of 200 mM NaCl were able to flower and set fruit. While the leaves accumulated high concentrations of sodium, the tomato fruits continued to contain only low concentrations of sodium. In tomato, expression of gene A1DREB1A showed tolerance to drought and BcZAT12 gene showed tolerance to heat and drought stress.(IIVR,2012). Research on the physiology of stress tolerance has demonstrated that tolerance to a specific stress is determined by several component traits and controlled by corresponding genes. A combination of a genome-wide scan of expression, using DNA arrays, and QTL analysis could provide important information in identifying the major genes association with stress tolerance.

Spices

In spice also, abiotic stress is a major limiting factor of plant growth and crop yields. With a goal to raise spices with better suitability towards rapidly changing environmental inputs, physiological, biochemical and molecular tools are being used to improve tolerance to abiotic stresses. A few genotypes tolerant to moisture stress were short listed from germplasm in black pepper, cardamom and a few seed spices which can grow in rain fed/reduced moisture conditions. The black pepper cultivar 'Kalluvally' is tolerant to moisture stress. These lines are being evaluated for their yield potential and a few genotypes were promising in cardamom. Inter varietal hybridization involving the shortlisted accessions are in progress to converge the genes for drought tolerance with other resistance characters and quality attributes.

In black pepper mapping populations were also developed to tag stress resistance genes. Phenotyping, genotyping and molecular profiling of these populations are in progress, this will help in future Marker Assisted Selection and reduce breeding time.

In vitro co-culture of plant tissue explants with beneficial microorganisms induces developmental and metabolic changes in the derived plantlets which enhance their tolerance to abiotic and biotic stresses. The induced resistance response caused by the inoculants is referred to as "biotization." There is enough experimental evidence with bacteria (bacterization) and vesicular arbuscular mycorrhiza (mycorrhization) inoculations to recommend utilization of this technology in commercial micropropagation. Use of *Trichoderma*, Fluorescent *Pseudomonas* and VAM at hardening stage helped TC Black pepper plants perform better compared to conventionally propagated plants due to better root system, better establishment and better growth. Use of Plant Growth Promoting Rhizobacteria (PGPR) is another approach which is being attempted especially in seed spices. Seed coating with IISR strains gave positive indications of better germination, establishment and yield under reduced water usage. In perennial crops like nutmeg and clove, grafting with root stocks of wild species have given initial indication of better establishment.

Exploiting somaclonal variations is another approach to develop stress tolerant varieties especially in ginger, turmeric and seed spices. Reports are also available on *in vitro* selection for salt tolerance in fenugreek, *Trigonella foenum-graeceum*.

Plant biotechnology appears to be an attractive alternative in respect of the possibility for direct introduction of single genes into crops. Employing transgenic technology, functional validation of various target genes like 'Osmotin' are also being considered and a few putative transgenics were developed and are being evaluated for drought tolerance. Accumulation of osmoprotectants through gene transfer is another approach which can be attempted. With the advent of genome sequencing, transcriptome approach is also being attempted in black pepper to annotate and isolate genes and transcription factors involved in abiotic stress resistance.

In general breeding for abiotic stress tolerance has made a good beginning and soon may yield lasting results. With the availability of many novel genes for stress tolerance, development of transgenics and use of root stocks use of will be the shortest way of developing abiotic stress tolerant varieties in vegetables and spices.

References

1. Abdalla A A and Verderk K.(1968). Growth, flowering and fruit set of tomato at high temperature. *The Neth.J.Agric.Sci* 16: 71-76.

2. Abdul-Baki A A and Stommel, J (1995)Pollen viability and fruit set of tomato genotypes under optimum and high-temperature regimes. *Hort Science* 30: 115-117.

3. Ahn, S. J., Im, Y. J., Chung, G. C., Cho, B. H. and Suh, S. R. (1999). Physiological responses of grafted cucumber leaves and rootstock roots affected by low root temperature. *J. Scientia horticulture*. 81: 397-408.

4. Ali M (2000) Dynamics of vegetables in Asia: A synthesis. (In) Ali M (ed) Dynamics of vegetable production, distribution and consumption in Asia. AVRDC, Shanhua, Taiwan, pp 1-29.

4. Ali, Q., Ashraf, M. and Athar, H.R. (2007). Exogenously applied proline at different growth stages enhances growth of two maize cultivars grown under water deficit conditions. *Pakistan J. Bot.* **39**: 1133-144.

5. Arao, T., Takeda, H. and Nishihara, E. (2008). Reduction of cadmium translocation from roots to shoots in eggplant (*Solanum melongena*) by grafting onto *Solanum torvum* rootstock. *Soil sci. plant nutr.* **54**: 555-59.

6. Ashraf,M.and Foolad,M.R.(2007).Roles of gycinebetaine and proline in improving plant abiotic stress.*Environ.Expt.Bot*.59: 206-216.

7. AVRDC [Asian Vegetable Research and Development Centre]. (2001). *Report 2001* [on line]. Asian Vegetable Research and Development Centre, Taiwan.

8. Black, L. L., Wu, D. L., Wang, J. F., Kalb, T. Abbas, D. and Chen, J. H. (2003). Grafting Tomatoes for production in the Hot-Wet season. AVRDC publication number 03-551.

9. Cardarelli, M., Rouphael, Y, and Colla, G. (2010). Mitigation of alkaline stress by arbuscular mycorrhiza In zucchini plants grown under mineral and organic fertilization. *J. Plant Nutr. Soil Sci.* **173**: 778-787.

10. Chaves MM and Oliveira.(2004).Mechanisms underlying plant resilience to water deficits: prospects for water-saving agriculture. *J.Exp.Bot.*55: 2365-2384.

11. Cheeseman JM.(1988).Mechanisms of salinity tolerance in plants. *Plant Physiol* 87: 57-550.

12. Colla, G., Rouphael, Y., Leonardi, C. and Bie, Z. (2010c). Role of grafting in vegetable crops grown under saline conditions. *Scientia Hort.* **127**: 147-155.

13. Cuartero,J.,Bolarin,M.C.,Asins,M.J. and Moreno,V.(2006)Increasing the salt tolerance in tomato.*J.Exp.Bot.Plants and salinitySpecial Issue,*57: 1045-1058.

14. Déjardin A, Sokolov LN and Kleczkowski,LA.(1999). Sugar/osmoticum levels modulate differential ABA-independent expression of two stress responsive sucrose synthase genes in Arabidopsis. *Biochem. J.* 344: 503-509.

15. Drew,MC.(1979). Plant responses to anaerobic conditions in soil and solution culture. *Curr. Adv.Plant Sci.*36: 1-14.

16. Erickson AN and Markhart, AH.(2002). Flower developmental stage and organ sensitivity of bell pepper (*Capsicum annuum* L.) to elevated temperature.*Plant Cell Environ.*25: 123-130.

17. FAO,(2001).Climate variability and change: A challenge for sustainable agricultural production. Committee on Agriculture, Sixteenth Session Report, 26-30 March, 2001. Rome, Italy.

18. Flowers,TJ.(2004). Improving crop salt tolerance. *J.Exp.Bot.* 55: 307-319.

19. Foolad,MR.(2004). Recent advances in genetics of salt tolerance in tomato. *Plant Cell Tissue Organ Culture* 76: 101-119.

20. Foolad MR, Zhang LP and Subbiah,P.(2003). Genetics of drought tolerance during seed germination in tomato: inheritance and QTL mapping. *Genome* 46: 536-545.

21. Gur A and Zamir,D.(2004). Unused natural variation can lift yield barriers in plant breeding. *PLoS Biol* 2: 1610-1615.

22. Hanson, PM., Chen, JT and Kuo, CG.(2002).Gene action and heritability of high temperature fruit set in tomato line CL5915. *Hort.Science* 37: 172-175.

23. Hazra P, Samsul HA, Sikder D and Peter,KV.(2007). Breeding tomato (*Lycopersicon esculentum* Mill) resistant to high temperature stress. *Int. J. Plant Breed.* 1(1).

24. Hsieh TH, Lee JT, Charng YY and Chan,MT.(2002).Tomato plants ectopically expressing Arabidopsis CBF1 show enhanced resistance to water deficit stress. *Plant Physiol.*130: 618–626.

25. Huang, Y., Tang,R.,Cao,Q.Land Bie,Z.L.(2009).Improving fruit yield and quality of cucumber by grafting on to the salt tolerant rootstock under NaCl stress. *Scientia Hort.,122: 26-31.*

26. IIVR (2012).Annual Report 2011-2012. Indian Institute of Vegetable Research, Varanasi.pp.93-107.

27. Kawase, M.(1981). Anatomical and morphological adaptation of plants to waterlogging. *Hort.Science* 16: 30-34.

28. Kuo, DG, Tsay JS, Chen, BW and Lin, PY. (1982). Screening for flooding tolerance in the genus *Lycopersicon. Hort.Science* 17(1): 76-78.

29. Lin KH, Lo HF, Lee SP, Kuo CG and Chen,JT.(2006). RAPD markers for the identification of yield traits in tomatoes under heat stress via bulk segregant analysis. *Hereditas* 143: 142-154.

30. Liu Q, Kasuga M,SakumaY,Abe H,Miura S,Yamaguchi-Shinozaki K and Shinozaki,K. (1998). Two transcription factors, DREB1 and DREB2, with an EREBP/AP2 DNA binding domain separate two cellular signal transduction pathways in drought- and low-temperature-responsive gene expression, respectively, in Arabidopsis. *Plant Cell* 10: 1391-1406.

31. Ma, Q.F., Turner, D.W., Levy, D. and Cowling,W.A.(2004). Solute accumulation and osmotic adjustment in leaves of Brassica oilseeds in response to soil water deficit. *Australian J. Agric. Re*s. **55**: 939-945.

32. Manuel, C. P. and Deng-Lin, W. (2009). Grafting Sweet Peppers for production in the Hot-Wet season. AVRDC publication number 09-722.

33. Masle J, Gilmore SR and Farquhar,GD.(2005). The ERECTA gene regulates plant transpiration efficiency in *Arabidopsis. Nature* 436: 866-870.

34. Nirmal Babu K, George JK, Bhat AI, Prasath D, and Parthasarathy, VA.(2011). Tropical Spices *in* HP Singh, VA Parthasarathy and K Nirmal Babu (eds) *Advances in Horticulture Biotechnology - Vol 5 – Gene cloning and Transgenics.*Westville Publishing House, New Delhi. pp. 529- 542.

35. Opena RT, Green SK, Talekar NS and Chen,JT.(1989).Genetic improvement of tomato adaptability to the tropics: Progress and future prospects. In: Green SK (ed) Tomato and pepper production in the tropics. AVRDC, Shanhua, Taiwan pp. 70-85.

36. Ort DR (2001). When there is too much light. *Plant Physiol* 125: 29-32.

37. Oztekin, G.B., Tuzel, Y. and Tuzel,I.H.(2012). Does mycorrhiza improve salinity tolerance in grafted plants? *Scientia Hort.* (in press).

38. Park S, Li J, Pittman JK, Berkowitz GA, Yang H, Undurraga S, Morris, J, Hirschi KD and Gaxiola, RA.(2005). Up-regulation of a H+-pyrophosphatase (H+-PPase) as a strategy to engineer drought –resistant crop plants. *PNAS* 102: 18830-18835.

39. Peter KV and Nirmal Babu,K.(2012). Role of plant tissue culture in biotic and abiotic stress tolerance – in spice crops, National Symposium on Impact of Plant Tissue Culture on Advances in Plant Biology, St. Xavier's College, Ahmadabad, 21Jan 2012. Abstracts, p. 63-64.

40. Rick,C.M.(1973).Potentila genetic resources in tomato species: clues from observations in native habitats.In: SrbAM(ed.)Genes,enzymes and populations,Plenum Press,New York,p.255-269.

41. Reymond P, Weber H, Damond M and Farmer,EE.(2000). Differential gene expression in response to mechanical wounding and insect feeding in *Arabidopsis. Plant Cell* 12: 707-720.

42. Sakuma Y, Liu Q, Dubouzet JG, Abe H, Shinozaki K and Yamaguchi-Shinozaki,K.(2002). DNA-binding specificity of the ERF/AP2 domain of *Arabidopsis* DREBs, transcription factors involved dehydration- and cold-inducible gene expression. *Biochem. Biophys. Res. Comm.* 290: 998-1009.

43. Sato S, Peet MM and Thomas,JF.(2002). Determining critical pre- and post-anthesis periods andphysiological process in *Lycopersicon esculentum* Mill. Exposed to moderately elevated temperatures. *J. Exp. Bot.* 53, 1187-1195.

44. Savvas, D., Giotis, D., Chatzieustratiou, E., Bakea, M. and Patakioutas, G.(2009). Silicon suppy in soilless cultivations of zucchini alleviates stress induced by salinity and powdery mildew infection. *Environ.Exp. Bot.* **65**: 11-17.

45. Savvas, D., Ntatsia, G. and Barouchas, P.(2012). Impact of grafting and rootstock genotype on cation uptake by cucumber (*Cucumis sativus* L.) exposed to Cd or Ni stress. *Scientia. Hort.* (in press).

46. Seki M, Narusaka M, Abe H, Kasuga M, Yamaguchi-Shinozaki K, Carninci P, Hayashizaki Y and Shinozaki,K.(2001).Monitoring the expression pattern of 1300 Arabidopsis genes under drought and cold stresses by using a full-length cDNA microarray. *Plant Cell* 13: 61-72.

47. Singh,H.P.(2012). Ongoing research in abiotic stress due to climate change in Horticulture.1-23.

48. Stevens, MA and Rudich, J.(1978). Genetic potential for overcoming physiological limitations on adaptability, yield, and quality in tomato. *Hort Science* 13: 673-678.

49. Sung DY, Kaplan F, Lee KJ and Guy,CL.(2003). Acquired tolerance to temperature extremes. *Trends Plant Sci.* 8: 179-187.

50. Wei, G.P., Yang, L.F., Zhu, Y.L. and Chen, G. (2009). Changes in oxidative damage, antioxidant enzyme activities and polyamine contents in leaves of grafted and non-grafed eggplant seedlings under stress by excess of calcium nitrate. *Scientia. Hort.* **120**: 443-451.

51. Yamaguchi T and Blumwald,E.(2005). Developing salt-tolerant crop plants: challenges and opportunities. *Trends Plant Sci.* 10(12): 616-619.

52 Yildirim E and Guvenc,I.(2006). Salt tolerance of pepper cultivars during germination and seedling growth. *Turk. J. Agric. Forestry* 30: 347-353.

53. Zhang HX and Blumwald,E.(2001). Transgenic salt-tolerant tomato plants accumulate salt in foliage but not in fruit. *Na.t Biotechno.l* 19: 765-768.

54. Zhang JZ, Creelman RA and Zhu, JK.(2004). From laboratory to field. Using information from Arabidopsis to engineer salt, cold, and drought tolerance in crops. *Plant Physiol.* 135: 615-621.

2015, **Horticulture for Nutrition Security**   *Pages* **345–366**
Editor: **Prof. K.V. Peter**
Published by: **DAYA PUBLISHING HOUSE, NEW DELHI**

Chapter 16

Azadirachtin:
Its Structure and Insect Activity

Pathipati Usha Rani

Neem *(Azadirachta indica* A juss) of the family Meliacea is a tree with miracle powers and wonder chemicals often considered as divine tree in many parts of Asia. Its medicinal value is traditionally known for centuries. As a response to the feeding by multiple pests, the neem tree might have evolved over the evolutionary period with several defense mechanisms/tactics resulting in the production of the chemical azadirachtin. The role of plant defensive chemicals is to discourage pest herbivory. Understanding the chemistry of plant defensive compounds will lead to the identification of possible insect feeding deterrent substances that could be isolated in sufficient quantities or synthesized for application on plants as crop protective agents. Neem seeds actually contain more than a dozen azadirachtin analogs, in which azadirachtin A is the major form and the remaining minor analogs also might contribute up to a considerable extent. Hundreds of chemicals are being extracted and identified from neem which not only cure human ailments but also aid them in getting rid of harmful insect pests and pathogens. These properties are due to a large number of secondary plant metabolites found in various parts of the tree. The most important of all these is Azadirachtin, which belongs to an organic molecule class called tetranortriterpenoids ($C_{35}H_{44}O_{16}$, Molecular weight: 720), is a highly oxidized, tetranortriterpenoid from the neem kernel and has been rated as the most potent naturally occurring crop protection chemical (Figure 16.1). It is a subject of study ever since its discovery in the year 1968. The insecticidal effect of the compound azadirachtin was first reported by Ruscoe (1972). It boasts a plethora of oxygen

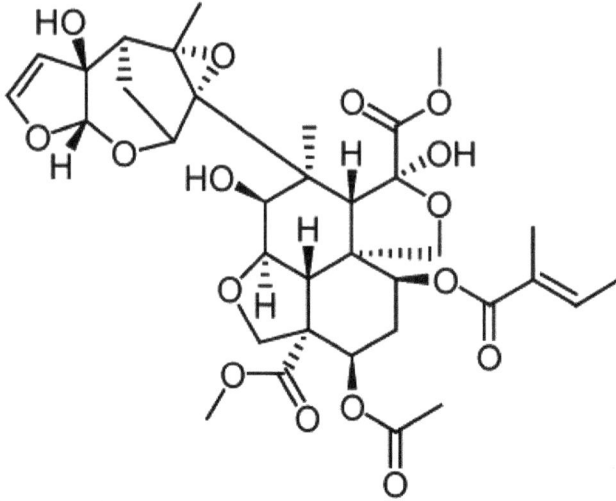

Figure 16.1: Structure of Azadirachtin A.

functionality, comprising an enol ether, acetal, hemiacetal, and tetra-substituted oxirane as well as a variety of carboxylic esters. The seed after extraction with hexane yields a paste and oil. This paste is first extracted with ethanol and then with chloroform to yield crude Azadirachtin which is about 20gm for 1 kg of seeds. The active ingredients biodegrade rapidly in sunlight (it degrades within 100 hours when exposed to light) and within a few weeks in the soil. The half life of, azadirachtin was estimated to be approximately 20 h on olives growing in Italy (Caboni *et al.,* 2002). It breaks down rapidly (in 50-100 hrs) in water also, hence it may not cause long -term effects on fish populations (Martineau, 1994) and no significant effects on other wildlife were reported.

It is also reported to be safe on beneficial arthropods (bio-control agents) and the effects are generally considered to be minimal. It is a valuable pesticide and is used as a grain protectant for centuries without apparent harm to humans and other mammals (the LD_{50} in rats is > 3,540 mg/kg). The presence of this amazing chemical not only protects the neem tree from hundreds of pests existing around but also when used on crop plants or in human dwellings and stored food prevents the pests from destroying. Azadirachtin because of its abundance and unique mode of action. It does not knock down or kill the insect instantaneously like most neurotoxic insecticides. Instead, it elicits physiological and behavioural responses in insects, which lead to their death. It can repel, disrupt the growth and reproduction or cause severe morphological or physiological abnormalities. Azadirachtin interferes with the neuro endocrine system of several insects which controls the synthesis of ecdysone and juvenile hormone, and have a direct role in moulting inhibition. This knowledge indicates that azadirachtin may have more than one mode of action. It alters the organelle formation, causes disruption of cell division, block the transport and release of neurosecretory peptides and inhibit the sperm formation (Mordue *et al.,* 2010). The treatment might also cause inhibition of digestive enzymes in mid gut cells. Azadirachtin at molecular level may act by preventing transcription or translation of proteins expressed at particular stages of the cell cycle (Mordue *et al.,* 2010). Neem seeds actually contain more than a dozen of azadirachtin analogs, in which azadirachtin A is the major form and the remaining minor analogs also might contribute up to a considerable extent. The overall effects of azadirachtin can be described as follows. It may repel the approaching herbivore thus preventing the insect from initiating the feeding, it may stop the feeding of the insect that initiated feeding as an antifeedant or it may cause growth defects after ingestion due to secondary hormonal or physiological substances disrupt normal development interfering with chitin synthesis. Susceptibility to the various effects of neem differs by species.

Lepidoptera are among the organisms which are the most sensitive to neem extracts (Schmutterer and Singh, 1966). Azadirachtin have multiple modes of action on insects. It exerts strong negative influence on feeding, fecundity and post embryonic development of several insect species; thus causing significant reduction in its further development and growth. Azadirachtin exhibit several biological effects like feeding deterrence, repellence, oviposition deterrence, hindering the flying ability, mating disruption, insect growth regulating effect on metamorphosis, reduced fecundity and viability of eggs, sterilant activity to eggs, ovicidal and insecticidal action are known against a large variety of insects. These varied effects are age and stage dependent (Schmutterer, 1995). Research over the past a few decades has shown that it is one of the most potent growth regulators and feeding deterrents ever discovered. By ingesting a small quantity of this chemical, the insects become inactive and stop feeding. Depending on insect and dosage applied azadirachtin can be residual insecticide for up to 10 days in some cases. The potential applications of azaditachtin in human and veterinary medicine are numerous (Schmuttere,r 2002; Talwar *et al.,* 2002; Kleeberg and Strang, 2009; Van der Esch *et al.,* 2009; Ketkar, 2009). Extensive variation in azadirachtin content of neem was attributed to the variations in the ecotypes (Kumar and Parmar, 1997), local environmental conditions such as humidity, rainfall, temperature, or season (Ermel, 1995, Venkateswarlu *et al.,* 1997) and seasons (winter or monsoon) (Sidhu and Behl, 1996). Often it is reported in the literature about variations of azadirachtin content in neem from even different countries or different regions within a country (Siddhu *et al.,* 2003) and has found wide variations in oil and azadirachtin contents among different provinces, while examining intra province variability among forty-three provinces of India.

Azadirachtin: Insect Antifeedant Activity

The major competence of azaditachtin is repelling and/or preventing the feeding of many species of pest insects as well as some nematodes. Azadirachtin, the predominant biologically active chemical is also known as the 'most potent insect antifeedant discovered to date' (Miller *et al.,* 2006). It is an excellent antifeedant to a range of lepidopterans and it is a boon to farmers as most of the destructive pests particularly those which cause damage during feeding stages are deterred from feeding further due to the treatment. It is now reported as a feeding deterrent against almost 200 species of insects, and as the antifeedant its effects are structural dependent.

It was initially shown as feeding inhibitor to dessert locust (*Schistocerca gregaria*) (Butterworth and Morgan, 1968). The increased neural activity of the chemoreceptors is correlated with the antifeedant responses. A combined neural input from the insect's chemical senses such as taste receptors on tarsi, mouthparts and oral cavity and a central nervous integration of this sensory input is generally responsible for the feeding behaviour of any insect. Azadirachtin stimulates specific 'deterrent' cells in chemoreceptors and also block the firing of 'sugar' receptor cells, which normally stimulate feeding (Simmonds and Blaney, 1984; Blaney *et al.,* 1990; Simmonds *et al.,* 1990). This causes the feeding inhibition thus leading to starvation death of the insect (Koul and Wahab, 2004). In several insect species, it is shown that the biological activity of Azadirachtin is dose dependent, increasing concentration increases the deterrence index, thus achieving a good dose response and Lack of specific components, particularly nitrogen or water, leads to a lower growth rate and lower metabolic efficiency (Schoonhoven *et al.,* 2005). It was suggested that the differential toxicity of azadirachtin is at least partly due to difference in uptake rate. The dessert locust, *Schistocerca gregaria*, will not even touch a plant treated with azadirachtin in trace amounts, whereas, the Mexican bean beetle, *Epilachna varivestis*, can tolerate relatively high doses of azadirachtin treated diet without any negative effects. However, this insect is highly responsive to the growth–inhibitory effect of these compounds. No feeding inhibition was recorded when azadirachtin is tested against three holometabolous insects belonging to different orders: Mexican bean beetle, *Epilachna varivestis* Muls.; Mediterranean flour moth, *Ephestia kuehniella* Zell.; and honeybee, *Apis mellifera* L. However, there was a great reduction in larval development and their transformation into adults and the reduction was dose-dependent; pupal mortality was also observed (Rembold *et al.,* 1982). Results obtained by treating the dessert locust, *Schistocerca gregaria* with azadirachtin are interesting. Azadirachtin effects on development, food consumption and utilization and haemolymph constituents of final instar *Schistocerca gregaria* was studied by Rao and Subrmnayam (1986). The treatment caused severe moulting deformities. Its effect on females was more drastic as they showed negative growth rate revealing the catabolic state. Maximum changes in haemolymph protein and amino acid levels were observed particularly in females between 5th and 8th day after treatment. Neoliya *et al.* (2005) through a laboratory study demonstrated the effects of azadirachtin on head protein levels in *H. armigera*. They found that medium to low dose of azadirachtin significantly influenced the total head protein profile in larvae of *H. armigera*. The chemical applied topically to the 2nd instar larvae efficiently reduced the protein concentration. Similar decrease in protein concentration has been reported in heamolymph system of larval *S. litura* (Ayyangar *et al.,* 1990). Their work also implies that azadirachtin may not exist in the neuroendrocrine system up to 72 h when administered topically. However, contrary to this ingested azadirachtin has been found accumulated in the head and other tissues of *Peridroma sausia* (Kaul, 1993) indicating different mode of azadirachtin action.

The structural dependency of the antifeedant effect is well dealt with. Most of the antifeedant studies are made using lepidopteran insects. Since these are the most damaging pests, consuming large quantities of leaf materials causing considerable economic damage. Hence majority of work reported constituted major lepidopteran pests. If the compound treated with is effective in preventing the larval feeding it is great advantageous in plant protection. A study on the nutritional physiology and gut enzyme activity of the lesser mulberry pyralid, *Glyphodes pyloalis* Walker, a monophagous and dangerous pest of mulberry in Guilan province, northern Iran treated with neem formulation containing azadirachtin revealed interesting effects of this compound (Khosravi and Jalali Sendi, 2013). All nutritional indices, except approximate digestibility, decreased in exposed insects. The effects of azadirachtin on consumption and inhibition of growth of *G. pyloalis* could be attributed to the lowered activities of both the proteases and α-amylase in the midgut, and the inhibition of feeding behaviour. The biochemical compounds in the hemolymph, such as protein, lipid, and glucose were decreased in the larvae fed on treated leaves but the amount of uric acid was increased when compared with the control (Khosravi and Jalali Sendi, 2013). Azadirachtin's efficiency in inhibiting peristalsis, reducing enzyme production as food moves through gut, inhibiting midgut cell replacement, and feeding reduction is reported with other insects too (Anuradha and Annadurai, 2008).

Huang *et al.* (2004) showed that azadirachtin significantly influenced the protein level in *Spodoptera littoralis* Fabricius. Other research where azadirachtin interfered with protein synthesis is in the desert locust (Annandurai and Rembold, 1993), and hemolymph of *S. litura* (F.) (Li *et al.*, 1995). Adequate level of nutrient ingestion, assimilation and conversion of food into energy and biomass are related to growth effects (Rees, 1978). Azadirachtin has been shown to interfere with digestibility and efficiency of conversion of ingested food in several species of insects (Fagonee, 1984; Arnason *et al.*, 1985; Ayyangar and Rao 1989; Timmins and Reynolds, 1992). *Spodoptera littoralis* (Boisduval) is sensitive to the azadirachtin treatments just like other Lepidopterans. Feeding of the third instar larvae with sub lethal doses of azadirachtin incorporated artificial diet resulted with reduced food intake, a reduced weight gain in their instars immediately after treatment and prolonged larval life (Martinez and van Emden, 1999). According to them, Azadirachtin did not influence digestion efficiency but diminished the ability of the larvae to convert both ingested and digested nutrients into growth, particularly immediately after treatment. Secondary antifeedant effects exerted by azadirachtin varied with the insect species. It was not observed in M. *sexta* (Timmins and Reynolds, 1992) or in *P. saúda* (Koul and Isman, 1991). Higher doses may lead to the growth disruption.

It is shown to interact with peritrophic membrane, and damage cellular surfaces of the midgut. In Lepidoptera the antifeedant response is also correlated with increased neural activity of the chemo receptors. It is interesting to know that both monophagous and polyphagous insects responded in similar manner to azadirachtin treatments, whereas, several other plant compounds showed different activities to different lepidopteran larvae. Percentage feeding inhibition in the polyphagous pest, *S. litura* treated with the leaf extracts of *Terminalia catappa* was more (about 65 per cent), but the percentage feeding inhibition of mono phagous pest, *A. janatha* was only 35 per cent indicating differential susceptibility of the two major lepidopteran pests of Castor, *Ricinus communis* (L.) (Devanand and Usha Rani, 2008). *Sterulia fitida* too did not show any toxic or antifeedant effects to *A. janata* at 2 mg cm^{-2} but was good feeding deterrent as well as toxicant to *S. litura* at this dosage (Usha Rani and Rajasekharreddy,

2009). A few amino acid derivatives of Plumbagin showed excellent antifeedant and toxic effects to *S. litura* but the same compounds failed to show any such activities to *A. janata* (Sreelatha *et al.,* 2009). Similarly, butyl and decahexyl derivatives of plumbagin were good feeding deterrents to *S. litura* but the activity was totally absent in the monophagous pest, *A. janatha* (Prasad *et al.,* 2012).

Azadirachtin has systemic action in certain crop plants, greatly enhancing its efficacy and field persistence (Schmutterer, 2002). Neem has some systemic activity in plants and potential against several sucking pests. Certain chemicals isolated are most effective against actively growing immature insects too. Azadirachtin was significantly found to decrease the number of feeding site by the pentatomid *N. viridula* on peacan nuts (Seymour *et al.,* 1995). Systemic Antifeedant activity of Azadirachtin was analyzed in adult *Myzus persicae* (Sulzer) (Homoptera: Aphididae) on *Nicotiana clevelandii* (Gray) seedlings by electrical penetration graph (EPG) method (Nisbet *et al.,* 1993). Azadirachtin treatment significantly reduced the percentage of probes that reached sieve elements and increased non-penetration activity before and after the first period of ingestion from the sieve elements. The number of probes initiated, and the numbers of sieve tube penetrations were also increased with increased azadirachtin concentration. Petioles of intact cabbage leaves placed in azadirachtin solutions at 1-10ppm level were translocated systemically in to the plant and reduced the food consumption of the *Pieris brassicae* larvae (Arpaia and Van Loon, 2011). Seedlings of winter barley (*Hordeum marinum*) treated with different concentrations of azadirachtin when exposed to cereal aphid, *Rhopalosiphum padi* (L.), were less preferred by the aphids than normal plants. The compound has systemic activity towards the aphids and prevented them from feeding on the azadirachtin treated plants (West and Mordue, 2011).

The presence of azadirachtin was shown to be responsible for highly deterrent and growth regulatory activity of the neem seed extract to rose aphid, *Microsiphum comosae* (L.) and Chrysanthemum aphid, *Macrosiphoniella sanbornii* (Gillete) and the activity was related to the concentration of the compound in the extract (Opender Koul, 1999). Growth regulatory effects of azadirachtin are influenced by the host plant and the stage of treatment. Topical application of azadirachtin at 1, 2 and 5 µg on eggs and larvae of *Triboleum castaneum* (Herbst) (Col., Tenebrionidae) did not have any adverse effects but it reduced emergence of normal adults when applied on less than 6 hour old pupae (Mukherjee and Ramachandran, 1989). The secondary antifeedant effect of azadirachtin was reported in *Spodoptera littoralis* which consumed less food after feeding on diet treated with azadirachtin for two days as third instar larvae than the control larvae. This confirmed a secondary antifeedant effect of azadirachtin, as treated insects were transferred back to normal diet that does not contain azadirachtin (Martinez and Emden, 1999). However, the ability to convert food into biomass was reduced and the growth was affected.

Azadirachtin: Insect Growth Disruptor

Among all the analogs and related compounds synthesized till today, azaditachtin is popular because it is structurally similar to insect hormones called 'ecdysones' that regulates the process of metamorphosis. Several hormones and other physiological changes synchronize the process of metamorphosis and azadirachtin seems to block the insect's production and release of these vital hormones, so that the process of moulting is inhibited, thus breaking their life cycle. Effects on the hormone physiology of insects have been described (Mordue, 2002; Rembold, 2004). Studies with *Ostrinia furnacalis* Guenée (Lepid., Pyralidae) (feeding azadirachtin in artificial diet) showed a gradual decrease in the body weight and even the lipid content became less than that of control (Shinfoon, 2009). Several histopathological changes occurred in brain, corpora cardiaca, corpora allata and the prothoracic gland of these treated larvae indicating involvement of neuroendocrine system of insect. A comparison of larval growth inhibitory effects of azadirachtin on six species of Noctuid moths, the black army cutworm, *Actebia fennica* (Tausch.), the bertha armyworm, *Mamestra configurata* Walker, the variegated cutworm, *Peridroma saucia* Hubner, the zebra caterpillar, *Melanchra picta* (Harr.), the Asian armyworm, *Spodoptera litura* (Fab.), and the cabbage looper, *Trichoplusiani* Hubner showed that the feeding of artificial diet containing azadirachtin at various doses inhibited neonate larval growth of all species in a dose dependent fashion (Isman, 1993). Interference of azadirachtin with ovarian and testes development in turn affects the fecundity and fertility of insects as an important component of azadirachtin effects on insects (Mordue, 2000). Azadirachtin impede spermatogenic meiosis in Locusts testes, arresting meiosis at metaphase I (Limton *et al.*, 1997).

Azadirachtin inhibited the growth of rice moth, *Corcyra cephalonica* Staint, a serious pest of stored grains and stored food product when applied topically (Sharma, 1992). The compound prevented larval development at higher doses applied and disturbed both larval-pupal and pupal-adult transformation, which is interpreted as interference with the morphogenetic hormone pool size. Azaditachtin also induces typical effects on the insect endocrine system. The systemic application of seed extracts of neem containing Azadirachtin as main ingredient severely disrupted the development of coniferophagous bark beetles, *Dendroctonus ponderosae* Hopkins (Duthie Holt and Borden, 1999). The fungal associates of Mountain pine beetle, *Ophiostoma clavigerum* (Robbins Jeff and Davids) and *O. ips* were also killed due to the application of azadirachtin mixed in malt extract agar. In fact, pine trees can be protected against mountain pine beetles by application of an emulsifiable concentrate of azadirachtin in tree trunks. Complete molt inhibition was shown by radioimmunoassay in two more beetles, large milkweed bug, *Oncopeltus fasciatus* and the mexican bean beetle, *Epilachna varivestis* due to blockage of ecdysteroid synthesis and release (Redfern *et al.*, 1982, Dorn, 1986). A single dose of azadirachtin had reduced hemolymph ecdysteroid titers to level which is too low for induction of ecdysis (Garcia *et al.*, 1990). Azadirachtin attacks a highly specific target molecule which is a highly specific receptor in the neurolemma of the brain and this receptor is absent in mammals (Rembold, 1989). Another unique, but undefined mode of action exhibited by Azadirachtin is its growth inhibitory effects on endoparasites as compared with the other IGRs. Azadirachtin treated *Rhodnius prolixus*, the vector for transmittance of Chagas disease, is no longer attractive to parasite flagellate, *Trypanosoma cruzi* due to the treatment (Rembold and Garcia, 1989). Azadirachtin has a highly complicated molecular structure which is involved in its biological activity. It affects the bug's hormone system and also disrupts the carefully synchronized arrangement between the parasite and its intermediate host in a lasting way. The treated insect loses its biosemiotic identity and is no longer attractive or viable for the parasite (Rembold and Garcia, 1989). Even at a minimum dosage applied, injection of azadirachtin into *Manduca sexta* larvae that were parasitized by *Cotesia congregata* adversely affected subsequent endoparasite development, when the compound was administered prior to the first larval ecdysis of the wasps (Beckage *et al.,* 1988). Parasites in azadirachtin-treated hosts never grew, though the host larvae could survive up to two weeks time, the parasites failed to extricate themselves from the exuviae that lead to their death. The characteristic hemolymph polypeptides of the terminal stage larvae were disappeared or not found in these parasitized larvae. The azadirachtins seem to have a higher repelling effect against hemi- than against holo metabolous insects.

Different species of genus Spodoptera are susceptible to azadirachtin applications, which often cause abnormal growth and reproduction (Figures 16.2 and 16.3). Injecting 1-2 µg Azadirachtin to the fifth stage larvae of *Spodoptera mauritia* Boisd resulted in molting abnormalities and larval-pupal transformation of the treated insects. The larval duration was enhanced and produced sixth stadium larvae that failed to pupate. The azadirachtin treatment completely prevented normal pupation (Jagannadh and Nair, 1992). Being a potent insect growth inhibitor, azadiractin A appears to work by blocking the synthesis and release of molting hormones (ecdysteroids) from the prothoracic gland (Mordue and Blackwell, 1993). Final instar larvae of the African armyworm, *S. exempta* Walker (Lep: Noctuidae) topically applied with azadirachtin reacted to the treatment. The application of this compound adversely affected oogenesis and reproductive maturation in subsequent female moths. It is suggested that azadirachtin interfere with vitellogenin synthesis and/or its uptake by developing oocytes. Such larval treatment also caused substantial decrease in fecundity and although fertility in affected females was not decreased significantly, emerging larvae were less viable, less than 40 per cent reaching the fourth instar

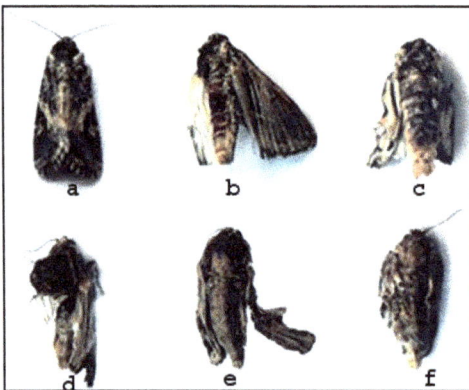

Figure 16.2: Growth Disruption in *Spodoptera litura*, Deformed Adults Due to the Treatment (b to f). Normal adult (a).

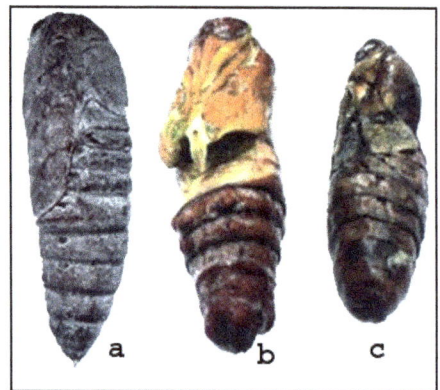

Figure 16.3: Abnormal Pupae of *Achoaea janata* Resulted Due to the Treatment (b-c) Normal pupa-a.

(Tanzubil, 1990). Injection of azadirachtin in to final (fifth)-instar larvae of the tobacco budworm, *Heliothis virescens* (Fabr.) reduced whole body and haemolymph titres of the moulting hormones, ecdysone and 20-hydroxyecdysone, thereby preventing normal development to the pupal stage. It also reduced brain levels of prothoracicotropic hormone (PTTH)-like activity, which is the stimulation of *in vitro* moulting hormone synthesis by the prothoracic glands (Barnby and Klocke, 1990).

The topical administration and abdominal injections of azadirachtin to the adults of *Dysdercus koenigii* F (Het: Pyrrochoridae) resulted with 50 per cent mortality but in the remaining survived insects, embryogenesis was severely impaired and trophocytes got damaged (Koul, 2009). The treatment evoked various specific and nonspecific effects during the course of development. Prolonged development, wing deformities, unplasticisation of wing lobes, development of wingless adults and larval mortality were the characteristic features.

Many studies showed the importance of content of azadirachtin in formulations. In a study using three sucking pests of tea: *Helopeltis theivora, Waterhouse, Scirtothrips dorsalis* Hood and *Empoasca flavescens* Fabricius, Roy and Gurusubramanian (2011) determined the dose-mortality response to formulations having varying azadirachtin content. They found a variation among the activity of the test insects. LC50 decreased as the azadirachtin content of neem formulation was increased.

Histological and electrophoretical analysis showed that the storage protein formation in the fat body, necessary for pupation, did not occur in higher dose treated (1.0 µg) individuals, but it was not affected by low dose (0.1 µg) treatment. It is assumed that azadirachtin causes metabolic defects at higher concentrations. Interestingly the dose of azadirachtin required to obtain growth retardant activity to different insects varies with the insect species exposed. Azadirachtin applied to the soil at 1 and 2. 5 mg to potted potato (*Solanum tuberosum*) and tomato (*Lycopersicon esculentum*), remarkably retarded the growth of *Manduca sexta* larvae and strongly disrupted the pupation of *Leptinotarsa decemlineata* larvae, respectively. However, much higher quantity (e"25 mg) of azadirachtin is required on sorghum (*Sorghum biocolor*), to slow the population growth of *Schizaphis graminum*, and prolong the survival of treated plants (Hu *et al.,* 1998), but even 150 mg did not affect the aphid growth. Castor semilooper, *Achaea janata* (Linn) (Lep: Noctuidae) is a severe defoliator of castor, *Ricinus communis* plants. Larval food consumption, faecal pellet production and weight gain of *this insect was* influenced by the level of food deprivation, age of the leaf and concentration of azadirachtin. Differences in food consumption due to starvation were affected by the concentration of azadirachtin (Ramachandran *et al.,* 1989).

Azadirachtin has a profound influence on reproduction of the green stink bug, *Nezara viridula* (Heteroptera: Pentatomidae). The fifth instar nymphs showed varied effects with different treatment doses. A lower dose (2-50 ng) applied resulted with only about half of the treated population molting in to normal adults, while 20 ng treated nymphs that developed in to normal looking adults, whose fecundity is drastically affected. It is also evident that when males obtained from these treated nymphs after mating with untreated or normal female *N. viridula* had also resulted with reduction in fecundity (Riba *et al.,* 2003). Ovipositional repellent effect was noted when the ovipositional surface is treated with azadirachtin. Almost all the nymphs of stinkbugs either died during the process of nymphal–adult moulting or developed in to nymphal-adult intermediates due to the azadirachtin treatment at higher doses (200-500 ng) (Riba *et al.,* 2003). Topical application of neem seed extract containing azadirachtin to third instar nymphs of a pentatomid bug *Perillus bioculatus* (F) delayed moulting and caused moulting deformities (Goldstein, 1991). Another heteropteran susceptible for azadirachtin treatment is *Eurygaster integriceps* Put (Het: Scutelleridae) a Palearctic species and the most destructive pest of wheat and other Graminae. A commercial formulation containing azadirachtin caused delayed effects on these bugs, apart from inhibiting the egg hatching (Kivan, 2005). A different result was obtained by (Suderland *et al.,* 2002) by treating small rice stink bug *Oebalus poecilus* (Hem: Pentatomidae) in Guyana. The application of azadirachtin reduced the feeding activity but did not affect the oviposition of this Pentatomid. The effects on eggs of another Heteropteran were assessed by Dorn (1986) who found that acetone diluted solution of azadirachtin had no effect on egg hatchability of *Oncopeltus fasciatus* Dallas (Het: Lygaeidae). A continuous dietary exposure of azadirachtin interfered with the ability of vine weevil to produce eggs, decreased the number and proportion of viable eggs. Several weeks of adult feeding with azadirachtin has increased their mortality slightly (Cowles, 2004).

The systemic action of azadirachtin is evident in the larvae of *Pieris brassicae* too. The insertion of the cabbage, *Brassica oleracea* leaves in azadirachtin treated solutions for 24 hour time period caused the systemic uptake by the plant leaves and the translocation of the active metabolites in to the plant resulted in lethal action with ecdysis in combination with the effect on developmental rate of *P. brassicae* (Arpaia and Van Loon, 2011). Larvae fed for 24 h on rice leaf cuts dipped in different solutions of the partially-purified fractions and methanolic extracts exhibited pronounced developmental abnormalities and mortalities in succeeding larval instars and in pupal and adult stages

Mosquitoes are not exempted from the effects of this wonder plant chemical. The oral administration of a commercial formulation (Neem Azal® (NA)), containing azadirachtin A at 34 per cent, through artificial blood feeding to a laboratory strain of *Anopheles stephensi,* resulted in causing a delay in oocyte development in vitellogenesis and choriogenesis phases of mosquitoes (Lucantoni *et al.,* 2006). Distinct structural modifications indicative of a complete block of oogenesis, impairment of vitellogenesis and vitelline envelope formation and a severe degeneration of follicle cells occurred due to the Azadirachtin feeding in these insects which is revealed by ultrastructural studies on ovaries. However, feeding of this chemical through an artificial membrane even at a high dose (up to 200 ng per female) did not affect the feeding behaviour of the yellow fever mosquitoe, *Aedes aegypti* (Ludlem, 1988). High doses of ingested azadirachtin fail to inhibit or delay oviposition in this insect, but, significant, transient retardation of oocyte growth is observed for up to 72 h after feeding. The azadirachtin efficiency in inhibiting molting in larvae of the face fly, *Musca autumnalis* DeGeer, was studied and was found that delayed lethal action will affect the pupation, adult *emergence,* and changes in size of pupae occurs due to the treatment (Ibrahim and Dora, 1984).

Azaditachtin Structure and its Activity Relation with Insects

Structure

Azadirachtin is a highly interesting compound both for its chemical structure as well as its varied biological activity. Azadirachtin was first extracted from the Indian neem tree in 1968, but its structure was not correctly determined until 1985. For the next two decades, about 40 organic synthetic chemists and several teams worked hard for the molecule's synthesis. Neem seeds consists of quiet a lot of allelochemicals and among these, only azadirachtin, a tetranortriterpenoid, has been extensively studied as an antifeedant, growth inhibitor and growth regulator (Koul 1992; Mordue and Blackwell, 1993). Azadirachtin is a highly complex molecule contains 16 stereocenters, of which 7 of them are quaternary and is also one of the most highly oxidized limonoids known (Veitch *et al., 2007*). Azadirachtin has an interesting as well as a complex chemical structure which required 18years to solve and another 22 years to synthesize (Morgan, 2009). The level of azadirachtin content varies in neem ecotypes in relation to climate, soil type and altitude where it is grown (Rengasamy *et al.,* 1996). Azadiradione and epoxy azadiradione are feeding deterrents to a lesser extent than azadirachtin (Govindachari *et al.,* 1995).

As Veitch *et al.* (2007) explained, 'azadirachtin's structure is immensely complex, with 16 adjacent chiral centres, and a central bond congested by neighbouring atoms. We call it a who's who of oxygen atoms as it contains almost every type of oxygen functionality, he said. Azadirachtin is also sensitive to acid, base, and unstable in light: its tendency to rearrange has scuppered many attempted syntheses.

Azadirachtin-H, which belongs to a group of C-*seco*-tetranortriterpenoids (C-*seco*-limonoids) of great interest for their insect antifeedant and ecdysis-inhibiting activity, has some unusual features: the absence of a carbomethoxy group at C (11); the presence of a cyclic hemiacetal function at C(11); the-orientation of the hydroxyl group on C(11), opposite to that in all other known azadirachtins with a hydroxyl group on C(11), except azadirachtin-I (Govindachari *et al.,* 1996). Substantial evidence is available to show that azadirachtin inhibits cell proliferation in Sf9 cells. Using MTT assay and Flow Cytometry the effect of azadirachtin on proliferation of Sf9 and MCF7 cells were evaluated by Salehzadeh and Abbasalipourkabir (2011). Nisbet *et al.* (1997) showed that (3H) dihydroazadirachtin binds to the nuclei of Sf9 cells. Blaney *et al.* (1990) assessed the antifeedant activity of about 40 compounds of azadirachtin, azadirachtin-derivatives and related limonoids against four major lepidopteran insects, *Spodoptera littoralis, Spodoptera frugiperda, Heliothis virescens* and *Heliothis armigera.* Azadirachtin and dihydroazadirachtin were the most potent of the 40 compounds tested. The results showed that hydrogenation of the C-22, 23 double bond did not decrease antifeedant activity and the nature of the substitutes at C-1, C-3 and C-11 were important. Molecules with bulky substitutes at either C-22 or C-23 were usually ineffective antifeedants as were compounds lacking an epoxide. The highly reactive epoxide ring of azadirachtin is stable during transport inside the insect which makes the molecule unmetabolized even weeks after application. Methylation of the hydroxy substitutions on the azadirachtin molecule usually resulted ιν α decrease in the antifeedant activity, as did the addition of bulky groups to the dihydrofuran ring (Simmonds *et al.*, 1995).

Extensive research has been done on the structure activity relations and biological activity of neem and its products particularly tetranortriterpenoids isolated from various parts of neem tree. Among them, azadirachtin-A has been shown to be a potent insect antifeedant and ecdysis inhibitor. Tetrahydro azadirachtin concentrates was obtained by Catalytic hydrogenation of the 60 and 90 per cent azadirachtin concentrates and this Tetrahydro azadirachtin and Dihydro azadirachtin that formed during the first 5h hydrogenation were identified by LC-ESI-MS on the basis of their unique mass fragmentation pattern (Vandana Sharma *et al.*, 2006). Their feeding and growth inhibitory activity to *Helicoverpa armigera* (H¨ ubner) larvae was analyzed and compared to that of azadirachtin concentrates. Almost all doses tested showed good activity, while tetrahydroazadirachtin-A (90 per cent) displayed higher antifeedant activity, Tetrahydro azadirachtin-A (90 per cent) and azadirachtin- A (90 per cent) were effective as growth retardants (Vandana Sharma *et al.*, 2006).

Decalin and dihydrofuranacetal fragments related to those in azadirachtin exhibited antifeedant activity against larvae of the African leaf-worm *Spodoptera littoralis* Boisd. All the decalin fragments tested were methoxy (C11) derivatives of azadirachtin. The most active decalin fragment had a ketone substitution at C7. Overall, the compounds were more active when tested in combinations of one decalin fragment and one dihydrofuranacetal fragment than when tested singly. Although some of these combinations did show significant levels of antifeedant activity, sometimes coupled with a synergistic effect, they were not as active as either azadirachtin or dihydroazadirachtin (Blaney *et al.*, 1994). Methylation of the hydroxy substitutions on the azadirachtin molecule usually resulted decrease in the antifeedant activity, as did the addition of bulky groups to the dihydrofuran ring.

An extremely efficient antifeedant and growth disrupting agent was synthesized by Ley, (1990) by incorporation of the hydroxydihydrofuran portion of azadirachtin. The compound formed has antifeedant activity comparable to that of the natural product. Structur-activity relationship studies on the two fragments of azadirachtin, *i.e.*, dihydrofuranoacetal and decalin moieties, revealed that the former imparted antifeedant activity and the latter caused disruption of insect growth and development.

Activity

It is intriguing to know that azadirachtin unlike many other herbal compounds has profound, fascinating and varied effects on insects. Most of the plant derived chemicals exert limited effects on pests (Usha Rani 2009, 2011). Either they are directly toxic or repellent or deterrent or growth inhibitor. Very limited number of plants has combined or multiple effects. Since Neem has all these above or many of the above characteristics it acquired a high status among botanical pesticides. Its reliable efficacy is linked to the physiological action of azadirachtin as an insect growth regulator; the antifeedant effect is highly variable among pest species.

Though azadirachtin is known to contain different properties, it at first reduces or prevents the feeding of the insect that has approached by strongly acting on the insect's chemo and gustatory receptors as an antifeedant. Even after this, if the insect resists and consume the plant then azadirachtin at physiological level stops peptide hormone release thus blocking the synthesis and release of molting hormones (ecdysteroids) from the prothoracic gland, which leads to incomplete ecdysis in immature insects. (Aerts, 1997 and Mordue, 1993). This effects the growth and development of the insect leading to the reduction of the pest populations. Finally azadirachtin has a damaging effect on most of the insect's tissues, including muscle, fat and gut cells. It may cause toxic effects killing them. Different insect species have different responses to individual compounds. The compound is likely to reduce the potential for development of genetic resistance or development of behavioural desensitisation, similar to other mixtures of plant essential-oil allelochemicals (Hummel Brunner and Isman 2001). Azadirachtin is unstable to heat, light, water, pH, microbes, etc., but its reduced derivatives, namely, dihydro- and tetrahydro-azadirachtin are more stable to light, heat, moisture, etc., and also retain the bioactivity (Ley, 1990).

Using behavioural and electrophysiological bioassays Simmonds *et al.* (1995) analyzed the antifeedant activity of azadirachtin and 56 azadirachtin analogues, including 22, 23 dihydroazadirachtin, against larvae of *Spodoptera littoralis*. Among all the analogues azadirachtin exhibited best activity, although many showed significant antifeedant activity at high concentrations. The activity of the majority of the analogues stimulated a dose-dependent response from a neurone in the medial styloconic maxillary sensilla which correlated with the behavioural activity (Simmonds *et al.*, 1995). Structur-activity relationship studies on the two fragments of azadirachtin, *i.e.*, dihydrofuranoacetal and decalin moieties, demonstrated that the former imparted antifeedant activity and the latter caused disruption of insect growth and development.

Studies on the structure—activity relation of a few azadirachtin analogues revealed the specificity of the activity against certain insects (Huang *et al.*, 2003). The analysis of toxicities and antifeedant activities of thirteen asymmetrical 1,3,4-oxadiazoles containing a 2H-pyridazin-3-one group were shown to possess considerable growth inhibitory activity of larvae of a number of Lepidoptera, but not against the insects belonging to other orders such as Homoptera, Diptera and Acarina. They possessed a mode of action that might be similar to the chitin-synthesis inhibition of oxadiazole compounds and/or the juvenile hormone effect of pyridazinone compounds (Huang *et al.*, 2003). Govindachaeri *et al.* (1995) evaluated insect antifeedant and growth-regulating activities of 22 limonoids (both natural and their derivatives) against *Spodoptera litura*. The C-seco limonoids (azadirachtins A, B, D, H, and I) were the most effective compounds exhibiting antifeedant as well as growth regulatory properties. They suggested that the cyclohexenone A ring and the *á*-hydroxy enone group in the B ring may be important for antifeedant activity, while the presence of a cyclohexenone or 1,2-epoxide in the A ring coupled with an *á*-hydroxy enone in the B ring for growth regulatory activity and an acetoxy at C-7 instead of *á*-hydroxy enone and perhaps the carbonyl at C-16 increase growth regulatory activity.

Fungi

Azadirachtin the major secondary active metabolite of Neem oil extract is a natural product which almost ideally represents an insect–specific control agent.

However, it is also known for its excellent bioefficacy on nematodes and fungi too. Azadirachtins A, B, and H, the three major bioactive constituents of neem seed kernel, were purified from this methanolic concentrate by employing reverse phase medium-pressure liquid chromatography (MPLC), using methanol-water solvent system as an eluant (Sharma *et al.* (2003). Azadirachtin A concentrates and azadirachtin A, B, and H were evaluated for their potential growth inhibitory activity towards *Rotylenchulus reniformis*, a reniform nematode and a significant plant pathogenic pest by Sharma *et al.* (2003). The nematicidal activity of all these compounds is evident and this activity of the azadirachtin concentrates was directly related to the azadirachtin A content. Azadirachtins B and H contributed to the activity of azadirachtin concentrates. It is reported to exhibit significant nematicidal activity against the root knot nematode *Meloidogyne incognita* (Devkumar and Goswami, 1992).

The azadirachtin concentrates showed a similar trend in fungicidal activity as that in the case of nematodes tested. It is effective against several plant pathogenic fungi such as *Rhizoctonia solani* Kuhn. and *Sclerotium rolfsii* Sacc, that cause severe damage to several crops and agricultural produce. Again the activity against two test fungi was directly related to the azadirachtin A content. However, in this case of fungi toxicity, Azadirachtins B and H were considerably more fungitoxic than azadirachtin A (Sharma *et al.,* 2003). Azadirachtins represent a major group of bio rationals which had impact on plant and stored product protection (Saxena, 2002).

Conclusion: Azadirachtin is a wonder chemical and a valuable natural pesticide having multiple effects on major insect pests and with low mammalian toxicity coupled with less environmental impacts. Its mode of action, structure–activity relationships, and its biosynthesis needs are to be explored more thoroughly. Compounds which deter insects from feeding (antifeedants) are attracting special attention owing to their potential use in integrated pest control management systems. Ecologically acceptable methods have led to a rapidly growing interest in behaviour-modifying chemicals from natural sources. It is now considered to hold a great potential as pest control agent for use in agriculture.

Acknowledgements

I gratefully acknowledge the support of Director, Indian Institute of Chemical Technology, Hyderabad, India and several scientists in this world who inspired me with their wonderful scientific knowledge and discoveries and also the research grant from Ministry of Environment and Forests, New Delhi. I am also thankful to my students Jois Madhusudanamurthy for the pictures and Salma for her assistance in preparing the manuscript.

References

1. Aerts R J and Mordue Luntz A. J. (1997). Feeding deterrence and toxicity of neem triterpenoids. *Journal of Chemical Ecology* **23**: 2117- 2132.

2. Arnason, J T, Philogene B J R, Donskov N, Hudon M, McDougall C, Fortier C G, Morand R, Gardner D, Lambert, Morris C, and Nozzolillo C. (1985). Antifeedant and insecticidal properties of azadirachtin to the European corn borer, *Ostrinia nubilalis. Entomologia Experimentalis et Applicata* **38**: 29-34.

3. Arpaia S and Van Loon J J A. (2011). Effects of azadirachtin after systemic uptake into *Brassica oleracea* on larvae of *Pieris brassicae* The Netherlands Entomological Society DOI: 10.1111/j.1570-7458.1993.tb00690.x.

4. Annadurai R S and Rembold H. (1993). Azadirachtin A modulates the tissue-specific 2D polypeptide patterns of the desert locust, *Schistocerca gregaria*. *Naturwissenschaften* **80**: 127–130.

5. Anuradha A and Annadurai R S. (2008). Biochemical and molecular evidence of azadirachtin binding to insect actins. *Current Science* **95**: 1588–1593.

6. Ayyangar G. and Rao P J. (1989). Azadirachtin effects on consumption and utilisation of food and midgut enzymes of *Spodoptera litura* (Fabr). *Indian Journal of Entomology* **51**: 373- 376.

7. Ayyangar N R and Nagasampagi B. A. (1990). Role of botanical in integrated pest management. PP: 54-61.In: Chari, M. S. and Ramaprasad, G. (eds.). Proc. Symp. Botanical Pesticides in IPM, Rajahmundry.

8. Beckage N E, Metcalf J S, Barbara D. Nielsen, and. Nesbit D J. (1988). Disruptive Effects of Azadirachtin on Development of *Cotesia congregata* in Host Tobacco Hornworm Larvae. *Archives of Insect Biochemistry and Physiology* **9**: 47-65.

9. Blaney W. M, Simmonds M. S. J, Ley S. V, Anderson J. C and Toogood P. L. (1990). Antifeedant effects of azadirachtin and structurally related compounds on lepidopterous larvae *Entomologia Experimentalis et Applicata* **55**: 149–160.

10. Blaney W M, Simmonds M S J, Ley S V, James C A, Stephen C S and Anthony W. (1994). Effect of azadirachtin-derived decalin (perhydronaphthalene) and dihydrofuranacetal (furo[2, 3-fo]pyran) fragments on the feeding behaviour of *Spodoptera littoralis*. *Pesticide Science* **40**: 169–173.

11. Barnby M A and Klocke J A. (1990). Effects of azadirachtin on levels of ecdysteroids and prothoracicotropic hormone-like activity in *Heliothis virescens* (Fabr.) larvae, *Journal of Insect Physiology* **36:** 125-131.

12. Butterworth J and Morgan E. (1968). Isolation of a Substance that suppresses Feeding in Locusts". *Chemical Communications London* **23**: DOI: 10.1039/ C19680000023.

13. Caboni P., Cabras M., Angioni A, Russo M and Cabras P. (2002). Persistence of azadirachtin residues on olives after field treatment. *Journal of Agricultural and Food Chemistry* **50**: 3491–3494.

14. Cowles, Richard S. (2004). Impact of azadirachtin on vine weevil (Coleoptera: Curculionidae) reproduction, *Agricultural and Forest Entomology* **6**: 291–294.

15. Devakumar C. and Goswami B K. (1992). Nematicidal principals from neem (*Azadirachta indica* Juss) Part III Isolation and bioassay neem (*Azadirachta indica* A. Juss) Part III Isolation and bioassay of some neem meliacins. *Pesticide Research Journal* **4**: 81-86.

16. Dorn A. (1986). Effects of azadirachtin on reproduction and egg development of the heteropteran *Oncopeltus fasciatus* Dallas. *Journal of Applied Entomology* **102**: 313-319.

17. Devanand P and Usha Rani P. (2008). Biological potency of certain plant extracts in management of two lepidopteran pests of *Ricinus communis* L. *Journal of Biopesticides* **1**: 170 – 176.

18. Duthie Holt M A. and Borden J H. (1999). Treatment of lodgepole pine bark with neem demonstrates lack of repellency or feeding deterrency to the mountain pine beetle, *Dendroctonus ponderosae* Hopkins (Coleoptera: Scolytidae). *Journal of the Entomological Society of British Columbia* pp. 21-24.

19. Ermel K, Pahlich E, Schmutterer H. (1987). Azadirachtin content of neem kernels from dierent geographical locations and its dependence on temperature, relative humidity and light. In: Schmutterer H, Ascher K R S, eds. Natural pesticides from the Neem tree (*Azadirachta indica* A. Juss) and other tropical plants. pp.171-184. Eschborn, Germany: GTZ.

20. Fagonee I. (1984). Effect of azadirachtin and of a neem extract on food utilisation by Crocidolomia binotalis. pp. 211-223 in Schmutterer H. and Ascher, K R S. (Eds) Natural pesticides from the neem tree (*Azadirachta indica* A. Juss.) and other tropical plants. Proceedings of the Second International Neem Conference, Rauischholzhausen, Eschborn, GTZ.

21. Garcia E S, Feder D, Gomes L and Rembold H. (1990). Short and long term effects of Azadirachtin A on development and egg production of *Rhodnius prolixus Mem. Inst. Oswaldo Cruz* **85**: 11-15.

22. Govindachari T R, Gopalakrishnan G, Rajan S S, Kabaleeswaran V and Lessinger L. (1996). Molecular and crystal structure of azadirachtin-H, Acta Crystallographica. *Structural Science* **52**: 145-150.

23. Govindachari T R, Narasimhan N S, Suresh G, Partho P D, Gopalakrishnan G, and Krishna K G N. (1995). Structure-Related Insect Antifeedant And Growth Regulating Activities Of Some Limonoids. *Journal of Chemical Ecology* **21**: 1585-1600.

24. Goldstein H J, and Keil C B. (1991). Prospects for integrated control of the Colorado potato beetle (Coleoptera: Chrysomelidae) using *Perillus bioculatus* (Hemiptera: Pentatomidae) and various pesticides. *Journal of Economic Entomology* **84**: 1645-1651.

25. Hu M, Klocke J A and Barnby M A. and Shinfoon C. (1998). Systemic Insecticidal Action Of Azadirachtin 9 Neem Seed And Chinaberry Seed Extracts Applied As Soil Drenches To Potted Plants. *Entomologia Sinica* **5**: 177-188.

26. Huang Z W, Shi P, Dai J Q and Du J W. (2004). Protein metabolism in *Spodoptera litura* (F.) is influenced by the botanical insecticide azadirachtin. *Pesticide Biochemistry and Physiology* **80**: 85–93.

27. Huang Q, Quain X, Song G and Cao S. (2003). The toxic and anti feedent activity of 2H-pyridazin-3-one-substituted 1,3,4-oxadiazoles against the armyworm *Pseudaletia separate* (Walker) and other insects and mites. *Pest Management Science* **59**: 933-939.

28. Hummel brunner L. A and Isman M B. (2001). Acute, sublethal, antifeedant and synergistic effects of monoterpenoid, essential oil compounds on the tobacco cut worm *Spodoptera litura* (Lepidoptera: Noctuidae). *Journal of agricultural and food chemistry* **49**: 715-720.

29. Ibrahim A G and Dora K H. (1984). Biological Activity of Azadirachtin, Component of the Neem Tree Inhibiting Molting in the Face Fly, *Musca autumnalis* De Geer (Diptera: Muscidae) *Environmental Entomology* **13:** 803-812.

30. Isman M B. (1993). Growth inhibitory and antifeedant effects of azadirachtin on six noctuids of regional economic importance. *Journal of Pesticide Science* **38**: 57–63.

31. Jagannadh V and Nair V S K. (1992). Azadirachtin-induced effects on larval-pupal transformation of *Spodoptera mauritia*, *Physiological Entomology* **17**: 56–61.

32. Ketkar C M. (2009). Veterinary applications of neem (*Azadirachta indica* A. Juss.) and its products. In: Biological control of plant, medial and veterinary pests. Kleeberg H and Strang R. Eds. Proceedings of the 14th workshop, ISBN 3-925614-29-X. 27-30.Wetzlar, Germany.

33. Kivan M. (2005). Effects of azadirachtin on the sun pest, *Eurygaster integriceps* Put (Het: Scutelleridae) in the laboratory. *Journal of Central Europian Agriculture* **6**: 157-160.

34. Kleeberg H and Strang R. (2009). Biological control of plant, medical and veterinary pests. Proceedings of the 14thworkshop, ISBN 3-925614-29-X. 274 P. Wetzlar, Germany.

35. Khosravi R and Jalali Sendi J. (2013). Effect of Neem Pesticide (Achook) on Midgut Enzymatic Activities And Selected Biochemical Compounds In The Hemolymph of Lesser Mulberry Pyralid, *Glyphodes Pyloalis* Walker (Lepidoptera: Pyralidae). *Journal of Plant Protection Research* **53** No.3.

36. Koul O and Wahab S. (2004). Neem: Today and in the New Millenium. pp. 244–291. In: "Present Concepts of the Mode of Action of Azadirachtin from Neem" (R.J. Mordue, ed.). Kluwer Academic Publishers pp. 291.

37. Koul O, and Isman M B. (1991). Effects of azadirachtin on the dietary utilisation and development of the variegated cutworm *Peridroma saucia*. *Journal of Insect Physiology* **37**: 591-598.

38. Koul O. (1992). Neem allelochemicals and insect control. *Allelopathy* pp. 389-41.

39. Koul O. (1999). Insect Growth Regulating and Antifeedant Effects of Neem Extracts and Azadirachtin on Two Aphid Species of Ornamental Plants. *Journal of Biosciences* **24**: 85-90.

40. Koul O. (2009). Azadirachtin. II. Interaction with the reproductive behaviour of red cotton bugs **98**: 221–223.

41. Kumar J and Parmar B S. (1997). Neem oil content and its key chemical constituents in relation to the agro-ecological factors and regionsof India. *Pesticide Biochemistry and Physiology* **9**: 216-225.

42. Ley S V. (1990). Synthesis of antifeedants for insects: novel behaviour-modifying chemicals from plants. Ciba Foundation Symposium **154**: 80-87.

43. Li X D, Chen W K and Hu M Y. (1995). Studies on the effects and mechanisms of azadirachtin and rhodojaponin- on *Spodoptera litura* (F.). *Journal of South China Agricultural University.* **16**: 80–85.

44. Linton Y M, Nisbet A J, and Mordue (Luntz) A J. (1997). The effect of Azadirachtin on testes of the Desert Locust, *Schistocerca gregaria* (Forskal) *Journal of Insect Physiology* **43**: 1077-1084.

45. Lucantoni L, Giusti F, Cristofaro M, Pasqualini L, Esposito F, Lupetti P and Habluetzel A. (2006). Effects of a neem extract on blood feeding, oviposition and oocyte ultrastructure in *Anopheles stephensi* Liston (Diptera: Culicidae) pp. 361-371.

46. Ludlum C, and Sieber K. (1988). Effects of azadirachtin on oogenesis in Aedes aegypti. *Physiological Entomology* **13**: 177-184.

47. Martineau Jess. (1994). Agri Dyne Technologies, Inc. MSDS for Azatin-EC *Biological Insecticide.*

48. Martinez S S and Van Emden H F. (1999). Sublethal concentrations of azadirachtin affect food intake, conversion efficiency and feeding behaviour of *Spodoptera littoralis* (Lepidoptera: Noctuidae) *Bulietin of Entomological Research* **89**: 65-71.

49. Miller T A, Lampe D J and Lauzon C R. (2006). Transgenic and paratransgenic insects in crop protection. p.87–103. In "Insecticide Design Using Advanced Technologies" (I. Ishaaya, R. Nauen, R. Horowitz, eds.). Springer-Verlag, Heidelberg, pp. 267. Germany.

50. Mordue (Luntz) A J. (2002). The cellular actions of azadirachtin. In: The neem tree *Azadirachta indica* A. Juss. and other meliaceous plants (Schmutterer, H. ed.). *Neem Foundation, Mumbai* pp. 266-274.

51. Mordue (Luntz) A J and Nisbet A J. (2000). Azadirachtin from the Neem Tree Azadirachta indica: its Action against Insects. *Anais da Sociedade Entomológica do Brasil* **29**: 615-632.

52. Mordue A J (Luntz), Morgan E D and Nisbet A J. (2010). Azadirachtin, a natural product in insect control. In Insect Control: Biological and Synthetic Agents.ed by Lawrence I. Gilbert, Sarjeet S. Gill.

53. Mordue A J and Blackwell A. (1993). Azadirachtin: an update. *Journal of Insect Physiology* **39**: 903–924.

54. Morgan E D. (2009). Azadirachtin, a scientific gold mine, *Bioorganic and Medicinal Chemistry* **17**: 4096-105.

55. Mukherjee S N. and Ramachandran R. (1989). Effects of azadirachtin on the feeding, growth and development of *Tribolium castaneum* (Herbst) (Coleoptera, Tenebrionidae). *Journal Applied Entomology and Zoology* **10**: 145-149.

56. Neoliya N K, Singh D and Sangawan R S. (2005). Azadirachtin influences total head protein content of *Helicoverpa armigera* Hub. Larvae. *Current Science* **88**: 1889-1890.

57. Nisbet A J, Woodford J A T, Strang R H C and Connolly J.D. (1993). Systemic antifeedant effects of azadirachtin on the peach-potato aphid *Myzus persicae*. *Entomologia Experimentalis et Applicata* **68**: 87–98.

58. Nisbet A J, Mordue (Luntz) A J, Grossman R B, Jennens L, Ley S V and Mordue W. (1997). Characterisation of azadirachtin binding to Sf9 nuclei *in vitro*. *The Journal of Physiology and Biochemistry* **34**: 461-473.

59. Prasad K R, Babu S K, Rao R, Suresh G, Rekha K, Murthy J M, Usha R P, and Rao J M. (2012). Synthesis and Insect Antifeedant Activity of Plumbagin Derivatives. *Medicinal Chemistry Research* **21**: 578-583.

60. Rao P J and Subrahmanyam B. (2009). Azadirachtin induced changes in development, food utilization and haemolymph constituents of *Schistocerca gregaria* Forskal. *Angew Entomology* **102**: 217-224.

61. Ramachandran R, Mukherjee SN and Sharma R N. (1989). Effects of food deprivation and concentration of azadirachtin (from *Azadirachta indica*) on the performance of *Achaea janata* and *Spodoptera litura* on young and mature leaves of *Ricinus communis*. *Entomologia Experimentalis Et Applicata* **51**: 29-35.

62. Reese J C. (1978). Chronic effects of plant allelochemicals on insect nutritional physiology. *Entomologia Experimentalis et Applicata* **24**: 625-626.

63. Redfern R E, Kelly T J, Borkovec A. B and Hayes D. K. (1982). Ecdysteroid titres and molting aberrations in last stage *Oncopeltus* nymphs treated with insect growth regulators. *Pesticide Biochemistry and Physiology* **18**: 351–356.

64. Rembold, H. (2004). Der Niembaum: Quelle für eine neue Strategie im Pflanzenschutz *Entomologie Heute* **16**: 235-243.

65. Rembold H, Sharma G K, Czoppelt C H, and Schmutterer H. (1982). Azadirachtin: A potent insect growth regulator of plant origin Article first published online: 26 AUG 2009 DOI: 10.1111/j.1439-0418.1982.tb03564.x1982 *Blackwell Verlag GmbH*.

66. Rembold H. (1989). Azadirachtins: Their structure and mode of actionInsecticides of Plant Origin,pp.150. American Chemical Society, Washington, DC.

67. Rembold H and Garcia E.S. (1989). Azadirachtin inhibits *Trypanosoma cruzi* infection of its Triatomine host, *Rhodnius prolixus*. *Naturwissenschaften* **76**: 77-78.

68. Rengasamy S, Kaushik N, Kumar J, Koul O and Parmar B.S. (1996). Azadirachtin content and bioefcacy of some neem ecotypes of India. In: Neem and Environment. (eds RP Singh, MS Chari, AK Raheja and W Kraus) Oxford and IBH Publishing, New Delhi, India **1**: 207–217.

69. Riba M, Martí J and Sans A. (2003). Influence of azadirachtin on development and reproduction of *Nezara viridula* L. (Het., Pentatomidae) *Journal of Applied Entomology* **127**: 37–41.

70. Roy S and Gurusubramanian G. (2011). Bioefficacy of azadirachtin conten to neem formulations against three major sucking pests of tea in Sub Himalayan tea plantation of North Bengal, India. *Agricultura Tropica et Subtropica* **44**: 134-143.

71. Ruscoe (1972). Growth disruption eects of an insect antifeedant. *Nature New Biology* **236**: 159-160.

72. Salehzadeh A and Abbasalipourkabir R. (2011). Differential Uptake of Azadirachtin by Sf9 and MCF7 Cell Lines *Indian Journal of Fundamental and Applied Life Sciences* 1: 146-*152.*

73. Saxena, R.C. (2002). Pests of stored products. In: The Neem Tree (Schmutterer H. ed) pp. 524-537.

74. Schmutterer H and Singh R.P. (1966). List of insect pests suscep-tible to neem products. *in* Schmutterer, H. (*Ed.*) *The neem tree: source of unique natural products for integrated pest management, medicine, industry and other purposes. Weinheim VCH* pp. 326-365.

75. Schmutterer H. (Ed.) (1995). The Neem Tree: Source of Unique Natural Products for Integrated Pest Management, Medicine, Industry and Other Purposes, *Weinheim VCH* pp. 696.

76. Schmutterer H. (2002). The neem tree *Azadirachta indica* A. Juss. and other meliaceous plants. Sources of unique natural products for integrated pest management, medicine, industry and other purposes. 2nd edition, *Neem Foundation*, Mumbai.

77. Schoonhoven L M, van Loon J J A and Dicke M. (2005). *Insect-Plant Biology*. 2nd ed. Oxford University Press. pp 409.

78. Seymour J, Bowman G and Crouch M. (1995). Effect of neem seed extract on feeding frequency of *Nezara viridula* L. (Hemiptera: Pentatomidae) on pecan nuts. *Journal of the Australian Entomological Society* **34**: 221-223.

79. Shinfoon C, Xing Z, Siuking L and Duanping H. (2009). Growth disrupting effects of azadirachtin on the larvae of the Asiatic corn borer (Ostrinia furnacalis Guenée) (Lepid., Pyralidae)Laboratory of Insect Toxicology, South China Agricultural College, Guangzhou *Journal of Applied Entomology* **99**: 276 - 284.

80. Sharma G K. (1992). Growth-inhibiting activity of azadirachtin on *Corcyra cephalonica*, *Phytoparasitica* **20**: 47-50.

81. Sharma V, Walia S, Kumar J, Nair M G. and Parmar B S. (2003). An efficient method for the purification and characterization of nematicidal azadirachtins A, B, and H, using MPLC and ESIMS. *Journal of Agricultural and Food Chemistry* **51**: 3966-3972.

82. Sidhu O P, Kumar V, and Behl H M. (2003). Variability in Neem (*Azadirachta indica*) with Respect to Azadirachtin Content. *Journal of Agricultural and Food Chemistry* **51**: 910–915.

83. Sidhu O P and Behl H M. (1996). Seasonal variations in azadirachtin in seeds of Azadirachta *Journal of Current Chemical and Pharmaceutical Sciences* **70**: 1084-1086.

84. Simmonds M S J and Blaney W M. (1984). Some effects of azadirachtin on lepidopterous larvae. p. 163-180. In: "Natural Products from the neem Tree and Other Tropical Plants"Proceedings 2nd International Neem conference, pp. 345, Germany.

85. Simmonds M S J, Blaney W M, Ley S V, Anderson J C and Toogood P L. (1990). Azadirachtin: structural requirements for reducing growth and increasing mortality in lepidopterous larvae. *Entomologia Experimentalis Et Applicata* **55**: 169–181.

86. Simmonds M S J, Blaney W M, Ley S V, Anderson J C, Banteli R, Denholm A A, Green P C W, Grossman R B, Gutteridge C, Jennens L, Smith S C, Toogood P L and Wood A. (1995). Behavioural and neurophysiological responses of *Spodoptera littoralis* to azadirachtin and a range of synthetic analogues, *Entomologia Experimentalis et Applicata* **77**: 69–80.

87. Sreelatha T, Hymavathi A, Murthy J M, Usha Rani P and Rao J M. (2009). Synthesis of Insect Antifeedant compounds from Plumbagin derivatives against *Achaea janata* and *Spodoptera litura. Journal of Agricultural and Food Chemistry.* **57**: 6090–6094.

88. Talwar G P, Raghuvanshi P and Jacobson M. (2002). Neem for control of fertility and sexually transmitted pathogens of the reproductive tract. In: The Neem Tree (Schmutterer, H. ed) pp. 666-677.

89. Tanzubil P B and McCaffery A R. (1990). Effects of Azadirachtin on reproduction in the African armyworm (Spodoptera exempta), *Entomologia Experimentalis et Applicata.* **57:** 115–121.

90. Timmins W A. and Reynolds S E. (1992). Azadirachtin inhibits secretion of trypsin in midgut of Manduca sexta caterpillars: reduced growth due to impaired protein digestion. *Entomologia Experimentalis et Applicata* **63**: 47-54.

91. Usha Rani P, Venkateshwaramma T and Devanand P. (2011). Bioactivities of *Cocos nucifera* L. (Arecales: Arecaceae) and *Terminalia catappa* L. (Myrtales: Combretaceae) leaf extracts as post-harvest grain protectants against four major stored product pests. *Journal of Pest Science.* 84: 235-247.

92. Usha Rani P and Rajasekharreddy P. (2009). Toxic and antifeedant activities of *Sterculia foetida* (L.) seed crude extract against *Spodoptera litura* (F.) and *Achaea janata* (L.). *Journal of Biopesticides* **2:** 161-164.

93. Vandana S, Suresh W, Swaran D, Jitendra K and Balraj S P. (2006). Azadirachtin-A and tetrahydroazadirachtin-A concentrates: preparation, LC-MS characterization and insect antifeedant/IGR activity against *Helicoverpa armigera* (Hubner) *Pest Management Science* **62**: 965.

94. Van der Esch S A, Carnevali F and Amici A. (2009). Effect of neem derived products on gastrointestinal nematodes in vitro and in vivo in sheep. In: *Biological control of plant, medical and veterinary pests* (Kleeberg, H. and Strang, R.eds.), Proceedings of the 14th workshop, ISBN 3-925614-29-X. Wetzlar, Germany.

95. Veitch G E, Beckmann E, Burker B J, Boyer A, Maslen S L and Ley S V. (2007). Synthesis of azadirachtin: a long but successful journey. *Angewandte chemie* **46**: 7629-7632.

96. Venkateswarlu B, Katyal J C, Choudhari J and Mukhopadhyay K. (1997). Azadirachtin content in the neem seed samples collected from different dry land regions. *Neem Newsletters.* (IARI) **14**: 7-11.

97. West A J, and Mordue (Luntz) A. J. (2011). The influence of azadirachtin on the feeding behaviour of cereal aphids and slugs. *Entomologia Experimentalis et Applicata.* DOI: 10.1111/j.1570-7458.1992.tb00644.x

2015, Horticulture for Nutrition Security
Editor: Prof. K.V. Peter
Published by: DAYA PUBLISHING HOUSE, NEW DELHI

Pages 367–383

Chapter 17

Nanotechnologies for Crop Production

J.C. Tarafdar

Nanotechnology is the creation and utilization of materials, devices and systems through the control of the properties and structure of the matter at the nanomatric scale (1-100 nm). Materials reduced to the nanoscale show some unusual properties which are different from what they exhibit on a macro scale, enabling unique systematic applications. The interesting and sometimes unexpected properties of nanoparticles are broadly due to the large surface area of the material, which dominates the contributions made by the very small quantities of the materials. Nanoparticles, thus, take advantage of their dramatically increased surface area to volume ratio.

Nanotechnology is emerging out as the sixth revolutionary technology in the current era after the Industrial Revolution of Mid 1700s, Nuclear Energy Revolution of the 1940s, the Green Revolution of 1960s, Information Technology Revolution of 1980s and Biotechnology Revolution of the 1990s. It is an emerging and fast growing field of science which is being exploited in a wide spectrum of disciplines such as physics, chemistry, biology, material science, electronics, medicine, energy, environment and health sectors.

The nanotechnology-aided applications have the potential to change agricultural production by allowing better management and conservation of inputs of plant and animal production. Currently, such nanotechnologies have been reported to be in use in the form of (i) Nanosensors, nanopesticides[2] (ii) Nanoscale adjuvants for pesticides[3] (iii) Bio synthesized nanoparticles for agricultural use[4,5,6,7] (iv) Smart delivery systems for nanoscale pesticides and fertilizers[2] (v) Feed additives[8,9] (vi) Veterinary medicines[10] (vii) Aqua culture[11] as bio sensors[12] (viii) Plant growth regulators[13] (ix) Plants to synthesize nanoparticles[14,15] and (x) Nano induced materials[5].

In order to gain the advantage of such development, more and more countries join the club of nanotechnology. A survey by Salamanca – Buentella[16] predicts availability of several nanotechnology applications for agricultural production for developing countries within next 10 years. These included - (i) Nanoforms zeolites for slow release and efficient dosage of water and fertilizers for plants; drugs for livestock; nanocapsules and herbicide delivery (ii) Nanosensors for soil quality and for plant health monitoring; nanosensors for pests detection (iii) Nanomagnets for removal of soil contaminants and (iv) Nano-particles for new pesticides, insecticides, and insect repellents.

Synthesis of Nanoparticles

The nanoparticles may be synthesized by physical, chemical, biological and aerosol techniques. Physical synthesis method includes sedimentation process, rotor speed mill, high energy ball mill and pot mill. In general, particles which have natural deposites may be purified and undergo grinding for physical synthesis for example phosphorus (P) nanoparticles are prepared by purifying rock phosphate and grinding with high energy ball mill or pot mill (Table 17.1).

Table 17.1: Synthesis of P Nanoparticles by Physical Means

Method Used	Average Particle Size (nm)
High energy ball mill	28.0
High energy pot mill	70.0

Mainly chemical synthesis of nanoparticles involved precipitation technique and sol gel method as well as Poly Vinyl Pyrrolidene (PVP) technique. Preparation of Zn nanoparticles from zinc acetate with potassium hydroxide in methanol is a perfect example of chemical precipitation of nanoparticles. It is possible to prepare nanoparticles of size ranges between 18-22 nm (Figure 17.1).

PVP technique is a solution phase synthesis of many nanoparticles. The hydroxyl end groups of PVP are a well-suited reductant for the aqueous synthesis of circular, triangular, and hexagonal nanoparticles. The morphology can be changed by adjusting the molar ratio of PVP to the salt precursor and by altering the molecular weight of PVP. For example, the preparation of palladium (Pd) nanoparticles, sodium paladium (II) tetrachloride (Na_2PdCl_4) is used with the molecular weight of PVP ranging between 10,000 and 55,000 (Table 17.2).

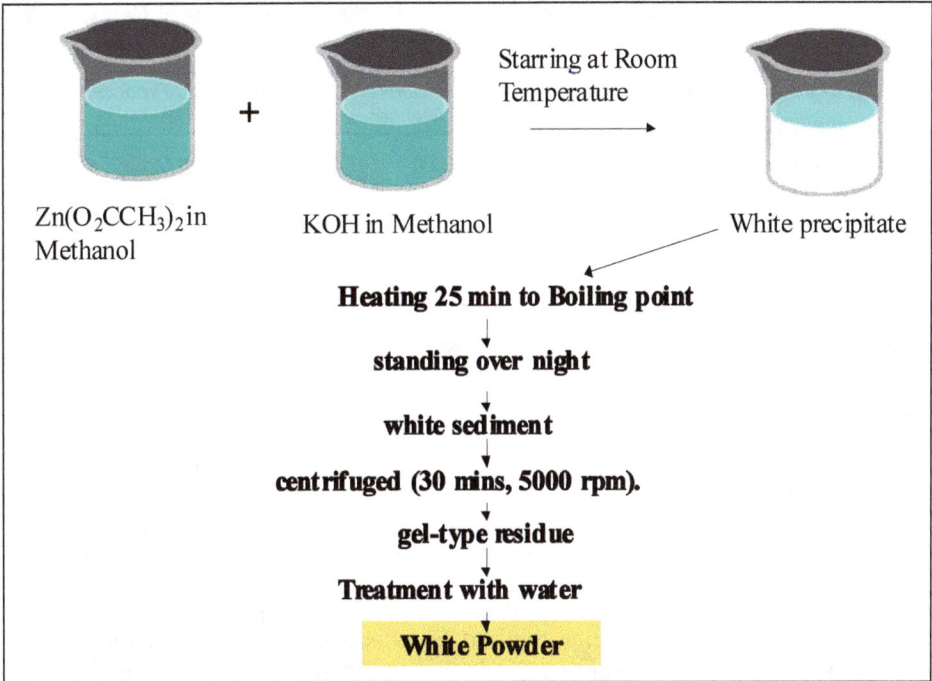

Figure 17.1: Preparation of Nano ZnO by Chemical Precipitation Method.

Table 17.2: Production of different Size and Shape of Pd Nanoparticles by PVP Technique

Metal Precursor	MW of PVP	Molar Ratio, PVP : Metal Precursor	Temp. °C	Time	Shape, Average Size, and Yield
Na$_2$PdCl$_4$	55000	1.5	80	5 h	Hexagonal, 40 nm, 10 per cent
Na$_2$PdCl$_4$	55000	5	80	5 h	Hexagonal, 45 nm, 70 per cent
Na$_2$PdCl$_4$	55000	15	80	5 h	Triangular, 25 nm, 30 per cent
Na$_2$PdCl$_4$	29000	15	80	5 h	Triangular, 10 nm, 40 per cent
Na$_2$PdCl$_4$	10000	15	80	5 h	Triangular, 50 nm, 70 per cent
Na$_2$PdCl$_4$	55000	5	95	5 h	Hexagonal, 30 nm, 30 per cent

Biological nanoparticles can be prepared by selecting microorganisms to grow in a particular salt solution after preparation of microbial balls. Enzymes are separated from the mycelium and used for nanoparticle synthesis. Biosynthesized nanoparticles are relatively stable and eco-friendly as they are naturally encapsulated by fungal protein. Lists of some organisms are given (Table 17.3), which can successfully be produced 100 per cent nanoparticles from the respective salt solutions.

Table 17.3: A List of some Organisms Efficient in Producing 100 per cent Nanoparticles from Respective Salts

Name of the Microorganisms	NCBI Gene Bank Accession Number	Nano Particle Produced
Aspergillus terreus CZR1	JF 681300	Zn and Mg
Aspergillus flavus CZR2	JF681301	Zn, Fe, Ag and Au
Aspergillus tubengensis TFR3	JF126255	Fe
Aspergillus japonicas AJP01	JF770435	Fe
Rhizoctonia bataticola TFR-	JQ675307	Ag, Au and Zn
Emericella variecolor TFR-16	KC175551	Mg and P
Bacillus megaterium JCT 13	JX442240	P
Pantoea tarafdar JCT 14	KC806057	N
Aspergillus ochraceus TFR-23	KC806053	K
Fusarium solani CZF-4	KC142125	Zn and Mg
Aspergillus flavus TFR- 12	JQ675295	Zn, Mg, P, Au, Ag, Fe, Pt and Ti

There are five different aerosol methods in use for nanoparticle synthesis. They are (i) furnace method (ii) flame method (iii) electron spray (iv) chemical vapor deposition and (v) physical vapor deposition method. By using furnace method, it is very difficult to produce the particle size less than 100 nm. Flame method can be used only with suitable precaution and can successfully produce TiO_2 nanoparticles. Electro-spray is a very useful method where exact size and shape of nanoparticles can be produced but yield is very low (1 g per year). Chemical vapor and physical vapour deposition methods are very popular for production of nanoparticles. To produce nanoparticles by aerosol technique one has to accurately control the gas flow rate, heater size and diffusion drier size. The instrumental set up for aerosol synthesis of nanoparticles is presented as Figure 17.2.

Characterization of Nanoparticles

Particle Size Analyser is generally used for initial size characterization of nanoparticles. Nanoparticle size analyzer uses the dynamic light scattering method to measure the size and distribution of particles undergoing Brownian motion. Dynamic Light Scattering (DLS) can basically measure suspensions and emulsions from 1 nm to 1 µm. Both lower and upper limits are sample dependent. The lower limit is influenced by concentration and how strongly the particle scatters light. The upper size limit is determined mainly by density of the particles. DLS algorithms are based on all particle movement coming from Brownian motion. The size distribution of Fe nanoparticles through particle size analyzer (DLS) is presented as Figure 17.3.

Transmission Electron Microscopy (TEM) is a microscopy technique whereby a beam of electrons is transmitted through on ultra thin specimen, interacting with the specimen as it passes through. An image is formed from the interaction of the electrons transmitted through the specimen; the image is magnified and focused onto an imaging device, such as a fluorescent screen, on a layer of photographic film, on to be

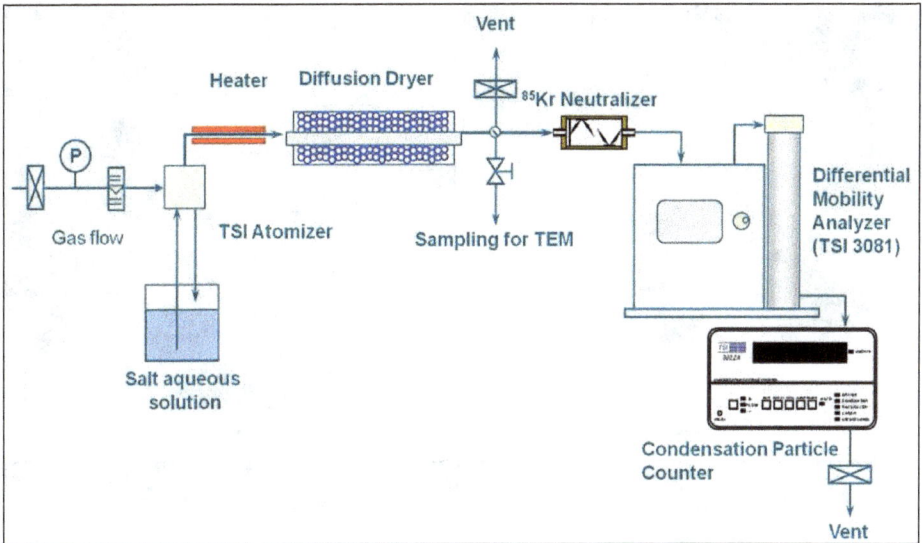

Figure 17.2: Instrumental Set-up for Aerosol Synthesis of Nanoparticles.

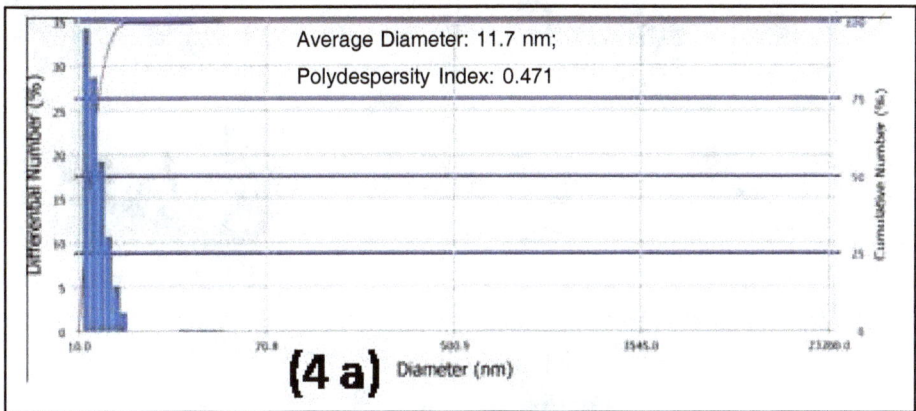

Average Diameter: 11.7 nm;

Polydespersity Index: 0.471

(4 a) Diameter (nm)

Figure 17.3: Particle Size Analysis Data of Biosynthesized Fe Nanoparticle.

detected by a sensor such as a CCD camera. TEM images of some of the nanoparticle are shown as Figure 17.4.

The scanning electron microscope (SEM) is a type of electron microscope that images the sample surface by scanning it with a high energy beam of electrons in a raster scan pattern. The electron interacts with atoms that make the sample's surface topography, composition and other properties such as electrical conductivity. The SEM images of Ag nanoparticle samples are shown as Figure 17.5.

Atomic force microscopy (AFM) is a very high-resolution type of scanning probe microscopy, with demonstrated resolution on the order of fractions of a nanometer, more than 1000 times better than the optical diffraction limit. The AFM is one of the

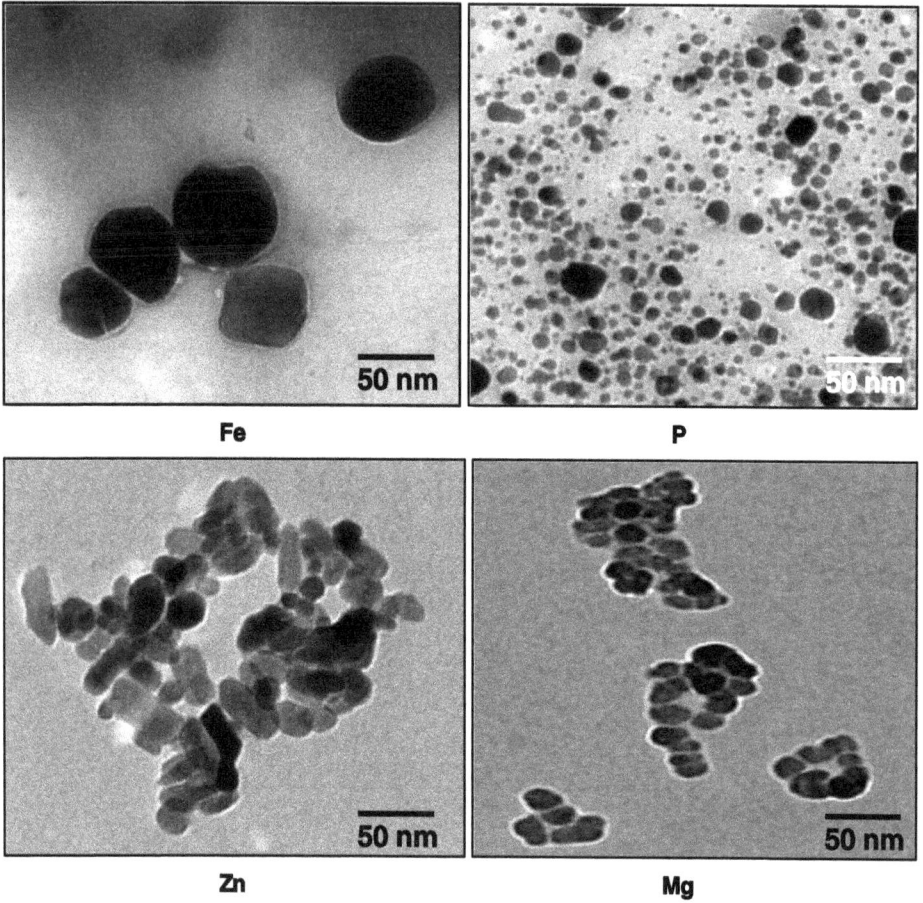

Figure 17.4: TEM Images of Biosynthesized Nanoparticles.

Figure 17.5: SEM Images of Ag Nanoparticle.

foremost tools for imaging, measuring and manipulating matter at the nanoscale. It offers the capability of 3D visualization and both qualitative and quantitative information on many physical properties including size, morphology, surface texture and roughness. Statistical information, including size, surface area, and volume distributions, can be determined as well. A wide range of particle sizes can be characterized in the same scan, from 1 nanometer to 8 micrometers. It offers visualization in three dimensions. AFM 3D images of Zn nanoparticles are shown as Figure 17.6.

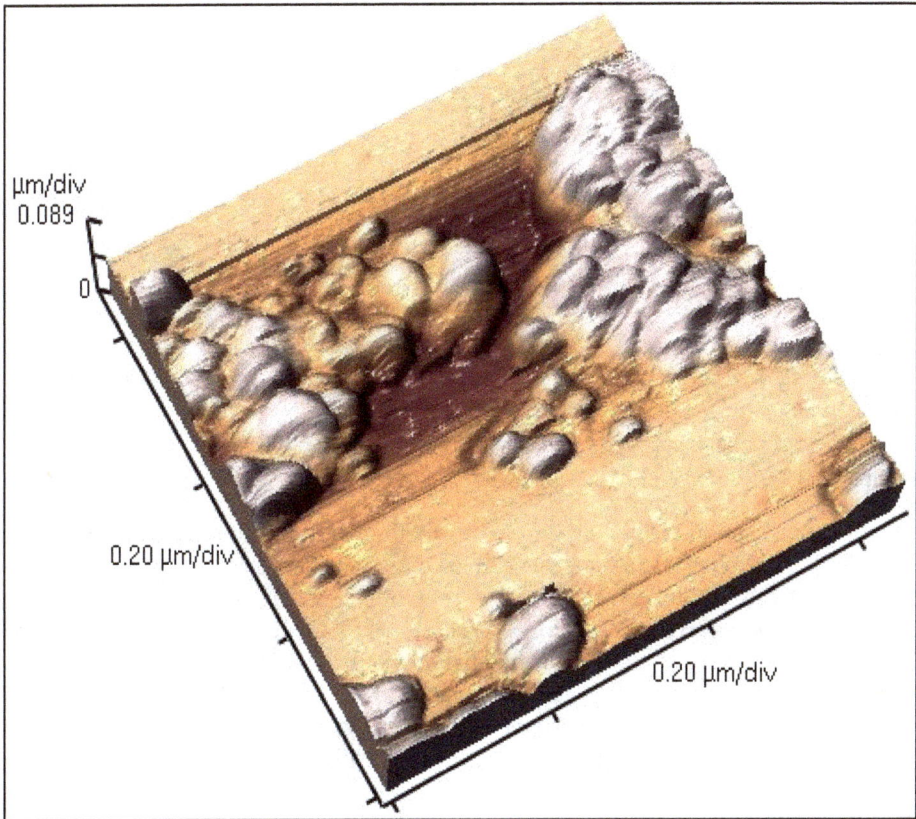

Figure 17.6: 3D AFM Image of Zn Nanoparticle.

Infrared (IR) spectroscopy is an important and popular tool for structural elucidation and compound identification. It is the absorption measurement of different IR frequencies by a sample positioned in the path of an IR beam. The main goal of IR spectroscopic analysis is to determine the chemical functional groups in the sample. Thus, IR spectroscopy is an important and popular tool for structural elucidation and compound identification. At temperature above absolute zero, all the atoms in molecules are in continuous vibration with respect to each other. When the frequency of a specific vibration is equal to the frequency of IR radiation directed on the molecule, the molecule absorbs the radiation. FTIR Spectra of Zn nanoparticle is shown as Figure 17.7.

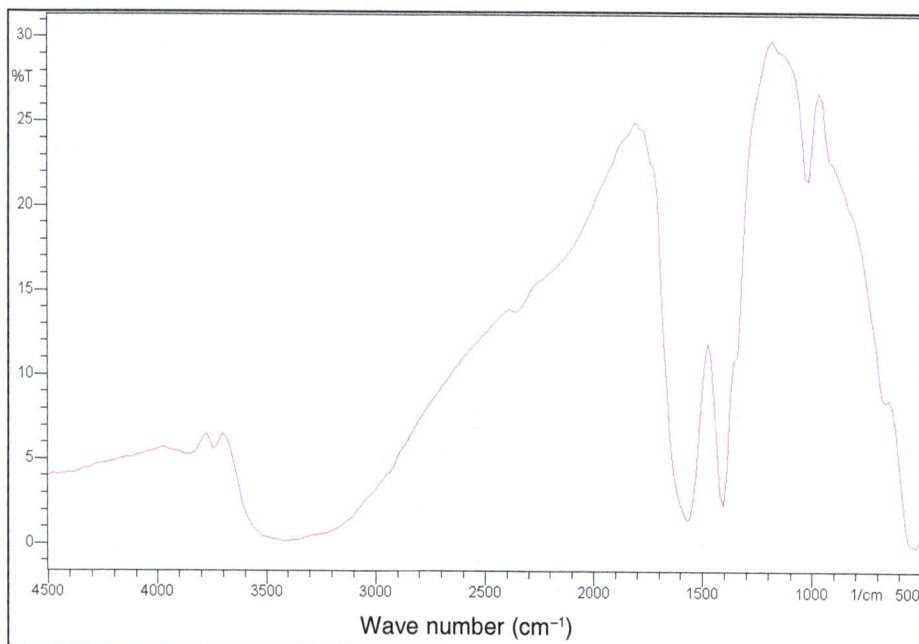

Figure 17.7: FTIR Image of Zn Nanoparticle.

X-ray diffraction or crystallography is a method of determining the arrangement of atoms within a crystal, in which a beam of X-rays strikes a crystal and diffracts into many specific directions. From the angles and intensities of these diffraction beams, a crystallographer can produce a three dimensional picture of the density of electrons within the crystal. From this electron density, we may get information on the mean positions of the atoms in the crystal, their chemical bonds and their disorders besides other information. Beside this, other characteristics techniques used are Energy Dispersive X-ray Spectroscopy, Lithography, ICP-MS and ICPOES.

Application of Nanoparticles

Aerosol spray (with the help of nebulizer) is much superior than traditional spray for application of nanoparticles to plants and microorganisms[6]. Lower concentration (5 ppm or less) can absorb and penetrate better through plants. Nanoparticle size (20 nm or less) may be better to apply for more benefit. Nanocube is found the better shape for more penetration both in plants and microorganisms.

Pathways of Nanoparticle Penetration

Nanoparticles are adsorbed to plant surfaces and taken up through natural nano-or micrometer-scale plant openings. Several pathways exist or are predicted for nanoparticle association and uptake in plants. Uptake rates depend on the size and surface properties of the nanoparticles. Very small sized nanoparticles can penetrate through cuticle. Larger nanoparticles can penetrate through cuticle-free areas, such as hydathodes, the stigma of flowers and stomata. Nanoparticles must traverse the cell wall before entering the intact plant cell protoplast. Result suggests that only nanoparticle less than 5 nm in diameter will be able to traverse the cell wall of undamaged cell efficiently. The penetration of nanoparticles and its translocation in plants are presented as Figure 17.8.

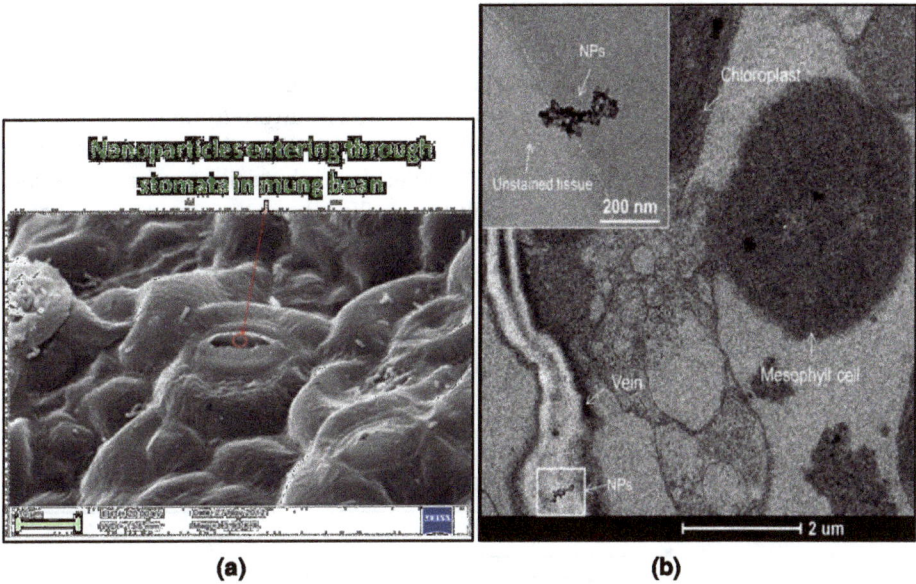

(a)   **(b)**

Figure 17.8: (a) Zn nano particle of 22 nm size entering through stomata in mung bean under arid environment (b) TEM images of NPs inside the leaf after applying nanoparticles for three day.

Effect of Nano Particles on Crop Production

Nano particles helped in seed germination. It was found that carbon nano tubes (CNTs) serve as new pores for water permeation by penetration of seed coat and act as a gate to channelize the water from the substance into the seeds. These processes facilitate germination which can be exploited in rainfed agriculture system. Nanofertilizers can enhance the fertilizer use efficiency, take care of imbalanced fertilization and multi-nutrient deficiencies. An enhanced production has been observed by foliar application of nano particles as fertilizer[17,18]. It was found that 640 mg ha^{-1} foliar application (40 ppm concentration) of nanophosphorus gives 80 kg ha^{-1} P fertilizer equivalent yield under arid environment of clusterbean and pearl millet. Currently, research is underway to develop nano-composite to supply all the required essential nutrients in suitable proportion through smart delivery system. Preliminary results suggest that balanced fertilization may be achieved through nanotechnology. The impact of nano-fertilizer products on physiological, biochemical, nutritional and morphological changes in plants and the fate of nano-products in soil and plant systems have to be studied. In addition, the effects of nano-fertilizer products on rhizosphere microorganisms and biogeocycling of nutrients have to be explored under natural field conditions.

Nanofertilizers have the opportunity to profoundly impact energy, the economy, and the environment by reducing nitrogen loss due to leaching, emissions and long-term incorporation by soil microorganisms[19]. Currently, the nitrogen use efficiency of plants is low due to the loss of 50 per cent and 70 per cent of the nitrogen supplied in conventional fertilizers. New nutrient delivery systems which exploit the porous nanoscale parts of plants could reduce nitrogen loss. Fertilizers encapsulated in nanoparticles will increase the uptake of nutrients. In the next generation of nanofertilizers, release of the fertilizer can be triggered by an environmental condition or simply be time released. Slow, controlled-release fertilizers have the potential to increase the efficiency of nutreint uptake. Nanofertilizers that utilize natural materials for coating and cementing granules of soluble fertilizer have the advantage of being less expensive to produce than those fertilizer that rely upon manufactured coating materials. Slow, controlled-released fertilizers may also improve soil by decreasing toxic effects associated with over application of fertilizer. Zeolites have been used as fertilizer delivery mechanism. The effect of nanoparticles/nanofertilizers under farmers field condition has resulted between 17 and 55 per cent improvement in crop yield (Table 17.4).

Table 17.4: Effect of Nano Phosphorus and Zinc on Yield of different Crops (Concentration of Zn nanoparticle 10ppm of 22 nm size and P nano particle 40 ppm of 18 nm size)

Treatments	Grain Yield (kg ha⁻¹)		Flower[1]
	Clusterbean	*Pearl Millet*	*Cauliflower*
Control	325	625	550
Ordinary Zn	338 (4.0)*	647.5 (3.6)	–
Nano Zn	438.7 (35.0)	731.2 (17.0)	–
Ordinary P	338 (4.0)	771.2 (23.4)	–
Nano P	471.2 (45.0)	921.7 (47.5)	–
Nano P + Nano Zn	–	–	850 (54.5)
LSD (p = 0.05)	27.5	47.2	23.7

* Per cent improvement over control.

Developing a target specific herbicide molecule encapsulated with nanoparticle is aimed for specific receptor in the roots of target weeds, which enter into system and translocated to parts that inhibit glycosis of food reserve in the root system. This will make the specific weed plant to starve for food and gets killed[20]. One nanosurfactant based on soybean micelles claims to make glyphosate-resistant crops susceptible to glyphosate when it is applied with the 'nanotechnology-derived surfactant'[3]. Less herbicide is required to achive the weed reduction effects desired. If the active ingredient is combined with a smart delivery system, herbicide will be applied only when necessary according to the conditions present in the field.

Several pesticide manufacturers are developing pesticides encapsulated in nanoparticles[21]. These pesticides may be time released or released upon the occurance of an environmental trigger (for example, temperature, humidity, light). It is unclear whether these pesticide products will be commercially available in the short term.

Generally, there are three kinds of controlled release systems (CRS), *viz.*, zero-order release, first-order release and square-root-time release which may be selected in view of the environmental condition and pest/pathogen biology. Biodegradable microbial polymers like polyhydroxyl alkanates have proved effective for controlled release of pesticides[22,23]. Temperature sensitive polymers *e.g.* Intelimer can regulate the release of pesticide with the progress of season depending on the temperature fluctuations, thereby protecting the active ingredient from unwanted leaching and degradation[24]. Some important controlled release devices are microcapsules, microspheres, coated granules and granular matrices that may be used in formulating the nanopesticides. A microcapsule is a reservoir system where active ingredient of the pesticide is contained within the core which is surrounded by a membrane. Micro-encapsulation has a potential scope in formulating nanopesticides. A microsphere is a monolithic system consisting of spherical or irregular shape particles (20 nm-2000 µm size) in which the active agent is dissolved or dispersed in a polymer matrix[25]. Various naturally occurring polymers (polysaccharides and proteins[26,27]), synthetic polymers (polystyrene, polyacrylamide, polyamides, polyesters etc.[28,29,27]), and inorganic materials (silica, zeolites, inorganic oxides, glass beeds, ceramic etc.[30,31,32,33]) can be explored to test their suitability in formulating nanopesticides helpful in crop protection.

Nano-based viral diagnostics, including multiplexed diagnostics kit development, have taken momentum to detect the exact strain of virus and the stage of application of some therapeutic to stop the disease. Detection and utilization of biomarkers, that accurately indicate disease stages, is also an *emerging area* of research in bio-Nanotechnology. Measuring differential protein production in both healthy and diseased states leads to the identification of the development of several proteins during the infection cycle. Needless to mention, that such Nano-based diagnostic kits would not only increase the speed of detection, but also fine-tune the degree of detection. Clay nanotubes (Halloysite) are developed as carriers of pesticides for low cost, extended release and better contact with plants, and they will reduce the amount of pesticides by 70-80 per cent, hence reducing the cost of pesticide and also the impact on water streams.

Water filtration may be improved with the use of nanofiber membranes and the use of nanobiocides, which appear promisingly effective. Biofilms are mats of bacteria wrapped in natural polymers. These can be difficult to treat with antimicrobials or other chemicals. They can be cleaned up mechanically, but at the cost of substantial down-time and labour. Work is in progress to develop enzyme treatments that may be able to break down such biofilms.

Nanoscale carriers can be utilized for the efficient delivery of fertilizers, pesticides, herbicides, plant growth regulators, etc. The mechanisms involved in the efficient delivery, better storage and controlled release include: encapsulation and entrapment, polymers and dendrimers, surface ionic and weak bond attachements among others. These mechanisms help to improve stability against degradation in the environment and ultimately reduce the amount to be applied, which reduces chemical runoff and alleviates environmental problems. These carriers can be designed in such a way that they can anchor the plant roots to the surrounding soil structure and organic matter. This can only be possible through the understanding of molecular and conformational mechanisms between the delivery nanoscale structure and targeted structures and matter in soil. These advances will help in slowing the uptake of active ingredients, thereby reducing the amount of inputs to be used and also the waste produced.

Biosensors provide high performance capabilities for use in detecting contaminants in food or environmental media. They offer high specificity and sensitivity, rapid response, user-friendly operation, and compact size at a low cost[34]. Several nano-based biosensors on direct enzyme inhibition have been developed to detect contaminants[34]. While the direct enzyme inhibition sensors currently lack the analytical ability to discriminate between multiple toxic substances in a sample (such as simultaneous presence of heavy metal and pesticide), they may prove useful as a screening tool to determine when a sample contains one or more contaminants. These methods are amenable to deployment in single-use test strips (making them useful to those in the field).

Nanobiosensor is a gold nanoparticle-based sensor that detects crystal violet or malachite green- prohibited dyes used to reduce fungal and parasitic infection – at concentrations as low as 2 ppb in seafood[35]. A technique that requires less use of specialized equipment is voltammetric detection using a nanocomposite film of zirconia-gold ZrO_2/Au nanoparticles[36]. Such a technique was used to detect parathion residues at a detection limit of 3 ng/mL. This technique is still dependent on the use of standard laboratory equipment and electricity. According to Hu[37], detection of multiple residues of organophosphorus pesticides has been accomplished using a nanomagnetic particle in an enzyme-linked immunosorbent assay (ELISA) test. The authors suggest that ELISA is more cost-effective than analytic tests requiring expensive laboratory equipment and that it does not require the same high levels of skill as tests using analytical chemistry.

Identity Preservation (IP) is a system that creates increased value by providing customers with information about practices and activities used to produce a particular crop or other agricultural products. Certifying inspectors can take advantage of IP as a mere way of recording, verifying, and certifying agricultural practices. Through IP, it is possible to provide stakeholders and consumers with access to information, records and supplier protocols. Quality assurance of agricultural products safety and security could be significantly improved through IP at the nano-scale. Nano-scale IP holds a possibility of the continuous tracking and recording of the history which a particular agricultural product experiences. The nano-scale monitors linked to recording and tracking devices to improve identity preservation of food and agricultural products. The IP system is highly useful to discriminate organic versus convnetional agricultural products.

Nanoparticles can be used for the bioremediation of resistant or slowly degradable compounds like pesticides. These harmful compounds tend to join the positive holes, are degraded and converted into non-toxic compounds. Otherwise these harmful compounds enter the food chain resulting in serious problems for the body. The electron hole pair, especially the negative electrons resulting from the excitation of nanoparticles, can also be used as a disinfectant of bacteria, as when bacteria make contact with nanoparticles, the excited electrons are injected into their bodies, which result in the bacterial removal from the object concerned, as in fruit packaging and food engineering.

In modern environmental science, removal of wastewater is an emerging issue due to its effects on living organisms. Many strategies are applied for wastewater treatment with little success. Photocatalysis can be used for purification, decontamination and deodorization of air. It is found that semiconductor sensitized photosynthetic and photocatalytic processes can be used for the removal of organics, destruction of cancer cells, bacteria and viruses. Application of photocatalytic degradation has gained popularity in the area of wastewater treatment.

The impact of nanotechnology is huge, ranging from basic food to food processing, from nutrition delivery to intelligent packaging. It is estimated that the nanotechnology and nano-bio-info convergence will influence over 40 per cent of the food industries by 2015. There is a strong need to develop nanofood through nano-engineering of food ingredients. Under this, texture, taste, flavour and color of food ingredients can be modified using nanoengineering without losing their nutritional value or with improved nutritional quality. For example, plant proteins may be given meat flavour by modifications. Efforts are required to develop smart nanocarrier out of fruit and vegetable wastes of Industries. This has two fold applications; (i) waste can be a continuous source of nanotechnology produce and large scale demand of substrate for nanoproducts development can be met. (ii) Utilization of waste protects environmental pollution. Nanofilms may be developed for extending shelf life of perishables like fruits, vegetables, and flowers during transportation to prevent loss.

Crop growth and field conditions like moisture level, soil fertility, temperature, crop nutrient status, insects, plant diseases, weeds, etc. can be monitored through advancement in nanotechnology. This real-time monitoring is done by employing networks of wireless nanosensor across cultivated fields, providing essential data for agronomic intelligence processes like optimal time of planting and harvesting the crops. It is also helpful for monitoring the time and level of water, fertilizers, pesticides, herbicides and other treatments. These processes are needed to be administered given specific plant physiology, pathology and environmental conditions and ultimately reduce the resource inputs and maximize yield. Scientists and engineers are working from dawn to dusk in developing the strategies which can increase water use efficiency in agricultural productions, *e.g.* drip irrigation. This has moved precision agriculture to a much higher level of control in water usage, ultimately towards the conservation of water. More precise water delivery systems are likely to be developed in the near future. These factors critical for their development include water storage, *in situ* water holding capacity, water distribution near roots, water absorption efficiency of plants, encapsulated water released on demand, and interaction with field intelligence through distributed nano-sensor systems. Nanotechnology enabled sensors may able to detect and identify harmful chemical or biological agents in the air and soil with much higher sensitivity than possible today.

Soils of the arid region are generally course textured, contain low soil organic carbon and have poor structure. As a result, these soils dry rapidly and are extremely prone to wind erosion. These twin factors of wind erosion and rapid drying adversely affect the survival of microorganisms in arid soils. Exo-polysaccharides produced by microorganisms can absorb moisture and can also bind the soil particles together to improve soil aggregation. Thus increasing production of exo-polysaccharides can help in reducing erodibility of soil, increasing soil moisture retention, survivability of microorganisms and C build up into the system. Nanoparticles Zn (10 ppm) and Fe (30 ppm) may help to increasing more polysaccharide release by microorganisms to increasing moisture retention capacity through better aggregate formation (Table 17.5) and C as well as microbial build up in the soils.

Table 17.5: Effect of Nanoinduced Polysaccharide Powder on Soil Aggregation

Treatment	Per cent Improvement of Aggregate Size Over Control After 30 Days (1 per cent w/v)		
	1.0 mm	*0.5 mm*	*0.18 mm*
Polysaccharide from *Bacillus coagulans*	80.7	No change	No change
Polysaccharide from *Alcaligenes faecalis*	33.4	82.9	56.4

Plants exude about 25 to 30 per cent of the photosynthates through roots. These exudates consist of low molecular weight amino acids, amino sugars, organic acids and polysaccharides that can provide energy and C skeleton for synthesis of exo-polysaccharides. Mg and Ti nanoparticles may help to absorb more solar radiation by plant leaves resulted more photosynthesis and enzyme release by plant roots for more native nutrient mobilization for plant nutrition.

References

1. Mukal, D., Sexena, N. and Dwivedi, P. D. (2009). Emerging trends of nanoparticles application in food technology: Safety paradigms. *Nanotoxicology* **3,** 10–18.

2. Bio-Based. (2010). Soy soap research update.< http: //www.biobased.us/ media/BioBased-2009-Research-Report.pdf >.

3. Tarafdar, J.C., Raliya, R. and Rathore, I. (2012). Microbial synthesis of phosphorus nanoparticles from Tri-calcium phosphate using *Aspergillus tubingensis* TFR-5. *Journal of Bionanoscience* 6, 84-89.

4. Tarafdar, J.C., Agrawal, A., Raliya, R., Kumar, P., Burman, U. and Kaul, R.K. (2012). ZnO nanoparticles induced synthesis of polysaccharides and phosphatases by *Aspergillus* fungi. *Advanced Science, Engineering and Medicine* 4, 1-5.

5. Tarafdar, J.C., Xiang, Y., Wang, W.N., Dong, Q. and Biswas, P. (2012). Standardization of size, shape and concentration of nanoparticle for plant application. *Applied Biological Research* 14, 138-144.

6. Mahajan, P., Dhoke, S. K., Khanna, A. S. and Tarafdar, J. C. (2011). Effect of nano-ZnO on growth of mung (*Vigna radiata*) and gram (*Cicer arietinum*) seedlings using plant agar method. *Applied Biological Research* **13**, 54-61.

7. Shi, Y. H., Xu, Z. R., Feng, J. L., and Wang, C.Z. (2006). Efficacy of modified montmorillonite nanocomposite to reduce the toxicity of aflatoxin in broiler chicks. *Animal Feed Science and Technology* **129,** 138–148.

8. Spriull, J. C. (2006). Novel pre-harvest approaches to control enteric food-borne bacteria in poultry. M.S. thesis, North Carolina State University, Raleigh.

9. Ochoa, J., Irache, J. M., Tamayoa, I., Walz, A., DelVecchio, V. G. and Gamazoa, C. (2007). Protective immunity of biodegradable nanoparticle-based vaccine against an experimental challenge with *Salmonella enteritidis* in mice. *Vaccine* **25,** 4410–4419.

10. Kumar, S.R., Ahmed, V. P. I., Parameswaran, V., Sudhakaran, R., Babu, V. S. and Hameed, A. S. (2008). Potential use of chitosan nanoparticles for oral delivery of DNA vaccine in Asian sea bass (*Lates calcarifer*) to protect from *Vibrio* (*Listonella*) *anguillarum*. *Fish and Shellfish Immunology* **25,** 47–56.

11. Food Safety Authority of Ireland (FSA) (2008). The Relevance for Food Safety of Applications of Nanotechnology in Food and Feed. <http://www.nanowerk.com/nanotechnology/reports/reportpdf/report119.pdf>.

12. Choy, J.H., Choi, S.J., Oh, J.M. and Park, T. (2006). Clay minerals and double layered hydroxides for novel biological applications. *Applied Clay Science* **36**, 122–132.

13. Gardea-Torresdey, J. L., Parsons, J. G., Dokken, K., Peralta-Videa, Troiani, H. E., Santiago, P. and Jose-Yacaman, M. (2002). Formation and growth of Au nanoparticles inside live alfalfa plants. *Nano Letter* **2**, 397–401.

14. Gardea-Torresdey, J.L., Gomez, E., Peralta-Videa, J., Parsons, J. G., Troiani, H. E. and Jose-Yacaman, M. (2003). Alfalfa sprouts: A natural source for the synthesis of silver nanoparticles. *Langmuir* **19,** 1357–1361.

15. Salamanca-Buentello, F., Persad, D. L., Court, E. B., Martin, D. K., Daar, A. S. and Singer, P. A. (2005). Nanotechnology and the developing world. *PLoS Medicine* **2,** 383–386.

16. Raliya, R. (2012). Application of nanoparticles on plant system and associated rhizospheric microflora. *Ph.D. Thesis*, Jai Narian Vyas University, Jodhpur, India, pp. 199.

17. Tarafdar, J.C. (2012). Perspectives of nanotechnological applications for crop production. *NAAS News* 12, 8-11.

18. DeRosa, M.C., Monreal, C., Schnitzer, M., Walsh, R. and Sultan, Y. (2010). Nanotechnology in fertilizers. *Nature Nanotechnology* **5,** 91.

19. Chinnamuthu, C. R. and Kokiladevi, E. (2007). Weed management through nanoherbicides. In: *Application of Nanotechnology in Agriculture*. C.R. Chinnamuthu, B. Chandrasekaran, and C. Ramasamy (Eds.) Tamil Nadu Agricultural University, Coimbatore, India.

20. OECD and Allianz (2008). *Sizes that matter: Opportunities and risks of nanotechnologies*. Report in cooperation with the OECD International Futures Programme. <http://www.oecd.org/dataoecd/32/1/44108334.pdf>.

21. Pepperman, A. B., Kuan J. C. W. and McCombs, C. (1991). Alginate controlled release formulations of metribuzin. *J. Controlled Release* 17, 105-111.

22. Lee, P. I. and Good, W. R. (1987). Overview of controlled-release drug delivery. *American Chemical Society*, 348.

23. Deborah, C., Meyers, H., Greene, P. A. and Lawrence, C. (1992). Temperature-activated release of trifluralin and diazinon. *ASTM Special Technical Publication, STP 1112 (Pestic. Formulations Appl. Syst)*, 11: 57-69.

24. Sotthivirat, S., Haslam, J. L., Lee, P. L. and Rao, V. M. (2007). "Controlled Porosity-Osmotic Pump Pellets of a Poorly Water-Soluble Drug Using Sulfobutylether-b-Cyclodextrin, (SBE)7M-b-CD, as a Solubilizing and Osmotic Agent, *J. Pharm. Sci.*, 98: 2364-2374.

25. Saravanan, K. M. and Panduranga, R. (2010). Pectin -gelatin and alginate-gelatin complex coacervation for controlled drug delivery: Influence of anionic polysaccharides and drugs being encapsulated on physicochemical properties of microcapsules. *Carbohydr. Polym.*, 80: 808-816.

26. Pong, C. C., Kai, L. T., Ming, L. S. and Chang, H. C. (2006). Release properties on gelatin-gum arabic microcapsules containing camphor oil with added polystyrene, *Colloids Surf.*, 50: 136-140.

27. Shukla, P. G., Sivaram, S. and Mohantya, B. (1992). Structure of carbofuran in crosslinked starch matrix by carbon-13 NMR: correlation of release and swelling kinetics with the dynamic behavior of polymer chains. *Polymer*, 33: 3611-3615.

28. Pong, C. C., Chen, C. J. Kimio, I. and D. Toshiaki (2005). Permeability of dye through poly(urea-urethane) microcapsule membrane prepared from mixtures of di- and tri-isocyanate. *Colloids Surf., B*, 44: 187-190.

29. Arshady, R. (1999). *Microspheres, Microcapsules and Liposomes: Prepartions and Chemical Applications*, Vol. 1, Citus Books, London, United Kingdom.

30. Pong, C. C. Miho, K., Takao, Y., Masahiro, N. and Toshiaki, D. (2003). Effect of dispersing medium.

31. Pong, C. C. and Toshiaki, D. (2003). Preparation of alginate complex capsules containing eucalyptus essential oil and its controlled release. *Colloids Surf., B*, 32: 257-262.

32. Chuan, H. W., Pong, C. C. and Lin, G. Y. (2006). Controlled release properties of Chitosan encapsulated volatile Citronella Oil microcapsules by thermal treatments," *Colloids Surf., B*, 53: 209-214.

33. Amine, A., Mohammadi, H., Bourais, I. and Palleschi, G. (2006). Enzyme inhibition-based biosensors for food safety and environmental monitoring. *Biosensors and Bioelectronics* **21,** 1405-1423.

34. He, L., Kim, N.J., Li, H., Hu, Z. and Lin, M. (2008). Use of a fractal-like gold nanostructure in surface-enhanced Raman spectroscopy for detection of selected food contaminants. *Journal of Agricultural and Food Chemistry* **56,** 9843–9847.

35. Wang, M. and Li, Z. (2008). Nano-composite ZrO_2/Au film electrode for voltammetric detection of parathion. *Sensors and Actuators B: Chemical* **133,** 607-612.

36. Hu, Y., Shen, G., Zhu, H. and. Jiang, G. (2010). A class-specific enzyme-linked immunosorbent assay based on magnetic particles for multiresidue organophosphorus pesticides. *Journal of Agricultural and Food Chemistry* **58,** 2801–2806.

2015, **Horticulture for Nutrition Security**
Editor: **Prof. K.V. Peter**
Published by: **DAYA PUBLISHING HOUSE, NEW DELHI**

Pages **385–410**

Chapter 18

Environment-Sensitive Male-Sterility in some Food Crops

K.B. Saxena and M. Bharathi

Flowering is an indispensable phenomenon of natural reproduction. To enter in the reproductive phase, most plant species get critical signals from different environmental factors which facilitate the conversion of their vegetative buds into reproductive tissues. The major factors that influence the appearance of flowers are moisture stress, temperature, photoperiod and irradiation. According to Bernier *et al.* (1993) these factors are perceived by different plant parts. For example, leaves are affected by photoperiod and irradiation; while temperature is perceived by all plant parts. Similarly, vernalization affects shoot apex and moisture availability is perceived by roots. These external determinants do not act independently and their interaction is inevitable and a particular factor may alter or substitute the direct effect of the other factors; and this may change the threshold level of individual factor in inducing flowering. Since at a given point of time, these environmental factors may act on different parts of the plant in their own way, the interaction among their gene products (proteins) may decide the transition of the vegetative buds into reproductive buds. Once flowering is induced in the plants, the further development of reproductive parts leads to differentiation into male/female gametes which participate in the natural reproduction. In this processes also, certain environmental factors play an active role and any departure from the normal process of microsporogenesis/megasporogenesis leads to disorders such as male or female sterility. In this chapter the authors have not attempted a thorough review of male sterility systems of food cops but restricted to environment-sensitive male sterility with brief description of their origin, possible variants, maintenance and utilization in two-parent hybrid breeding.

Basics of Male Sterility Systems in Plants

Any biological abnormality during microsporogenesis in plants,caused by either natural or artificial factors,leads to the appearance of male sterility in their flowers. Such events do not allow plants to reproduce through normal sexual mating. J. G. Kolreuter, a German scientist, is credited for recording the first ever incidence of male sterile plants while studying hybrid progenies of crosses involving different *Nicotiana* species in the middle of 18th century (Roberts, 1929). Darwin (1877) postulated that the loss of reproducing ability of plant helps in evolution and enhanced adaptation through gene transfer from diverse related and unrelated individuals through cross pollination.

The emergence of male sterile plants in nature is through spontaneousmutation (mostly recessive)of nuclear fertility alleles, it is popularly called as **'genetic male sterility'**. These abnormal plantsare exposed as homozygous recessive (*msms*) in the self-pollinated progenies of heterozygote (*Msms*) individuals.Generally, such mutants appear in low frequency and are often lost in the self-pollinated crops. However, in both cross- as well as partially cross-pollinated crops, therecessive mutant allele is protected by its dominant male fertile counterparts as heterozygotes.The other form of male sterility, commonly called as **'cytoplasmic male sterility' (CMS)**, is caused by certain deleterious interaction between nuclear and cytoplasmic genomes.In these genotypes the male sterility is conditioned by the factors housed in the cytoplasm and it is always inherited through maternal parent. These genotypes are maintained by the individuals carrying fertile or normal cytoplasm and identical nuclear genes. In certain cases the CMS genotypes with sterile cytoplasm,carry male fertility restoring dominant nuclear gene (*FrFr*) that makes the plant male fertile. The male sterile plants carry the corresponding recessive nuclear alleles (*frfr*) and sterile cytoplasm. These types are designated as '**cytoplasmic nuclear (or genetic) male sterility'**

In 20th century, the male sterility systems were reported in a number of plant species (Kaul, 1988) of economic importance. The evolution of the concept of hybrid vigour by Shull (1908) and subsequently by others, helped in understanding the potential benefits of male sterility in enhancing productivity of crops. Stephens (1937) for the first time utilized male sterility in hybrid seed production in sorghum. At the same time Jones and Emsweller (1937) also demonstrated its use in hybrid seed production of onion.These developments triggered male sterility based commercial hybrid breeding programmes in a number of food and horticultural crops. During this period massive research projects were undertaken globally at a number of plant breeding institutes to understand different types of male sterility systems with respect to their physiology, biochemistry, and genetics.

Male sterility in the plants can arise due to reasons such as defective growth anddifferentiation of anthers, impaired microsporogenesis, and failurein the release ofmature pollen grains and/or inability of pollen grains to germinate on stigma. In general these abnormalities do not adversely affect female reproductive system; and if such plants are pollinated they produce normal seeds.Based on the type of defects in androecium, the male sterility systems have been classified by Kaul (1988) as structural (absence or deformity of anthers), sporogenous (defective microsporogenesis), and functional (failure of mature pollen to germinate). In addition, on the basis of genetic control mechanisms, it has also been classified as genetic, cytoplasmic, and cytoplasmic nuclear (or genetic) male sterility.

Genetic Male Sterility

It is the most common form of reproductive abnormality found in both monocots as well as dicots coveringover 300plant species (Kaul 1988). In this system the male sterility is controlled by nuclear genetic factors with no influence of cytoplasmic. According to Kaul (1988) in most cases genetic male sterility is controlled by one or two pairs of recessive alleles. However, there are a few exceptions where male sterility is controlled by dominant genes.Theserecessive mutant male sterile plants generally arise spontaneously, and are lost if not maintained as heterozygotes (*Msms*). In cases where male sterility is controlled by dominant allele, its maintenance through reproductive means is difficult.

Cytoplasmic Male Sterility

It arises due to the presence of abnormal (defective) mitochondrial genome. Such cytoplasmis designated as 'sterile' (S); and it can arise spontaneously or through wide hybridization. These male sterile plantsdo not produce pollen grains because their nucleus contains a pair of recessive non-restoring (*msms*) alleles. The cytoplasmic male sterility is maintained by the genotypes which carry 'fertile' (F) or 'normal' (N) cytoplasm and non-restoring recessive nuclear alleles. Over 150 plant species have been reported to carrythis type of male sterility (Kaul 1988).This male sterility system cannot be used for field crops due to difficulties in producing large quantities of hybrid seed; but it is suitable for horticultural crops where seeds are non-commercial entity and only fruits are consumed.

Cytoplasmic Nuclear Male Sterility

This male sterility system is the most popular as far as its utility in plant breeding is concerned. In this system the expression of male sterilityis conditioned by a strong interaction between cytoplasmic and nuclear genomes. The fertility restoration of this male sterility mechanisms is controlled by dominant nuclear genes and depending on the type of fertility restoring gene, the expression of male fertility restoration could be total or partial. The hybrid technology based on this male sterility is popularly called as "three-line hybrid system" and include 'A' line-the male sterile female line with 'S' cytoplasm and recessive fertility nuclear alleles (*frfr*),B-line (maintainer of the female parent) with fertile 'N' cytoplasm and recessive nuclear alleles (*frfr*). This line when crossed with 'A' line, the entire progeny is male sterile. The third parent(designated as 'R' - line) contains dominant fertility restoring gene (*FrFr*). This line has ability to restore the male fertility of the hybrid plants produced by crossing with 'A' line.

Environment–Induced Male Sterility

This male sterility system is unique and has a recent identification. In this case the expression of male sterility/fertility is determined by nuclear genes whose expression is controlled by specific environmental factors; of these, temperature and photo-period are the major male sterility inducing agents. These two factors may or may not influence the plants independently. It is evident that every gene conferring male sterility in the plant kingdom is not prone to environmental changes and only few environment - sensitive gene(s) have so far been reported; and these are present in both genetic as well as cytoplasmic nuclear male sterility systems. In certain cases the 'converted fertile' plants revert back to male sterility under conducive environments.

Genetic Control of Male Sterility Systems

The molecular factors determiningCMS is still unclear. Studies have shown that its expression is linked to certain rearrangements of mitochondrial genome and consequently specific toxic proteins are produced that inhibit fertility restoration. In fact a number of explanations have been put forward from time to time, but still the genetic basis of this male sterility has not been properly understood in most crops. Recent studies have shown that the male sterility is associated with chimeric mitochondrial ORFs (open reading frames). Wang *et al.* (2006) demonstrated that in rice the ORF encodes a cytotoxin peptide which determines the expression of male sterility. Iwabuchi*et al.* (1993) showed that an abnormal copy of a mitochondrial gene produced aberrant mRNA transcripts containing an additional ORF. Hanson and Bentolila (2004) reported that male sterility may also be associated with alterations in promoter regions and portions of coding regions of mitochondrial ATP synthase. The genomic studies on mitochondria of A_4 cytoplasm in pigeonpea recognized 13 ORFs which can trigger male sterility (Tuteja*et al.* 2013). Further, Pallavi Sinha (unpublished) recorded10 bpdeletion in *nad7a* gene that was responsible for producing male sterile plants in pigeonpea.

Fertility Restoration of Male Sterility

The restoration of male fertilityin the hybrids involving cytoplasmic nuclear male sterile lines is an integral part of hybrid breeding programmes. Once an 'R' line is crossed with 'A' line, the dominant *Fr*nuclear gene of 'R' line overcomes the ill-effects of defective mitochondrial genome in the hybrid plant. According to Kaul (1988) the *Fr* gene produces certain proteins which repair the damage and make the hybrid plant male fertile. In most crops the pollen fertility is controlled by one or two dominant genes (Kaul, 1988). Saxena *et al.* (2011) reported that in pigeonpea two dominant genes were responsible for fertility restoration and the hybrids with a single gene were also fertile but they produced small quantity of pollen and exhibited instability over diverse environments. Similarly in maize also, four fertility restoring genes are reported and of these, two are major genes, and the other two genes yieldonly partial restoration (Wise *et al.*, 1999). Further, it was also reported that one of the major fertility restoring gene reduced sterility causing protein by 80 per cent (Kennel *et al.*, 1987).The fertility restoration has also been associated with genes encoding pentatrico peptide repeat proteins (Hanson and Bantolila, 2004).

Origin of Male Sterility Systems

Natural Occurrence

Nature has provided unlimited variability and in the past it has yielded a number of economic traits in different crops. There are numerous examples of it and among these, various male sterility systems are unique and these have benefitted millions through the cultivation of high yielding hybrids. Genetic male sterility arises due to mutation of male fertility nuclear gene (*MsMs*) to its recessive form to produce heterozygote (*Msms*) individuals. Self-pollination of such plants reveals male sterile segregants. Under natural conditions in the self-pollinated crops the male sterile mutants are generally lost but, in cross pollinated or partially cross pollinated crops such mutants are preserved by natural hybridization. According to Kaul (1988) genetic male sterility arising due to spontaneous variation has been reported in over 175 plant species. The natural frequency of cytoplasmic and male sterile mutants in nature is relatively less, because it require natural mutation in mitochondrial genome to make its cytoplasm male sterile. Similarly, the natural occurrence of cytoplasmic nuclear male sterility system is also low since it requires simultaneous mutations both in the mitochondria and nucleus. According to Kaul (1988) so far only 46 plant species are credited to have produced cytoplasmic nuclear male sterility under natural conditions.

Chance Recombinants

In search of promising recombinants with useful traits, not available in the primary gene pool, the use of wild species representing secondary gene pool in breeding is quite common. There are numerous examples where duringevaluation of inter-specific F$_1$ or F$_2$ population some male sterile plants have emerged. This may be due to partial incompatibility of the two nuclear genomes or interaction between nuclear and cytoplasmic genomes. In this way male sterility systems have been successfully developed in some cereal, oil seeds, and various other groups of crops (Kaul, 1988).

Targeted Breeding

Cytoplasm Substitution with Related Wild Species

The development of male sterility through cytoplasm substitution is based on the concept of bringing cytoplasmic and nuclear genomes of diverse origins within a single genotype. This is achieved by crossing a wild relative of a crop as female parent with a cultivated line as male parent. This combination integrates the cultivated nucleus into the cytoplasm of wild species and brings together the two diverse entities in a new genotype; and in many cases it has yielded male sterile plants. The selection of an appropriate maintainer genotype is also important and it should be done with care. In most cases the fertility restorers can also be selected germplasm; but in case it is not available, then the breeders need to select fertile segregants originating from the same cross. Besides this, in crops like pearl millet, soybean, and cotton cytoplasmic nuclear male sterile recombinant segregants have also been selected from inter-varietal crosses (Kaul 1988). The frequency of such useful recombination, however, is very low.

Mutations

In addition to above, specific mutagenic agents can also be used to breed male sterility systems in a number of crops. According to Kaul (1988) over 35 plant species have been tried successfully to develop male sterility systems through mutagenesis. Among the mutagenic agents tried, gamma rays and Ethium Bromide have been found most effective. Soybean and pearl millet are good example where cytoplasmic nuclear male sterile lines have been developed through mutagenesis (Burton and Hanna 1976; Kaul 1988).

Chemical Hybridizing Agents

Some of the chemicals are also known to have gametocide properties, and these as a group, are called as 'chemical hybridizing agents' (CHA). Moore (1950) and Naylor and Davis (1950) were the first to induce male sterility by spraying maleic hydrazide in maize. Soon other chemicals (alpha naphthalene acetic acid and beta indole acetic acid) were reported to have induced female flowers in cucumber (Laibach and Kriben 1951). According to Colhoun and Steer (1982) the ideal CHA must be very specific and should not affect other parts of the plants; and at the same time should not be transmitted to the progeny in any form. The major advantage of this system is that it does not require any maintainer line. Tu and Banga (1998) reported that chemicals like 'Dalapan' can cause male sterility in cotton, pearl millet, wheat, linseed, sesame, capsicum and some other crops. 'Ethrel' is effective in barley, mustard, oat, pear millet, rice, and wheat. Similarly, 'gibberellic acid' has been found effective in inducing male sterility in rice, maize, barley, oats, sunflower, and onion. 'Maleic hydrazide' produces male sterility in capsicum, cotton, oats, sorghum, and onion.

Genetically Engineered Plants

Recent advances in DNA recombination technology has made it possible to synthesize male sterile lines and their restorers. Mariani*et al.* (1990) were the first to develop such a genotype. This was achieved by transferring tobacco and rapeseed plants with a chimeric dominant gene from *Bacillus amyloliquefacience.* This gene disrupts the normal process of pollen formation and causes male sterility. Besides this, some other technologies such as induction of modified glucanase gene (Worrall *et al.,* 1992) and hormone engineering (Schimulling*et al.,* 1988) have been explored in the past. The cytoplasmic nuclear male sterility can also be produced through asexual recombination. Their use, however, has not found favor with plant breeders in any commercial hybrid crop.

Environment-Sensitive Male Sterility (ESMS)

In this form of male sterility changes in some environmental factors are responsible for the expression of male sterility. In fact all the reported male sterility genes do not respond to such changes. These environment-sensitive genes are spread across the species and genera and may be identified in both genetic as well as cytoplasmic nuclear male sterile genotypes. According to Kaul (1988) the major non-genetic factors which are known to cause male sterility in plants are temperature, photo-period, duration and qualityof light, and different soil-borne stresses. Of these, temperature and photo-period are the major male sterility inducing environmental agents. These two important factors may or may not function independently. There is a lot of literature which shows a significant role of interaction between these factors in the expression of male sterility and its reversal to male fertility. It is also found that the genotypic differences within a species can also alter the expression of male sterility/fertility of the plants.Photo-periodism is a developmental response of plants to the length and frequency of dark period for specified durations. The plants in general use a photo-period receptor protein (phytochrome or cryptochrome) and provide signals to start the process of flowering or to remain vegetative. Recently, Jarillo*et al.* (2008) have published a detail model to explain how photoperiod controls flowering event in the plants. The first true ESMS was identified by Shi (1981), a Chinese researcher in rice and it was a spontaneous photo-sensitive male-sterile mutant. Similarly, the first case of temperature sensitivity was recorded in rice by Young and Wang (1990) and Sun *et al.* (1989).This line was male sterile under high temperature (28-33°C) and male fertile under low temperature (22-27°C). Subsequently, these types of male sterility systems were detected in a number of crops across the genera.The threshold for sex change and its reversal, however, may be different for different species or genotypes. It is also established fact that the expression of the sensitive genes may also be influenced by genetic background of the genotype. Besides these, there are reports in literature about the male sterility that is induced by micro-nutrient (*e.g.* copper, boron etc.) deficiency. This type of male sterility induction is of academic interest or for some specific glass house experiments. This aspect is not covered in this chapter.There is a huge literature on various aspects of environment related induction of flowering in dozens of crops representing different genera and species. The authors, therefore, have not made attempts to review the bulky literature and only important aspects of some basic contents have been covered with appropriate examples.

Cereal Crops

Rice

Among cereals,riceis the most researched and utilized crop with respect to environment induced male sterility system. The first rice ESMS mutant was reported by Shi (1981) in a late maturing japonica variety 'Nongken 58' in China in 1973; and it was found to be sensitive to photo-period changes; and after selection it was designated as 'Nongken 58S'.In this mutant the male sterility was induced by daylength of ≥14 h. On the contrary, the plants grown under the photo-periods of ≤13 h 45m restored their male fertility (Shi, 1981, 1985; Shi and Dong, 1986; Lu and Wang, 1988). Another photo-sensitive rice segregants was isolated from anF_1 population of a cross between *O. glaberrima* and *O. sativa* (Sano, 1993). Similarly, Satoh *et al.* (1992) identified yet another photo-sensitive male sterility system from an inter-variety cross. This material was also fertile under in the environmental conditions with photoperiod ≤ 13.5 h. Yuan *et al.* (1993) reported the existence of two photo-period reactions that govern growth and development of rice. The first reaction was responsible for acceleration (or delay) in panicle differentiation and heading; while the second photo-period reaction determinedthe formation of fertile (or sterile) pollen. They further mentioned that the second photo-period reaction required more critically timed daylength, greater light intensity, and relatively higher temperature than that of the first photo-period reaction.

A thermo-sensitive male sterile natural mutant was identified in rice by Tan *et al.* (1989). This mutant expressed male sterility under high temperature and male fertility under low temperature. Subsequently, temperature sensitive rice genotypes were also reported by Zhou *et al.* (1988), Sun *et al.* (1989), Young and Wang (1990), Virmani and Voc (1991). These genotypes were male sterile under high temperature (28-33°C) and male fertile under low temperature (22-27°C).Maruyama *et al.* (1991) identified a rice mutantthat was derived through irradiation using 20 kr gamma rays. This mutant was completely male sterile at 31/21°C; partial male-fertile at 28/15°C; and complete male fertile at 25/15°C. On the contrary, Jiang (1988) and Zhang *et al.*(1991)reported temperature-sensitive mutants with a reverse response to variations in temperature; and with male sterility at 24°C and male fertility at 27°C. Satake and Yoshida (1977) reported that high temperature (35-41°C) during the period of anthesis in rice produced sterile pollen. According to Sun *et al.* (1989) and Maruyama *et al.* (1991) the genes that control the response to temperature were simply inherited. On the contrary, Zhang*et al.* (1991) and Siddiq *et al.*(1995) reported that the male sterility gene in ricewastightly linked to temperature sensitive nuclear gene. In summary, rice crop has both types of responses in expressing male sterility/fertility*i.e.* some genotypes produce male sterility reaction under high temperature or long photo-period regime; and male fertility under low temperature or short photo-period; while the other group of materials responses in the reverse way.

The discovery of Nongken 58S, a natural recessive photoperiod sensitive male sterile line, served as a starting point for development of two-line hybrid rice in China. The original PGMS rice Nongken 58S is a spontaneous mutant, and many studies have shown that fertility segregation in crosses between 58S and its wild-type progenitor is conditioned by a single Mendelian locus. It is thus interesting that a second locus has become involved in this system, and homozygosity of recessive alleles at both loci is required for expression of male sterility. This implies that the cultivar Nongken 58 was already homozygous for the recessive allele at the second locus before it mutated to become PGMS rice. In this connection it should be noted that, in a previous study of a cross between two japonica lines in which the fertility also displayed typical two-locus segregation, Zhang *et al.* (1990) reported a linkage between a PGMS gene and a locus for dwarfism located on chromosome 5. A major difficulty presently encountered in the utilization of PGMS rice in two-line hybrid breeding is the temperature-mediated fertility variation observed in many newly developed PGMS lines. The substantial amount of self-pollinated seeds produced by these male sterile lines under some environmental conditions prevented their use in hybrid seed production. The seed-setting rates of these male sterile lines under long-day conditions varied from zero under favourable growing conditions, to low in average growing conditions, and to 30-40 per cent in cooler-than-usual conditions. The extent of such fertility variations depends on the genetic background of the genotypes and, in general, it is more serious in *Indica* than in *Japonica* genetic backgrounds. There may be several reasons for such temperature-mediated fluctuations in the male sterility. These may include (i) a possibility of the gene at the "second locus" may differ from the one in PGMS line (ii) the possibility that there may be a presence of multiple alleles at the second locus conditioning male sterility, as is the case in soybean and pea (21) and for a locus governing fertility restoration in corn (22), so that different lines may carry different alleles with varying degrees of temperature sensitivity; and (iii) the possibility that additional modifying genes with thermo-sensitive expression may be involved in the system.

Wheat

Among winter cereals relatively more research has been carried out in wheat and both GMS and CMS genotypes have been reported to be influenced by environmental factors. Fisher (1972) recorded amazing results of variable photo-periods on reproductive organs in wheat. A short day (10 h) treatment at initial reproductive stage converted stamen into ovaries and ovules that were formed on the anther lobes. In another report the early and late formed tillers in wheat were found to have different pollen fertility levels and thus, signified the role of temperature on the male fertility (Jan, 1974). Luo *et al.* (1998) also reported the selection of photo-sensitive wheat lines. Murai and Tsunewaki (1993) reported identification of true photo-sensitive wheat genotypes which produced male fertility reaction during the daylength of ≤14.5 h and male sterility during photo-periods of ≥ 15 h.

Utilization of a two-line breeding system via photoperiod-thermo sensitive male sterility has a great potential for hybrid production in wheat (*Triticum aestivum* L.). 337S is a novel wheat male sterile line sensitive to both short daylength/low temperature and long daylength/high temperature. The first long daylength-sensitive D2 type CMS wheat line was discovered by Sasakuma and Ohtsuka (1979), and thereafter a series of photoperiod-thermo sensitive male sterile lines have been identified in wheat (Tan *et al.*, 1992; Murai and Tsunewaki, 1993; Luo *et al.*, 1998; Murai, 1998; Xu and Yan, 1998). Most of these sterile lines are difficult to use for hybrid wheat production due to their requirements for extreme daylength or temperature. Recently, a novel wheat male sterile line 337S was identified, which shows good male sterility under both short daylength/low temperature and long daylength/high temperature (Guo *et al.*, 2006a). The photoperiod-thermo sensitive male sterility in 337S is governed by two recessive genes located on chromosomes 2B and 5B, respectively, under long daylength/high temperature (Guo *et al.*, 2006b). There are two sowing windows for this line to be used as a male sterile line. Under an appropriate sowing time, it becomes fertile with the self-fertility rate >50 per cent, and thus it can also be used as a maintainer line. 337S is the first wheat male sterile line sensitive to both short daylength/low temperature and long daylength/high temperature, providing two sowing time windows for the expression of male sterility. This male sterile line has no harmful cytoplasmic effect and is controlled by recessive genes (Guo *et al.*, 2006a). The inheritance of male sterility under short daylength/low temperature was detected to be monogenic.

Other Cereals

In **maize** only the lines with S-cytoplasm exhibit thermo-sensitivity. Duvick (1966) while studying the stability of male sterility system in maize observed that some genotypes were male sterile in hot and dry environments, while other genotypes expressed partial male fertility in cool and humid environment. He *et al.* (1997) selected a thermo-sensitive male sterile maize mutant in 1992 in China and it was designated as 'Qiong 6 Qms'. This mutant was insensitive to change in the lengths of photoperiod and expressed male sterility in summer and male fertility in winter sowings. Qiong 6 Qms is being used to develop two-parent maize hybrids in China.Tang *et al.* (1997) reported a male sterility system in **sorghum** that was controlled by some specific interaction between temperature and photo-period. Kidd (1961), perhaps, was the first to report the induction of male fertility in the sterile population by exposing sorghum plants to high (40° C) temperature regime. These observations were later confirmed by Kontian and Hongyi (1981) and Zhang and Fu (1982). On the contrary, Downes and Marshall (1971) reported development of male sterility under cool (13° C) nights during meiosis. Murty (1986) reported male sterility during short days and low temperature in A_2 cytoplasm of sorghum. In the warm weather, however, the expression of male sterility was not complete with existence of partial fertile flowers. The exposure of **barley** plants to short days for a period of two weeks significantly reduces pollen fertility. Sharma and Reinbergs (1976) reported the identification of male sterile mutants in barley that were male sterile at high ($\geq 30°$ C) temperature and male fertile at low ($\leq 15°$ C) temperatures. By evaluating barley genotypes at two diverse altitudes, Ahokas and Hockett (1977) reported differential photo-period sensitivity for flowering.

Legume Crops

In **faba bean** only limitedsources of heritable male sterility have been reported. Berthelem and Le Guen (1975) and Duc (1980) reported significant effects of light intensity and temperature on the expression of male sterility. With plants converting to male fertility when the temperature ranged between 17-27°C. In **soybean**, Caviness and Fagala (1973) documented the effect of both photo-period and temperature on pollen fertility, with no pod set recorded under high temperature. Wei *et al.* (1994, 1997) reported the first true photo-period sensitive male sterile mutant from a local soybean cultivar.

In **pigeonpea** (*Cajanus cajan*) bothtemperature and photo-period are known to influence the initiation and appearance of floral buds, but their role in determining the male fertility/sterility has not been established. In the studies conducted by Saxena (2014),temperature was found to influence the fertility status of plants. Under the temperature regime of $\geq 25°$ C the plants were completely male-sterile. In contrast when daily mean temperatures dropped down to $< 24°$ C, the male-sterile plants turned fully fertile and produced self-pollinated pods (Table 18.1). In early generations of breeding this material, Saxena *et al.* (2004)observed that some male sterile pigeonpea plants converted to male fertility much earlier than the rest, and these male sterile plants were classified as 'early' and 'late' converters. This suggested the presence of more than one gene with different temperature thresholds to produce fertile plants. All the 'converted male fertile' plants reverted back to male sterility when these plants encountered high temperatures (Table 18.1).

Vegetable Crops

In **cabbage**, Rundfeldt (1960) identified a natural mutant that was male fertile under low temperature and expressed male sterility under warmer environments.

Timin and Dobrutskaya (1981) reported significant effects of environment on male sterility in **carrots**. In **watermelon** and **muskmelon** among cucurbits,the sex expression of the plants is greatly influenced by prevailing environmental factors (Rudich and Peles, 1976; Kaul, 1988).

Table 18.1: Field Observations Recorded in Three Months on Male-Sterility and Fertility in Four Temperature-Sensitive Selections

Year	Selection	September		November		February	
		Sterile Plants	Fertile Plants	Sterile Plants	Fertile Plants	Sterile Plants	Fertile Plants
2007	Envs S-1	13	1	2	12	11	0
	Envs S-2	11	2	1	12	11	2
	Envs S-3	9	1	2	8	8	0
	Envs S-5	13	3	2	14	12	2
	Total	46	7 (13.2 per cent)	7	46 (86.8 per cent)	42	4 (8.7 per cent)
2008	Envs S-1	22	0	1	21	22	0
	Envs S-2	8	0	1	7	8	0
	Envs S-3	10	0	0	10	7	0
	Envs S-5	18	0	3	15	16	0
	Total	58	0 (0.0 per cent)	5	53 (91.4 per cent)	53	0 (0.0 per cent)
2009	Envs S-1	37	0	0	37	37	0
	Envs S-2	32	0	0	32	32	0
	Envs S-3	27	0	0	27	25	0
	Envs S-5	23	0	0	22	21	0
	Total	119	0 (0.0 per cent)	0	118 (100.0 per cent)	115	0 (0.0 per cent)

() per cent fertile plants

Source. Saxena, 2014.

Among vegetables relatively good work has been done in onion and tomato. Temperature variability has been reported to influence pollen fertility/sterility in **onion.** Barham and Munger (1950) reported that no viable pollen was produced at the temperature <21°C. van der Meer and van Bennekom (1978) reported genotypic variability for response to variable temperatures. In one population the pollen sterility was found to be controlled by temperatures of 14°C and below; while the other population did not show any response. In **tomato** the influence of temperature on genetic male sterility system has been reported with no influence of photoperiod. Rick and Boynton (1967) reported a tomato mutant that was male sterile at 30-32°C. Stevens and Rudich (1978) reported a mutant where temperature of 38°/27°C resulted in the reduction in pollen production. Sawhney (1983) also reported temperature control of male sterility in tomato. An exposure of the mutant line at 23°/18°C produced sterility and under low temperature regime of 15-18°C the plants produced fertile pollen grains.

Other Crops

Brar (1982) reported a mutant in **sesame**, that was male sterile under glass house and fertile under field conditions. In **sugarbeet** also, low temperature was found to induce male sterility (Kinoshita, 1971). Xi *et al.* (1997) reported selection of a thermo-sensitive mutant in **brassica**. Fan and Stefensson (1986) found that the expression of male sterility was controlled by low (22°/16°C) temperature in brassica; and the full pollen fertility was expressed at 30°/24° C. In the middle temperature range the plants exhibited variable response to pollen fertility. According to Myer and Myer (1965) the male sterility in **cotton** is expressed at high (32-38° C).

Besides male sterility,fertility restoration of some F_1 hybrid combinations is also adversely affected by environmental factors, mainly temperature and humidity; and in different environments produces different levels of pollen fertility in the plants. In pigeonpea also, this has been a problem in A_2(*C. scarabeoides*) cytoplasm based hybrids. Saxena *et al.* (2011) attributedthe variability in fertility restoration due to low temperature stress in A_4 cytoplasm (*C. cajanifolius*) based hybrids.

Breeding Elite ESMS Lines

The environment-sensitive male sterility system is unique and if explored seriously, then it can be used for enhancing crop productivity through its use in hybrid breeding programmes. Besides its low cost, it is easy to implement and large seed quantities of female parent and hybrid can be produced with good quality control. Although this type of male sterility has been found in a number of crops (see section 4) but China is the sole leader in this field with ESMS-based hybrids commercialized in rice, sorghum, maize and brassica. In rice this technology has been exploited on a very large scale with high adoption to benefit the Chinese farmers. Some other countries such as India, Vietnam, Malaysia, Philippines, and Bangladesh etc. have also started using ESMS hybrid rice technology; but China is the real champion.

For a sustainable ESMS hybrid breeding programme it is essential that new parent material is generated on a regular basis. In this endeavour, it is essentialthat the sensitive male sterility gene is not lost during breeding process and each selected plant/progeny should carry the sensitive gene. To achieve this, it is important that in each generation the breeding materials should be planted under inductive environment that would allow the expression of male sterility gene. The seed from the selection should be harvested (as a ratoon crop) when the selected male sterile plants convert to full fertility and produce normal seed set. To launch a breeding programme for hybrid breeding using EGMS lines, the first important activity is to characterize the line with respect to its behavior under different temperature/

photoperiod regimes to find out the critical fertility and sterility points. Once these threshold points are confirmed, then the next important activity will be to create or identify locations with specific requirements of the environments. There should two specific sites, one site should be male fertility inducing (environment I), while the other should be able to induce male sterility (environment II). These distinct sites/ environments should be used in breeding new EGMS cultivars. For rice breeding in China, two sites with specific temperature and photo-period requirements are being used on a large scale (Virmani *et al.*, 1997). For a typical TGMS crop breeding/selection, a programme has been outlined in Table 18.2 using the summer (hot) and winter (cool) environments.

Table 18.2: A Procedure for Breeding New Thermo-Sensitive A-lines

Year	Generation	Season	Activity
1	–	Rainy (warm)	Plant parents, select male sterile plants, make crosses with elite male line
2	F1	Rainy (warm)	Grow F1, examine sterility of each plant, harvest F2 seed
3	F2	Rainy (warm)	Grow 2000 plants, examine each plant for sterility, reject the fertile plants, number the sterile plants. Carry them to winter season.
		Winter (cool)	Observe each plant for fertility and pod set. Reject poor pod setting plants. Harvest about 200 single converted plants for evaluation in F3 progeny rows.
4	F3	Rainy (warm)	Grow F3 rows, evaluate them for male sterility. Reject off- type progenies. Take them to the cool season.
		Winter (cool)	Select progenies on the basis of conversion to male fertility. Record data on the rate of conversion. Bulk or single plant harvest after rouging.
5	F4	Rainy (warm)	Grow F4 rows, evaluate them for male sterility. Reject off- type progenies. Take them to the cool season.
		Winter (cool)	Select progenies on the basis of conversion to male fertility. Record data on the rate of conversion. Bulk or single plant harvest after rouging.
6	F5	Rainy (warm)	Grow F5 rows, evaluate them for male sterility. Reject off- type progenies. Take them to the cool season.
		Winter (cool)	Select progenies on the basis of conversion to male fertility. Record data on the rate of conversion. Bulk harvest and record yield.
7	F6 (New A- lines ready; test them for combining ability before using in hybrid program		Grow F6 rows, evaluate them for male sterility and other traits. Reject inferior progenies. Take them to the cool season. Select progenies on the basis of conversion to male fertility. Record data on the rate of conversion andyield. Bulk harvest for use in hybrid breeding program.

Molecular Aspects of ESMS Systems

Wheat

To date, only a few temperature, photoperiod, or photoperiod-thermo sensitive male sterile genes inwheat have been mapped (Xing *et al.,* 2003; Cao *et al.,* 2004). The mapping analysis indicated that the male sterile gene *wptms3* located on chromosome 1B, flanked by *Xgwm413* and *Xgwm182*, differed from those reported by Guo*et al.* (2006b) and thus is a new gene. To date, several fertility restoring genes against CMS in wheat have been mapped on chromosome 1B (Ahmed *et al.,* 2001; Li*et al.,* 2005; Zhou *et al.,* 2005). Therefore, there are regions on chromosome 1B related to fertility performance. The SSR marker *Xgwm413* was identified to be closely linked to the male sterile gene and was found to be linked to yellow rust resistance genes in earlier reports (Peng *et al.,* 1999; 2000a; 2000b; Ma *et al.,* 2001). These studies indicated that the genes are not randomly distributed over the genome of a species, but they are rather frequently clustered on particular chromosomes (Peng *et al.,* 1999). The clustering of genes coding a trait may be the result of the co-evolution of plant species and their adaptation to environments (Peng *et al.,* 1999). With the identification of the molecular markers linked to the male sterile genes *wptms3, wptms1* and *wptms2*, they could be very useful for developing and improving new male sterile lines via marker-assisted selection.

In wheat breeding programmes, we can use the linked markers to distinguish the sterile genotypes earlier, which can help to shorten the breeding time and facilitate the whole procedure. As the long daylength/high temperature induced male sterility in 337S is controlled by two complementary genes (Guo*et al.,* 2006a; 2006b), and only the plants with genotype *aabb* are sterile and the ratio of sterile plants is low. Thus, mapping of *wptms3* provides a more feasible and reliable method for identifying male sterile plants in practical breeding. Therefore, it is also helpful for breeders to use these linked markers to identify those male sterile recessive genotypes in early generation and period with high accuracy. It is clear that the fertility of photoperiod sensitive sterile lines is determined by the interaction of genes and environments and the mechanism is complex. Isolation, cloning, and characterization of the male sterile gene can promote molecular study of the genetic male sterile trait and improve its application in hybrid breeding. Even though three male sterile genes (*wptms1, wptms2,* and *wptms3*) have been identified in the novel wheat male sterile line 337S, the linkage between markers and target genes is not tight enough for gene cloning. Therefore, fine mapping and cloning the male sterile genes in 337S under different environments are underway and will further improve understanding of the genetic mechanism of the male sterile system in utilization of wheat heterosis.

Rice

ESMS line is useful in hybrid rice seed production. It is not possible to identify the locus of sterility gene using morphological marker even if it is single gene, because the evaluation of fertility which is easily influenced by the environmental factor and is expressed as quantitative trait is difficult (Yoshiahi and Yamaguchi, 1997). Zhang *et al.* (1994) reported that expression of PGMS is more stable in *japonica* than in *indica* genetic background leading to the assumption that TGMS expression is also variable according to the genetic background.Although two-line hybrids developed using this EGMS germplasm have made great impact in improving rice yield in China during the past two decades, people knew less about the molecular mechanism of how the daylength and temperature co-ordinately regulate the fertility transition of EGMS in rice.

In the genetic analysis it was reported that *pms3*waslocated on chromosome12. This was the original mutation which converted Nongken58 to become the PGMS rice NK58S. Recently, *pms3* was cloned and shown to encode a long non-coding RNA (lncRNA) named LDMAR. A sufficient amount of LDMAR is required for male fertility under long-day conditions. A spontaneous G-C mutation causing a SNP between NK58 and NK58S, eventually brings about heritable increased methylation in the promoter region of LDMAR, which reduces the level of LDMAR expression. This then results in premature programmed cell death (PCD) in the anther development under long days, and hence express male sterility(Ding *et al.*, 2012a). In addition, Ding*et al.* also reported that RNA-dependent DNA methylation (RdDM) is involved in the regulation of PGMS. Promoter siRNA of LDMAR derived from *AK111270* is associated with the DNA methylation level of LDMAR, which reduces the expression level of LDMAR, and therefore male sterility in Nonken58S under long-day conditions (Ding *et al.*, 2012b). *P/TMS12-1*, which confers PGMS in the japonica rice line NK58S and TGMS in the indica rice line PA64S, encodes a unique non-coding RNA, which produces a 21-nucleotide small RNA named *osa-smR5864w*. This RNA shares identity with the product of *pms3* at the nucleotide level, which is responsible for the fertility of the pollen of NK58S and PA64S (Zhou *et al.*, 2012). Taken together, these findings suggest that this non-coding small RNA gene is an important regulator of male development controlled by cross-talk between the genetic networks and the environmental conditions.

The studies by Zhou (2012) suggest that a non-coding small RNA gene *p/tms* 12-1 is an important regulator of male development controlled by cross talk between the genetic networks and environmental conditions. They further indicated that a point mutation in this gene probably leads to loss-of-function for a small RNA namely *osa-smR5864m* constituting a common cause for PGMS and TGMS in the *japonica* and *indica* lines, respectively. To date a number of loci that control PGMS or TGMS in different lines have been mapped in different chromosomes. Photoperiod-sensitive genic male sterile (PSGMS) rice has a number of desirable characteristics for hybrid rice production.

Great efforts have to be invested to understand the biological functions of long non- coding RNAs (lncRNAs). In PGMS, how the lncRNAs (pms3) sense the different photoperiod and alter their expression is also very interesting and awaits much deeper research. The studies have shown that an lncRNA of 1,236 bases in length, referred to as long-day specific male-fertility–associated RNA (LDMAR), regulates PSMS in rice. The sufficient amount of the LDMAR transcript was required for normal pollen development of plants grown under long-day conditions. A spontaneous mutation causing a single nucleotide polymorphism (SNP) between the wild-type and mutant altered the secondary structure of LDMAR. This change brought about increased methylation in the putative promoter region of LDMAR, which reduced the transcription of LDMAR specifically under long-day conditions, resulting in premature programmed cell death (PCD) in developing anthers, thus causing PSMS (Ding *et al.*, 2011). Thus, an lncRNA could directly exert a major effect on a trait like a structure gene, and a SNP could alter the function of an lncRNA similar to amino acid substitution in structural genes. Molecular elucidating of PSMS has important implications for understanding molecular mechanisms of photoperiod regulation of many biological processes and also for developing male sterile germplasm for hybrid crop breeding.

According to Deninang(2012)the C-to-G mutationin a non-coding RNAis responsible for the temperaturesensitivemutation in *indica* background, suggesting thatit is the genetic background, but not thencRNA locus *per se*, that is responsiblefor the photoperiod to temperature sensitivemale sterility change.

Elucidation of the genetic and molecular bases of PSMS, although still far from completion at this stage, provides important implications for the studies of other photoperiod-regulated processes. It indicates that each of the processes may have a distinct genetic and molecular control at least at the lower level of regulatory hierarchy of photo-periodism. Therefore, the processes have to be investigated individually to understand them. We also speculate that at higher levels, such as the perception of day length, time keeping, and photoreceptors, among other things, these processes should be subjected to the same regulatory machineries as photoperiod flowering.

Characterization of genes and proteins related to male sterility aims to understand how and why the male sterility occurs, and which proteins are the key players for microspores abortion. Recently, a series of genes and proteins related to cytoplasmic male sterility (CMS), photoperiod-sensitive male sterility, self-incompatibility, and other types of microspores deterioration have been characterized through genetics or proteomics. Especially the latter, offers us a powerful and high throughput approach to discern the novel proteins involving in male-sterile pathways which may help us to breed artificial male-sterile system. This represents an alternative tool to meet the critical challenge of further development of hybrid rice. Taken together, concerted efforts from multiple angles would pave the road to rapid progress in understanding the complex regulatory networks and finally enable us to attain a holistic concept for the development of rice male reproductive system. All the data harvested in these studies will definitely help us to freely manipulate fertility in rice and other crop plants to facilitate hybrid breeding in the future.

Use of Environment-Sensitivity in Hybrid Technology

To make practical use of the environment-sensitive genotypes in hybrid seed production programmes, the selection of two production sites with strict temperature regimes and least fluctuations is essential. For the multiplication of quality seed of female parent and hybrid, the maximum safe mean temperature during crop growth, particularly reproductive phase, should not exceed the limits prescribed through field and/or laboratory experiments. These temperature/photoperiod bars would allow complete expression of the gene(s) responsible for the unique behavior of the genotypes and maintain pollen sterility/fertility status of the plants for quality seed production.

The salient features of ESMS-based hybrid seed production of rice, as used in China, is summarized in Table 18.3. This methodology can also be explored in pigeonpea as described by (Saxena, 2014) and given in Table 18.4. For sowing of the crop the months of June (high temperature) and September (low temperature) were selected. The data (Table 18.4) indicated that the two crops behaved differently with respect to their pollen fertility. In the June sown crop, over 92 per cent plants were male-sterile and this population can be used for hybrid seed production. In contrast, the September-sown crop appeared like a normal pure line variety with 98 per cent plants being male-fertile, and it can be used for the multiplication of A-line without involving B-line. Therefore, the seed system strategy involving temperature-sensitive pigeonpea genotypes would require two distinct sites, each with characteristically different temperature regime. These will allow complete expression of the gene(s) responsible for the unique behavior of the genotypes. For multiplication of female parent, the maximum safe temperature during crop growth, particularly reproductive phase, should be in the range that does not allow conversion of male fertile plants to sterility. This temperature bar will maintain pollen fertility status of the plants and allow production of fertile flowers and normal pod set. The seed produced from such isolated plots will remain genetically pure. For hybrid (A x R) seed production involving temperature-sensitive the male-sterile lines should be grown a season when the temperatures are high.

Table 18.3:Techniques Involved in Hybrid Seed Production of Rice in China Using TGMS System

Technique	Number/Quantity
Row ratio	2: 14-16
Width of female parent (m)	2.5
Plants/hill	2
Hills/ha	
A-line	530,000
R-line	27,000
Plants/ha	
A-line	1060,000
R-line	54,000
GA_3 (g/ha)	120-180
GA_3 Concentration	100
Seed set (per cent)	45-50
Hybrid yield (kg/ha)	2,600

Source: SS Virmani, IRRI, Philippines.

The pigeonpea experiment (Table 18.4) also showed that in some special environments the seed production of both the female parent and hybrid can be done at a single location. In this system the hybrid seed production plot is sown in isolation in early rainy season (June) to produce male sterile flowers for hybridization.The multiplication of female parent can be taken up in another isolation in September sowing; and this will produce male fertile flowers and a good harvest of female parent (A-line) can be taken without the use of B-line and pollinating insects.

Table 18.4: Segregation for Male-Sterility and Fertility as Affected by Date of Planting at Patancheru during 2008

Date	June 15 Sowing				September 30 Sowing			
	Sterile Plants	Fertile Plants	Per cent Fertile Plants	Yield/ Plant (g)	Sterile Plants	Fertile Plants	Per cent Fertile Plants	Yield/ Plant (g)
August 28-29	164	13	7.9	–	–	–	–	–
November 25-30	3	148	–	68-113	8	204	–	12-53
March 21-24	113	8	7.1	–	168	13	7.7	–

Source: Saxena, 2014.

General Discussion

Exploitation of two basic genetic phenomenon in commercial crops, identified as cytoplasmic nuclear male sterility and hybrid vigour, have not only saved the earth from hunger but also spared space for the cultivation of food crops. In the cultivation of hybrids seed is the most expensive input because every year a new seed stock is required to realize the maximum benefit from the technology. To limit the seed cost, the cytoplasmic nuclear male sterility based three-parent hybrid technology is always preferred over genetic male sterility based two-parent hybrids. To further ease the situation the hybrid technology based on environment-sensitive male sterility system is being advocated, particularly for rice in China. This system not only eliminates the use of B-line but also makes the selection of heterotic hybrid combinations much easier and faster.

The environment-sensitive male sterile mutants have been recovered in about three dozens of crops but its use in commercial hybrid breeding is limited to rice only. Yuan (1987) proposed its use in rice hybrid breeding programme. Since it eliminates the requirement of maintainer 'B' line, this hybrid system, is popularly called as 'two-parent hybrid breeding'. The hybrids rice based on this technology cover over 300,000 ha areas with yields as high as 8-9 t/ha. Now considering the pressure of providing food to ever increasing population, it would be appropriate that this technology be extended to other crops also. For example in a new crop like pigeonpea, this male sterility system is now available (Saxena, 2014) and breeding programmes to develop hybrids can be launched.This will a collaborative effort that would involve identification ofideal seed production sites each for female parent and hybrids. The temperatures at Patancheru during the months of August, November, and March arevastly different; and thesecan be used to provide the required fertility status of pigeonpea hybrid parents (Table 18.5). Based on prevailing temperatures one canalso identify the large scale seed production sites and then the process of commercialization of hybrids can begin. The future research in this endeavour should now be concentrated on the issues such as identification of high yielding hybrids with known threshold

Table 18.5: Mean Temperatures and Photo-periods during Critical Standard Weeks Recorded at Patancheru

Std. Week	Period	Day Length (h)	Average Air Temperature (°C)		
			2007-08	2008-09	2009-10
34.0	20-26 Aug	12.6	26.0	26.0	25.8
35.0	27 Aug-02 Sep	12.5	26.0	26.7	25.2
Mean	12.6	26.0	26.4	25.5	
47.0	19-25 Nov	11.2	19.1	23.9	22.9
48.0	26 Nov-02 Dec	11.2	20.3	22.3	20.4
Mean	11.2	19.7	23.1	21.7	
12.0	19-25 Mar	12.1	25.8	27.0	29.0
13.0	26 Mar-01 Apr	12.2	26.5	28.3	30.3
Mean	12.2	26.2	27.7	29.7	

Source: Saxena, 2014.

points of fertility conversion of different A- lines with respect to temperature and photo-period. Simultaneously, attempt should also be made to understand the molecular basis of sex reversion under different environments. Finally, it will be important to identify genes/QTLs responsible for controlling this trait that will facilitate quick transfer of these genes into heterotic hybrid parents.

References

1. Ahmed, T.A., Tsujimoto, H. and Sasakuma, T. (2001). QTL analysis of fertility-restoration against cytoplasmic male sterility in wheat. *Genes Genet. Syst.*, 76(1): 33-38. [doi: 10.1266/ggs.76.33].

2. Ahokas, H. and Hockett E. A. (1977). Male sterile mutant of barley. IV. Different fertility levels of *msgqci* (cv. Vantage), an ecoclinal response. *Barley Genet. Newsl.*, 7: 10-11.

3. Barham, W.S. and Munger, H.M.(1950).The stability of male sterility in onions. *Proc Am SocHorticSci*, 56: 401-409.

4. Bernier, G., Havelange, A., Claude, H., Petitjean, A. and Lejeune, P. (1993). Physiological signals that induce flowering. *The Plant Cell*,5: 1147- 1155.

5. Berthelem, P. and Le Guen J. (1975). Rapport d'activite-station d'amelioration des plantes. INRA Rennes (France): 1971-74. (As reported in Kaul, 1988.).

6. Brar, D. S. 1982. Male sterility in sesame. *Indian J. Genet.*,42: 23-27.

7. Burton, G.W. and Hanna, W.W.(1976). Ethidium bromide induced cytoplasmic male sterility in pearlmillet. *Crop Sci.*, 16: 731-732.

8. Cao, S.H., Guo, X.L., Liu, D.C., Zhang, X.Q. and Zhang, A.M. (2004). Preliminary gene-mapping of photoperiod-thermo sensitive genic male sterility in wheat (*Triticum aestivum* L.). *Acta Genet. Sin.*, 31(3): 293-298 (in Chinese).

9. Caviness, C.E. and Fagala, B.L.(1973). Influence of temperature on a partially male sterile soybean strain. *Crop Sci.*, 11: 564-566.

10. Colhoun, C.W. and Steer, M. W. (1981). Microsporogenesis and the mechanism of Cytoplasmic Male Sterility in maize. *Annual Botanic*, 48: 417-424.

11. Danmeng Zhu and Xing Wang Deng. (2012).A non-coding RNA locus mediates environment-conditioned male sterility in rice. *Cell Research*, 1: 1-2.

12. Darwin, C. (1877) The different forms of flowers on plants of the same species, Murray, London.

13. Ding, J., Lu, Q., Ouyang, Y., Maoa, H., Zhanga, P., Yaob, J., Xua, C. and Lia, X. (2012a).A long noncoding RNA regulates photoperiod- sensitive male sterility, an essential component of hybrid rice. *Proc Natl Acad Sci USA,*109: 2654-2659.

14. Ding, J., Shen, J., Mao, H., Xie, W., Li, X. and Zhang, Q. (2012b). RNA-directed DNA methylation is involved in regulating photoperiod-sensitive male sterility in rice. *Mol. Plant* 5, 1210–1216.

15. Downes, R.W. and Marshall, D.R. (1971). Low temperature induced male sterility in Sorghum bicolor. *Australia J. Expt. Agr. Anim. Husb.*, 11: 352–356.Duc, G.1980. Effect of environment on the instability of two sources of cytoplasmic male sterility in faba beans. *FEBS Lett.*, 2: 29-30.

16. Duvick, D.N. (1966). Influence of morphology and sterility on breeding methodology. (In) Frey KJ, editor. *Plant breeding*. Vol. I. Ames, Iowa (USA): Iowa State University Publications. p 85-138.

17. Emsweller, S.L. and Jones, H.A.(1935). An interspecific hybrid in *Allium.Hilgardia*, 9: 265–273.

18. Fan, Z. and Stefansson, B. R. (1986). Influence of temperature on sterility of two cytoplasmic male sterility systems in rape (*Brassica napus* L.) *Can. J. Plant Sci.*, 66: 221-227.

19. Fisher, J.E. (1972). The transformation of stamens to ovaries and of ovaries to inflorescences in *L. aestivum* L. under short day treatment. *Bot. Gaz.*, 133: 78-85.

20. Guo, R.X., Sun, D.F., Cheng, X.D., Rong, D.F. and Li, C.D. (2006a). Inheritance of photoperiod-sensitive male sterility in wheat. *Aust. J. Agric. Res.*, 57(2): 187-192.

21. Guo, R.X., Sun, D.F., Tan, Z.B., Rong, D.F. and Li, C.D. (2006 b).Two recessive genes controlling thermophotoperiod-sensitive male sterility in wheat. *Theor Appl Genet.,*112(7): 1271–1276.

22. Zhou, H., Liu, Q., Li, J., Jiang, D., Zhou, L., Wu, P., Lu, S., Li, F., Zhu, L., Liu, Z., Chen, L., Liu, Y. and Zhuang, C. (2012). Photoperiod- and thermo-sensitive genic male sterility in rice are caused by a point mutation in a novel noncoding RNA that produces a small RNA.*Cell Research*, 22: 649-660.

23. Hanson, M.R. and Bentolila, S. (2004). Interactions of mitochondrial and nuclear genes that affect male gametophyte development. *Plant Cell*, 16: S154–S169.

24. Iwabuchi, M., Kyozuka, J. and Shimamoto K. (1993). Processing followed by complete editing of an altered mitochondrial atp6 RNA restores fertility of cytoplasmic male sterile rice. *EMBO J.,* 12: 1437-1446.

25. Jan, C.C. (1974). Genetic male sterility in wheat (*Triticum aestivum* L.): expression, stability, inheritanceand practical use. Ph.D. thesis. University of California, Davis, Calif., USA.

26. Jarillo, J.A., Pineiro, M., Cubas, P. and Martinez-Zapater, J.M. (2009). Chromatin remodeling in plant development. *Inter J Dev Biol* 53: 1581–1596.

27. Jiang, Y. M. (1988). Studies on the effect of high temperature on fertility of the female sterile line in Dain-type hybrid rice.*J. Yunnah. Agric. Univ.,* 3 (2): 99-107.

28. Kaul, M.L.H. (1988) Male sterility in higher plants. Springer Verlag Berlin, Heidelberg. New York,1805 pp.

29. Kennell, J.C., Wise, R.P. and Pring, D.R. (1987). Influence of nuclear background on transcription of a maize mitochondrial region associated with Texas male sterile cytoplasm. *Mol. Gen. Genet.,* 210: 399–406.

30. Kinoshita, T. (1971).Genetical studies on the male sterility of sugarbeets (*Beta vulgaris*) and its related species. *J Fac Agric Hokkaido Univ.,* 56: 435-541.

31. Kongtian, Z. and Hongyi, F.(1981). Effect of high temperature on fertility of the male sterile lines in Sorghum. *Annu Rep Inst Genet Acad Sin*, p. 120.

32. Laibach, F. and Kribben, F.J. (1951). Der Einfluss von Wuchstoff auf das Geschlecht der Bliiten bei einer monoezischen Pflanze (*Cucumis sativus* L.). *Beitr Bioi Pflanz.,* 28: 64-67.

33. Li, X.L., Liu, L.K., Hou, N., Liu, G.Q. and Liu, C.G. (2005). SSR and SCAR markers linked to the fertility-restoring gene for a D2-type cytoplasmic male-sterile line in wheat. *Plant Breeding*, 124(4): 413-415.

34. Lu, Q., Li, X.H., Guo, D., Xu, C.G. and Zhang, Q.(2005). Localization of *pms3*, a gene for photoperiod-sensitive genic male sterility, to a 28.4-kb DNA fragment. *Mol Genet Genomics.,*273: 507- 511.

35. Lu, X. and Wang, J. (1988). Fertility transformation and genetic behavior of Hubei photoperiod-sensitive genic male sterile rice. In: Hybrid rice. Los Baños (Philippines): International Rice Research Institute. p 129-138.

36. Luo, H.B., He, J.M., Dai, J.T., Liu, X.L. and Yang, Y.C. (1998). Studies on the characteristics of seed production of two ecological male sterile lines in wheat. *J Hunan Agric Univ.,*24(2): 83–89. (in Chinese).

37. Luo, Z., Zhong, A. and Zhou, Q. (1988). Effect of a new non-toxic chemical gametocide in rice. In: Hybrid rice. Manila (Philippines): International Rice Research Institute.p. 147-156.

38. Ma, J.X., Zhou, R.H., Dong, Y.S., Wang, L.F., Wang, X.M. and Jia, J.Z. (2001). Molecular mapping and detection of the yellow rust resistance gene *Yr26* in wheat transferred from *Triticum turgidum* L. using microsatellite markers. *Euphytica*, 120(2): 219-226.

39. Mariani, C., De Beuckeleer, M., Truettner, J., Leemans, J. and Goldberg R.B. (1990). Induction of male sterility in plants by a chimeric ribonuclease gene. *Nature*, 347: 737-741.

40. Maruyama, K., Araki, H. and Kato, H. (1991). Thermosensitive genetic male sterility induced by irradiation. In: Rice genetics II. Manila (Philippines): International Rice Research Institute. p. 227-235.

41. Meyer, V.G. and Meyer, J.R.(1965). Cytoplasmically controlled male sterility in cotton. *Crop Sci.*, 5: 444-448.

42. Moore, R.H. (1950). Several effects of maleic hydrazide on plants. *Science.*, 112: 52-53.

43. Murai, K. (1998). Two-Line System for Hybrid Wheat Production Using Photoperiod-Sensitive Cytoplasmic Male Sterility. *In*: Zhang, A., Huang, T.C. (Eds.), Proceedings of the First International Workshop on Hybrid Wheat. China Agricultural University Press, Beijing, p.37-39 (in Chinese).

44. Murai, K. and Tsunewaki, K.(1993). Photoperiod-sensitive cytoplasmic male sterility in wheat with *Aegilops crassa* cytoplasm. *Euphytica*, 67(1-2): 41-48. [doi: 10.1007/BF000 22723].

45. Murty, U.R. (1986). Effect of A$_2$ cytoplasm on the inheritance of plant height in temperate × tropical sorghum crosses. *Sorghum Newsletter,* 29: 77.

46. Naylor, A.W. and Davis, E.A.(1950). Maleic hydrazide as a plant growth inhibitor. *Bot Gaz.*, 112: 112-116.

47. Pallavi Sinha., Saxena, K.B.,Rachit Kumar Saxena., Vikas Kumar Singh.,Suryanarayana, V.,Sameer Kumar, C.V., Mohan Katta, A.V.S., AamirKhan, W. and Varshney. R.K. (2014). *nad7a*: a candidate gene associated with cytoplasmic male sterility in pigeonpea (*Cajanas cajan*). Unpublished.

48. Peng, J.H., Fahima, T., Röder, M.S., Li, Y.C., Dahan, A., Grama, A., Ronin, Y.I., Korol, A.B. and Nevo, E. (1999). Microsatellite tagging of the stripe-rust resistance gene *YrH52* derived from wild emmer wheat, *Triticum dicoccoides*, and suggestive negative crossover interference in chromosome 1B. *Theor. Appl. Genet.*, 98(6-7): 862-872. [doi: 10.1007/s001220051145].

49. Peng, J.H., Fahima, T., Röder, M.S., Huang, Q.Y., Dahan, A., Li, Y.C., Grama, A. and Nevo, E. (2000a). High-density molecular map of chromosome region harboring stripe-rust resistance genes *YrH52* and *Yr15* derived from wild emmer wheat, *Triticum dicoccoides*. *Genetica*, 109(3): 199-210. [doi: 10.1023/A: 1017573726512].

50. Peng, J.H., Fahima, T., Röder, M.S., Li, Y.C., Grama, A. and Nevo, E. (2000b). Microsatellite high-density mapping of the stripe rust resistance gene *YrH52* region on chromosome 1B and evaluation of its marker-assisted selection in the F2 generation in wild emmer wheat. *New Phytol.*, **146**(1): 141-154. [doi: 10.1046/ j.1469-8137.2000.00617.x].

51. Rick, C.M. and Boynton, J.E. (1967). A temperature sensitive male sterile mutant of the tomato. *Am. J. Bot.*, 54: 601-611.

52. Roberts, H.F. (1929). Plant Hybridization Before Mendel, New Jersey: Princeton University Press.

53. Rudich, J. and Peles, A. (1976) Sex expression in watermelon as affected by photoperiod and temperature. *Sci. Hort.*, 5: 339-344.

54. Rundfeldt H. (1960). Untersuchungen zur zuchtung des Kopfkohls (*B. olerecea* L. var. *Capitata*). *Z. Pflanzenzuchtung* 44: 30-62. (As reported in Kaul, 1988.).

55. Sano, Y. (1983). A new controlling gene sterility in F_1 hybrids of two cultivated rice species: Its association with photoperiodic sensitivity. *J Hered.*, 74: 435-439.

56. Sasakuma, T. and Ohtsuka, I. (1979). Cytoplasmic effects of *Aegilops* species having D genome in wheat. I. Cytoplasmic differentiation among five species regarding pistilody induction. *Seiken Ziho*, 27: 59–65.

57. Satake, T. and Yoshida, S. (1977). Mechanism of sterility caused by high temperature at flowering time in *indica* rice. *JARQ Japan*, 11: 127-128.

58. Satoh, K. (1992). Influence of different seeding times on fertility performance of rice strain X 88 and Chunkanbohan nou 12in Okinava. Jpn. J. breeding (Suppl.1).

59. Sawhney, V.K. (1983). Temperature control of male sterility in a tomato mutant. *J. Hered.*, 74: 51-54.

60. Saxena, K.B., Sultana, R., Saxena, R.K., Kumar, R.V., Sandhu, J.S., Rathore, A. and Varshney, R.K. (2011). Genetics of fertility restoration in A_4 based diverse maturing hybrids in pigeonpea [*Cajanus cajan* (L.) Millsp.]. *Crop Science*, 51: 1 - 5.

61. Saxena, K.B. (2014). Temperature sensitive male-sterility system in pigeonpea. *Current Science*, 7 (2): 277-281.

62. Saxena, K.B., Tikka, S.B.S. and Mazumdar, N.D. (2004). Cytoplasmic genic male-sterility in pigeonpea and its utilization in hybrid breeding programme. (*In*) Pulses in New Perspective. Ali M, Singh, B.B., Kumar, S., and Dhar, V. (eds.). Indian Institute of Pulses Research, Kanpur 208024, UP, India, p 132-146.

63. Schimalling, C., Schell, J. and Spena, A. (1988). Single genes from Agrobacterium rhizogenes influence plant development. *EMBO J.*, 7: 2621-2629.

64. Sharma, R.K. and Reinbergs, E. (1976) Male sterility genes in barley and their sensitivity to light and temperature intensity. *Ind. J. Genet. Plant Breed.*, 3 2: 408-410.

65. Shi, M.S. and Deng, J.Y. (1986). The discovery, determination and utilization of Hubei photosensitive genic male sterile rice (*Oryza sativa* L. subsp. *japonica*). *Acta Genet. Sin.*, 13(2): 107-112.

66. Shi, M.S. (1981). Preliminary report of breeding and utilization of late japonica natural double purpose line. *J. Hubei Agric. Sci.*, 7: 1-3.

67. Shi, M.S. (1985). The discovery and the study of the photosensitive recessive male sterile rice (*Oryza sativa* L. subsp. *japonica*). *Sci. Agric. Sin.*, 2: 44-48.

68. Shull, G. H. (1908). The composition of a field of maize. *Am. Breeders Assoc. Rep.*, 4: 296–301.

69. Siddiq, E.A., Ahmad, I., Viraktmath, B.C., Jauhar, Ali. and Hoan, T. (1995). Status of hybrid rice research in India. *In*: Rai,M Mauria S (eds). Hybrid research and development. Indian Society of Seed Technology. New Delhi, India., Pp 139 - 158.

70. Stevens, M.A. and Rudich, J. (1978). Genetic potential for overcoming physiological limitations on adaptability, yield and quality in tomato. *Hort. Sci.*, 13: 673- 678.

71. Sun ZX, Min SK and Xiong ZM. (1989). A temperature sensitive male sterile line found in rice. Rice Genet. Newsl. 6: 116-117.

72. Tan, C.H., Yu, G.D., Yang, P.F., Zhang, Z.H., Pan, Y. and Zheng, J. (1992). Preliminary study on sterility of thermo-photosensitive genetic male sterile wheat in Chongqing. *Southwest China J. Agric. Sci.*, **5**(4): 1-6 (in Chinese).

73. Tan, Z. C., Li, Y.,Chen,B. and Zhou,G. Q. (1990). Studies on ecological and adaptability of dual purpose line Annong 1S. *Hybrid Rice*, 3: 35-38.

74. Tang, H.V., Pring, D.R., Shaw, L.C., Salazar, F.A. and Muza, F.R. (1996). Transcript processing internal to a mitochondrial open reading frame is correlated with fertility restoration in male-sterile Sorghum. *Plant J.*, 10: 123-133.

75. Timin, N. I. and Dobrutskaya, *E.G.* (1981). Cytoplasmic male sterility under different environmental conditions. *Ekol Genet Moldavian USSR*, 147: 1-17.

76. Tu, Z.P. and Banga, S.K. (1998). Chemical hybridizing agents. Narosa Publishing House, New Delhi. *In*: Hybrid cultivar development p.160.

77. Tuteja, R., Saxena, R.K., Davila, J., Shah, T., Chen, W., Xiao, Y.L., Fan, G., Saxena, K. B., Alverson, A. J., Spillane, C., Town, C. and Varshney, R. K. (2013). *Cytoplasmic male sterility-associated chimeric open reading frames identified by mitochondrial genome sequencing of four Cajanus genotypes.DNA Research*, 20 (5): 485-495.

78. van der Meer, Q.P. and van Bennekom, J.L. (1978). Improving the onion crop (*Allium cepa* L.) by transfer of characters from *A. fistulosum. Biuletyn Warzywniczy*, 22: 87-91.

79. Virmani, S.S., Viraktamath, B.C. and Lopez, M.T. (1997). Nucleus and breeder seed production of thermosensitive genic male sterile lines. *Int. Rice Res. Newsl.* 22(3): 26-27.

80. Virmani, S.S. and Voc, P.C. (1991). Induction of photo- and thermo-sensitive male sterility in *indica* rice. *Agron. Abstr.* 119.

81. Wang, Z., Zuo, Y., Li, X., Zhang, Q., Chen, L., Wu, H., Su, D., Chen, Y., Guo, J., Luo, D., Long Y.,Zhong, Y. and Liu, Y.G. (2006). Cytoplasmic male sterility of rice with Boro II cytoplasm is caused by a cytotoxic peptide and is restored by two related PPR motif genes via distinct modes of mRNA silencing. *Plant Cell* 18, 676–687.

82. Wei, B.G., Sun, G.C., Chang, J.W. and Jiang, C.X. (1994). The determination of photoperiod-temperature sensitive male sterile soybean. Proceedings of Third National Youth Symposium on Crop Genetics and Breeding. Edited by China Agricultural Scientech Press. p 185-189.

83. Wei, G., Zhang, Y., Zhang, D. and Liu, J.(1997). Mt DNA heterogeneity of cytoplasmic male sterility in maize. *Acta Genet Sin.,* 24: 66-77.

84. Wise R.P., Gobelman-Werner K., Pel D., Dill C.L. and Schnable P.S. (1999). Mitochondrial transcript processing and restoration of male fertility in T-cytoplasm maize *J. Heredity* 90: 380-385.

85. Worrall, D., Hlrd, D.L., Hodge, R., Paul, W., Draper, J. and Scott, R. (1992). Premature dissolution of the microsporocyte callose wall causes male sterility in transgenic tobacco. *Plant Cell,* 4: 759-771.

86. Xi, D.W., Cheng, W.J., Yi, D.L., Yu, D.R., Ning, Z.L., Deng, X.X. and Li, M. (1997). Genetic analysis on the TGMS line in rape of *B. napus* Xiangyou 91S. In: Lu RL, Cao XB, Liao FM, Xin YY, editors. Proceedings of the International Symposium on Two-Line System of Heterosis Breeding in Crops, 6-8 September 1997, Changsha, China. p 215-220.

87. Xing, Q.H., Ru, Z.G., Zhou, C.J., Xue, X., Liang, C.Y., Yang, D.E., Jin, D.M. and Wang, B. (2003). Genetic analysis, molecular tagging and mapping of the thermo-sensitive genic male-sterile gene (*wtms1*) in wheat. *Theor Appl Genet.,* 107(8): 1500–1504.

88. Xu, N.Y. and Yan, J.Q. (1998). Studies on photoperiod-sensitive cytoplasmic male sterility in wheat. *J Wuhan Bot Res.,*16(2): 97–105. (in Chinese).

89. Yamaguchi, Y., Ikeda, R., Hirasawa, H., Minami, M. and Ujihara, A. (1997). Linkage analysis of thermosensitive genic male sterility gene, *tms-2* in rice (*Oryza sativa* L.) *Breed. Sci.,* 47: 371-373.

90. Yang, D. and Wang, N.Y. (1990). The breeding of themosensitive male sterile rice R 59TS. *Scientia Agric. Sinica,* 23: 90.

91. Yuan, L. (1986). Hybrid rice in China. *Chin J Rice Sci.,* 1(1): 8–18.

92. Yuan, S.C., Zhang, Z.G., He, H.H., Zen, H.L., Lu, K.Y., Lian, J.H. and Wang, B.X. (1993). Two photoperiodic reactions in photoperiod-sensitive genic male sterile rice. *Crop Sci.,* 33: 651-660.

93. Zhang, K.T. and Fu, H.Y. (1982). Effect of high temperature treatment on male sterility in sorghum. *Acta Genet.* (China) 9: 71-77.

94. Zhang, D., Deng, X., Yu, G., Lin, X., Xie, Y. and Li, Z. (1990). Chromosomal location of the photoperiod sensitive male genic sterile gene in Nongken 58S. *J HuazhongAgric Univ.,*9: 407–419.

95. Zhang, M., Liang, C., Huang, Y. and Chen, B. (1994). Fertility expression of photoperiod temperature sensitive genic male sterile rice in Guangzhan and the response to photoperiod and temperature treatments. *J. Trop. Subtrop. Bot.,* 2(4): 100-107.

96. Zhang, Z.G., Yuan, S.C., Zen, H.L., Li, Y.Z., Li, Z.G. and Wei, C.L. (1991). Preliminary observations of fertility changes in a new type temperature sensitive male sterile rice IVA. *Hybrid Rice,* 1: 31-34.

97. Zhou, T.B., Xiao, H.C., Lei, D.Y. and Duan, Q.Z. (1988). The breeding of *indica* photoperiod male sterile lines. *J. Hunan Agric. Sci.,* 6: 16-18.

98. Zhou, W.C., Kolb, F.L., Domier, L.L. and Wang, S.W. (2005). SSR markers associated with fertility restoration genes against *Triticum timopheevii* cytoplasm in *Triticum aestivum. Euphytica,*141(1-2): 33–40. doi: 10.1007/s10681-005-5067-5.

2015, Horticulture for Nutrition Security
Editor: **Prof. K.V. Peter**
Published by: **DAYA PUBLISHING HOUSE, NEW DELHI**

*Pages **411–423***

Chapter 19

Indigenous Leaf Vegetables

P.G. Sadhankumar, S. Nirmaladevi and K.V. Peter

Vegetables especially green leaf vegetables are the most sustainable and affordable source of micro nutrients and vitamin A in the diet. They are much in minerals like iron and calcium. Leaves are departed urea more carotenoids then tuber vegetables and fruits. Green leaf vegetables are probably the most important group of foods as they have the best source of alkaline minerals. Daily intake of 100g fresh green leaf vegetable is recommended.

Palak (*Beta vulgaris* var. *bengalensis*)

Origin and Distribution

Palak (Beat leaf) is an important leaf vegetable grown in tropical and sub-tropical regions. Palak probably originated in the Indo-Chinese region (Nath, 1976).The Romans used this crop as food for humans and animals. The crop was introduced into USA in 1800 and was known as garden beet (Campbell, 1976). It is a popular crop in Bangladesh. The important palak growing states in India are Uttar Pradesh, West Bengal, Punjab, Haryana, Delhi, Madhya Pradesh, Maharashtra, Gujarat and Rajastan. It is becoming popular in South India. and is also known as Indian spinach, Spinach beet,Garden beet, Palong, Palang sag, Teegabatchali, Vusavyeley Dumpsbucchale, and Pasalai. It belongs to family Chenopodiaceae, genus *Beta* and species *vulgaris*, Chromosome number is 2n = 18. Leaves and stem are used after cooking. Leaves and tender stem are used for making pakoda. It possesses medicinal properties also. It is mildly laxative, diuretic, used in case of fever and inflammations of lung and bowels.

Methods of Propagation

Palak is a seed propagated crop. Seeds are sown directly in the field.

Varieties

There are a number of varieties released from IARI,New Delhi.They are

All Green

It is a high yielding variety (12.5 t/ha) Leaves are uniform green. Leaf margin is entire. 6 to 7 cuttings can be taken at an interval of 15 to 20 days.

Pusa Palak

This is a selection from a cross between Swiss Chard and local palak. It is a late bolting variety and produces uniform green leaves without any purple pigmentation.

Pusa Jyothi

Plants are vigorous, quick growing and regenerate quickly after taking cutting. It yields 29 t/ha in 6-8 cuttings.

Pusa Bharati

Leaves are cordate measuring 25 cm long and 14 cm in breadth. Leaves are smooth, tender and green without red pigmentation. First harvest can be done 30-40 days after sowing. The duration is 156 days and suited for both Kharif and summer. It is a selection from S 44-1.

Pusa Harit

This was released from IARI Regional Station, Katrain and is suited for cultivation in hilly regions throughout the year. Plants are upright, vigorous with green, slightly crinkled giant sized leaves. It is a late bolting variety that tolerates alkaline soil conditions

Arka Anupama

The variety Arka Anupama released from IIHR, Bangalore, is a multicut, late bolting variety. Leaves are dark green, thick, medium large and succulent. It is moderately resistant to *Cercospora* leaf spot under field conditions (Varalakshmi *et al.*, 2004).

Jobner Green

This variety was released from University of Udaipur, Jobner,Gujarat. It produces uniform green, large, thick succulent tender leaves with entire margin. It gives an average yield of 29.6 t/ha

Ooty 1

This variety was released from Tamil Nadu Agricultural University, Coimbatore. This is a selection from a local type. It is a high yielding variety yielding 15 t/ha. This is suited for year round cultivation and can withstand frost. The carotene content is high.

Banerjee's Giant

This is a popular variety in West Bengal. It produces large thick leaves on a succulent stem.

Pant Composite

This variety was released from GB Pant University of Agriculture and Technology, Pantnagar. This is a high yielding, composite variety tolerant to *Cercospora* leaf spot.

HS 23

This variety was released from Haryana Agricultural University, Hisar. This produces large, dark green, thick leaves. The first cutting can be taken 30 days after sowing. A total of 6-8 cuttings can be taken at an interval of 15 days.

Agronomic Practices

Palak can be successfully grown in fertile, well drained soil.It tolerates slightly alkaline soil, but performs best in neutral soils. This crop can be grown both in tropical and sub tropical conditions. It withstands frost and high temperature conditions. In plains of India, the crop is sown in June-July and September-October. In hills, it is sown from April to June. In South India, the crop is sown during November-December.

Apply farmyard manure @ 25 t/ha and NPK @ 80:60: 60 kg/ha. The entire quantity of P and K and half N are given as basal. Remaining N is applied in 3-4 splits after each harvest. Irrigation depends on soil type and season. Frequency of irrigation is more in light soils than in heavy soils. Immediately after sowing, give one irrigation. Give another irrigation on the third day. Subsequent irrigations may be given once in seven days depending on soil moisture. For removing weeds, 2-3 hoeing cum weeding will be required.

Harvest and Post-harvest Processing

Crop is ready for harvest in about 3-4 weeks after sowing. Subsequent harvests can be taken at 15-20 days interval. After harvest, the leaves cannot be stored for longer time and hence it should be sent to the market immediately.

Pests and Diseases and their Management

Pests

Aphids

Aphids are seen on the tender parts of the plant. They suck sap from the plant tissues.They can be controlled by spraying 0.1 per cent malathion.

Leaf Eating Catterpillars

Catterpillars of *Laphygma exigua* feed on leaves and make holes on the leaves. They can be controlled by spraying 0.1 per cent malathion.

Diseases

Cercospora Leaf Spot

Small, circular, light brown coloured spots with ash coloured centre appear on the leaves. Later the spots turn black with a papery centre. As it is a leafy vegetable, systemic fungicides cannot be used. Spray 0.2 per cent mancozeb for controlling the disease.

Downy Mildew

This disease is caused by *Perenospora spinaciae.* Symptoms appear on the outer leaves as pale yellow spots which are irregular in shape. Petioles may also be effected. Spray 0.2 per cent mancozeb for controlling the disease.

Spinach (*Spinacia oleracea* Linn.)

Spinach is a cool season vegetable grown for its edible leaves. It belongs to family Chenopodiaceae, genus *Spinacia* and species o*leracea*. It has 2 n = 2 x = 12 (Ryder, 1979). It is a rich source of dippa-carotene, vitamin C and minerals like calcium and iron. Spinach is dioecious. A few monoecious plants with varying proportions of male and female flowers are also seen. Male flowers are seen in clusters on the spike whereas female flowers are borne in clusters in leaf axils. Male plants flower earlier than female plants.Spinach varieties can be classified into 3 groups.

1. *Savoy*: They produce dark green, crinkled and curled leaves, mostly fresh market types. They are mostly hand harvested.

2. *Flat or smooth leaved*: They produce broad, smooth leaves and are suited for processing and mostly machine harvested.

3. *Semi-savoy*: It is a hybrid type. They produce slightly crinkled leaves having texture of Savoy type. It is grown both for fresh market and processing.

In Spinach, cultivars are either prickly seeded or round seeded. Prickly seeded cultivars perform well in hilly areas and round seeded ones in plains.

Based on growth habit, spinach varieties are classified as prostrate, semi-erect and upright.

Rosa (1925) described four sex forms in spinach.

1. *Extreme males*: These are smaller in size bearing staminate flowers. Seed stalk is characterised by presence of small leaves and normally flower earlier than other three vegetative males. These also bear staminate flowers. These are larger in size.

2. *Vegetative males*: Plants are larger in size.

3. *Female plants*: These bear only pistillate flowers. They are larger in size and remain vegetative for longer period.

4. *Monoecious plants*: These bear varying proportions of staminate and pistillate flowers.

Extreme males are not preferred due to their small sized leaves and early bolting nature. Monoecious plants are the most desirable types both for market and seed production.

Origin, Distribution, Evolution and Variation

Spinach is a native of Central Asia, most probably Persia. This crop was first cultivated by the Arabs. Spinach is included in the diet for centuries. It was introduced into Europe in 15ᵗʰ Century. Americans started growing it in 19ᵗʰ century.It reached North Africa through Syria and Arabia. The seed type in spinach does not indicate whether it is smooth leaved or Savoy type. Asian types are mostly prickly seeded and have triangular shaped leaves.

Methods of Propagation

Spinach is a seed propagated crops. Seeds are sown directly in the field.

Varieties

☆ **Virginia Savoy**: It is a prickly seeded cultivar having blistered dark green large leaves with a round tip. The plants are upright and vigorous.

☆ **Early smooth leaf**: It is a smooth seeded exotic cultivar producing small light green leaves. Leaves have a pointed apex.

☆ **Marathon**: It is a savoy type. Plants are large and produce dark green leaves.

Cultivation

Sandy loam soil with good amount of organic matter are preferred for the crop. A pH of 6.5 to 7 is ideal. Spinach grows poorly, if pH is below 6.Spinach grows the best in a cool weather and it withstands frost better than any other vegetable.

Sowing season varies with location. In plains, crop is sown during September-October. In hills, it is sown during August. Seed rate is 40-45 kg/ha and spacing is 60 cm x 25 cm.NPK @ 150:40:60 kg/ha should be given (Veeraraghavathatham *et al*,1998). Immediately after sowing,one irrigation should be given. Another irrigation on third day and subsequent irrigations at an interval of 10-15 days should be given.

The first harvest can be taken 25-30 days after sowing. Subsequent harvesting can be done at an interval of 15-20 days.

Diseases

☆ **White rust**: Causal organism is *Albugo occidentalis*. Initially, on the lower side of leaves white, blister like pustules are seen. Later, white lesions are seen on upper side of leaves. Upper surface will be chlorotic.

☆ **Downy mildew**: Causal organism is *Peronospora effuse*. Initially, yellowish areas are seen on the upper side of leaves. On underside, a gray to violet fungal growth can be seen.

☆ **Anthracnose**: Causal organism is *Colletotrichum spinaciae*. Initially, small, dark olive coloured spots appear on leaves. Later, they enlarge in size and become tan in colour. Then they coalesce.

☆ *Cercospora* **leaf spot**: Causal organism is *Cercospora beticola*. Initially small white spots appear on leaves. Later, they enlarge in size and sometimes coalesce.

☆ **Fusarium Decline**: Causal organism is *Fusarium oxysporum* f. sp. *spinaceae*. Infected plants turn yellow and wilt. Vascular system of the root is darkened.

☆ **Blight:** Causal organism is cucumber mosaic virus. The plants become chlorotic and the crown leaves become narrow and crinkled and the inward rolling of leaf margins occur.

☆ **Tobacco ringspot virus**: Small, indistinct, chlorolic spots appear on young foliage. Later, coppery brown chlorosis appear on the leaves. Affected plants become yellow and stunted.

☆ **Beet Curly Top:** This is a viral disease. Infected plants produce a rosette of tightly curled, small leaves in the centre of plant. Later the plant dies. The virus is transmitted by beet leaf hopper.

Basella (*Basella alba* and *B. rubra*)

Basella (Syn. Malabar spinach, Ceylon spinach, Indian spinach, Vine spinach, Poi, Chinese spinach, Malabar climbing spinach and East Indian spinach) is a popular leaf vegetable grown in almost all parts of India. Fresh tender leaves and stems are consumed as leaf vegetable after cooking. The mucilaginous qualities of plant make it an excellent thickening agent in soups, stews etc. Apart from its use as a leaf vegetable, the colouring matter contained in the leaves, stem and ripe fruits of red cultivars are used for coloring food. Chinese sell the bright red juice in the form of a powder called "gintjoo". Because of the mucilaginous nature, the leaves and stem are used as poultice.The juice of leaves is prescribed against constipation especially for children and pregnant women (Burkill, 1935). A paste of root is applied to swellings and is also used as a rubefacient. The leaf juice is a demulcent, diuretic, febrifuge and laxative. A paste of leaves is applied externally to treat boils. The flowers are used as an antidote to poison.

Origin and Distribution

Basella originated in Asia and more particularly in India. It is widely grown in North East and South India. It is also cultivated in Tropical Asia and Africa.

Methods of Propagation

Basella is propagated either by seed or stem cuttings or by root cuttings.

Cultivation

Basella grows well in hot and humid climates. Low temperature slows down growth rate and results in small leaves. Growth is also limited at altitudes higher than 500 m. Partial shading produces larger leaves compared to growing under full sunlight. Day lengths shorter than 13 h result in flowering. Although the crop is adapted to many soils, sandy loam is the most suitable. The soil should be moist, fertile and well supplied with organic matter. The crop comes up well when soil pH is between 5.5 to 8.0.

In Northern and Eastern plains of India, seeds are sown from March to May while in Southern parts, it is sown in July and again in October-November. Basella is planted either by direct seeding or by transplanting. It can also be propagated through stem cuttings or by root cuttings. Choice of planting method depends on availability of seeds, labour and growing season. A seed rate of about 12.5 to 16.0 kg will be required to sow a hectare of land. When direct seeding is done, seeds are sown in rows in well prepared beds. Furrows of 1.0 - 1.5 cm depth are spaced at 10-15 cm apart on bed. The seeds are sown 5 cm apart in rows. The seeds are covered with a layer of compost. The seedlings are thinned to stand 10-15 cm apart at two to three true leaf stage.

The seedlings can also be raised in seedling trays or seed beds. Seedling trays are filled with potting mixture having good water holding capacity. eg. peat moss, potting mixture, rice hulls, vermiculite and sand. Two or three seeds are sown per tray cell at 1.0 to 1.5 cm depth. Thinning is done to retain one seedling per tray cell at two to three true leaf stage. If the seedlings are raised in seed beds, soil should be partially sterilized by burning dry organic matter. Seeds are then sown in furrows spaced at 5 cm apart and then covered with soil. Seed beds are then covered with insect proof nets to provide shade. Later they are hardened by slowly exposing to direct sunlight prior to transplanting. Seedlings are ready for transplanting when they have five true leaves.

Recommended spacing varies with variety and harvest method. For once over harvest, raised beds of 20-30 cm height at 90 cm width at top are taken at convenient length. Rows are spaced 10-15 cm apart with 15 cm between plants within rows. Irrigate immediately after transplanting to establish good root to soil content.

Stem cuttings of 20-25 cm length with three to four internodes are soaked in water overnight and planted at spacing of 20-30 cm between rows and 15-20 cm within a row at a depth of 5-10 cm. Two or three stem cuttings are planted per hole. Stem or root cuttings should be planted during monsoon month or in early summer

A basal dressing of 20-30 tonnes of farm yard manure and 60:60:40 kg NPK/ha has to be applied before transplanting or sowing. Frequent irrigation promotes growth of plant and makes plant succulent. Water stagnation should be avoided. Water stress induces early flowering and may lead to thin winy stems and small leaves. The crop requires irrigation at an interval of 5-6 days in summer and 7-8 days during cool season. When plant starts trailing, it should be trained on supports.

Harvest

Crops raised from seeds will be ready for harvest with edible stems and leaves 8-10 weeks after sowing. The plants raised from root or stem cuttings will be ready for harvest in about 6 weeks after planting. Yield varies from 14-19.5 t./ha.

Pests and Diseases

The most important diseases affecting the crop are damping off (*Pythium aphanidermatum*), leaf spot (*Acrothecium basellae, Fusarium moniliforme* and *Cercopsora* sp.) and mosaic. The crop is almost free from insect attack.

Agathi (*Sesbania grandiflora* Pers)

Agathi (*Sesbania grandiflora*) is a perennial tropical, quick growing and soft wooded tree belonging to family Fabaceae. It has ornamental, food and fodder values. The flowers are also eaten as vegetable. Flowers are good for making " bajji".The tree can be used as a standard for pepper and betelvine. In Tamil Nadu, agathi is grown around banana as a wind-break and around coconut seedlings as a shade plant. The bark yields good fibre and a gum and the juice of flowers is said to improve the sight, when squeezed into the eyes.Leaves and flowers of agathi have nutritional and medicinal properties.

Origin and Distribution

It is a native of Malaysia. It is not grown on large scale for vegetable purpose. It is grown in parts of Punjab, Delhi, Bihar, Orissa, Assam, Bengal, Tamil Nadu and Kerala. There are two forms of agathi, one with red flowers and the other with white flowers. The white flowered agathi is suitable for kitchen garden.

Methods of Propagation

Agathi is propagated through seeds.

Cultivation

It grows the best in black cotton soils and comes up quickly when the surface soil is loose and uneven. It is resistant to drought. The seeds are first sown in the nursery.They are transplanted in the mainfield when they are 30-45 cm height in pits.

Pits are taken at a spacing of 90-100cm either way.

In Tamil Nadu, two months after sowing/transplanting ammonium sulphate is applied to seedlings. A tree yields 4.5 - 9.1 kg of leaves / year. The plants come to flower by September - December and to fruiting during summer.

Harvest

Leaves are produced throughout the year. First harvest can be taken 3-4 months after planting. Tender leaves are harvested.

In early stages of crop growth, it suffers severe attack by seedling blight caused by *Colletotrichum capsici*. The disease can be effectively controlled by 1 per cent Bordeaux mixture spray.

Chekkurmanis (*Sauropus androgynous* Merr.)

Chekkurmanis (*Sauropus androgynous* Merr.) (Syn. *S. albicans* Blume, *S. gardnerianes* wight., *S. sumatrans* Mig.) is a perennial leaf vegetable. It also called as multi vitamin and multi mineral packed leaf vegetable. It has unique position in the list of leaf vegetables because of its high nutritive value and multivarious uses. Its leaves are very rich in protein, minerals and vitamins A, B and C. The crop is being grown in Southern India, Indonesia and Singapore. The leaves can be cooked like other greens.The leaves are used to give a light green colour to pastry and to fermented rice in the Dutch East Indies. Leaves are used for preparation of soup in Java. The plant is also useful for growing as a hedge around home gardens. In Java, it is often planted in live fences and in midst of garden beds to provide light shade for other vegetables planted in beds.Besides other uses, leaves are used as cattle and poultry feed in certain parts of the country. In some other places, plants are planted as a soil binder to prevent soil erosion.

Chekkurmanis has several medicinal properties. The juice of leaves pounded with roots of pomegranate (*Punica granatum*) and leaves of jasmine (*Jasminum sambac*) is used against eye troubles. A decoction of its roots is often recommended for fever in rural areas. Pounded roots and leaves are used as poultice for ulcers in the nose.

Origin and Distribution

Chekkurmanis is native of India and Burma. It is found in Sikkim Himalayas, Khasi, Abor and Arka Hills at 1200 m elevation.It is also seen in the Western Ghats of Kerala from Wynad Northwards at altitude of 300 to 1200 m. It was introduced into Kerala from Malaya in the year 1953. In Malayalam, it is known as 'Madhura Kheera'. The plant grows wild in the evergreen forests of the Western Ghats and in the southern parts of Kerala. It was introduced in to Tamil Nadu through the Agricultural Research Stations of the state during 1955-56 where it became popular in the name 'Thavasi Murungai'.

Methods of Propagation

It is propagated by stem cuttings which root easily and also through fresh seeds.

Cultivation

The plant grows well in all types of soil. A warm humid climate with good rainfall is the best suited for the luxuriant, succulent growth of leaves and twigs. It tolerates shade to some extent. The crop comes up well in mild humid locations also.

It is propagated by stem cuttings which root easily and also through fresh seeds. Both hardwood and semihard-wood cuttings of 20-30 cm length after trimming the leaves are used as planting materials. They are planted in shallow furrows spaced 30 cm and at a distance of 10-15 cm in 2-3 rows. Rooted cuttings can be made in nursery beds also or in pots and then transplanted around the kitchen garden.

The land is ploughed and leveled well. Well rotten FYM is added @ 2 kg/sq.m. The cuttings are planted in shallow furrows atleast 15 days earlier to the onset of monsoon during April-May. Subsequently frequent irrigations are given until root initiation takes place.

After the onset of monsoon, it does not require much irrigation. The cuttings come up very well and would be ready for harvest within 3 - 4 months of planting. Apex of the plant is nipped off which enables the plants in putting forth new branches. The tender shoots and leaves can be harvested intermittently for several subsequent years.

Eventhough it can withstand the hot dry weather for a long period, watering of plants in such condition is desirable for getting constant appearance and growth of new leaves.

Curry Leaf (*Murraya koenigii* Linn.Sprengal)

Curry leaf is a spicy leaf vegetable belonging to family Rutaceae and chromosomer number is 2n=18. It used in culinary preparation to enhance flavour and taste of food and is a very rich source of iron and Vit.A. The leaves are slightly pungent,bitter and feebly acidic in taste.Curry leaf loses flavour when cooked.To retain fresh flavour,the leves should not be removed from branches until ready for use.

Origin and Distribution

Curry leaf is a spicy leaf vegetable originated in Tarai tract in foothills of Himalayas from Kumaon to Sikkim in Bengal, Assam, Deccan plateau and in western ghats. It is a backyard crop in many south Indian homesteads. On a commercial scale, it is cultivated in Tamil Nadu, Andhra Pradesh, and Karnataka.Curry leaf grows throughout India including Andaman and Nicobar Isands upto an altitude of 1500m.Leaves are rich source of vit.A and minerals. The peculiar aroma is due to sulphur containing essential oil.

Propagation

Curry leaf is propagated by seeds as well as one year old root suckers (Regeena and Ravi,1995).Raising seedlings in nursey and transplanting in main field is normal practice. Well ripe fruits are harvested from high yielding mother trees seeds are extracted and sown in nursery beds or poly bag within 3-4 days. Each fruit contains 2-3 seeds. Nursery beds of size 1x1 m with 30cm height are prepared and adequate farm yard manure is incorporated. Seeds are sown at a spacing f 10cm. and they germinate in three weeks.

Varieties

DWD 1 (Suwasini)

Clone developed by University of Agricultural Sciences,Bangalore has dark green highly aromatic shining leaves with oil content of 5-22 percent.The clone is sensitive to low temperature.

DWD 2

It is an open pollinated seedling progeny developed by UAS, Dharwad with pale green leaves having lesser aroma (5.09 per cent).It is not sensitive to low temperature and is superior in bud burst, internodal length, shoot length and weight of new shoots. Due to winter insensitive nature, farmers get extra income.

Senkaampu

It is local cultivar with purplish petiole and highly aromatic leaves.

Agronomic Practices

Curry leaf is a hardy crop which can tolerate high temperature. It can tolerate maximum temperature ranging from 26-37^0 C. But when temperature falls below 16^0 C, the vegetative buds become dormant. Curry leaf can establish very well in red sandy loam with good drainage. The main field is ploughed to fine tilth and well rotten farmyard manure is added @ 20t/ha.Pits of size 30cm^{3}are dug at spacing of 1.2 to1.5m and one year old seedlings are planted at the centre of each pit. Pits are irrigated immediately after planting. Second irrigation is given on third day and thereafter at weekly intervals. The interspace is kept free of weeds by periodical hoeing and in the first year,one intercrop of a pulse crop can be taken.A fertilizer mixture comprising of 150g N,25g P_2O_5 and 50g K_2O per plant along with 25 kg FYM is essential for curry leaf. The method of application does not influence growth. Plants should be pruned at the age of 8-10 months after planting at a height of 30cm from ground level (Krishnarajan,1991). This will help to maintain plant in a bushy state. In total 5-6 branches are retained per plant.

Harvest and Post-harvest Processing

Harvesting starts 10-12 months after planting.At the end of first year 250-400 kg leaves can be harvested from 1 ha. In the second and third years harvesting can be done once in 4 months and total yield per year is approximately 5400 kg/ha. In fourth year harvesting is done once in 3 months and 2500 kg leaves will be obtained in each harvest. From fifth year onwards harvesting is done at three months interval and average yield will be 20t/ha green leaves. After each harvest,farm yard manure is applied @ 20kg/plant and mixed with soil. The leaves retain flavour even after drying and hence these are marketed both as fresh and dried form.

Pests and Diseases and their Management

Citrus Butterfly

The early instar larvae of citrus butterfly feed on leaves and causes severe defoliation. Application of Malathion @1ml/l is effective

Psyllid Bug and Scale Insects

Both these cause severe defoliation. Removal of affected branches and spraying dimethoate @ 1ml/l controls the pests.

Leafspot

The disease cause severe defoliation. Spraying carbendazim @1g/l controls the disease

Basal stem rot caused by *Sclerotinia rolfsi* has been reported from some parts of the country

Value Added Products

Curry leaves ground with mature coconut kernel and spices form an excellent preserve. Leaves yield volatile oil and crystalline glucoside- 'Koenigiin'. A glucoside 'Murrayin' is obtained from flowers. Spice powder is prepared for export. Oil used as fixative for heavy type of soap and perfume is extracted from leaves by solvent extraction method (Bhat,2007).

Bathua (*Chenopodium album* L.)

Bathua,also known as 'Common lambs quarters' a nutritious leafy vegetable belonging to family Chenopodiaceae.The chromosome number 2n=36or 54.It is a fast growing upright annual plant common in temperate regions. The leaves have a waxy coating. The tender stem and leaves are used as leafy vegetable. It occurs throughout Britain but is less frequent in north and west. It grows best in fertile soils and common on sandy loams and frequent in clay.

Propagation

The crop is propagated through seeds which are slightly smaller than mustard seeds. Seeds can be directly sown in field or seedling can be raised in nursery and transplanted. Brown seeds germinate rapidly but black seeds persists in soil for longer time. Plants from brown seeds produce same proportion of black an brown seeds as those from black seeds. Immature seeds are capable of germination.

Varieties

Pusa Bathua No. 1

It is the first improved variety of the crop released from IARI, new Delhi.Plants grow to 2.25 m height with attractive purplish green leaves. Leaves along with tender shoots may be picked 45 days after sowing.Edible portion is 60 per cent more rich in vit.C and 10 per cent more in dippa carotene than local by available types

Ootty 1

Tamil Nadu Agricultural University, Coimbatore has developed this variety. It grows to a height of 38-40 cm and yield about 28.9t/ha in 55 days in hills and 17.0 tonnes in 50 days in plains.Leaves are dark green and unfolded leaves are pinkish in color. It shows resistance to *Cercospora* leaf spot, *Colletotrichum* leafspot and *Macrophomina* foot rot.

Seeds are either sown directly or seedlings are transplanted in main field at spacing of 30x15cm.At wider spacing plant grows vigorously.Crop growth increase with increase in Mg content in soil. After about 35-40 days either seedlings can be uprooted or clipped near ground level and used as green.

Portulaca (*Portulaca oleracea* Linn)

Portulaca belongs to the family Portulacaceae, genus *Portulaca* and species *oleracea*. This is an annual succulent herb with fleshy oval leaves,and yellow flowers.The chromosome No.2n= 54. It is rich in beta carotene, folic acid,vit.C and essential fatty acids.

Common purslane is a native of India and the middle East. It was naturalised elsewhere and is considered as an invasive weed in some regions.It was highly esteemed in ancient Egypt and cultivated in Europe as far back as the Middle Ages.

Purslane can be grown in any type of soil. It is propagated by seeds. The seeds are sown directly from April to September in hills and March to June in plains. Regular irrigation is required for raising good crop. The plant becomes ready for first harvest in about a month after sowing. It matures quickly. Individual leaves and young shoots are harvested for use in salads in summer. The leaves do not remain fresh for any length of time. So for continuous supply of leaves seeds are sown at 15 days interval (Bhat, 2007).

References

1. Bhat K.L (2007). Minor vegetables Untapped Potential. Kalyani Publishers, Ludhiana.

2. Burkill IH (1935). A dictionary of the economic products of the Malay Peninsula, Crown Agents, London.

3. Campbell GKG (1976). Sugar beet. In Evolution of Crop Plants (Simmonds NW, Ed). Longman, London.

4. Krishnarajan, I. (1991). Curry leaf cultivation, Spice India 3: 5-6.

5. Nath, P. (1976). Vegetable for the Tropical Region. ICAR, New Delhi, Low Priced Book Series No. 2.

6. Peter, K.V. (Ed) (2007, 2008, 2009). Underutilised and Underexploited Horticultural Crops, Vols. I to V. New India Publishing Agency, New Delhi.

7. Regeena S and Ravi S 1995.Curry leaf cultivation in the homesteads of South Kerala. Spice India, 9: 11-13.

8. Rosa JT (1925). Sex expression in spinach. Hilgardia, **1**: 258.

9. Ryder EJ (1979). Leafy salad vegetables. AVI Publishing Company, Westport, Connecticut, USA.

10. Varalakshmi B, Naik G and Pratapreddy VV (2004). Improved leafy vegetable varieties from IIHR, Bangalore. Proceedings of the Indian Horticultural Congress, New Delhi (6-9 Nov). p. 118-119.

11. Veeraraghavathatham D, Jawaharlal M and Seemanthini Ramadas (1998). A guide on vegetable culture. Suri Associates, Coimbatore.

2015, Horticulture for Nutrition Security
Editor: **Prof. K.V. Peter**
Published by: **DAYA PUBLISHING HOUSE, NEW DELHI**

Pages **425–444**

Chapter 20

Potatoes for Nutritional Security

Brajesh Singh, B.P. Singh and Pinky Raigond

Food security in broader sense may be realized when all people at all the times have access to sufficient, safe and nutritious food to maintain a healthy and active life. Food security depends on availability of food, affordability and proper utilization of food. Generally it is considered that surplus food grains leads to nutritional security, but it is not always true to all the population, as affordability is another major issue. There are a few other indicators of food and nutritional security as well like proportion of undernourished population, <5 years old underweight children and mortality of <5 years old children (Singh and Rana 2013). Undernourished population and less than 5 years old underweight children are more in Bangladesh and in India also the status remained alarming over the years (Figures 20.1 and 20.2). Globally India is second in child malnutrition after Bangladesh (Jain and Pandey 2012). In India still more than 40 per cent children are underweight, which is a great concern. However, the rate of mortality of <5 years old decreased over the period, but is still 6.3 per cent in India, which needs to be reduced further through nutritious and sufficient food access to pregnant women and children (Figure 20.3). To overcome this situation those fruits and vegetables should be popularized, which are available throughout the country in all the seasons and are in reach of all income groups, especially poor.

Potato: A Ray of Hope

Potato (*Solanum tuberosum*) which is locally called as 'alu' is one such candidate that can solve the problem of food security as well as malnutrition. Potato was originally domesticated about 8,000 years ago by communities of hunters and gatherers in the Andes mountain range of South America. Sea voyagers from Europe

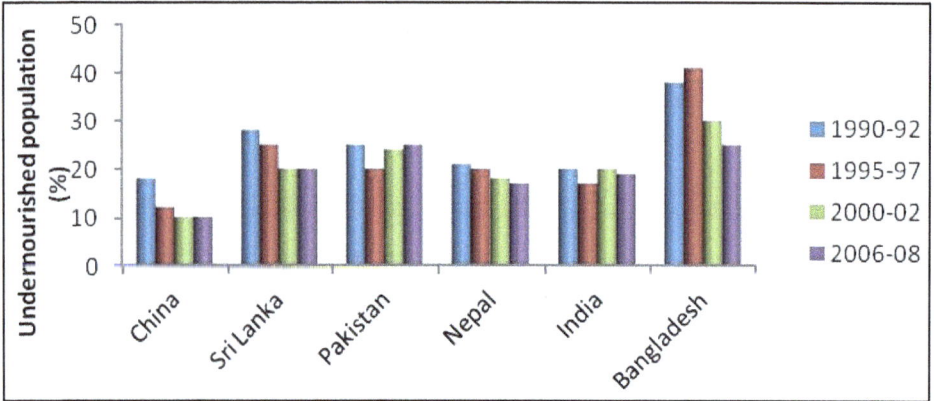

Figure 20.1: Undernourished Population in different Countries (*Source*: Singh and Rana, 2013).

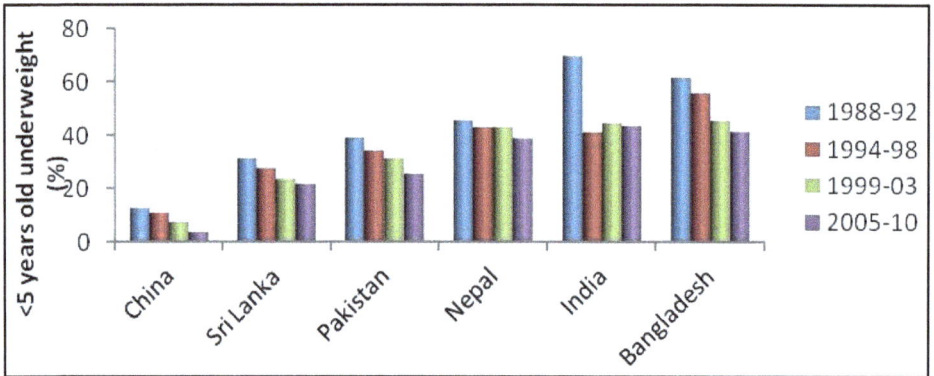

Figure 20.2: <5 years Old Underweight as an Indicator of Food and Nutritional Security (*Source*: Singh and Rana, 2013).

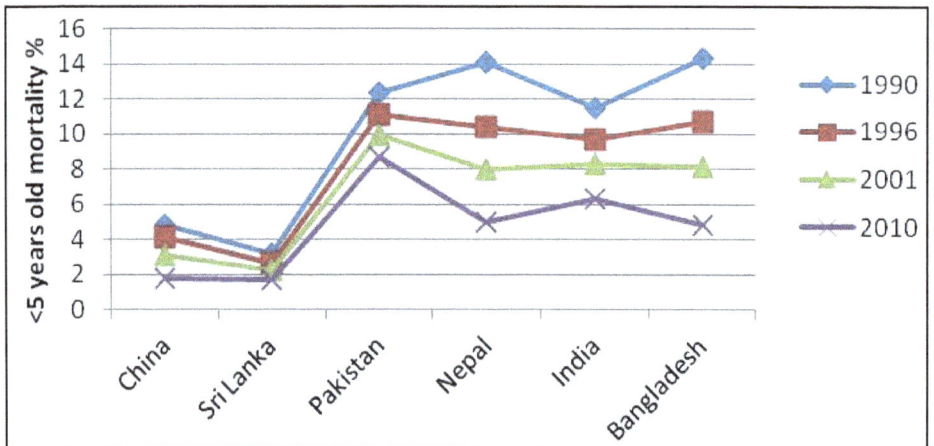

Figure 20.3: <5 Years Old Mortality as an Indicator of Food and Nutritional Security (*Source*: Singh and Rana, 2013).

who came in search of Asian treasure, took potato with them and introduced potato to ancient agrarian sub-continent only about 400 years ago. Though Europeans did not accept potato readily, it became the favourite food of the sailors who used to take potato tubers for consumption on ocean voyages, specifically to avoid scurvy. Potato was introduced to India most likely by Portuguese traders during late 16th to early 17th century. Potato was accepted as a primary vegetable supplement because of its mild flavour and its utilization in combination with other foods. Nutritional value of potato was known since long, specially its high content of ascorbic acid to prevent scurvy (Pandey and Chakrabarti 2008). One of the prominent publications of the Food and Agriculture Organization (FAO 2008) has emphatically considered and recommended potato as a potential crop for the poorest of the poor, to ensure global food, nutritional and income security in future.

Potato is a flexible crop compared to other vegetables and can be grown under conditions where other crops may fail to grow. Moreover its short and flexible life cycle brings the yield within 100 to 120 days and is hence also suitable for double cropping and intercropping systems. Potato is a good option for food and is capable of producing nutritious food more quickly on lesser land compared to any other major food crops. Potatoes yield more edible energy, protein and dry matter per unit area and time compared to other crops due to its high protein-calorie ratio (17g protein: 1000 kcal) and short life cycle (Singh and Rana 2013). Farmers can harvest up to 80 per cent of biomass as edible, nutritious food in case of potato, whereas in case of cereals only 50 per cent can be harvested as grains. Growth of potato in terms of production and productivity was higher in comparison to maize, rice and wheat during 2008 to 2012. Serious food security problem will appear in future due to stagnation of crop yields, exhausting soils and increasing population in the country. In such a scenario, potato provides a ray of hope due to its highest per hectare, per day production of edible dry matter and vital nutrients (Singh and Rana 2013).

Potato: A Wholesome Food

Potato is considered as the most productive vegetable and provides a major source of nutrition and income to many population and communities. Due to its versatility in way of cooking *viz.* boiling, baking, deep frying etc. potato became popular over the period of time and is being consumed by one and all. Potato is popularly known as the "Vegetable King". It may be consumed in the form of snacks (chips, fries and dehydrated products) by the rich, though most of the undernourished households consume potato, as primary or secondary source of food and nutrition. In India, potatoes have been utilized largely for consumption as fresh potatoes and the major part of potato harvest (approx. 68.5 per cent) goes to domestic table consumption. The nutritional value of potato is well acclaimed and is known as a versatile, carbohydrate-rich and low-fat food. Freshly harvested potatoes contain about 80 per cent water and 20 per cent dry matter, out of this dry matter approximately 60-80 per cent is constituted in the form of starch. Its content of dry matter, edible energy and edible protein makes it a good choice for nutrients availability. On dry weight basis, potato protein content is similar to that of cereals and is very high in comparison to other roots and tubers. Potato is known to everyone as a supplier of energy but its ability to supply vital nutrients is vastly underestimated. Potato is an excellent source of complex carbohydrates, dietary fibres and vitamin C. It also contains a variety of health-promoting compounds, such as, phytonutrients that have antioxidative activity. Among these, important health-promoting compounds are carotenoids, flavonoids, and caffeic acid, as well as unique tuber storage proteins, such as patatin, which exhibit activity against free radicals. Potato is also a substantial source of ascorbic acid, thiamine, niacin, pantothenic acid and riboflavin (Table 20.1). Due to the nutritional value of potato, it is highly desirable in human diet. The nutritive value of a potato containing food depends on the other components served with it and on the method of preparation. By itself, potato is not fattening and the feeling of satiety that comes from eating potato can actually help people to control their weight. However, preparing and serving potatoes with high-fat ingredients raises the caloric value of the dish. Since the starch in raw potato cannot be digested by humans, they are prepared for consumption by boiling (with or without the skin), baking or frying. Each preparation method affects potato composition in a different way, but all reduce fiber and protein content, due to their leaching into cooking water and oil, destruction by heat treatment or chemical changes such as oxidation. Table 20.2 showed the contribution made by consumption of 100g of boiled potato to recommended dietary allowances (RDA) for energy, protein, thiamine, niacin, pyridoxine and folic acid.

Table 20.1: Composition of Potato and its Products per 100g of Edible Portion

	Raw	Boiled (in skin)	Fried	Potato Chips	French Fries	Potato Flour
Water (per cent)	79.8	79.8	46.9	1.8	44.7	7.6
Carbohydrate (g)	17.1	17.1	32.6	50.0	36.0	79.9
Protein (g)	2.1	2.1	4.0	5.3	4.3	8.0
Fat (g)	0.1	0.1	14.2	39.8	13.2	0.8
Iron (mg)	0.6	0.6	1.1	1.8	1.3	17.2
Calcium (mg)	7	7	15	40	15	33
Ascorbic acid (mg)	20	16	19	16	21	19
Thiamine (mg)	0.10	0.09	0.12	0.21	0.13	0.42
Riboflavin (mg)	0.04	0.04	0.07	0.07	0.08	0.14
Niacin (mg)	1.5	1.5	2.8	4.8	3.1	3.4

Source: Ezekiel *et al.*, 1999.

Potato: A Low Energy Diet

Raw potato (on dry weight basis) provides about 80 kcal energy whereas, a boiled potato provides about 69 kcal energy per 100 g of weight. When eaten without added fat as in case of frying, potato is a good food for weight conscious people because of its low energy density. The energy value of potato is less than major food crops like rice, wheat, maize and sorghum (Figure 20.4). The energy value is lower than other tuber and root crops as well as food products from animal origin (Figure

Table 20.2: Percentage of Recommended Dietary Allowances (RDA) of some Major Nutrients Provided by 100g of Boiled Potatoes

Nutrients	Infant (up to 6 months with 5.4 kg body weight)		Adult Man (Moderately active with 60 kg body weight)	
	RDA	Per cent of RDA Provided by 100g Boiled Potatoes	RDA	Per cent of RDA Provided by 100g Boiled Potatoes
Energy (k cal/d)	108/kg	12	2875	2
Protein (g/d)	2/kg	18	60	3
Thiamine (mg/d)	55µg/kg	34	1.4	7
Niacin (mg/d)	710µg/kg	31	18	7
Pyridoxine (mg/d)	0.1	100	2	12
Folic acid (µg/d)	25	28	100	7
Ascrbic acid (mg/d)	25	70	40	42

Source: Ezekiel and Khurana, 2003.

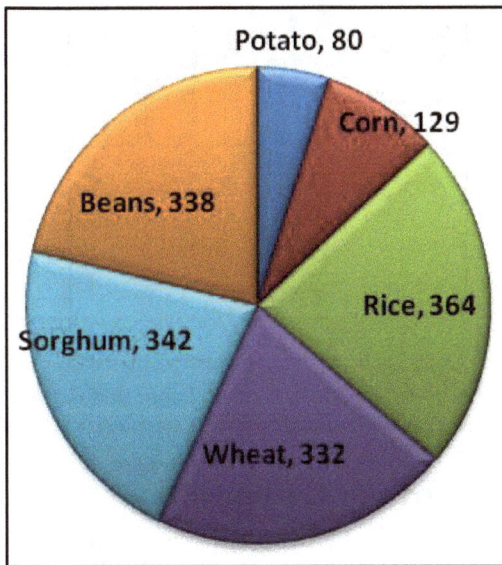

Figure 20.4: Comparison of Energy Provided by Raw Potato and other Plant Foods (kcal per 100g edible portion).

20.5). Potatoes are an excellent source of complex carbohydrates. These carbohydrates take longer time for break down into glucose and result in energy that lasts longer. Complex carbohydrates are longer chains of sugars, such as starches and fiber. In potato, starch is the major carbohydrate and sucrose, fructose and glucose are the main sugars. Carbohydrates are the body's primary source of fuel for energy. The energy produced through potato gets stored as glycogen in muscle and liver and functions as a readily available energy during prolonged, strenuous exercise.

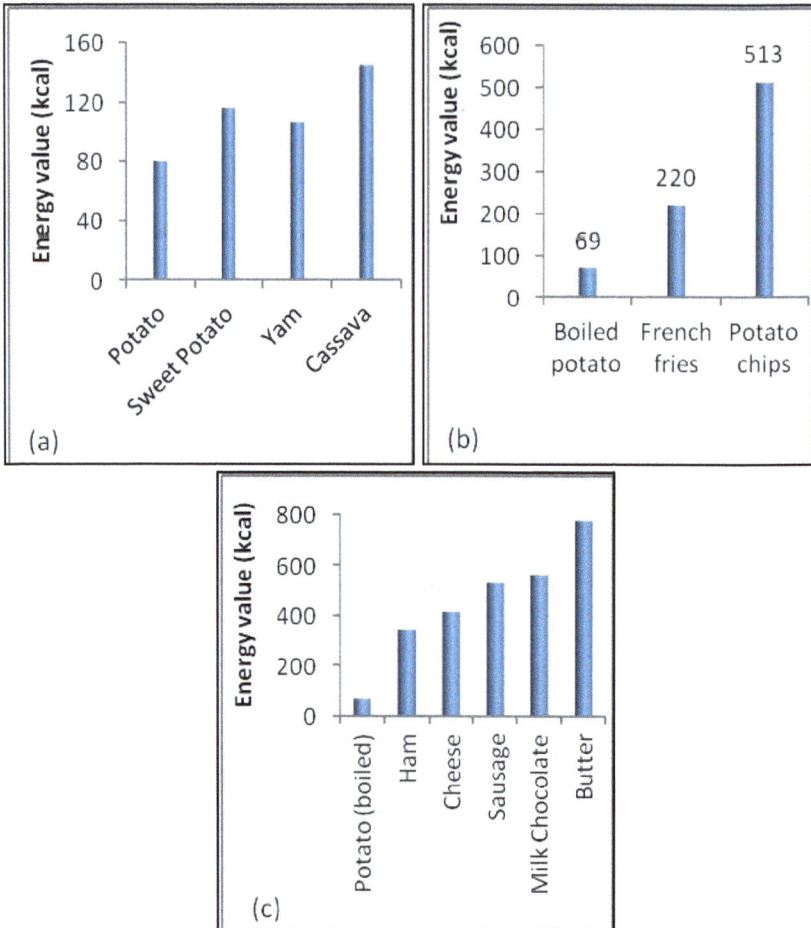

Figure 20.5: Energy Value of (a) Potato and other Tuber and Root Crops; (b) Potato and Potato Products; (c) Potato and other Food Products from Animal Origin (kcal per 100g).

Potato: A Good Source of Carbohydrates

The major role of carbohydrates in nutrition is to provide energy. The complex carbohydrates present in potato are important to a healthy diet. Carbohydrates in potato are mostly found in the form of starch. On an average, potato contains 14-16 per cent of starch on fresh weight basis. The total carbohydrate content of potato (about 18.5 per cent) is lower than some important crops such as wheat, rice, corn, sorghum and beans (Figure 20.6). Sugars are the most basic carbohydrates, the building blocks of complex carbohydrates. Starch furnishes most of the energy supplied by the potato. Digestibility of starch influences the energy value of the potato and hence also the bulk of potato which must be eaten to supply a given amount of energy. The digestibility of potato starch is low in raw state but improves considerably after cooking or processing. Sucrose, fructose and glucose are the main sugars in potatoes.

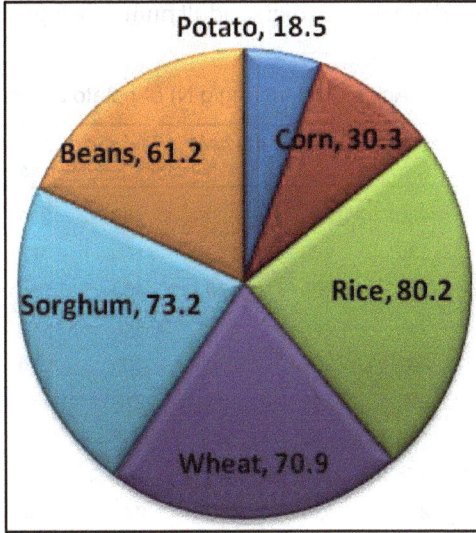

Figure 20.6: Carbohydrate Content of Raw Potato vis-a-vis other Plant Foods (g per 100g edible portion).

Potato: Source of High Quality Protein

Potato is a very good source of high quality protein. Average protein content of potato is 2 per cent on fresh weight basis and about 10 per cent on dry weight basis. Potato protein content is lower than wheat, rice, corn, sorghum, and beans but is higher than other major root and tuber crops like sweet potato, yam and cassava (Figure 20.7). The total nitrogen of potato tubers can be divided into soluble protein, insoluble protein and soluble non-protein nitrogen. The insoluble protein fraction is mainly present in the peel. Soluble potato protein contains substantial levels of the essential amino acids. Free amino acids present in potatoes are totally available for absorption. Potato protein has an adequate ratio of total essential amino acids to total amino acids and a balance among individual essential amino acid concentrations to meet the needs of infants and small children (Table 20.3 and Figure 20.8). However, the digestibility of potato protein is relatively low in infants. Potato protein has a very high biological value since all the essential amino acids are present in it in a good proportion (Table 20.4). The biological value in potato is higher than major cereals and higher than even proteins of animal origin like milk and beef. With its high lysine content, potato can supplement diets which are limiting in lysine (Figure 20.9). Potato has a clear advantage over cereals in India because of its ability to provide high quality protein. Diet which can fulfill only the energy requirement of body cannot support growth of children, if its protein content is below the recommended requirement. However, if a diet provides inadequate energy, its protein is metabolized as a source of energy rather than being used for growth. Therefore, diet should be well balanced in terms of energy and protein. Therefore, potato is a superb food which has correct balance between net protein calories and total calories adequate for all age groups. Nutritive value of potato protein is better than beef, wheat flour, rice and corn in terms of recommended nitrogen in diet of adult human beings (Ezekiel and Khurana 2003).

Table 20.3: Essential Amino Acids Content (mg/g N) of potato and other Food Crops

Essential Amino Acids	Potato	Rice	Maize	Wheat	Soybean	Sweet Potato	Yam	Cassava
Arginine	330	480	290	290	450	280	480	580
Histidine	100	130	160	130	150	90	120	110
Lysine	320	230	200	170	400	260	280	290
Tryptophan	100	80	40	70	80	110	70	80
Phenylalanine	270	280	290	280	300	270	300	180
Tyrosine	170	290	240	180	210	150	200	100
Methionine	90	150	120	90	80	100	100	50
Cystine	50	90	100	140	100	30	-	90
Threonine	220	230	280	180	240	280	220	200
Leucine	380	500	720	410	480	360	400	300
Isoleucine	270	300	240	220	320	290	230	250
Valine	310	380	300	280	320	380	290	240

Source: Ezekiel *et al.*, 1999.

Table 20.4: Nutritional Value of Potato Protein Compared to Egg Protein

Essential Amino Acids	Composition of Potato Protein as Compared to Egg Protein (mg/g N)		Concentration in Potato Protein Relative to Egg Protein (per cent)
	Egg Protein	Potato Protein	
Methionine	214	120	45-61
Lysine	448	441	60-96
Tryptophan	93	84	90-100
Isoleucine	423	261	62-80
Leucine	490	499	68-100
Phenylalanine	362	231	64-77
Threonine	325	204	63-80
Valine	463	293	63-79

Source: Ezekiel *et al.*, 1999.

Potato: Say No to Fat

There is a common misconception that eating potato may cause obesity due to its high fat content which is not at all a true statement, since potatoes contain very little quantity of fat. The average fat content of potato is 0.1 per cent on fresh weight basis which is too low to have any negative nutritional significance. Fat content in potato

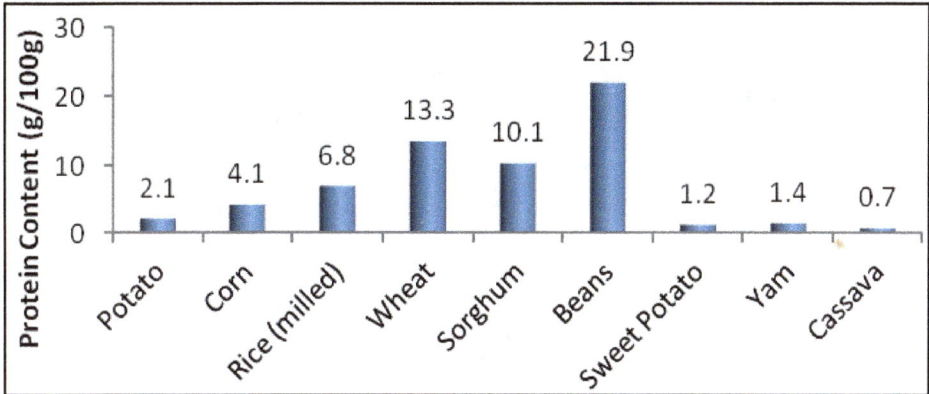

Figure 20.7: Protein Content of Raw Potato and other Plant Foods (g/100g edible portion) (*Source*: Ezekiel *et al.*, 1999).

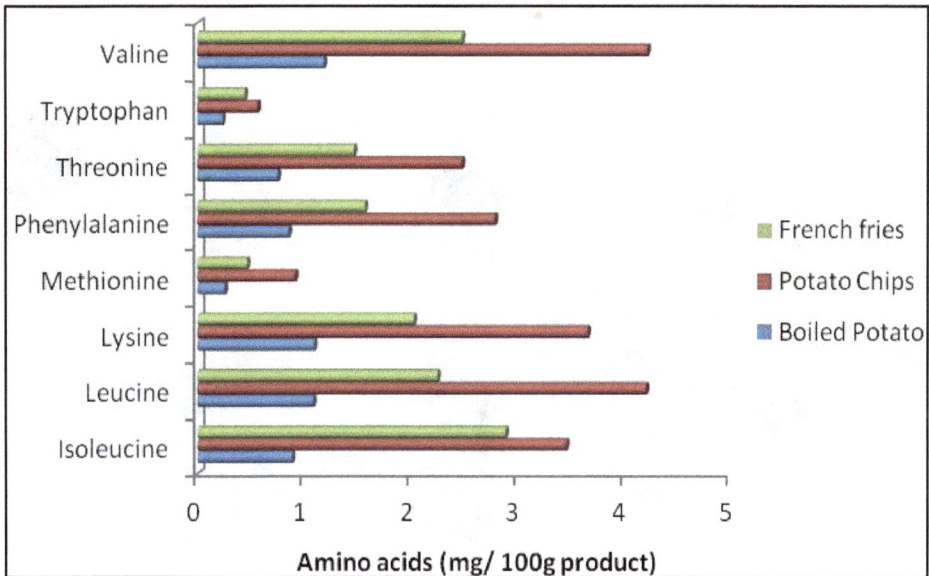

Figure 20.8: Amino Acid Content of Potato and Potato Products.

is lower than major cereals like rice, wheat, maize and sorghum (Figure 20.10). The little fat present in potato contributes towards potato palatability. Major proportion (i.e nearly 60-80 per cent) of potato fat consists of unsaturated fatty acids and linoleic acid is the predominant one. The high content of unsaturated fatty acids increases the nutritive value of the fat present in potato. When eaten without added fat, potato is good for weight conscious people because of its low energy density. However, when fat is added to the fried or processed potato products, it becomes rich in energy and may certainly become a cause of concern. Especially excessive consumption of processed potato products such as chips and French fries containing up to 40 per cent fat may cause obesity (Ezekiel and Khurana 2003).

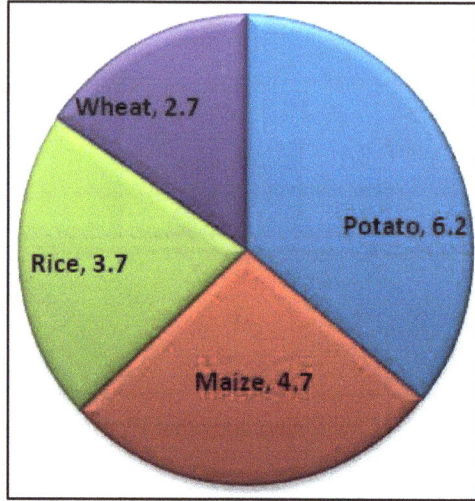

Figure 20.9: Lysine Content (g/100g Protein) of Potato is Higher than Major Cereal Crops (Right).

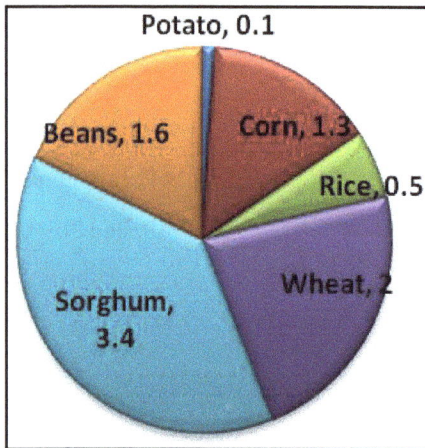

Figure 20.10: Fat Content of Raw Potato is Negligible Compared to other Plant Foods (g/100g edible portion); French Fries of CPRI (Right).

Potatoes: Good Source of Vitamins

Potato is one of the rich natural source of vitamin C or ascorbic acid as it contains 17-35mg of ascorbic acid per 100 g tuber. Potatoes have high quantities of vitamin C than other vegetables like carrots, onion and beet root (Figure 20.11). When consumed in sufficient quantities, potatoes itself can meet all the vitamin C requirements of an individual. Potato is an important source of thiamine, niacin and pyridoxine and its derivatives (vitamin B6 group). It also contains pantothenic acid (vitamin B5), riboflavin and folic acid. B-vitamins are essential for general health and growth, since they are water soluble. However it is recommended that potatoes should not be

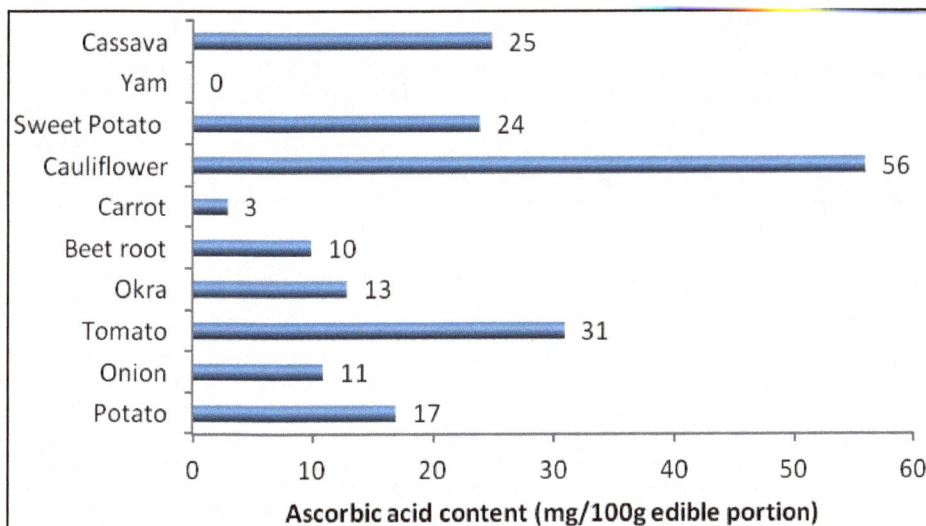

Figure 20.11: Ascorbic Acid Content of Potato and some other Plant Foods.

washed after peeling to prevent loss of vitamins. A small potato can deliver more than 20 per cent Daily Value of vitamin C and vitamin B 6.

Vitamin C

Potato contains 17-35mg/100g Vitamin C which is far high compared to corn, wheat, rice, sorghum and beans. This water-soluble vitamin acts as an antioxidant and stabilizes or eliminates free radicals, thus helping to prevent cellular damage. Potato as major source of Vitamin C provides protection against scurvy. That was why ancient Americans consumed potatoes to prevent scurvy. Vitamin C enhances the absorption of iron in human body and may help support the body's immune system. The ascorbic acid content of potato declines when potatoes are stored, cooked or processed. Though potato loses some of its vitamin C during storage, substantial amounts are retained until it sprouts. In Indian scenario, the consumption of fresh fruits and vegetables is much less and for poor it is almost negligible. Potato being in reach of all income groups may serve as the main source of vitamin C, especially for poor people.

Vitamin B Complex

Potato is an important source of vitamin B complex. All B vitamins help the body to convert food (carbohydrates) into fuel (glucose), which is used to produce energy. These B vitamins, often referred to as B complex vitamins, also help the body use fats and protein. B complex vitamins are needed for healthy skin, hair, eyes and liver. They are also required for proper functioning of nervous system. Potatoes are important source of thiamin, niacin and pyridoxine. They provide 0.10 mg vitamin B1 (thiamine) per 100 g of freshly harvested potatoes (Figure 20.12). Thiamine is needed to release energy from carbohydrates. Insufficient intake of thiamine results in a disease called 'beriberi' affecting the peripheral nervous system (polyneuritis) and/or the

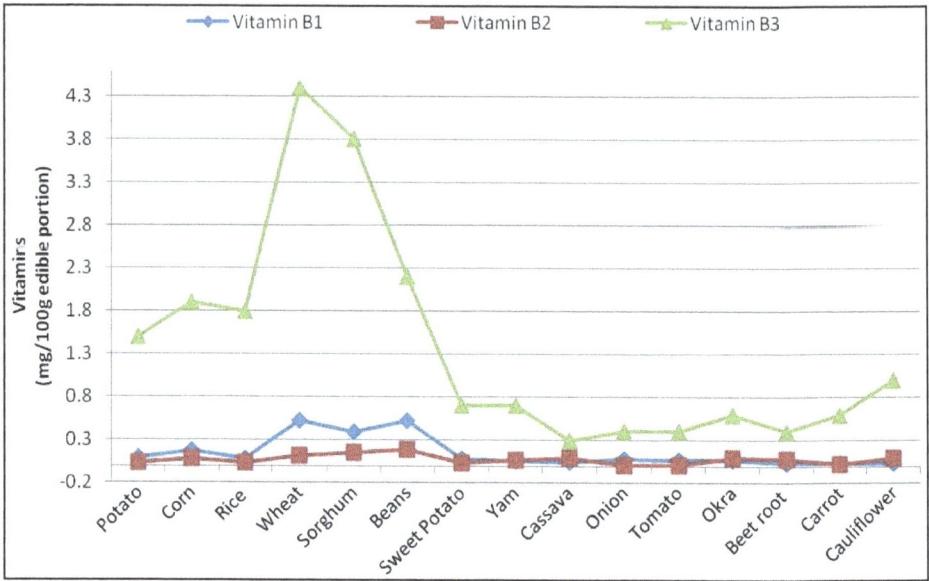

Figure 20.12: Some of B-complex Vitamins in Raw Potato and other Plant Foods.

cardiovascular system. Potato contains 0.01 mg riboflavin per 100 g freshly harvested potatoes. It is also used for treating riboflavin deficiency, acne, muscle cramps, burning feet syndrome, carpal tunnel syndrome and blood disorders. Being an important source of niacin (Vitamin B3) potato provide 1.5 mg niacin per 100g freshly harvested potatoes. Niacin is involved in both DNA repair, and the production of steroid hormones in the adrenal gland. Pantothenic acid is an essential nutrient. Potatoes contain 0.3mg pantothenic acid per 100g of freshly harvested potato. Pantothenic acid is required to synthesize coenzyme-A (CoA), as well as to synthesize and metabolize proteins, carbohydrates and fats. Potatoes are good source of vitamin B6 (pyridoxine). Vitamin B6 plays important roles in carbohydrate and protein metabolism. It helps the body to synthesize nonessential amino acids needed to synthesize various body proteins. It is also a cofactor for several enzymes involved in energy metabolism, and it is required for the synthesis of haemoglobin – an essential component of red blood cells. Potatoes contain 14 mg folic acid per 100 g of freshly harvested potatoes. Folic acid is essential to numerous body functions. Human body needs folate to synthesize, repair and methylate DNA as well as to act as a cofactor in biological reactions involving folate. It is especially important in aiding rapid cell division and growth, such as in infancy and pregnancy. Children and adults both require folic acid to produce healthy red blood cells and prevent anemia. The role of potato as a source of some vitamins of B complex is vastly underestimated. 100g of boiled potatoes (boiled with skin) can fulfill the daily requirement of thiamine, niacin, folic acid and pantothenic acid (Ezekiel and Khurana 2003).

Potato: A Cocktail of Minerals

Potato is a good source of important minerals and trace elements. 100g of potato contains approximately 40-65 mg of phosphorus. The phosphorous present in potato is more assimilable than the phosphorous present in other food crops because of the relatively small percentage of phytic acid in potato. The lower phytic acid content of potatoes enhances phosphorous bioavailability to human body and also helps in increased bioavailability of calcium, iron and zinc (Figure 20.13). The potassium content of potato is also relatively high *i.e* 247-455mg/100g fresh weight. Because of high potassium content, potatoes are not included in the diet of patients with renal failure. On the other hand, the sodium content of potato is very low and the content is 11mg/100g fresh weight. Potatoes are a good source of iron and their iron content is comparable to most of the other vegetables. 100g of cooked potato can supply between 6 and 12 per cent of daily iron requirement for children or adult. Moreover, due to high ascorbic acid content of potato, bioavailability of non-haem form of iron from potato is increased. Potatoes mixed with other food are also beneficial as it increases the bioavailability of iron from other foods also due to its high ascorbic acid content. Moreover iron availability from potato is higher compared to other foods such as kidney beans, wheat flour and bread, reason being the high proportion of iron from potato is soluble. Potatoes provide a good source of magnesium and magnesium content in potato is up to 22mg/100g fresh weight. Potato can be consumed with foods low in magnesium such as milk. Magnesium content of milk is one fifth to one tenth of potato. Hence potato is superior to milk in terms of magnesium, but consumed together they form best combination as milk is rich in calcium and potato provides magnesium. Zinc is an important trace element found in potato. Though zinc content of potato is not very high but its availability is high because of the low phytic acid

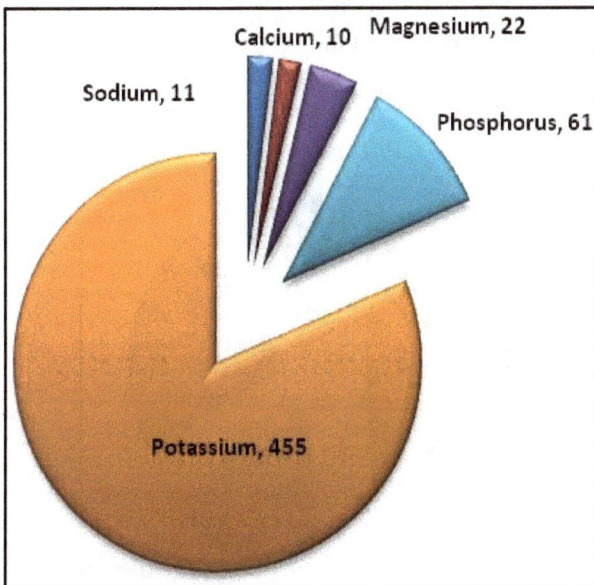

Figure 20.13: Mineral Content of Raw Potato (mg/100g edible portion).

content. Potato can supply at least part of daily requirements of trace elements like copper, manganese, molybdenum and chromium. Traces of boron, bromine, iodine, aluminium, cobalt and selenium are also present in potato. A small potato can deliver 10 per cent Daily Value of folate, manganese, magnesium and phosphorus. Therefore, potato being in reach of poorest of the poor can play a vital role in eradication of 'Hidden Hunger' which is also known as micronutrient malnutrition.

Potato: Its Dietary Fiber Helps in Digestion

Potato is a rich source of dietary fiber. Cellulose, pectin and pectin associated substances are higher in potatoes compared to cereal brans (Ezekiel and Khurana, 2003). Dietary fiber content in raw potato tuber ranges from 1-2 g/100g fresh weight. Unpeeled potatoes contain more dietary fibers than peeled potatoes. The dietary fiber from potato tuber comes mainly from its cell walls that constitute about 1.2 per cent of the fresh weight of the tubers (Figure 20.14). To increase the dietary fiber intake, potatoes must be consumed along with peel. More than half of the dietary fiber in potato is in the form of pectic substances which improves the quality of potato dietary fiber and thus helps in lowering cholesterol levels. One medium potato with the skin contributes 2 g of fiber or 8 per cent of the daily value (Figure 20.15). Dietary fiber is a complex carbohydrate and is the part of the plant material that cannot be digested and absorbed in the bloodstream. Dietary fiber has been shown to have numerous health benefits, including improving blood lipid levels, regulating blood glucose, and increasing satiety, which may help in weight management. The main components of dietary fiber are non-starch polysaccharides (NSP), lignin, resistant starch and non-digestible oligosaccharides. Potatoes also contain resistant starch which is known as 'starch and starch degradation products that escape digestion in the small intestine of healthy individuals' (Table 20.5). Resistant starch acts in similar fashion as fibers and is found naturally in foods such as legumes, bananas, potatoes and some unprocessed whole grains. In Indian potatoes resistant starch content is approximately 1.5 to 2 per cent in cooked and cooled tubers (Raigond *et al.*, 2014). Natural resistant

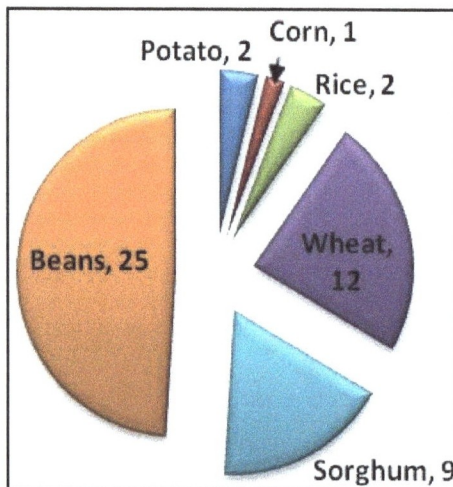

Figure 20.14: Dietary Fiber Content of Raw Potato and other Foods (g/100g edible portion).

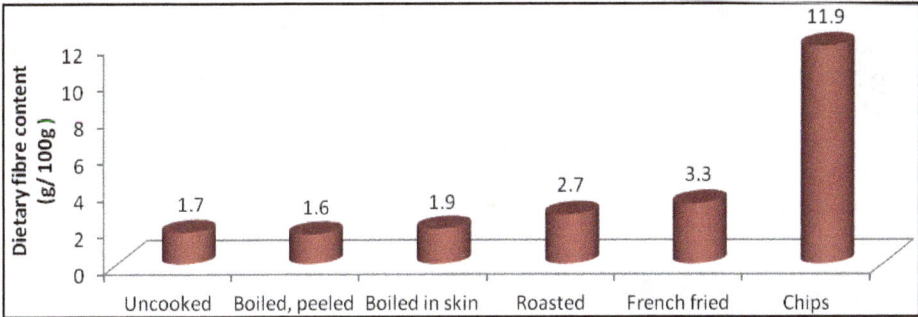

Figure 20.15: Dietary Fiber Content of Potatoes when Cooked by different Methods (*Source*: Ezekiel *et al.*, 1999).

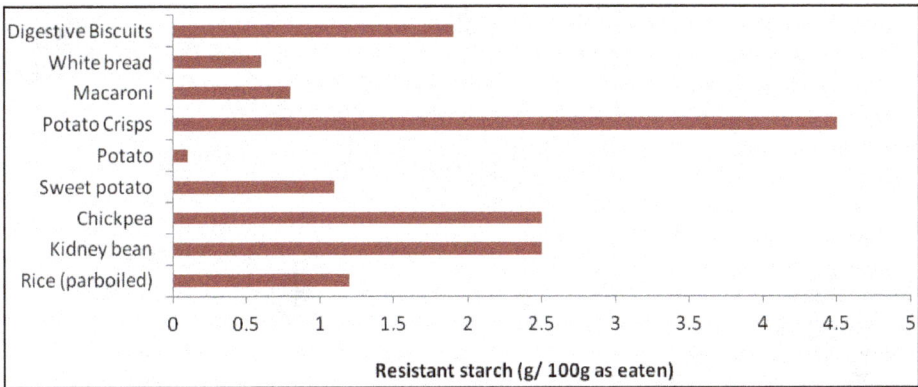

Figure 20.16: Resistant Starch Content of different Food Materials (*Source*: Englyst *et al.*, 1996).

Table 20.5: Resistant Starch Content of different Food Materials (Per cent dry matter)

Negligible (≤1.0 per cent)	Boiled potato (hot), boiled rice (hot), pasta, breakfast cereals containing a high proportion of bran, wheat flour.
Low (1.0–2.5 per cent)	Breakfast cereals, biscuits, breads, pasta, boiled potato (cooled), boiled rice (cooled).
Intermediate (2.5–5.0 per cent)	Breakfast cereals (corn flakes, rice crispies), fried potatoes, extruded legumes.
High (5.0–15.0 per cent)	Cooked legumes (lentils, chickpeas, beans), peas raw rice, autoclaved and cooled starches (wheat, potato, maize), cooked and frozen starchy foods.
Very high (> 15.0 per cent)	Raw potatoes, raw legumes, amylo-maize, unripe banana, retrograded amylose.

Source: Goni *et al.*, 1996.

starch is insoluble, fermented in the large intestine and a prebiotic fiber (i.e it may stimulate the growth of beneficial bacteria in the colon). Other types of resistant starch may be soluble or insoluble and may or may not have prebiotic properties.

Resistant starch appears to exert beneficial effects within the colon. The amount of resistant starch found in potatoes is highly dependent upon processing and preparation methods (Figure 20.16). For example, cooking and then cooling potatoes leads to nearly a two-fold increase in resistant starch. Even processed potatoes (*e.g.* potato flakes) appear to retain a significant amount of resistant starch with the potential to confer health benefits. Resistant starch is considered as the third type of dietary fiber, as it can deliver some of the benefits of insoluble fiber and some of the benefits of soluble fiber.

Potatoes: A Rich Source of Antioxidants

Along with vitamins and minerals, potatoes contain a number of small molecules, many of which are beneficial phytonutrients such as phenols, flavonoids, kukoamines, anthocyanins and carotenoids. Coloured potatoes may serve as a potential source of natural anthocyanin pigments and also a powerful source of antioxidant micronutrients. Potato antioxidants have potential role in immune function and disease prevention. Yellow pigmented potatoes are known to have high carotenoids content such as lutein, zeaxanthin, violaxanthin and antheraxathin (Ah-Hen *et al.*, 2012). Potato carotenoids are primarily oxygenated carotenoids which are also known as xanthophylls. Purple pigmented potatoes have health benefit against cardiovascular disease while consumption of yellow pigmented potatoes enhances immune response (Kaspar *et al.*, 2013). Lutein, zeaxanthin, violaxanthin and neoxanthin are the major carotenoids present in potatoes and β-carotene has been detected only in trace amounts. The orange colour of the tuber flesh is due to zeaxanthin, whereas the yellow colour is due to lutein. Generally high phenolic contents such as anthocyanin are present in dark colored potatoes. But white/cream fleshed potatoes also contain phenolics to some extent.

Antinutrients: Glycoalkaloids and Acrylamide

Potatoes when exposed to sunlight form some alkaloids called 'glycoalkaloids' and the main constituents are α-chaconine and α-solanine. Normally, potatoes contain less than 5 mg solanine per 100 g fresh weight which is far lesser than the safety limit of 20 mg/100 g and glycoalkaloids content of potato is so low that it is not even perceptible by taste (Morris and Lee 1984). However, if the potatoes become green, they might contain higher glycoalkaloids resulting in health threats and therefore, it is not recommended to consume greened potatoes. It is a fallacy to say that potato glycoalkaloids can be poisonous to human beings. Tubers have much lower glycoalkaloids content and the distribution is not uniform, with higher levels found in the periderm and cortex, decreasing markedly towards the pith (Dale *et al.*, 1998). Therefore, most of the glycoalkaloids (80 per cent) are found in outer layer and therefore can be easily removed. With precautions, bad effects of glycoalkaloids may be avoided.

Acrylamide which is classified as a Group 2A carcinogen (that is, probable human carcinogen), has been detected in common foods, such as potato chips, French fries, cookies, cereals, and bread, that are prepared or cooked at a temperature of over 120°C. Swedish researchers (Tareke *et al.*, 2002) shocked the food safety world when they presented preliminary findings of acrylamide in some fried and baked foods, most notably potato crisps and French fries, at levels of 30–2300 µg/kg. Since then lot of emphasis has been given on analysis of acrylamide in fried and baked products. Although acrylamide has probably been part of our diet since man first started cooking, because of concerns over safety, there is need to reduce the levels of acrylamide in foods. Acrylamide is generated during a side reaction of the Maillard reaction which occurs between amino acid (asparagine) and reducing sugars (fructose and glucose). Fried products prepared from Indian potato processing varieties *viz.*, Kufri Chipsona-1, Kufri Chipsona-2, Kufri Chipsona-3, Kufri Himsona and Kufri Frysona (63-101 µg/kg) contained low concentration of acrylamide (Singh *et al.,* 2010).

Nutritional Benefits Associated with Potato Consumption

Potatoes contain a complete range of nutrients, including those necessary for growth and development of human beings.

☆ They are rich in vitamin C and thus prevent scurvy.

☆ Potato is a good source of vitamin B6. Many of the building blocks of protein, amino acids require B6 for their synthesis, as do the nucleic acids used in the creation of our DNA. Because amino and nucleic acids are such critical parts of new cell formation, vitamin B6 is essential for the formation of virtually all new cells in the body.

☆ Potatoes being a natural source of phytochemicals such as carotenoids, phenolic compounds, flavonoids and anthocyanin help in reducing the risk of chronic diseases, including cancer, age-related neuronal degeneration or cardiovascular diseases.

☆ Raw potato juice is regarded as an excellent food remedy for rheumatism.

☆ More than half of dietary fiber in potato is in the form of pectic substances, which helps in lowering cholesterol levels in human. Moreover, the dietary fiber dilutes highly caloric components in food, stimulates peristaltic movement and improves digestion.

☆ Because of its low sodium content, potatoes can be used in diets given to patients with high blood pressure.

☆ Potato is a good source of resistant starch. Natural resistant starch helps maintain a healthy colon and a healthy digestive system via several mechanisms and prevents colorectal cancer and type 2 diabetes. Resistant starch may also help to burn fat and may lead to lower fat accumulation. Resistant starch contributes to oral rehydration solutions for the treatment of diarrhea. It is predicted to help maintain "regularity" with a mild laxative effect due to increased microbial activity in the large intestine. When resistant starches are included in a meal (e.g cooked and cooled potatoes), it slows down the absorption of sugars from other foods. That means there is more gradual rise and fall in blood sugar levels after eating. That's particularly helpful for diabetics, who need to keep their blood sugar levels steady.

Some Misconceptions on Potatoes: Far from Truth

There are many misconceptions prevalent in society concerning the nutritional value of potato.

☆ The most common misconception is that potatoes are fattening. With a fat content of less than 0.1 per cent and very low calorie content, by no means it can cause obesity. Potatoes are known to absorb considerable amount of fat while frying which is a common way of consuming potatoes in Indian recipes. Hence, potato is wrongly blamed of causing obesity, while the real culprit is the fat which it might have absorbed at the time of frying.

☆ Another common misconception is that potatoes can cause or worsen diabetes. Resistant starch present in potato helps in more gradual rise and fall in blood sugar levels after eating and hence potatoes are not an unhealthy food for diabetics. In western countries, potatoes are consumed by one and all.

☆ One more misconception is that all of the potato's nutrients are found in the skin. While the skin does contain approximately half of the total dietary fiber, the majority (> 50 percent) of the nutrients are found within the potato itself. As is true for most vegetables, cooking does impact the bioavailability of certain nutrients, particularly water-soluble vitamins and minerals, and nutrient loss is greatest when cooking involves water (boiling) and/or extended periods of time (baking). To protect most of the nutrition in a cooked potato, steaming and microwaving are the best methods of cooking potatoes.

"With the removal of misconceptions among populations, potatoes might play the role of nutritious source of food and become 'bread of life'"

Conclusion

'Global Hunger Index 2012' report of International Food Policy Research Institute showed serious concerns about food security in India. To overcome such serious and alarming situations potato provides a way out to tackle the problem of hunger and provides food and nutritional security (Table 20.2). Potato can be a perfect substitute for other cereals due to its high nutritional value and high production and productivity. Potato is a nourishing and wholesome food. It's low energy density is advantageous when eaten without much added fat. Potatoes contain high quality protein rich in essential amino acids. It is a rich source of vitamin C and is far superior in this respect to most other vegetables and cereals. Considerable quantities of some of the B group vitamins are also present in potato. Potatoes contain many minerals and trace elements and simultaneously are low in fats. Moreover, among the major food crops, bioavailability of minerals is potentially high in potatoes because of the presence of high concentrations of the compounds that stimulate micronutrient absorption such as ascorbate. Low concentration of absorption inhibitors such as phytate and oxalate also improve the bioavailability of minerals from potato. Hence, potato as such is a wholesome food and anyone can live by eating potatoes alone.

With ever increasing population, potatoes are destined to be very crucial for providing food and nutritional security to populations in the developing countries including Indian masses.

'If one had to live on one food alone, the POTATO would be better by far than any other major food crop available today"

Dr RL Sawyer, Former DG, CIP, Lima, Peru'

References

1. Ah-Hen, K., Fuenzalida, C., Hess, S., Contreras, A., Vega-Galvez, A. and Lemus-Mondaca, L. (2012) Antioxidant capacity and total phenolic compounds of twelve selected potato landrace clones grown in southern Chile. Chil J Agric Res 72 (1): 3-9

2. Dale, M.F.B., Griffiths, D.W. and Bain, H. (1998) Effect of bruising on the total glycoalkaloids and chlorogenic acid content of potato (*Solanum tuberosum*) tubers of five cultivars. J Sci Food Agric 77: 499-505.

3. Englyst, H.N., Veenstra, J.and Hudson, G.J. (1996). Measurement of rapidly available glucose (RAG) in plant foods: a potential *in vitro* predictor of the glycaemic response. Brit J Nutr 75:327-337.

4. Ezekiel, R. and Khurana, S.M.P. (2003) Nutritional and medicinal value of potatoes. Extension Bulletin no. 35. CPRI, Shimla, pp:14.

5. Ezekiel, R., Sukumaran, N.P. and Shekhawat, G.S. (1999) Potato: A wholesome food. Tach. Bull 49. CPRI.

6. FAO (2008) The state of food insecurity in the world 2008. Economic and Social Development Department, Food and Agriculture Organization of the UN. Rome.

7. Goni, L., Garcia-Diz, E., Manas, F. and Saura-Calixto (1996) Analysis of resistant starch: a method for foods and food products. Food Chemistry. 56, 445-449.

8. Jain, S. C. and Pandey, A. (2012) Improvement of nutritional quality of bakery products through fortification/enrichment. Processed Food Industry.16 (2): 18-25.

9. Kaspar, K.L., Park, J.S., Brown, C.R. and Mathison, B.D., Navarre, D.A., Chew, B.P. (2013) Pigmented potato consumption improves immune response in men: a randomized controlled trial. Am J Advanced Food Sci Technol 1: 15-25

10. Morris, S.C. and Lee, T.H. (1984) The toxicity and teratogenicity of Solanaceae glycoalkaloids, particularly those of potato (*Solanum tuberosum*). Food Technol. Australia 36: 118-124.

11. Pandey, S.K. and Charabarti, S.K. (2008) Twenty steps towards hidden treasure; technologies that triggered potato revolution in India. Pp281.

12. Raigond, P., Ezekiel, R. and Kaundal, B. (2014) Starch fractions of cooked potatoes at low temperature. Potato J 41(1): 58-67.

13. Singh B, Sharma S and Singh B.P. (2010). Acrylamide in French fries of Indian potato varieties grown in north-western hills. In: Book of Abstracts of the "National conference of Plant Physiology" held at BHU, Varanasi, India during Nov 25-27 2010. Organised by ISPP, New Delhi. Pp. 339.

14. Singh, B.P. and Rana, R.K. (2013) Potato for food and nutritional security in India. Indian Farming. 63(7): 37-43.

15. Tareke, E., P. Rydberg, P., Karisson, S. and Eriksson, M., Tornqvist. (2002) Analysis of acrylamide, a carcinogen formed in heated foodstuffs. *J. Agric. Food Chem.* **50**: 4998-5006.

2015, Horticulture for Nutrition Security
Editor: **Prof. K.V. Peter**
Published by: **DAYA PUBLISHING HOUSE, NEW DELHI**

Pages **445–450**

Chapter 21

Urban and Peri-urban Agriculture (UPA) for Food and Nutrition Security

K.V. Peter and Binoo Bonny

Urban and peri-urban agriculture (UPA) has evolved from a simple, traditional and informal activity into a commercial and professional initiative. It has become a key element in food security strategies officially recognized by the 15th FAO-COAG session in Rome, 1999 and at the World Food Summit, 2002. It is conceptualized as an agricultural production process located within (intra-urban) or on the fringe (peri-urban) of a town, a city or a metropolis, which grows, processes and distributes a diversity of agriculture products, using largely human, land and water resources, products and services found in and around that urban area.In developing countries migration to cities for employment is clogging space for habitation and associated basic entitlements like water,energy(electricity, liquid petroleum gas,health care, transport and above all availability and access to clean food.It is predicted that by 2020 that 60 per cent of India's population will be in cities leaving only 40 per cent in rural areas.

UPA provides the urban migrants who are constrained for the very necessities of life and that too at compromised quality standards and exorbitant rates, an opportunity for self provisioning of safe and diverse food. It serves as a means for vulnerable groups to minimize their food-insecurity problems by reducing their reliance on cash income for food by growing their own food on plots inside or outside the city, thus increasing their access to food. Furthermore UPA allows for saving on energy at various levels of food chain including packaging, transport, storage and distribution which will affect the final retail price of the food commodities. However, UPA cannot be expected to satisfy the urban demand for staple crops like cereals and tubers, but need to be appreciated as an important strategy for self-provisioning, especially for poor households. Moreover, the highly developed commercial urban farming that supplies urban residents with fresh vegetables, fruits and animal products also forms a major component of the urban food system. Thus it also helps in providing the diversity needed to ensure dietary quality – an important aspect of food security.

Given the urban constraints related to land, water and other amenities needed for farming, UPA generally prefers short duration crops like vegetables, annual flowers, ornamental trees, seasonal spices and herbs, aromatic plants and mushrooms. These can be grown for human consumption and ornamental use within and in the immediate surroundings of cities with rural agriculture continuing to be the primary source of basic food for urban dwellers also. Horticultural crops assume greater significance in the context of UPA and it is more of urban and peri-urban horticulture (UPH). Protected cultivation techniques are available to force fruits, vegetables and flower crops to grow off season. Hydroponics, aeroponics and aquaponics are possible methods of growing crops saving space, time, energy and labor.People have food security when they are able to grow enough food, or buy enough food, to meet their daily needs for an active and healthy life.In many of the 21st century's developing cities,all of those conditions of food security are threatened(www.fao.org).Poor households spend from 60 to 80 percent of their income on food, making them vulnerable when food prices rise or their income falls.Poverty(<$1.25 a day or less) and hunger levels in the developing world are on rise (Table 21.1.). Twenty nine percent of total population in developing countries are poor and 20 percent hungry- a shame and blot on humanity.Prevalence of undernourishmemt and progress towards the World Food Summit (WFS) and Millennium Development Goal (MDG) targets in developing countries are significantly widening and seem to be peace threatening (Table 21.2.). "A hungry man succumbs to temptations by evil and negatives"-Mahatma Gandhi.In India alone 251.5 million people are undernourished in 2004-2006 and Indias progress towards Millenium Development Goals is 0.9 while the target is only 0.5. Among different farm –size groups,agricultural labourers alone are poor to an extent of 26.4 percent and undernourished to the extent of 22.0 percent (Table 21.3). Extent of child malnutrition in selected countries was analyzed for the

Table 21.1: Poverty ($1.25 a day or less) and Hunger Levels in the Developing World, Percentages

Region	Per cent Poverty	Per cent Hunger (Undernourished)
Asia-Pacific	27	17
Latin America and Caribbean	8	10
Sub-Saharan Africa	51	32
Total Developing Countries	29	20

Table 21.2: Prevalence of Undernourishment and Progress Towards the World Food Summit (WFS) and the Millennium Development Goal (MDG) Targets in Developing Countries

World/Region/ Country	Total Population 2004-06 (Million)	Number of People Undernourished (Million)				Progress in Number Towards WFS (Target 0.5)	Progress in Prevalence Towards MDGs (Target 0.5)**
		1990-92	1995-97	2000-02	2004-06		
India	1134.4	210.2	193.5	223.0	251.5	1.2	0.9
China	1320.5	177.8	143.7	132.5	127.4	0.7	0.6
Brazil	186.8	15.8	15.6	16.6	11.9	0.7	0.6
Asia-Pacific	3518.7	585.7	528.5	552.1	566.2	1.0	0.8
Developed Countries	1269.5	19.1	21.4	18.7	15.2	0.8	Na
Developing Countries	5213.8	826.2	803.5	838.0	857.7	1.0	0.8
World	6483.3	845.3	824.9	856.8	872.9	1.0	0.9

Table 21.3: Proportion of Poor and Undernourished Persons in different Farm-Size Groups in Rural India, 2004

Farm Size	Share of each Group in Total Poor, Per cent	Share of each group in Total Under-nourished, Per cent
Agrl. Labourer	26.4	22.0
Marg. Farms	56.8	51.3
Small Farms	2.9	3.9
Med. Farms	1.3	2.1
Large Farms	0.4	0.6
Other Rural	12.2	20.1

Table 21.4: Extent of Child Undernutrition in Selected Countries

Indicators	India	Brazil	China	Russia	Nigeria
Low birth weight: 0 per cent of infants with LBW- 2000-2007	28.0	8.0	2.0	6.0	14.0
Children under 3 yrs. who are stunted, per cent	47.9	7.1	21.8		43.0
Children under 5yrs. who are underweight: weight for age- less than 2 S.D.: per cent	43.5	2.2	6.8	–	27.2

Table 21.5: Per Capita Production of various Food Items (Kg)

Year	Cereals	Pulses	Food-Grains	Oil-seeds	Sugar-cane	Milk	Fruits	Vegetables	Fish
1991-1995	192	15	207	23	283	67	33	64	5.0
1996-2000	191	14	205	24	297	75	45	83	5.6
2001-2005	177	12	189	20	258	82	43	88	5.8
2005-2006	176	12	188	25	254	88	50	99	6.0
2006-2007	180	13	193	21	281	90	51	100	6.2

criticality and seriousness of level of food, nutrition and health Low birth weight: 0 per cent of infants with LBW-200-2007, children under 3 years who are stunted (per cent) and children under 5 years, who are underweight: weight for age-less than 2 S.D.(per cent).Twenty eight percent of infants in India are low in weight,47.9 percent are stunted and 43.5 percent are underweight. The consequences of above figures are poor health, susceptibility to diseases, work loss and an ill healthy generation (Table 21.4). India alone has 42 percent of children below 5 years as underweight followed by Bangladesh (5 per cent), Pakistan (5 per cent), Nigeria (5 pe rcent) and other developing countries (43 per cent) (Figure 21.1). Access to nutritious food is a key dimension of food security.In Africa and Asia, urban households spend up to 50 percent of their food budgets on cheap "convenience" foods often deficient in vitamins

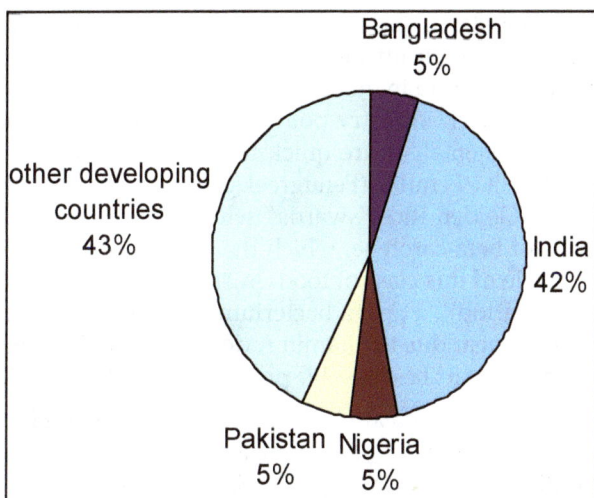

Figure 21.1: Share of Underweight Children under Five Years of Age.

and minerals essential for health.One study found that vitamin A deficiency, a cause of blindness, was more severe among Dhaka (Bengladesh) slum dwellers than among even the rural poor.Fruit and vegetables are the richest natural sources of micronutrients.In developing countries, daily consumption of fruits and vegetables is just 20 per cent of FAO/WHO recommendations.Urban meals rich in low-cost fats and sugars are also responsible for rising levels of obesity and overweight.In India, diet related chronic diseases like diabetes, are a growing health problem and mainly in urban areas. The percapita production of various food items are low compared to requirement (Table 21.5). Per capita production of fruits is only 51 kg/year and 100 kg/year and can be supplemented by production in urban and peri-urban areas.

Urban and peri-urban horticulture help developing cities to meet all those challenges.It boosts the physical supply of fresh, nutritious produce, available year round.It improves the urban poor's economic access to food when their household production of fruit and vegetables reduces their food bills and they earn a living by sale.The other dimension of urban and peri-urban horticulture is bio-cycling of garbage wastes into composts and manures.The urban sewage water is also purified for purpose of irrigating home gardens/nutrition gardens/kitchen gardens. The housewives are trained on gardening with recommended package of practices.No harvest is made immediately after use of pesticides, fruits and vegetables are washed in running water before cutting and use.State Agricultural Universities have come out with Package of Practices for guidance in Good Agricultural Practices (GAP) (KAU, 2011).

Biotechnology and Nutraceuticals

Nutraceuticals are functional foods based on the philosophy of using food as preventive medicine. Biotechnology provides tools to enhance active components in food. Any food or part of food or nutrient that provides health benefits, including prevention and treatment of a disease qualify to be called nutraceuticals. Since the introduction of the first genetically modified crop, the "Flavr Savr" tomato in 1994, "GM" products are now found in thousands of foods, from bagels to butter tarts to soy milk. While the biotech industry points to the safety and benefits of genetic modifications, environmentalists are quick to denounce it for potential harm to biodiversity. Products like Fenulife (Fenugreek galactomannon combination) which control blood sugar, Golden Rice "swarna" which is rice genetically modified to contain high levels of beta-carotene, which the body can convert into vitamin A indicate emerging role of this class of food. In "swarna" rice modification is done using genes from a daffodil, a pea, a bacterium and a virus. As many as 500,000 children go blind each year due to vitamin A deficiency, and this new rice is being touted as biotechnology at its best to solve problems of food insecurity.

It is estimated that 134 m ha are brought under transgenic crops since 1966 of which 46 per cent are in developing countries (China, India, Brazil, Argentina, Paraguay and South Africa) and in crops like Soybean, Maize, Cotton, and Canola. More than 50 crops and forestry trees are being targeted currently. Tissue culture propagation of plating material for horticultural crops is making waves in banana,ginger,cardamom and ornamentals.

The use of biotechnology in urban and periurban horticulture awaits anticipatory research for eco-friendliness and acceptance.

References

1. KAU (2011). Packages of Practices Recommendations (Crops) Directorate of Extension, Kerala Agricultural University, P O Mannuthy(www.kau.in)

2. www.fao.org.

Index

www.ingramcontent.com/pod-product-compliance
Lightning Source LLC
Chambersburg PA
CBHW050126240326
41458CB00122B/1434